To Make Amends, 2014, by Maxime Xavier

http://www.maximexavier.co.uk

The Turing Guide

THE TURING GUIDE

B. JACK COPELAND

JONATHAN P. BOWEN

MARK SPREVAK

ROBIN WILSON

and others

OXFORD
UNIVERSITY PRESS

OXFORD
UNIVERSITY PRESS

Great Clarendon Street, Oxford, OX2 6DP,
United Kingdom

Oxford University Press is a department of the University of Oxford.
It furthers the University's objective of excellence in research, scholarship,
and education by publishing worldwide. Oxford is a registered trade mark of
Oxford University Press in the UK and in certain other countries

Published in the United States of America by Oxford University Press
198 Madison Avenue, New York, NY 10016, United States of America

British Library Cataloguing in Publication Data
Data available

Library of Congress Control Number: 2016946810

ISBN 978–0–19–874782–6 (hbk.)
ISBN 978–0–19–874783–3 (pbk.)

Printed and bound by
CPI Group (UK) Ltd, Croydon, CR0 4YY

In memoriam
MAVIS BATEY (1921–2013)
CATHERINE CAUGHEY (1923–2008)
IVOR GRATTAN-GUINNESS (1941–2014)
PETER HILTON (1923–2010)
JERRY ROBERTS (1920–2014)

FOREWORD BY ANDREW HODGES

Author of the bestseller *Alan Turing, the Enigma*

This book celebrates Alan Turing's place in mathematics, science, technology, and philosophy, and includes chapters by a number of Turing's contemporaries. A glance at its pages will show the diversity of the contributions. Some are fastidious scholarship, bringing to life details of smudged typescripts and incomplete manuscripts from over 60 years ago. Some convey modern scientific developments. Some explore personal memories, or philosophical speculations. But they share a special concern to shed new light on hidden history.

Turing's centenary year reflected a general public sense that the issues of Alan Turing's life and work are as relevant as ever in the twenty-first century. One reason is obvious: the universality of the computer has invaded everyday consciousness. It has changed the relationship between the individual and the social world. The computer has made possible impassioned public campaigns for Turing's official recognition, demanding some remedy for his identity as a criminalized gay man. At the same time, the significance of the modern state's computer-based collection and analysis of information, of which he was the scientific founder, has made a new impact on the world political arena. Turing himself knew that the computer would involve everything: it was prefigured in his futuristic discussion of the meaning of mechanical intelligence, where his all-embracing discourse touched provocatively on topics from sex to cryptography.

Andrew Hodges
Mathematical Institute, University of Oxford

PREFACE

This book celebrates the many facets of Alan Turing, the British mathematician and computing pioneer who is widely considered to be the father of computer science. The book is written for general readers, and Turing's scientific and mathematical concepts are explained in an accessible way.

Each of the book's eight parts covers a different aspect of Turing's life and work. Part I is biographical: Chapter 1 contains a timeline of Turing's short but brilliant life, Chapter 2 is an appraisal by family member Dermot Turing, and Chapter 3, by Turing's close colleague and friend Peter Hilton, describes what it was like to work with a genius like Alan Turing, while Chapter 4 focuses on his trial—for being gay—and his shocking punishment. Part II deals with the early origins of the computer, and focuses in particular on Turing's 'universal computing machine', now known as the universal Turing machine. Part III explains exactly what Turing did as a codebreaker at Bletchley Park during the Second World War. The war was a disastrous interlude for many, but for Turing it provided an internationally important outlet for his creative genius. It is no overstatement to say that, without Turing, the war would probably have lasted longer, and might even have been won by the Nazis. The ultrasecret nature of Turing's wartime work meant that much of what he did was kept secret until recent times. Some remains classified to this day.

When the war was over, Turing left Bletchley Park and joined London's National Physical Laboratory. Part IV is about his post-war work on computing, first in London and then in Manchester: Turing had his own quirky but highly effective approach to designing hardware and software. Part V discusses artificial intelligence (AI), called 'machine intelligence' by Turing. He was AI's first major prophet and contributed a slew of brilliant concepts to the field that he founded. Part VI goes on to explain Turing's theory of morphogenesis, his final scientific contribution. This theory tries to unlock the secret of how shapes—such as the shape of a starfish or a daisy—are formed during biological growth. Turing's brilliant 1952 paper on morphogenesis made his reputation as a mathematical biologist, and was also the starting point of the modern field called 'artificial life'. Part VII describes some of Turing's contributions to pure and applied mathematics, including his 'Banburismus' method, used against the German Enigma code, and his work on the *Entscheidungsproblem* or 'decision problem', which gave rise to his universal computing machine. Part VIII is, as its title says, a finale: its topics range from speculations about the nature of the universe to a discussion of recent plays, novels, and music about Turing.

ACKNOWLEDGEMENTS

We thank the following institutions for supporting the research involved in creating this book: BCS-FACS Specialist Group; Birmingham City University; Bletchley Park; the British National Archives at Kew; the British Society for the History of Mathematics; Canterbury University, New Zealand; Det Informationsvidenskabelige Akademi, Copenhagen University, Denmark; Edinburgh University; the Israel Institute for Advanced Studies, Jerusalem; King's College, Cambridge; London South Bank University; Museophile Limited; Oxford University Department of Continuing Education; Queensland University, Australia; and the Swiss Federal Institute of Technology (ETH), Zurich, Switzerland.

Thanks also to Shaun Armstrong, Richard Banach, Will Bowen, Ralph Erskine, Terry Froggatt, Tula Giannini, Kelsey Griffin, Rachel Hassall, Michal Linial, Bob Lockhart, Keith MansField, Patricia McGuire, Francesca Rossi, Dan Taber, Narmatha Vaithiyanathan, and Billy Wheeler.

Special thanks to Graham Diprose for photo post-production and image editing.

CONTENTS

PART I

Biography

CHAPTER 1

Life and work

JACK COPELAND AND JONATHAN BOWEN

A few months after Alan Turing's tragically early death, in 1954, his colleague Geoffrey Jefferson (professor of neurosurgery at Manchester University) wrote what might serve as Turing's epitaph:[1]

Alan in whom the lamp of genius burned so bright—too hot a flame perhaps it was for his endurance. He was so unversed in worldly ways, so childlike it sometimes seemed to me, so unconventional, so non-conform[ing] to the general pattern. His genius flared because he had never quite grown up, he was I suppose a sort of scientific Shelley.

The genius who died at 41

After his short but brilliant career Alan Mathison Turing's life ended 15 days short of his forty-second birthday.[2] His ideas lived on, however, and at the turn of the millennium *Time* magazine listed him among the twentieth-century's one hundred greatest minds, alongside the Wright brothers, Albert Einstein, DNA busters Crick and Watson, and Alexander Fleming, the discoverer of penicillin.[3]

Turing's achievements during his short lifetime were legion. Best known as the mathematician who broke some of Nazi Germany's most secret codes, Turing was also one of the ringleaders of the computer revolution. Today, all who click, tap, or touch to open are familiar with the impact of his ideas. We take for granted that we use the same slab of hardware to shop, manage our finances, type our memoirs, play our favourite music and videos, and send instant messages across the street or around the world. In an era when 'computer' was the term for a human clerk who did the sums in the back office of an insurance company or science lab, Turing envisaged a 'universal computing machine', able to do *anything* that a programmer could pin down in the form of a series of instructions. He could not have foreseen this at the time, but his universal computing machine changed the way we live: it eventually caught on like wildfire, with sales of personal computers now hovering around the million a day mark. Turing's universal machine transported us into a world where many young people have never known life without the Internet.

There was more. In addition to his codebreaking, which saved untold millions of lives, and his remarkable theoretical and practical contributions to the development of the computer, Turing was also the first pioneer of what we now call artificial intelligence. Further, he made profound contributions to mathematics and mathematical logic, philosophy, mathematical biology, and the study of the mind. Jefferson said that Turing's genius 'shone from him': this book explains what that genius achieved.

Birth and early years

Turing's mother, Sara Stoney, came from a family of engineers and scientists, and his father Julius held a position in the Indian Civil Service, in the imperial city of Madras, now Chennai. Julius returned briefly to England with Sara in 1912 and Alan was born that year on 23 June (Fig. 1.1). He entered the world at what is now the Colonnade Hotel in Maida Vale, about half a mile from London's Paddington Station.

The young Alan grew up in the south of England, in a privileged world—cooks, maids, holidays abroad. But he lived the life of a near orphan, lodging with carers and seeing his parents only when they returned from India on leave. Sara described how she came back from one absence of many months to find Alan profoundly changed:[4]

From having always been extremely vivacious—even mercurial—making friends with everyone, he had become unsociable and dreamy.

Figure 1.1 Plaque marking Turing's birthplace at 2 Warrington Crescent, Maida Vale, West London.

Posted by Simon Harriyott to Wikimedia Commons, https://commons.wikimedia.org/wiki/File:Alan_Turing_ (5025990183).jpg. Creative Commons Licence.

As she departed again for her next stint in India, she was

left with the painful memory of his rushing down the school drive with arms flung wide in pursuit of our vanishing taxi.

The unsociable and dreamy child had been thrust into boarding school life at the tender age of 9. An adolescence of persecution and fagging awaited him. His first school was Hazelhurst, near Tunbridge Wells, a prep school for sons of the upper classes. At the age of 14 he moved on to Sherborne School in Dorset: founded in 1550 and built in the shadow of Sherborne Abbey, the place looked like a monastery. Turing arrived by bicycle, alone and dishevelled. 'I am Turing', he announced.[5]

Our timeline picks up the story from Turing's arrival at his new home.

Turing's timeline

1926   *The 14-year-old Turing bicycles more than 60 miles from Southampton to his new school.*

1927/8   *Aged 15½, he writes a précis of Einstein's book 'The Theory of Relativity'.*

1929   *His parents pay to put his name down for King's College, Cambridge.*

1930   *School friend Christopher Morcom dies of tuberculosis.*

1931   *Goes up to King's on a scholarship. Immerses himself in mathematics and in King's gay culture.*

Plays bridge and tennis, rows, skis, enjoys theatre and opera, practises his second-hand violin.

Figure 1.2 Alan Turing aged 16, 17, 18, and 19 in school photographs at Sherborne School.

Images provided courtesy of Sherborne School.

Figure 1.3 'Watching the daisies grow': Alan Turing drawn by his mother Sara.

Reproduced with thanks to the Turing family.

1934	Passes his final Cambridge exams with flying colours: first-class honours with distinction, or 'B Star wrangler' in Cambridge parlance.
1935	Elected a Fellow of King's College at the very young age of 22.
	Attends Max Newman's lectures on the foundations of mathematics. Is inspired to invent the universal Turing machine.
1936	Completes and publishes his famous paper 'On computable numbers, with an application to the Entscheidungsproblem', laying the foundations for modern computer science.
	Travels to the United States to study for a PhD at Princeton University under Alonzo Church. Works on designing ciphers in his spare time.
1937	Studies what mathematicians call 'intuition'.
	Investigates a possible way of circumventing Kurt Gödel's famous incompleteness theorem.
	Plays hockey and tennis.
1938	Finishes his PhD thesis 'Systems of logic based on ordinals'.
	Builds an electrical cipher machine.
	Leaves Princeton for Cambridge as war approaches. Resumes life at King's.
1939	Gives a course of lectures on mathematical logic.
	Attends classes on the foundations of mathematics taught by the eccentric Cambridge philosopher Ludwig Wittgenstein.
	Begins work on a German cipher machine called 'Enigma'. Pays regular visits to veteran codebreaker Dilly Knox at the Government Code and Cypher School (GC&CS) in London.
	Writes his paper 'A method for the calculation of the zeta-function'. Builds a mechanical analogue computer for calculating the zeta-function; pieces of the machinery hang about in attics for many years afterwards.
	War with Germany declared on 3 September. Turing reports to the wartime headquarters of GC&CS at Bletchley Park (aka 'Station X'). Works in Knox's 'Research Section'.

Starts work on German Naval Enigma. Deduces the special method of protection used for transmissions to and from North Atlantic U-boats.

Invents the 'Banburismus' method of attacking U-boat messages.

Completes the preliminary design for his Enigma-breaking machine, the 'bombe'.

1940   Travels to Paris and is present when Polish codebreakers decrypt the first Enigma message to be broken since the start of the war.

Breaks the Enigma key codenamed Light Blue, but this turns out to be only a training key.

Hut 8 is established at Bletchley Park, with Turing in charge (Fig. 1.4). His goal: to break into naval messages on a daily basis.

His first bombe, 'Victory', is installed in Hut 1.

Cycles to and from work wearing a gas mask to ward off hay fever.

Joins the Home Guard and learns to shoot.

German bombs fall close to his lodgings in the village of Shenley Brook End.

Figure 1.4 Reconstruction of Turing's desk and office in Hut 8 at Bletchley Park.
Copyright Shaun Armstrong/mubsta.com. Reproduced with permission of Bletchley Park Trust.

1941   Breaks into current U-boat transmissions for the first time, giving merchant convoys crossing the North Atlantic a means of avoiding the U-boats. Bletchley Park is now able to read U-boat messages almost as quickly as the Germans.

Summoned to Whitehall for official congratulations and £200 bonus.

Proposes marriage to Joan Clarke, his colleague in Hut 8, but later calls it off.

With three other leading cryptanalysts he writes to Prime Minister Churchill complaining about shortages and bottlenecks at Bletchley Park. Churchill memos his Chief of Staff: 'ACTION THIS DAY Make sure they have all they want on extreme priority'.

Explores the concept of machine intelligence; discusses chess-playing algorithms with Jack Good.

1942   Sees his bombes help defeat Rommel in North Africa.

Joins the Research Section's battle against a new German cipher codenamed 'Tunny'. Invents a method called 'Turingery' for breaking the priceless Tunny messages — some signed by Hitler himself.

Discusses machines for Tunny breaking with Max Newman.

Discusses machine intelligence with Donald Michie. Foresees machine learning. Further explores machine chess.

Leaves Bletchley Park for the United States and liaises with US Navy codebreakers in Washington, DC.

Visits the National Cash Register corporation in Dayton, Ohio, to advise on design of American bombes.

1943	Works on speech encryption in New York's Bell Labs. Meets Claude Shannon.
	Returns from the United States and takes on the role of high-level scientific advisor at Bletchley Park.
	Establishes a small lab at Hanslope Park, a few miles from Bletchley Park. Starts work on a portable speech-encryption system.
	Lives among soldiers stationed at Hanslope, eats in their mess.
1944	Just up the road at Bletchley Park, the world's first large-scale electronic computer, Tommy Flowers' Colossus, is breaking Tunny messages from February.
	Meets lifelong friends Robin Gandy and Don Bayley at Hanslope.
	Career as runner begins. Comes an easy first in the mile at regimental sports.
1945	Completes documentation for his 'Delilah' speech encryption system.
	Celebrates VE Day (victory in Europe) by taking a quiet country walk with Bayley and Gandy.
	Travels to Germany with Flowers to investigate German cryptological and communications systems. When the atom bomb is dropped on Hiroshima, explains to Flowers how the bomb works.
	Surprise visit from John Womersley of the National Physical Laboratory (NPL). Accepts Womersley's offer to join the NPL and design an electronic universal Turing machine.
	Moves to London.
	Completes the design of his Automatic Computing Engine (ACE). Specifies processor speed of 1 MHz.
	Says computers can be programmed to 'display intelligence' but at the risk of them making 'occasional serious mistakes'.
	Describes the concepts of self-modifying programs and of programs that modify other programs, presaging the compiler concept.

1946   Presents his report 'Proposed electronic calculator' to the NPL's Executive Committee. This contains the first detailed design of a stored-program computer.

His OBE (Officer of the Order of the British Empire), awarded by King George VI for wartime service, arrives in the post.

Regularly runs 15 miles from the NPL to Flowers' lab in north London, to consult on hardware. Joins Walton Athletic Club and turns in winning performances in long-distance races.

Begins a series of weekly lectures on ACE at London's Adelphi Hotel; the lectures end in February 1947.

1947   Visits America to attend a computer conference at Harvard University. He is not invited to lecture by organizer Howard Aiken, but dominates in post-lecture discussions, especially on memory design.

Gives the earliest known lecture to mention computer intelligence, at Burlington House in London's Piccadilly. This is the public debut of a field now called 'artificial intelligence' (AI).

Pioneers computer programming and builds up a large software library at the NPL.

Fumes at lack of progress on ACE hardware – wants to solder the computer together himself.

Comes fifth in a large qualifying event for the 1948 London Olympics. Decides to enter Olympic trials, but develops a hip problem, quashing his Olympic hopes.

Very fed up with the NPL. Returns to Cambridge for a 12-month sabbatical. Plans to play lots of tennis.

Does pioneering work on machine intelligence in Cambridge.

Invents the LU (lower–upper) decomposition method, used in numerical analysis for solving matrix equations.

1948	Writes AI's first manifesto, titled 'Intelligent machinery'. In this he invents the concepts of a genetic algorithm, neuron-like computing, and connectionist AI.

Hypothesizes that 'intellectual activity consists mainly of various kinds of search' (anticipating Al Newell's and Herb Simon's 'heuristic search hypothesis', saying that intelligence consists in search).

Proposes using 'television cameras, microphones, loudspeakers, wheels and "handling servo-mechanisms"' to build a robot.

Describes a version of the experiment now called the Turing test.

Accepts Max Newman's offer of a job in the Computing Machine Laboratory at Manchester University.

Before leaving Cambridge starts work on software for Manchester's 'Baby', the world's first electronic stored-program computer.

Designs a chess program 'Turochamp' with friend David Champernowne — arguably the first ever AI program.

Moves to Manchester and meets Baby, built by engineers Freddie Williams and Tom Kilburn. Improves hardware, adding input–output equipment.

Experiments with programming musical notes.

1949 Runs his 'Mersenne Express' program for finding prime numbers.

Pioneers debugging techniques.

Tells reporter from 'The Times' that there is no reason why computers will not 'eventually compete on equal terms' with the human intellect.

Gives a lecture titled 'Checking a large routine' at a computer conference in Cambridge, foreseeing what computer scientists now call 'software verification'.

Designs a random number generator for the Ferranti Mark I computer.

1950 Writes the world's first programming manual.

Takes delivery of the first Ferranti Mark I. Soon has the computer to himself for two nights every week.

1950   Runs a program for calculating values of the zeta-function. Reports that the program ran from mid-afternoon until breakfast time the next morning when 'unfortunately at this point the machine broke down and no further work was done'.

Buys a Victorian red-brick semi, Hollymeade, in Wilmslow, an affluent suburb of Manchester (Fig. 1.5). Tells his mother Sara 'I think I shall be very happy here'.

Publishes his farsighted paper 'Computing machinery and intelligence' in the philosophy journal 'Mind', securing his reputation as the father of AI.

Describes his 'imitation game', the full-fledged Turing test.

1951   Elected a Fellow of the Royal Society of London.

Gives a talk on BBC radio titled 'Can digital computers think?'. Says 'If a machine can think, it might think more intelligently than we do, and then where should we be?' Discusses the idea of computers having free will.

In a lecture predicts that thinking computers 'would not take long to outstrip our feeble powers'. Says we should 'expect the machines to take control'.

Meets Christopher Strachey (then working as a schoolteacher) at the Manchester Computing Machine Laboratory. Suggests Strachey try writing a program to make the computer check itself.

Picks up a young man named Arnold Murray in Manchester's Oxford Street.

1952   Broadcasts on BBC radio again, discussing whether computers can be said to think. Predicts it will be 'at least 100 years' before any computer stands a chance in the imitation game, with no questions barred.

Reports a burglary at his house to the police. Admits to a sexual relationship with Murray. Tried and convicted of 'gross indecency'. Is sentenced to 12 months of 'organo-therapy' – chemical castration.

Publishes 'The chemical basis of morphogenesis', pioneering mathematical biology and artificial life. His allied research note 'Outline of development of the daisy' reflects a childhood interest (Fig. 1.3).

1953   Uses Ferranti Mark I to model biological growth.

Also uses the computer as a primitive word processor.

Organo-therapy ends.

Appointed to Readership in the Theory of Computing at Manchester.

Publishes 'Chess', describing his chess-playing program.

1954   Publishes his last paper 'Solvable and unsolvable problems'.

Dies sometime on the night of Monday 7 June, at his home in Wilmslow. His body is found the next day by his housekeeper Eliza Clayton. Post-mortem shows death to be a result of cyanide poisoning (Fig. 1.6).

At an inquest on 10 June the coroner returns a verdict of suicide.

Service and committal on 12 June at Woking Crematorium near Sara's home in Guildford.

Turing left no note, and (going by the inquest transcripts) no evidence was presented to the coroner to indicate that Turing intended suicide: the modern guideline is that a verdict of suicide should not be recorded unless there is clear evidence, placing it beyond any reasonable doubt that the person intended to take his or her own life.[6] At the inquest the coroner said, not very plausibly:[7]

In a man of his type, one never knows what his mental processes are going to do next.

The poisoned apple

One thing that almost everyone has heard about Alan Turing is that he bit into a poisoned apple. Shortly after he died a story in the papers, reporting that a scientist working on an 'electronic brain' had taken cyanide, presented the now familiar image of the bitten apple by the bedside:[8]

On a table at the side of the bed was half an apple from the side of which several bites had been taken.

An article in the *Washington Post* on the morning of what would have been Turing's hundredth birthday reiterated the usual claim that the Allied codebreaker had 'committed suicide by biting into an apple laced with cyanide'; and in 2014 a two-page article in the British *Daily Mail* stated baldly that Turing 'killed himself by eating an apple coated in cyanide'.[9] An apple was indeed found in Turing's bedroom near his body. It was never tested for cyanide, however: the love of a good story filled in the part about the apple being poisoned. The police pathologist, who thought that Turing had drunk cyanide dissolved in water, said only that the apple might have been used to take away some of the taste.[10] In fact, though, the presence of a half-eaten apple on Turing's bedside table offers no clue about how he died, since it was his long-standing habit to eat a few bites of apple last thing at night.[11]

A small cramped laboratory adjoined Turing's bedroom: the 'nightmare room', he called it.[12] In it the police found a glass jam (jelly) jar containing cyanide solution.[13] Charles Bird, the pathologist who carried out the post-mortem examination, thought that Turing must have drunk this cyanide solution.[14] However, the police sergeant attached to the coroner's office, Leonard Cottrell, who examined Turing's body at the scene of death, reported that there was 'no sign of burning about the mouth'—as might have been expected if Turing had drunk the poison—and said that he smelled no more than a 'faint' trace of bitter almonds (cyanide) around Turing's mouth.[15] Quite possibly this faint odour came from the exhaled froth that Cottrell noted on the lips, rather than from a residue left by downing gulps of the strong-smelling contents of the jam jar.

Some sort of experiment was going on in the nightmare room. Cottrell found a pan full of bubbling liquid, with electrodes that were wired via a transformer to the central light fitting in the ceiling. He noted a 'strong smell' of cyanide in the nightmare room.[16] Sara suspected that Turing might have died from inhaling cyanide gas from the experiment.[17] This is a possibility. Illicit drug 'cooks' working in small confined drug laboratories can die from accidental exposure to cyanide gas emitted from their chemical stews.[18]

Figure 1.5 Plaque marking Turing's house 'Hollymeade' in Wilmslow.

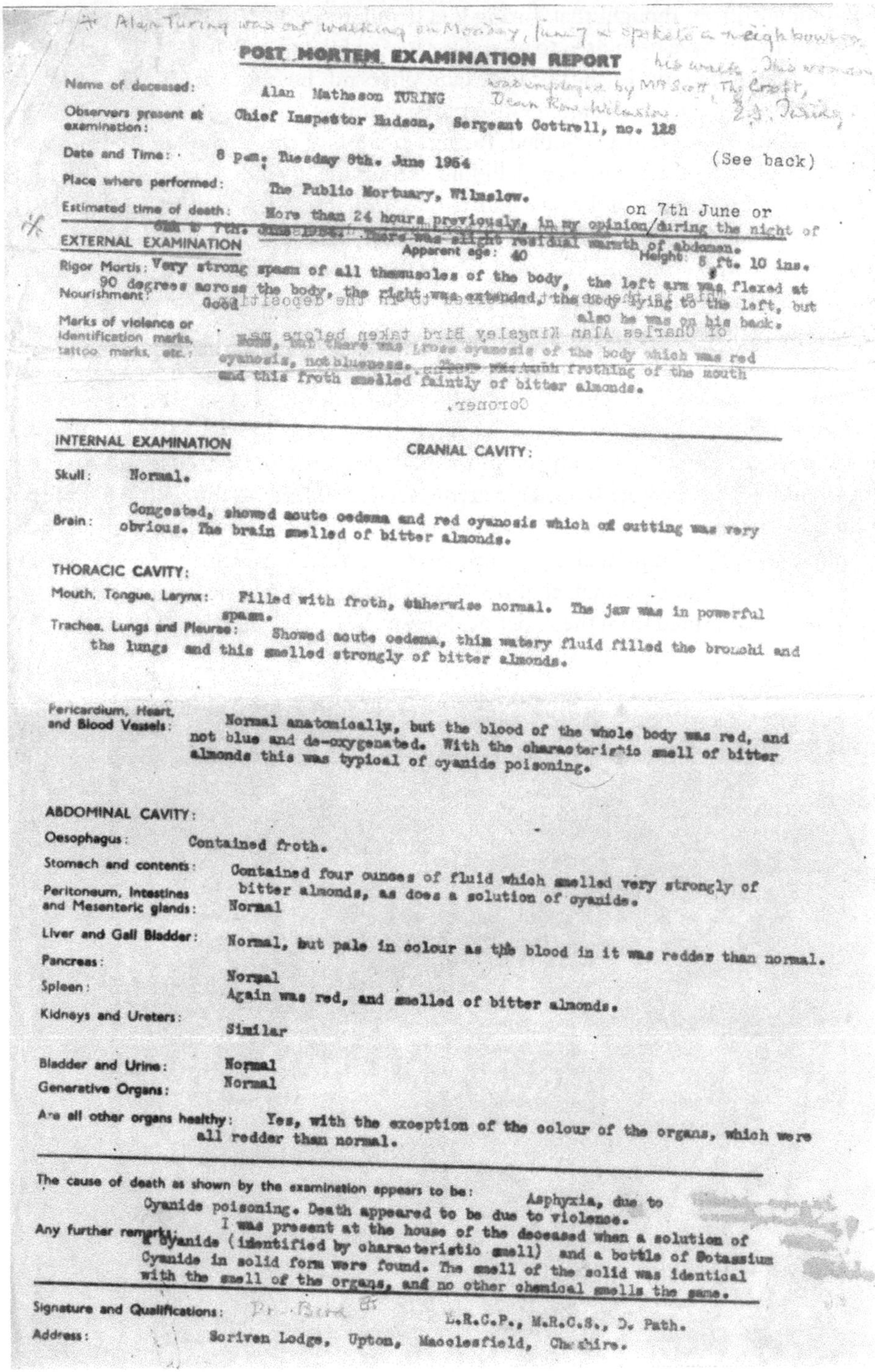

[handwritten top margin] At Alan Turing was out walking on Monday, [?] & spoke to a neighbour... his uncle... this woman was employed by Mrs Scott, The Croft, Dean Row Wilmslow. E.J. Turing

POST MORTEM EXAMINATION REPORT

Name of deceased: Alan Mathæson TURING

Observers present at examination: Chief Inspector Hudson, Sergeant Cottrell, no. 128

Date and Time: 8 p.m. Tuesday 8th. June 1954   (See back)

Place where performed: The Public Mortuary, Wilmslow.

Estimated time of death: More than 24 hours previously, in my opinion/during the night of ~~6th to 7th.~~ *on 7th June or* ~~There was slight residual warmth of abdomen.~~

EXTERNAL EXAMINATION
Apparent age: 40   Height: 5 ft. 10 ins.

Rigor Mortis: Very strong spasm of all the muscles of the body, the left arm was flexed at 90 degrees across the body, the right was extended, the body lying to the left, but also he was on his back.

Nourishment: Good

Marks of violence or identification marks, tattoo marks, etc.: None, but there was gross cyanosis of the body which was red cyanosis, not blueness. There was much frothing of the mouth and this froth smelled faintly of bitter almonds.

[handwritten faint:] was employed... Alan Kingsley Bird taken before... Coroner.

INTERNAL EXAMINATION
CRANIAL CAVITY:

Skull: Normal.

Brain: Congested, showed acute oedema and red cyanosis which on cutting was very obvious. The brain smelled of bitter almonds.

THORACIC CAVITY:

Mouth, Tongue, Larynx: Filled with froth, otherwise normal. The jaw was in powerful spasm.

Trachea, Lungs and Pleurae: Showed acute oedema, thin watery fluid filled the bronchi and the lungs and this smelled strongly of bitter almonds.

Pericardium, Heart, and Blood Vessels: Normal anatomically, but the blood of the whole body was red, and not blue and de-oxygenated. With the characteristic smell of bitter almonds this was typical of cyanide poisoning.

ABDOMINAL CAVITY:

Oesophagus: Contained froth.

Stomach and contents: Contained four ounces of fluid which smelled very strongly of bitter almonds, as does a solution of cyanide.

Peritoneum, Intestines and Mesenteric glands: Normal

Liver and Gall Bladder: Normal, but pale in colour as the blood in it was redder than normal.

Pancreas: Normal

Spleen: Again was red, and smelled of bitter almonds.

Kidneys and Ureters: Similar

Bladder and Urine: Normal

Generative Organs: Normal

Are all other organs healthy: Yes, with the exception of the colour of the organs, which were all redder than normal.

The cause of death as shown by the examination appears to be: Asphyxia, due to Cyanide poisoning. Death appeared to be due to violence.

Any further remarks: I was present at the house of the deceased when a solution of cyanide (identified by characteristic smell) and a bottle of Potassium Cyanide in solid form were found. The smell of the solid was identical with the smell of the organs, and no other chemical smells the same.

Signature and Qualifications: Dr. Bird   L.R.C.P., M.R.C.S., D. Path.

Address: Scriven Lodge, Upton, Macclesfield, Cheshire.

Figure 1.6 The post-mortem report, with handwritten notes by Sara Turing.

Turing might have committed suicide, but then again he might not. We shall most probably never know. Perhaps we should just shrug our shoulders, agree that the jury is out, and focus on Turing's life and extraordinary work.

Aftermath

In 2009 there was a handsome and long-awaited apology from the British Prime Minister, Gordon Brown:[19]

While Turing was dealt with under the law of the time, and we can't put the clock back, his treatment was of course utterly unfair, and I am pleased to have the chance to say how deeply sorry I and we all are for what happened to him.

Then, in 2013, Turing was granted a royal pardon by Britain's Queen Elizabeth II. At that time, the other men who were convicted under the same anti-gay legislation—about 75,000 of them—remained unpardoned (but see Chapter 42). Turing, who did nothing that needed pardoning, might perhaps have preferred to be left among the unpardoned, as a notable example of a man victimized by an unjust and wicked law.

The new fields that Turing had pioneered went from strength to strength in the years following his death—most visibly computer science and computer programming, but also artificial intelligence and mathematical biology. Fittingly, computing's equivalent of the Nobel Prize is called simply the A. M. Turing Award.

The Turing Guide is an introduction to the life's work of this shy, gay, witty, grumpy, courageous, unassuming, and wildly successful genius.

CHAPTER 2

The man with the terrible trousers

SIR JOHN DERMOT TURING

My uncle, Alan Turing, was not a well-dressed man. It is a tribute to those who employed him that he was able to flourish in environments that ignored his refusal to comply with social norms as much as he disregarded mindless social conventions. Social conventions, however, became an increasingly powerful influence over his life. Here I retell the story from the family perspective.

Caught on camera

There is an old photograph in the family album that shows Alan in his last years at Sherborne (Fig. 2.1). It was taken in June 1930—a few months after his friend Christopher Morcom's death—and Alan looks relaxed and happy. But his trousers are a complete disgrace. It is not clear who took the picture, but the timing suggests that it was done at Commemoration, the annual festival at Sherborne to which parents and dignitaries are invited, and where boys, particularly senior boys, should be smartly turned-out. Ordinarily, Alan's mother (my grandmother) would have intervened and spruced him up. But given that Alan was, like other boarding-school boys, responsible for his own clothes, she probably had no control over him any more, if indeed she ever had done.

Hazelhurst, 1922

My grandmother had had little direct control over Alan during his formative years. My grandfather was serving the Empire in India, and she, as a good memsahib, was expected to be with him to run his household. (From the distance of a century or so, this seems a waste of talent, for my grandmother had a formidable intellect as well as many other gifts, and in a later age

Figure 2.1 Alan Turing at Sherborne
School in 1930.

Reproduced with permission from Beryl Turing.

would probably have become a scientist of distinction.) So Alan was deposited in England with foster parents in St Leonards-on-Sea, and at nine years of age was sent off to a prep school called Hazelhurst, near Frant in Sussex.

School seems to have been a reasonably good experience for him—at least in his first term. There was the incident of the geography test. At that time my father, being four years older than Alan, was in the top form while Alan was in the bottom one. The whole school was made to do a geography test. Turing 1 (my father) got 59 marks and Turing 2 (Alan) got 77; my father considered this a thoroughly bad show.

At King's College, Cambridge, there is an archive that contains many things of interest relating to Alan Turing, including his letters home from Hazelhurst.[1] Curiously, these start 'Dear Mother and Daddy', which might give amateur psychologists something to chew on. A typical letter is this one, from 8 June 1924, when Alan was nearly 12:

Dear Mother and Daddy. I have started writing with my fountain pen again, please tell me if you think my writing is worse with it . . . I do not know whether I told you last week but once when I said how much I hated tapioca pudding and you said that all Turings hate tapioca pudding and mint-sauce and something else I had never tried mint-sauce but a few days ago we had it and I found out very much that your statement was true.

Alan's fountain pen was, of course, his own invention, and mint sauce is something of an aberration in the British culinary canon. The school reports groaned, inevitably, about Alan's terrible handwriting and untidiness. A famous drawing done by my grandmother, where Alan is studying daisies rather than watching the ball during a hockey game, dates from these Hazelhurst years (see Fig. 1.3).

When my father left Hazelhurst he was sent to Marlborough College, where he had a thoroughly miserable time. He concluded that Alan would wither in that environment and managed to persuade my grandparents to send Alan to Sherborne School, where things were a bit more liberal and Alan could indulge his passion for science. While at Sherborne, Alan won a handful of school prizes. As one might expect for the times, these were leather-bound classics with the school coat of arms nicely embossed on the front. They are now on display at the museum at Bletchley Park and are in mint condition: it is clear that Alan never touched them. From correspondence between Alan and the Morcom family, it seems that Alan wanted to receive for his prizes a collection of books on scientific subjects, but the school rules (even at Sherborne) were not liberal enough to allow that.

In 1931, after Sherborne, Alan went up to King's College, Cambridge, as an undergraduate. In an academic environment such as King's, or even Bletchley Park, he could get away with a lackadaisical approach to dress and social niceties. But while he was at King's Alan was not just buried away thinking about mathematics. His letters home show a vibrant interest in the anti-war movement: it was not possible to be cut off from politics in the early 1930s. The captain's book of the King's College boat club shows that by the time of his graduation Alan had become a reliable rower, and he helped the second eight to victory by stepping in to replace the number five man who became injured, perhaps in unedifying circumstances, mid-way through the May bumps races; the victory oar that Alan helped to win for his club is also on display at Bletchley. But there are few photographs—indeed, there are hardly any photographs of Alan as an adult.

At King's the photographs, like most of those from school, are the formally posed ones: graduation (Figs 2.2 and 2.3) and the boat club. In the family album is one of him at my father's wedding, and someone has clearly had a thorough go at him for this: his hair is shiny and well brushed and his trousers have something almost unique for him—a sharp crease. I suspect that my grandmother may have had a hand in that. Apart from the graduation shots, the other full-length pictures suggest that Alan disdained pointless workaday things such as having clean and nicely pressed trousers (Fig. 2.4).

Bletchley Park, 1939

Alastair Denniston was the head of the Government Code and Cypher School (GC&CS) in 1938. His plan was to have an emergency list of 'men of the professor type' to be called upon to re-create the successes of Room 40 which had broken German codes during the First World

Figure 2.2 Alan's graduation photograph in 1934.

Reproduced with permission from Beryl Turing.

War. It is unlikely that poor dress sense was a positive indicator for recruitment, but it says much for the foresight of the leaders of Bletchley Park—who were Royal Navy officers and might be expected to care about that sort of thing—that they were willing to look beyond the superficial and to make allowances for the eccentric, and sometimes aberrant, behaviour of their people. So Alan's name went on the Denniston list.

On 4 September 1939 Alan received his call and reported for duty at Bletchley Park, to where the GC&CS had moved. In those days Bletchley Park was a very different place from what it would later become. There were nine men of the professor type when Alan got there, and seven more arrived the next day. Over the next few years, codebreaking at Bletchley grew from a backroom approach modelled on the Age of Enlightenment into a huge and smoothly running intelligence machine, outgrowing the early ramshackle huts and moving into purpose-built accommodation, all of which one can still see there today.

Until the mid-1970s the family had little knowledge about what Alan had done at Bletchley Park. Although we knew there had been a codebreaking effort and that Alan had been part of it, we had no idea about its successes, its strategic or operational impact, or the significance of

Figure 2.3 Guildford in 1934: the lady in the hat may be wondering why the photographer is snapping a man with trousers like that.

Reproduced with permission from Beryl Turing.

his own role. It is well known that the Bletchley Park employees were sworn to secrecy about their work, even after the war had ended, so Alan was never in a position to tell his parents or my father what he had actually done. My grandmother's biography of Alan, *Alan M. Turing* (S. Turing 1959), refers to Alan's war years as 'work in the Foreign Office': she says that at first even his whereabouts were kept secret, although later Alan was able to say that he was working at Bletchley Park. But (as she wrote in 1959) 'no hint was ever given of the nature of his secret work, nor has it ever been revealed'.

The image of Alan as an eccentric man of the professor type grew out of his having to work at Bletchley with a large number of non-academic mortals and service personnel, who may have found it surprising that someone could be so highly regarded while so young, and with such a casual approach to dress and other social mores. One of his biographers, Andrew Hodges, has noted that there were stories going the rounds about Alan's trousers being held up with string

Figure 2.4 Julius Turing (Alan's father) and Alan in Guildford in 1938; Julius has a distinct crease in *his* trousers.

Reproduced with permission from Beryl Turing.

and his wearing a pyjama jacket under his sports coat. My grandmother was still forlornly try-ing to get him to tidy himself up:[2]

He lived at the Crown Inn, Shenley Brook End, some three miles out of Bletchley. Here his kind landlady, Mrs Ramshaw, took great care of him and generally mothered him and admonished him about his clothes. Someone on the staff at Bletchley Park reported to a relative of ours that Alan was 'wrapped up in his theories and wild as to hair and clothes and conventions, but a dear fellow'. Alan himself deplored the shabby clothes of some other people at Bletchley Park and complained that they wore them 'not even patched'. It was, of course [she reminds us] the time when clothing coupons restricted outlay on clothes.

My grandmother died in 1976, at the age of 93, but at that time there was little public understanding of what Alan Turing had really done, either at Bletchley or in his wider contribution to computer science: how cruel it seems for her not to have been allowed an insight into Alan's work in unravelling the secrets of the Enigma. Sure, the rule of secrecy had been thrown out in 1974, when Group Captain Winterbotham wrote an exposé called *The Ultra Secret*,[3] but Winterbotham's book does not contain a single mention of Alan Turing. In the exhibition at Bletchley Park there is a copy of a letter written to my grandmother just before she died, which tells her briefly that Alan's contribution to the codebreaking effort was highly significant—but there is no detail. In 1977 the BBC made a television series, *The Secret War*, and at home we crowded round a small black-and-white set to find out more. They showed a photograph— fortunately, head and shoulders only—of Alan, but they really didn't explain what he had done or how. It took several more years, and the publication of Andrew Hodges' superb biography (Hodges 1983), for Alan's role to be better understood.

Manchester, 1952

At Cambridge Alan could lead his life in a way that suited him—and, to a large degree, the years at Bletchley were also reasonably liberal. But after Bletchley, social convention caught up with him. After the war, Alan joined one of the efforts in Britain to build a universal computing machine. Living in Britain in the austerity years was not easy, and working outside the academic system at the National Physical Laboratory (NPL) added frustration and anti-climax into the mix.

Alan's time at the NPL was destined to be limited. One of the factors behind his parting company with them was that he had written a rather visionary paper on what uses a computer could be put to, which had been dismissed by the NPL's director, Sir Charles Darwin (grandson of *the* Charles Darwin), as a schoolboy's essay not suitable for publication.

When Max Newman rescued him with the offer of a post in Manchester it must have seemed a release, an opportunity to return to a more relaxed way of living and thinking. The development of a computing laboratory at Manchester provided Alan with an opportunity to use the machines in a thoroughly modern way: as a tool to try out ideas, not as advanced calculators for crunching through long arithmetical problems. The point for Alan was that actually building the thing and doing arithmetic on it were much less exciting than the wider opportunities that having a computer could open up.

The progress at Manchester caught the spirit of the times. This was the era of atomic bombs, space rockets, jet-engines, and televisions—all really exciting stuff, especially to schoolboys. As one may imagine, there was much public interest in the idea of an artificial brain. On 11 June 1949 *The Times* ran an article headed: 'The Mechanical Brain. Answer found to 300-year-old sum':

Experiments which have been in progress in this country and the United States since the end of the war to produce an efficient mechanical 'brain' have been successfully completed at Manchester University, where a workable 'brain' has been evolved . . . The Manchester 'mechanical mind' was built by Professor F. C. Williams, of the Department of Electro-Technics, and is now in the hands of two university mathematicians, Professor M. H. A. Newman and Mr A. W. [*sic*.] Turing. It has just completed, in a matter of weeks, a problem, the nature of which is not disclosed, which was started in the seventeenth century . . .

Breathless and heady stuff, and controversial. Alan had jumped in with both feet:

Mr Turing said yesterday: 'This is only a foretaste of what is to come, and only the shadow of what is going to be . . . It may take years before we settle down to the new possibilities, but I do not see why it should not enter any one of the fields normally covered by the human intellect, and eventually compete on equal terms. I do not think you can even draw the line about sonnets, though the comparison is perhaps a little bit unfair because a sonnet written by a machine will be better appreciated by another machine.'

The business about sonnets began with Sir Geoffrey Jefferson, who was the professor of neurosurgery at the University of Manchester and had spoken on the subject 'The mind of mechanical man' in his Lister Oration, the day before *The Times* interview. Sir Geoffrey was unenthusiastic about the idea of thinking machines, and among his remarks he had said:

not until a machine can write a sonnet or compose a concerto because of thoughts and emotions felt, and not by the chance fall of symbols, could we agree that machine equals brain.

For his peroration, Sir Geoffrey decided to quote several lines from *Hamlet*, including:

What a piece of work is a man! How noble in reason! How infinite in faculty . . .

and so forth. Of course Alan knew *Hamlet*. Alan's brother John, my father, says this:[4]

I think it must have been when Alan was due to take the School Certificate examination that he read Hamlet in the holidays. My father was delighted when Alan placed the volume on the floor and remarked 'Well, there's one line I like in this play'. My father could already see a burgeoning interest in English literature. But his hopes were dashed when Alan replied that he was referring to the final stage direction—Exeunt, bearing off the bodies.

Battle was joined, in the most civilized fashion: Sir Geoffrey for the establishment and for those who saw machines as calculators, and Alan for the visionaries. In 1952 the BBC brought them to face each other in a debate on the wireless about whether machines could be said to think, and to ensure fair play the referee was Professor Newman.

Intellectually, then, Manchester was a great place to be for Alan, but from a personal angle the move to Manchester was to be more complex. The story is well known. My grandfather had left Alan his gold watch and this, along with some minor items, was stolen from Alan's house by a local petty thief acting in cahoots with a young man that Alan had picked up in a pub. Being pretty unworldly, as well as deeply upset about the watch, Alan had gone to the police, who were far more interested in the sexual practices of a Fellow of the Royal Society who was working on the mechanical brain and had been giving talks about it on the wireless.

From the family viewpoint, all this was socially difficult. Alan's brother John, a solicitor, was consulted. He felt that the best thing was to minimize the amount of publicity, not least to protect their mother from the scandal, and pleading guilty was the best way to do that. In court Alan did as he was told, but defiantly: he was the one who had been morally wronged. The result was that the press coverage was minimal, and his mother was duly protected—a success of sorts. Alan did not go to prison or lose his job—another victory—but the Turing boys were both casualties in different ways.

As we all know, the court's sentence was a probation order coupled with the condition that he take a course of hormone 'therapy'; this caused Alan to grow breasts and (I imagine) mixed him up in all sorts of ways. Like John, I am a lawyer by profession, and there is a legal aspect to Alan's case which seems, from the perspective of some sixty years later, almost incredible.

Alan Turing was sentenced under the terms of the Criminal Justice Act 1948.[5] This was in fact quite a progressive piece of legislation: Section 1 abolished penal servitude and hard labour, and Section 2 abolished whipping, while Section 3 modernized the ability of the sentencing court to award probation as an alternative to a prison sentence. Until the 1948 Act the courts had not really been able to attach conditions to probation orders, but the reformed law allowed the judge to 'require the offender to comply with such requirements as the court considers necessary for securing the good conduct of the offender or preventing repetition of the same offence'.

Rather naively I had imagined that Alan was dealt with under this last section, and I wondered whether, when passing the 1948 Act, Parliament had intended that the 'requirements' would involve much more than residential conditions or supervision. These days, I suspect that an involuntary course of hormone injections might be challengeable under the Human Rights Act,[6] and even in 1952 it might have been unconstitutional under the Bill of Rights 1689.[7]

However, I was overlooking Section 4, which gave the court power 'if it makes a probation order, to include therein a requirement that the offender shall submit, for such period not extending beyond twelve months from the date of the order as may be specified therein, to treatment by or under the direction of a duly qualified medical practitioner'. The reason I had overlooked this is that this power applied only 'where the court is satisfied . . . that the *mental condition* of an offender is such as requires and as may be susceptible to treatment', and that the treatment would be carried out 'with a view to the improvement of the offender's mental condition'.

Alan was not just regarded as a criminal, he was treated as a mental case. A study carried out at Oxford University's Department of Criminology looked at 636 cases decided in 1953—not Alan's year, but close enough—in which men had been made subject to probation orders with Section 4 conditions attached.[8] Of these men, sixty-seven had been convicted of homosexual offences. There is a telling chart that sets out the medical diagnoses in respect of the twenty offenders whose homosexual offence had been with an 'adult victim':

Psychopathic personality,	10
Low intelligence,	2
Schizophrenia,	2
Illness with physical basis,	2
Anxiety states,	1
Ill-defined neurotic illness,	2
Depression,	1
No abnormality,	0

It seems that, in 1952, a gay man stood no chance of being thought sane. Even so, Alan's luck was out. The Oxford study shows that the treatment was hormonal in only three cases out of 414 where treatment was prescribed under Section 4.

Another consequence of the case was a deterioration in Alan's relations with the family. There was no way that social convention would allow for a discussion about Alan's sexuality at home—to use an anachronism, it was something that his mother would not be able to compute. Alan had found his brother's exasperated handling of the case insensitive, with unworthy words exchanged, which both of them regretted. These wounds needed time to heal, but time was running out. Within two years Alan was dead, apparently by his own hand. And social convention had a continuing role to play even then.

At the time of Alan's death in 1954, his mother was in Italy on holiday. John once again was the one to receive the call from Manchester, and took on the management of what had plenty of potential for more tabloid sensationalism. Worse, when he arrived in Manchester, John learned from Alan's psychiatrist that one of Alan's 'diaries' was missing: in the diaries Alan had committed to paper a range of emotions and catalogued his tortured assessments of other people. To have the diaries adduced in evidence in the coroner's court, with his mother and the nation's press listening to his unfilial remarks, was wholly unthinkable. John searched the house and at last found the diary. Nowadays there are still those who argue that the inquest was a whitewash, or at least superficial, but avoiding the distraction of this unwanted material coming onto the record was what John believed to be the right thing to do. And you did not have to look far to find plenty of other evidence, both physical and psychological, to explain the causes of Alan's death.

Alan Turing was evidently not the only gay man to be prosecuted in the 1950s. The law—and the resources being devoted to headline-grabbing prosecutions of celebrities—were in disrepute. Only two months after Alan's death, in the last days of Churchill's final premiership, a committee was set up under Sir John Wolfenden to consider the law and practice relating to homosexual offences and the treatment by the courts of persons convicted.[9] The committee recommended decriminalizing homosexual acts between consenting adults in private, but unfortunately their findings came too late to help Alan. At least the law was changed in the 1960s; but resetting social attitudes, so that gay people might not be thought of as having a 'mental condition such as requires treatment', has taken much longer.

Great Britain and beyond, 2012

In contrast to 1952, the year 2012 brought much to celebrate. How pleasing that, along with the Diamond Jubilee of Her Majesty Queen Elizabeth II and the successes of the UK-hosted Olympic Games, 2012 was also about Alan's ideas: the influence that he had on the foundations of mathematics, on computing, on cryptology, and on developmental biology, not to mention linguistic philosophy concerning the verb 'to think'—each continues to stimulate new thought and research.

The conferences and events that took place in 2012 allowed academics and others to stand back and draw lessons from Alan's life and work. He now stands as a positive role model. In this context, one thing of particular note was a speech in Leeds in October 2012 by the director of the Government Communications Headquarters (GCHQ). He paid an important centenary tribute to Alan Turing, but the keynote of his address was the importance of finding talent and skills in diversity. That was something that Bletchley Park had succeeded in doing in wartime, even if those values were lost sight of in post-war society. Bletchley's wartime culture of diversity is now regarded as exemplary, and has been drawn on as a model for a civil service recruiting drive. Social convention has moved to a different direction, and one in which future Alan Turings might find it easier to thrive.

Consider one further example. Perhaps, outside academic circles, the least known of Alan's studies are his contributions in the field of morphogenesis: visitors find it curious that the Turing building at Sherborne—not two minutes' walk from where the photo was taken of him in 1930—houses the school's biology department. One might expect that the most visual of his ideas would grab the imagination of many, but this chapter of his discoveries draws less attention than the more abstract work on codes and mathematics. Maybe this is because of an immense discovery in Cambridge towards the end of Alan's short life: the famous paper by Watson and Crick[10] on the structure of DNA was published in *Nature* on 25 April 1953, and Alan's work on diffusion in embryonic cells was eclipsed. After Alan's death, developmental biology became 'molecular biology', an idea so big that it swallowed an entire generation of biological scientists (including myself, I should confess, in a short career detour before I became a lawyer). Alan's centenary allows us a new perspective on this: we are reminded that there is much more behind the growth of organisms than their DNA.

The centenary stimulated a wealth of new thinking. Barry Cooper's *Alan Turing Year* website lists over 250 events of various descriptions that commemorated Alan's centenary year,[11] ranging from performances of new musical works to academic symposia of all flavours—my favourite was one held on the island of Hai Nan Do in China, with all presentations delivered in Esperanto. There seem to have been no fashion shows, but maybe that is for the best. Although Alan would have abhorred the fuss in 2012, it is far better that the fuss should be about these interesting things than about trivia such as his trousers.

CHAPTER 3

Meeting a genius

PETER HILTON

I had the good fortune to work closely with Alan Turing and to know him well for the last 12 years of his short life. It is a rare experience to meet an authentic genius. Those of us privileged to inhabit the world of scholarship are familiar with the intellectual stimulation furnished by talented colleagues. We can admire the ideas they share with us and are usually able to understand their source; we may even often believe that we ourselves could have created such concepts and originated such thoughts. However, the experience of sharing the intellectual life of a genius is entirely different; one realizes that one is in the presence of an intelligence, a sensitivity of such profundity and originality that one is filled with wonder and excitement.[1]

Leading light

Alan Turing was such a genius, and those, like myself, who had the astonishing and unexpected opportunity created by the strange exigencies of the Second World War to be able to count Turing as colleague and friend will never forget that experience, nor can we ever lose its immense benefit to us.

Before the war, in 1935–36, Turing had done fundamental work in mathematical logic and had invented a concept that has come to be known as the 'universal Turing machine' (see Chapter 6). His purpose was to make precise the notion of a computable mathematical function, but he had in fact provided a blueprint for the most basic principles of computer design and for the foundations of computer science.

I joined the distinguished team of mathematicians and first-class chess players working on the Enigma code in January 1942. Alan Turing was the acknowledged leading light of that team. However, I must emphasize that we were a team—this was no one-man show! Indeed, Turing's contribution was somewhat different from that of the rest of the team, being more concerned with improving our methods, especially the machines we used to help us, and less concerned with our daily output of deciphered messages. It was due to the efforts of Turing and the entire team that Churchill was able to describe our work as 'my secret weapon'. Turing could never have said—as Derek Jacobi, the actor who so dramatically played the role of Turing did in Hugh Whitemore's play *Breaking the Code*—'I broke the German code'. Neither Turing's modesty nor his honesty would have permitted such grandiosity. Turing's role was decisive, but he was not alone in playing that role.

When I presented myself at the gates of Bletchley Park in January 1942 I was greeted by a somewhat strange individual whose first question was 'Do you play chess?'—the questioner was none other than Turing. Fortunately I was able to answer 'Yes', and much of my first day of war service was spent in helping Turing to solve a chess problem that was intriguing him. I established an easy and informal relationship with him from virtually that first day at Bletchley Park, and was fortunate to maintain that relationship thereafter.

I left Bletchley Park in the summer of 1945, shortly after the end of the European war, and after a year at the Post Office Engineering Research Station I was demobilized and went back to Oxford University to work on my doctorate. In 1948 I took up my first academic appointment, at Manchester University, where the head of department was Max Newman, the great topologist who had played a decisive role as head of a key section at Bletchley Park (see Chapter 14). Max had done me the honour of appointing me as an assistant lecturer; he had also brought off the tremendous coup of luring Alan Turing away from the National Physical Laboratory to Manchester—with special responsibility for designing a computer to be built by Ferranti (see Chapters 20 and 23). Turing's collaboration with Newman was crowned with success. I can recall vividly the hours that Turing spent explaining to me how the computer functioned and how to use it. I left Manchester for Cambridge in 1952, but I remained in touch with Turing until his death.

Warm, friendly human being

Alan Turing was a warm, friendly human being. He was obviously a genius, but he was an approachable and friendly genius. He was always willing to take time and trouble to explain his ideas, but he was no narrow specialist, so that his versatile thought ranged over a vast area of the exact sciences; indeed, at the time of his death his dominant interest was in morphogenesis. He had a very lively imagination and a strong sense of humour—he was a fundamentally serious person, but never unduly austere.

We did not know during the war that Turing was a homosexual. This is not because Turing took elaborate steps to conceal his predilections. He was, characteristically, wholly honest about this and not ashamed, although he was never ostentatious about his preference. After the war, the law against the expression of male homosexuality was upheld with rigorous fervour in Britain, and in January 1952 Turing, then a Reader at Manchester University and a Fellow of the Royal Society, was arrested and charged with committing 'an act of gross indecency' with his friend Arnold Murray. Of course, he didn't deny the charge, but he did not agree that he had done anything wrong. He was bound over on condition that he submit to hormonal treatment designed to diminish his libido; the only obvious effect was that he developed breasts. Even in the straightened circumstances in which he found himself following his trial and conviction, he retained his enormous zest for life and for the free exchange of ideas.

As a by-product of his plea of guilty, he was no longer permitted to work as a consultant to GCHQ in Cheltenham where the codebreakers worked, nor to visit the United States. It is a tragic irony that the British security services should have been mobilized to exclude Turing—whose contribution to the work of GCHQ was of such inestimable value during the war—yet failed so conspicuously to detect the activities of the mole Geoffrey Prime, later convicted of spying for the Soviet Union while he was working at GCHQ. I. J. Good, a wartime colleague and friend, has so aptly remarked about Turing's tremendous contributions to the Nazi defeat:[2]

Fortunately, the authorities did not know Turing was a homosexual. Otherwise we might have lost the war.

Great orginality

If I were to attempt to characterize the nature of Turing's genius, I would say that it lay in his capacity for thought of great originality, so often going back to first principles for his inspiration. Even the most superficial examination of his *œuvre* shows how often he took up a new topic. His publications range very widely indeed; but to certain topics, in which he made fundamental advances, he made only one or two published contributions.

His very characteristic way of thinking can be illustrated by relating certain more light-hearted episodes in Turing's life, which also point up his quirky and infectious sense of humour. Let me describe first his approach to the game of tennis. Turing was a superb athlete—indeed, a marathon runner of great distinction. At Bletchley Park, he developed a real delight in playing tennis, and especially enjoyed playing doubles. He was very good up at the net, where his speed and good eye enabled him to make many effective interceptions. However, he was dissatisfied with his success rate—too often he intercepted a return from an opponent, but sent the ball into the net. Applying his remarkable thinking processes to a mundane problem, he reasoned as follows:

The problem is that, when intercepting, one has very little time to plan one's stroke. The time available is a function of the tautness of the strings of my racquet. Therefore I must loosen the strings.

Being Alan Turing, he then carried out the necessary alterations to his racquet himself. At this point, my recollection may be coloured by the great distance in time, but I seem to recall Turing turning up for his next game with a racquet somewhat resembling a fishing net. He was absolutely devastating, catching the ball in his racquet and delivering it wherever he chose—but plainly in two distinct operations and, therefore, illegally. He was soon persuaded to revert to a more orthodox racquet.

When the danger of an invasion of the UK by German paratroops seemed a really serious one, Turing volunteered for the Local Defence Volunteers, popularly known as the Home Guard, in order to become an expert at firing a rifle and hitting his target. In order to be enrolled it was, of course, necessary to complete a massive form full of irrelevant questions. As all those experienced in government bureaucracy know, when completing such forms the only essentials are to give one's name, to sign and date the form, and to answer every question. The answers themselves don't matter, because they are never likely to be read. One of the questions was:

Do you understand that, by enrolling in His Majesty's Local Defence Volunteers, you render yourself liable for military discipline?

Figure 3.1 Peter Hilton.

Photo taken by Marge Dodrill, reproduced courtesy of Meg Hilton.

Turing argued as follows: 'I can imagine no set of circumstances under which it would be to my advantage to answer this question "Yes" '. So he answered it 'No'. He was duly enrolled and soon became a first-class shot—he usually did very well the things that he set himself to do.

But then the danger of invasion receded and Turing began to find attending parades increasingly tedious, so he stopped going. In consequence, he started receiving nasty notes of increasing irritability. These culminated in a summons to his court-martial, presided over by Colonel Fillingham, Officer Commanding the Buckinghamshire Division of His Majesty's Local Defence Volunteers:

> Is it true, Private Turing, that you have attended none of the last eight parades?
> Yes, sir.
> Do you realize this is a very serious offence?
> No, sir.
> Private Turing, are you trying to make a fool out of me?
> No, sir, but if you look up my application for admission to the Home Guard, you will see that I do not understand I am subject to military discipline.

The form was produced, Colonel Fillingham read it, and became apoplectic. All he could say was:

> You were improperly enrolled. Get out of my sight!

CHAPTER 4

Crime and punishment

JACK COPELAND

I n 1952 Turing was arrested and tried for being gay. The court convicted him and sentenced him to chemical castration. It was disgraceful treatment by the nation that he had done so much to save. Turing faced this ordeal with his usual courage.[1]

Arnold

Turing wrote a short story.[2] Although only a few pages long and incomplete, it offers an intimate glimpse of its author. The central character—a scientist by the name of Alec Pryce, who works at Manchester University—is a thinly disguised Alan Turing.

Pryce, like Turing himself, always wore what Turing described as 'an old sports coat and rather unpressed worsted trousers'. Turing called this Pryce's 'undergraduate uniform', saying that it 'encouraged him to believe he was still an attractive youth'. At just the wrong side of 40, Turing must have been feeling his age. Pryce, whose work related to interplanetary travel, made an important discovery in his twenties, which came to be called 'Pryce's buoy'. The nature of the discovery is left unexplained, and Pryce's buoy is obviously a proxy for the universal Turing machine. 'Alec always felt a glow of pride when this phrase was used', Turing wrote revealingly.

'The rather obvious double-entendre rather pleased him too', Turing continued. 'He always liked to parade his homosexuality, and in suitable company Alec would pretend that the word was spelt without the "u"'. Pryce, we are told, has not had a sexual relationship since 'that soldier in Paris last summer'. Walking through Manchester, Pryce passes a youth lounging on a bench, Ron Miller. Ron, who is out of work and keeps company with petty criminals, makes a small income from male prostitution. He responds to a glance that Alec gives him as he passes, calling out uncouthly 'Got a fag?'. Shyly Alec joins him on the bench and the two sit together awkwardly. Eventually Alec plucks up courage to invite the boy to have lunch at a nearby restaurant. Beggars can't be choosers, Ron thinks meanly. He is not impressed by Alec's brusque approach and 'lah-di-dah' way of speaking, but says to himself philosophically, 'Bed's bed whatever way you get into it'.

'Ronald' is an anagram of 'Arnold', and it was in December 1951 that Turing first met Arnold Murray, the Ronald Miller of his short story. Turing picked up Murray in Manchester's Oxford Street and the two ate together.[3] Their first time was a few days later at Turing's house, Hollymeade, in Wilmslow. Afterwards Turing gave Murray a present of a penknife: probably the unemployed Murray would have preferred cash instead. The next time they had sex, Murray stole £8 from Turing's pocket as he left Hollymeade in the morning, and not long after this the house was burgled.

Even though the finger of suspicion pointed at Murray and his seedy friends, Turing spent the night with him one more time. In the morning he led Murray to the local police station. Turing went in, but not Murray. In the course of reporting the burglary he gave the police a wrong description and this, as the newspaper reporter covering his subsequent trial wrote luridly, 'proved to be his undoing'.

During questioning, Turing admitted to having had sex with Murray three times. The burglary dropped out of the picture, eclipsed by this sensational new information. As the police knew all too well, each of the three occasions counted as two separate crimes under the antique 1885 legislation still in force—the commission of an act of gross indecency with another male person, and the reciprocal crime of being party to the commission of an act of gross indecency. Six criminal offences.

After Turing made his statement, he said to a police officer: 'What is going to happen about all this? Isn't there a Royal Commission sitting to legalise it?' But not until 1967 was homosexuality decriminalized in the UK.

Trial

Three weeks later, at the end of February 1952, Turing and Murray appeared in court. The charges were read out and both men were committed for trial. The court granted Turing bail of £50, but refused to let Murray out of custody.

Following a distressing wait of more than four weeks, the trial was held in the quiet Cheshire town of Knutsford at the end of March. Turing's indictment began grandly 'The King versus Alan Mathison Turing', but George VI had recently died, and 'Queen' had been written above the hastily crossed out 'King'. Turing pleaded guilty on all six counts, as did Murray.

Putting on a brave face, Turing joked: 'Whilst in custody with the other criminals I had a very agreeable sense of irresponsibility'.[4] 'I was also quite glad to see my accomplice again', he admitted, 'though I didn't trust him an inch'. The mathematician Max Newman, Turing's long-time friend, was called as a character witness. 'He is completely absorbed in his work, and is one of the most profound and original mathematical minds of his generation', Newman said. It must have been good to hear these words, even on such a black day.

Murray's counsel attempted to shift the blame onto Turing, saying that Turing had approached Murray. If Murray 'had not met Turing he would not have indulged in that practice or stolen the £8', the barrister argued crassly. But his tactics worked: despite a previous conviction for larceny, Murray got off with 12 months' good behaviour. Turing's own counsel hoped to steer the court away from a prison sentence, and alluded to the possibility of organotherapy: 'There is treatment which could be given him. I ask you to think that the public interest would not be well served if this man is taken away from the very important work he is doing'.

The judge followed the barrister's lead, sentencing Turing to 12 months' probation and ordering him to 'submit for treatment by a duly qualified medical practitioner at Manchester Royal Infirmary'. It was not exactly the eulogy he deserved from the nation he had saved. Turing wrote in a letter 'No doubt I shall emerge from it all a different man, but quite who I've not found out'. He signed the letter 'Yours in distress'.[5]

The alternative of prison would probably have cost him his job, and with it his access to a computer. Already his arrest had cost him something else that mattered to him: as he told a friend, he would never be able to work for GCHQ again.[6] After the war GCHQ—Bletchley Park's peacetime successor—had offered him the astronomical sum of £5000 to break post-war codes (a figure more than six times his starting salary when he began work at the National Physical Laboratory in 1945).[7] One of his Bletchley Park colleagues, Joan Clarke, who stayed on as a peacetime codebreaker, confirmed that Turing visited GCHQ's Eastcote site after the war as a consultant.[8] But now Turing, the perfect patriot, had unwittingly become a security risk.

Probation and the 'treatment' from hell

The so-called treatment consisted of flooding his body with female hormones for a year.[9] 'It is supposed to reduce sexual urge whilst it goes on, but one is supposed to return to normal when it is over', Turing said, adding 'I hope they're right'.[10]

Turing seems to have borne it all cheerfully enough, enduring his trial and the subsequent abusive chemical 'therapy' with what his friend Peter Hilton described as 'amused fortitude'. He even regarded the hormone treatment 'as a laugh', his friend Don Bayley remembered.[11] Turing had led a resilient life and his resilience did not desert him now. The whole thing was an episode to be got through. 'Being on probation my shining virtue was terrific, and had to be', he said. 'If I had so much as parked my bicycle on the wrong side of the road there might have been 12 years for me'.[12]

Turing told his friend Norman Routledge that he was growing breasts. 'The fact that he had grown breasts as a result of the treatment he regarded as rather a joke', Routledge said.[13] The inhumane 'therapy' was evidently wreaking havoc with Turing's body, yet does not seem to have had the effect that the authorities desired. Even towards the end of the treatment, when the hormones would have been having their maximum impact, Turing was still finding young men that he met 'luscious', and he said so in a letter to his friend Robin Gandy.[14]

Despite the harshness of his personal life, Turing's career was enjoying a new crescendo. The logician turned codebreaker turned computer scientist turned artificial intelligence pioneer had now turned mathematical biologist. In August 1952, as the organo-therapy dragged on, the Royal Society published his groundbreaking paper describing his new theory of how things grow (see Chapters 33–35). It described the first phase of Turing's investigation into the chemistry of life.

During his probation, as alien hormones flooded through his body, Turing undertook the next phase of the investigation: working long hours at the console, he used the Manchester computer to simulate chemical processes described by his theory. In March 1953 two Cambridge researchers, Francis Crick and James Watson, cracked the chemical structure of DNA. Watson related that on the day of the discovery 'Francis winged into the Eagle to tell everyone within hearing distance that we had found the secret of life'.[15] Simultaneously, Turing was on the brink of discovering an even deeper secret: as we grow in our mother's womb, how does nature achieve that miraculous leap from microscopic twists of DNA to actual anatomy?

The movie *The Imitation Game*, starring Benedict Cumberbatch as Turing, had a scene showing Turing, his brain profoundly affected by the hormones, unable even to start a crossword puzzle, let alone solve it.[16] The scene is entirely fictitious, like much of the movie (*The New York Review of Books* described this as 'one of the film's most egregious scenes').[17] At the time the scene portrayed, the real Turing was busy with his own computer-assisted quest for the secret of life.

The Kjell crisis

A difficult spell blew up towards the end of Turing's probation with the arrival of a postcard announcing the visit of his Norwegian boyfriend, Kjell Carlsen.[18] Turing described his relationship with Kjell as one of 'Perfect virtue and chastity'.[19] 'A very light kiss beneath a foreign flag, under the influence of drink, was all that had ever occurred', he said. But the last thing he needed was Kjell turning up in Wilmslow during his probation. The postcard drew an astonishing response from the authorities, who must have been snooping on Turing's mail. Kjell never reached Turing. 'At one stage police over the N of England were out searching for him, especially in Wilmslow, Manchester, Newcastle etc.', Turing told Robin Gandy.[20] Kjell found himself back in Bergen.

The state was evidently keeping a very close watch indeed on Alan Turing. He knew Britain's best codebreaking secrets, and his arrest had come at exactly the wrong time. Guy Burgess and Donald Maclean had defected to Moscow in the middle of 1951, sparking a scandal that in the public mind associated treachery, Cambridge intellectuals, and homosexuality. MI5 and the Secret Intelligence Service would not want to be caught napping again. It was also in 1951 that the American scientists Ethel and Julius Rosenberg were sentenced to death for passing atomic weapons secrets to the Russians; their execution in the United States and the Kjell episode happened at virtually the same time. On both sides of the Atlantic anxiety about security was at white heat.

At that point in the Cold War the authorities must have seen Turing as one of the West's biggest security risks. During the war with Germany he had been a near-omniscient overlord of the Allied crypto-world—knowledge that had a long shelf life—and now, horror of horrors, this queer freethinking Cambridge intellectual had a foot in the atom world too. The Manchester computer was being used for atomic weapons work and Turing had been consulted at the outset, when the authorities approached the Manchester Computing Machine Laboratory about the need to carry out 'a series of very lengthy calculations'.[21] From that first contact in 1950 a permanent relationship quickly developed, and in October 1953 the British Atomic Weapons Research Establishment signed a formal agreement to purchase large blocks of time on the Manchester computer for top-secret work.[22]

What Turing called the 'Kjell crisis' seemingly passed, and a few weeks later in April 1953 his probation came to an uneventful end, well over a year before his death. He was rid of the organo-therapy, and in the warm sunny spring of 1953 the skies were blue again.

After the punishment

At the start of the summer of 1953 Turing packed his luggage for a holiday at the newly opened Club Méditerranée in Ipsos, on the Greek island of Corfu.[23] Sun, sea, men. He sent a postcard saying that he had met a lovely young man on the beach.

If Turing was suffering from life-threatening depression following the months of hormones, there is no evidence of it. So far as it is possible to tell from the evidence that survives, his mental state during the remaining 14 months of his life seems to have been unremarkable. His career, moreover, was at one of its highest points, with his research into growth going marvellously well and with the prospect of fundamentally important new results just around the corner. Turing talked about the topic of suicide with his friend Nick Furbank: 'It was the sort of thing I did discuss with him from time to time', Furbank told me.[24] But if Turing was talking obliquely about himself, this was not clear to Furbank—if he had an intention to take his own life, he did not tell his friend. Furbank felt no concern: caught up in other things, he unintentionally drifted out of Turing's circle, and there were no visits to Hollymeade during the 2 or 3 months prior to Turing's death. Furbank accepted the verdict of Turing's inquest—many did—and nearly 70 years later still believed that he had 'failed' Turing.

Another Turing movie, *Codebreaker*, was first shown on British television in 2011: much of the action of this well-told docudrama, produced by Patrick Sammon and Paul Sen, consists of fictitious conversations between Turing and his psychotherapist Dr Greenbaum.[25] It was in 1952 that Turing began seeing Franz Greenbaum, who lived a short bicycle ride away along the Wilmslow Road. In typical Turing fashion he soon formed friendships with not only the analyst but also his young family, often spending Sundays with the Greenbaums. Towards the end of the screenplay the Turing character discusses suicide with the Greenbaum character. In Sammon and Sen's follow-up article 'Turing committed suicide: case closed', they portrayed Greenbaum as feeling that 'suicide was a likely scenario' and as believing that, as Turing's therapist, he had 'insight into Alan's death as suicide rather than accident'.[26] These are not Greenbaum's own statements. In the year after Turing's death the real Greenbaum wrote in a letter to Sara (whether sincerely or not we cannot tell): 'There is not the slightest doubt to me that Alan died by an accident'.[27]

Sammon and Sen described the fact that, approximately four months before his death, Turing prepared a will as 'evidence of pre-meditation'.[28] Yet for a person to make a will at the age of 41 is not an unusual event. In any case, Turing's will seems poles apart from the last instructions of a man expecting to die soon. He included a small bequest to his housekeeper, Mrs Clayton, together with an additional annual bonus 'for each completed year in which she shall be in my employ from and after the thirtyfirst day of December One thousand nine hundred and fifty three' (see Fig. 4.1).[29] Never a man to waste words, it seems unlikely that Turing would have included this formula in his will unless he was actually envisaging Mrs Clayton's bonuses accruing during her future years of service to him.

Last days

The week before Turing died, he and Robin Gandy passed an enjoyable weekend together at Hollymeade: Turing's housekeeper Mrs Clayton said that 'they seemed to have a really good time'.[30] 'When I stayed with him the week-end before Whitsun', Gandy said, 'he seemed, if anything, happier than usual'.[31] Turing's neighbour, Mrs Webb, also found him perfectly cheerful. On Thursday 3 June, just four days before his death, he threw an impromptu party for her and her little boy Rob, making them tea and toast. 'It was such a jolly party', she recalled.[32] Gandy accepted the official verdict of suicide, but felt that there was no reason for Turing's action.

4. I GIVE the following pecuniary and specific legacies namely:-

(a) To the said Philip Nicholas Furbank in case he shall prove this my Will the sum of One hundred pounds

(b) To each of them my said brother John Ferrier Turing and my sister-in-law Joan and nieces Inagh Jean Shuna and Janet all of Glenthorne West Road Guildford who shall survive me the sum of Fifty pounds

(c) To Mrs Clayton of 6 Mount Pleasant Wilmslow in case she shall survive me the sum of Thirty pounds together with an additional sum of Ten pounds for each completed year in which she shall be in my employ from and after the thirtyfirst day of December One thousand nine hundred and fifty three

Figure 4.1 Extract from Turing's will, signed by him on 11 February 1954.

Another of Turing's wartime friends, Peter Hilton (the author of Chapter 3), was employed in the mathematics department at Manchester University and was seeing Turing regularly at this time. Hilton told me that, before going home that last time, Turing left a note in his university office: the note contained Turing's instructions to himself about the work he was going to do the following week. Another of Turing's colleagues at the university, computer engineer Owen Ephraim, who worked side by side with Turing at the computing laboratory from early 1954, told me: 'I was the last person to spend working time with Alan'. The two had 'said cheerio' as usual at the end of what turned out to be Turing's last week at the laboratory. 'Nobody from the police or elsewhere ever interviewed me to ask about his behaviour in those last days before his life ended', Ephraim said. He continued:

If I had been asked, I would have said that Alan Turing acted perfectly normally during those last days, and with as much dedication as ever.

Not long before his death Turing made an important appointment with Bernard Richards (the author of Chapter 35) to hear the details of this young scientist's exciting new confirmation of Turing's theory of biological growth. Tragically the appointment was for the very day Turing's body was discovered.[33] Richards says:[34]

I had indicated to Turing that I had obtained some startling three-dimensional results showing a biological solution to his Morphogenesis Equation. He said that our meeting could not come soon enough. Alas, he never came and I was shocked by the news of his death.

PART II

The universal machine and beyond

CHAPTER 5

A century of Turing

STEPHEN WOLFRAM

I never met Alan Turing; he died five years before I was born. But somehow I feel I know him well, not least because many of my own intellectual interests have had an almost eerie parallel with his. And by a strange coincidence, the 'birthday' of Wolfram Mathematica, 23 June 1988, is aligned with Turing's own.[1]

I think I first heard of Alan Turing when I was about 11 years old, right around the time I saw my first computer. Through a friend of my parents, I had got to know a rather eccentric old classics professor, who, knowing my interest in science, mentioned to me this 'bright young chap named Turing' whom he had known during the Second World War.

One of this professor's eccentricities was that, whenever the word 'ultra' came up in a Latin text, he would repeat it over and over again and make comments about remembering it. At the time, I didn't think much of it, although I did remember it. Only years later did I realize that 'Ultra' was the codename for the British cryptanalysis effort at Bletchley Park during the war. In a very British way, the classics professor wanted to tell me something about it, without breaking any secrets—and presumably it was at Bletchley Park that he had met Alan Turing.

A few years later I heard scattered mentions of Alan Turing in various British academic circles. I heard that he had done mysterious but important work in breaking German codes during the war, and I heard it claimed that after the war he had been killed by British Intelligence. At that time some of the British wartime cryptography effort was still secret, including Turing's role in it. I wondered why. So I asked around, and started hearing that perhaps Turing had invented codes that were still being used. In reality, though, the continued secrecy seems to have been intended to prevent its being known that certain codes had been broken, so that other countries would continue to use them.

I am not sure where I next encountered Alan Turing. Probably it was when I decided to learn all I could about computer science, and saw all sorts of mentions of 'Turing machines'. But I have a distinct memory from around 1979 of going to the library and finding a little book about Alan Turing written by his mother, Sara Turing.[2]

Gradually I built up quite a picture of Alan Turing and his work, and over the thirty or more years that have followed I have kept on running into him, often in unexpected places. In the early 1980s, for example, I had become very interested in theories of biological growth—only to find (from Sara Turing's book) that Alan Turing had done all sorts of largely unpublished work on that. And in 1989, when we were promoting an early version of Mathematica, I decided to make a poster of the Riemann zeta function, only to discover that Alan Turing had once held the record for computing its zeros (see Chapter 36)—earlier he had also designed a gear-based machine for doing this.

Recently I even found out that Turing had written about the 'reform of mathematical notation and phraseology'—a topic of great interest to me in connection with both Mathematica and Wolfram|Alpha—and at some point I learned that a high-school mathematics teacher of mine (Norman Routledge) had been a friend of Turing's late in his life. But even though my teacher knew of my interest in computers, he never mentioned Turing or his work to me. Indeed, thirty-five years ago, Alan Turing and his work were little known, and it is only fairly recently that he has become as famous as he now is.

Turing's greatest achievement was undoubtedly his construction in 1936 of a universal Turing machine—a theoretical device intended to represent the mechanization of mathematical processes—and in some sense Mathematica is precisely a concrete embodiment of the kind of mechanization that Turing was trying to represent.

In 1936, however, Turing's immediate purpose was purely theoretical—indeed, it was to show not what could be mechanized in mathematics but what could not. In 1931 Gödel's theorem had shown that there were limits to what could be proved in mathematics, and Turing wanted to understand the boundaries of what could ever be done by any systematic procedure in mathematics (see Chapters 7 and 37).

Turing was a young mathematician in Cambridge, and his work was couched in terms of the mathematical problems of his time. But one of his steps was the theoretical construction of a universal Turing machine capable of being 'programmed' to emulate any other Turing machine. In effect, Turing had invented the idea of universal computation, which was later to become the foundation on which all of modern computer technology is built.

At the time, though, Turing's work did not make much of a splash, probably largely because the emphasis of Cambridge mathematics was elsewhere. Just before Turing published his paper he learned about a similar result by Alonzo Church from Princeton, formulated not in terms of theoretical machines but in terms of the mathematics-like lambda calculus. As a result, Turing went to Princeton for two years to study with Church, and while he was there he wrote the most abstruse paper of his life.

Turing's next few years were dominated by his wartime cryptographic work. Several years ago I learned that during the war Turing visited Claude Shannon at Bell Labs in connection with speech encipherment (see Chapter 18). Turing had been working on a kind of statistical approach to cryptanalysis, and I am extremely curious to know whether Turing told Shannon about this and potentially launched the idea of information theory, which itself was first formulated for secret cryptographic purposes.

After the war Turing became involved with the construction of the first actual computers in England (see Chapters 20 and 21). To a large extent these computers had emerged from engineering and not from a fundamental understanding of Turing's work on universal computation. There was however a definite, if circuitous, connection. In 1943 Warren McCulloch and Walter Pitts in Chicago wrote a theoretical paper about neural networks that used the idea of universal Turing machines to discuss general computation in the brain. John von Neumann read this paper and used it in his recommendations about how practical computers should be

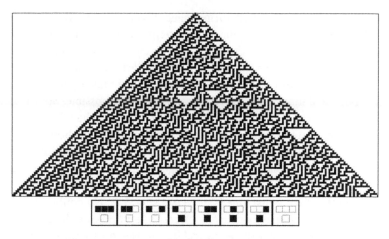

Figure 5.1 An example of a simple cellular automaton known as Rule 30 that I discovered in the early 1980s. Like many other systems in the computational universe, this generates remarkable complexity, even though its rules are very simple.

Photograph reproduced with the permission of Stephen Wolfram.

built and programmed. (Von Neumann had known about Turing's paper in 1936, but at the time did not fully recognize its significance, instead describing Turing in a recommendation letter as having done interesting work on the central limit theorem.)

It is remarkable that in just over a decade Alan Turing was transported from writing theoretically about universal computation to being able to write programs for an actual computer. I have to say, though, that from today's vantage point his programs look incredibly 'hacky'—with lots of special features packed in and encoded as strange strings of letters. But perhaps it is inevitable that to reach the edge of a new technology there has to be hackiness—and perhaps, too, it required a certain hackiness to construct the very first universal Turing machine. The concept was correct, but Turing quickly published an erratum to fix some bugs, and in later years it became clear that there were more bugs: at the time Turing had no intuition about how easily bugs can occur.

Turing also did not know just how general (or otherwise) his results about universal computation might be. Perhaps the Turing machine was just one model of a computational process, and other models—or brains—might have quite different capabilities. But gradually over the course of several decades it became clear that a wide range of possible models were actually exactly equivalent to the machines that Turing had invented.

It is strange to realize that Alan Turing appears never to have actually simulated a Turing machine on a computer. He viewed Turing machines as theoretical devices relevant for proving general principles, but he does not appear to have thought about them as concrete objects to be studied explicitly. Indeed, when Turing came to make models of biological growth processes, he immediately started using differential equations and appears never to have considered the possibility that something like a Turing machine might be relevant to natural processes.

Around 1980, when I became interested in simple computational processes, I also didn't consider Turing machines, and instead started off studying what I later learned were called 'cellular automata'. What I discovered was that even cellular automata with incredibly simple rules could produce incredibly complex behaviour, which I soon realized could be considered as corresponding to a complex computation (Fig. 5.1).

I probably simulated my first explicit Turing machine only in 1991. To me, Turing machines were built a little bit too much like engineering systems, rather than things that were likely to correspond to systems in nature. But I soon found that even simple Turing machines, just like simple cellular automata, could produce immensely complex behaviour.

In a sense, Alan Turing could easily have discovered this, but his intuition—like my original intuition—may have told him that no such phenomenon was possible. So it would probably only have been luck—and access to easy computation—that would have led him to find it.

Had he done so, I am quite sure that he would have become curious about what the threshold for his concept of universality would be, and how simple a Turing machine would suffice. In the mid-1990s I searched the space of simple Turing machines and found the smallest possible candidate, and after I put up a $25,000 prize a British college student named Alex Smith showed in 2007 that indeed this Turing machine is universal (Fig. 5.2). No doubt Alan Turing would very quickly have grasped the significance of such results for thinking about both mathematics and natural processes. But without the empirical discoveries, his thinking did not progress in this direction. Instead, he began to consider from a more engineering point of view to what extent computers should be able to emulate brains, and he invented ideas like the 'Turing test (see Chapters 25 and 27)'. Reading through his writings today, I find it remarkable how many of his conceptual arguments about artificial intelligence still need to be made—though some, like his discussions of extrasensory perception, have become quaintly dated (see Chapter 32).

Looking at Turing's famous 1950 article 'Computing machinery and intelligence',[3] one sees a discussion of programming into a machine the contents of *Encyclopaedia Britannica*—which he estimated as occupying sixty workers for fifty years. I wonder what he would think of Wolfram|Alpha which, thanks to progress over the past sixty years, and perhaps some cleverness, has so far taken at least slightly less human effort.

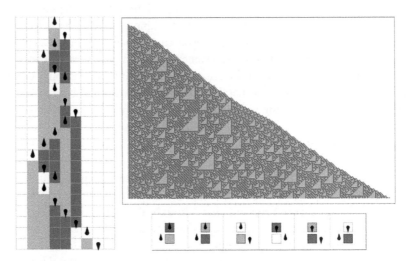

Figure 5.2 A Turing machine that is known to be the simplest possible universal Turing machine. I first identified it in the mid-1990s, and its universality was proved in connection with a prize that I sponsored in 2007. The machine has two states of the scanner (up or down) and three symbols (represented by colours or shades). The left-hand diagram shows twenty steps of evolution, while the compressed form on the right corresponds to a total of 129,520 steps.

In addition to his intellectual work, Turing has recently become something of a folk hero, most notably through the story of his death. It will almost certainly never be known for sure whether his death was intentional, but from what I know and have heard I rather doubt that it was (see Chapters 1 and 4). When one first hears that he died by eating an apple impregnated with cyanide, one assumes that it must have been intentional suicide. But when one later discovers that he was quite a tinkerer, had recently made cyanide for the purpose of electroplating spoons, kept chemicals alongside his food, and was rather a messy individual, the picture becomes a lot less clear.

I often wonder what Turing would have been like to meet. I do not know of any recording of his voice (though he did do a number of BBC radio broadcasts), but I gather that even near the end of his life he giggled a lot and talked with a kind of speech hesitation that seemed to come from thinking faster than he was talking. He seems to have found it easiest to talk to mathematicians. He thought a little about physics, but never seems to have got deeply into it. And he seems throughout his life to have maintained a childlike enthusiasm and wonder for intellectual questions.

Turing was something of a loner, working successively on his own on his various projects. He was gay and lived alone. He was no organizational politician and toward the end of his life seems to have found himself largely ignored, both by people working on computers and by those working on his new interest of biological growth and morphogenesis.

Turing was in some respects a quintessential British amateur, dipping his intellect into different areas. He achieved a high level of competence in pure mathematics and used that as his professional base. His contributions in traditional mathematics were certainly perfectly respectable, although not spectacular. But in every area he touched, there was a certain crispness to the ideas that he developed, even if their technical implementation was sometimes shrouded in arcane notation and masses of detail.

In some ways he was fortunate to live when he did. For he was at the right time to be able to take the formalism of mathematics as it had been developed and to combine it with the emerging engineering of his day, to see for the first time the general concept of computation.

It is a shame that he died over 20 years before computer experiments became widely feasible. I certainly wonder what he would have discovered when tinkering with Mathematica. I do not doubt that he would have pushed it to its limits, writing code that would horrify me. But I fully expect that, long before I did, he would have discovered the main elements of A New Kind of Science and begun to understand their significance.[4]

He would probably be disappointed that, 60 years after he invented the Turing test, there is still no full human-like artificial intelligence. Perhaps long ago he would have begun to campaign for the creation of something like Wolfram|Alpha, to turn human knowledge into something that computers can handle.

If Alan Turing had lived a few decades longer, he would doubtless have applied himself to half a dozen more areas. But there is still much to be grateful for in what he did achieve in his 41 years, and his modern reputation as the founding father of the concept of computation—and the conceptual basis for much of what I (for example) have done—is well deserved.

CHAPTER 6

Turing's great invention: the universal computing machine

JACK COPELAND

There is no such person as *the* inventor of the computer: it was a group effort. The many pioneers involved worked in different places and at different times, some in relative isolation and others within collaborative research networks. There are some very famous names among them, such as Charles Babbage and John von Neumann—and, of course, Alan Turing himself. Other leading names in this roll of honour include Konrad Zuse, Tommy Flowers, Howard Aiken, John Atanasoff, John Mauchly, Presper Eckert, Jay Forrester, Harry Huskey, Julian Bigelow, Samuel Alexander, Ralph Slutz, Trevor Pearcey, Maurice Wilkes, Max Newman, Freddie Williams, and Tom Kilburn. Turing's own outstanding contribution was to invent what he called the 'universal computing machine'. He was first to describe the basic logical principles of the modern computer, writing these down in 1936, 12 years before the appearance of the earliest implementation of his ideas. This came in 1948, when Williams and Kilburn succeeded in wiring together the first electronic universal computing machine—the first modern electronic computer.

A universal machine

In 1936, at the age of just 23, Turing invented the fundamental logical principles of the modern computer—almost by accident. A shy boyish-looking genius, he had recently been elected a Fellow of King's College, Cambridge. The young Turing worked alone, in a spartan room at the top of an ancient stone building beside the River Cam. It was all quite the opposite of a modern research facility—Cambridge's scholars had been doing their thinking in comfortless stone buildings, reminiscent of cathedrals or monasteries, ever since the university had begun to thrive in the Middle Ages.

A few steps from King's, along narrow medieval lanes, are the buildings and courtyards where, in the seventeenth century, Isaac Newton revolutionized our understanding of the universe. Turing was about to usher in another revolution. He was engaged in theoretical work in the foundations of mathematics. No-one could have guessed that anything of practical value would emerge from his highly abstract research, let alone a machine that would change all our lives.

As everyone knows, the way to make a computer do the job we want (word-processing, say) is simply to locate the appropriate program in its memory and start it running. This is the 'stored-program' concept and it was Turing's invention in 1936.[1] His fabulous idea, a product of pure thought, was of a single processor, a single slab of hardware, that could change itself seamlessly from a machine dedicated to one type of work into a machine dedicated to a completely different job—from calculator to word-processor to chess opponent, for example. It did this by making use of *programs*—sequences of coded instructions—stored inside its memory.

Turing called his invention the 'universal computing machine'; we now call it the *universal Turing machine*. If Isaac Newton had known about it, he would probably have wished that he had thought of it first. Nowadays, though, when nearly everyone owns a physical realization of Turing's universal machine, this idea of a one-stop-shop computing machine is apt to seem as obvious as the wheel and the arch. But in 1936, when engineers thought in terms of building different machines for different purposes, Turing's vision of a universal machine was revolutionary.

The hardware

For the next few years Turing's revolutionary ideas existed only on paper. He was interested right from the start in building his universal machine, but knew of no viable technology for doing so.[2]

The leading digital technology of the time, widely used in telephone exchanges, punched-card calculating equipment, and elsewhere, was the *relay*, an electromagnetically controlled switch. But relays were too slow, too bulky, and too unreliable to offer much promise as the basis for an engineered universal Turing machine. Nevertheless, the German computer pioneer Konrad Zuse used relays to build what were recognizably computers of a sort. His third try at building a relay-based computer—later dubbed Z3, for 'Zuse 3'—ran successfully in 1941. However, Zuse's machines did not incorporate the stored-program concept.[3] In the United States, too, Howard Aiken used relays to build a large-scale computer (the Harvard Mark 1), but again this incorporated no facility for storing programs inside the memory.

A crucial moment came in 1944, when Turing set eyes on Tommy Flowers' racks of high-speed *electronic* code-cracking equipment at Bletchley Park.[4] Unlike electromagnetic components, such as relays, electronic valves (tubes) have no moving parts except for a beam of electrons. As related in Chapter 14, Flowers (Fig. 6.1) had pioneered digital electronics during the 1930s, and as soon as he was introduced to the German 'Tunny' cipher system, in 1942, he realized that digital electronics was the technology that the codebreakers needed to make mincemeat of the new German cipher.

Flowers' giant Tunny-cracking 'Colossus' was certainly a computer—the world's first large-scale electronic digital computer, in fact—but it lacked the ultra-flexibility of the universal Turing machine. This was because Flowers had created Colossus for just one very specific purpose, breaking Tunny. Nor indeed did Colossus incorporate Turing's stored-program concept. While Flowers was designing the computer, Max Newman, Turing's teacher, friend, and code-breaking colleague, had shown him Turing's 1936 article about the universal machine, with its

Figure 6.1 Tommy Flowers.

The Turing Archive for the History of Computing. Photo restored by Jack Copeland and Dustin Parry.

key idea of storing programs in memory. But Flowers, who wanted to get Colossus working as quickly as possible, sensibly took no interest in including extra complications that went beyond what was strictly required for the job in hand.

As soon as he saw Colossus, Turing realized that electronics offered the means to build a miraculously fast universal stored-program computer.[5] Nevertheless, another four years would elapse before the first universal Turing machine in hardware ran the first stored program, on Monday 21 June 1948—the first day of the modern computer age. Based on Turing's ideas, and almost big enough to fill a room, this distant ancestor of our laptops, phones, and tablets was named simply 'Baby'. Thereafter, electronic stored-program universal digital computers became ever smaller and faster, until now they fit into coat pockets and school satchels, linking each of us to the whole wide world.

As explained in Chapter 20, two brilliant electronic engineers, Freddie Williams and Tom Kilburn, built Baby at Manchester University. Turing taught Kilburn the fundamentals of computer design during a course of lectures that he gave in London in 1946–47. By the end of the course, Kilburn was expertly applying the ideas that he had learned from Turing, and was soon drawing up preliminary schematics for Manchester's Baby. Baby was wired together in the Computing Machine Laboratory that Newman had recently established at the university. Newman had presided over Bletchley Park's giant installation of nine Colossi, in fact the world's first electronic computing facility.

Turing's ACE

At the end of the war, Newman—like Turing, gripped by the dream of building an electronic universal computing machine—had taken up the Fielden Chair of Mathematics at Manchester University.[6] Meanwhile, Turing joined London's National Physical Laboratory (NPL) to design an electronic universal stored-program computer—the Automatic Computing Engine, or ACE.[7] So began the race between London and Manchester to create the first universal computing machine in hardware.

Turing's plans caught the interest of the British press, with headlines such as ' "ACE" may be fastest brain', 'Month's work in a minute', and ' "ACE" superior to U.S. model'.[8] Behind the scenes, however, all was not well. Turing's plan was that Flowers and his assistants from the Colossus days should build ACE, but this idea soon ran into trouble. Flowers came under tremendous pressure to devote himself to the rebuilding of Britain's war-ravaged telephone system, and despite his willingness to collaborate with Turing, he rapidly became 'too busy', he said.[9] The NPL's slow-moving bureaucracy seemed unable to make alternative arrangements, and in April 1948 Turing's manager, John Womersley, reported ruefully that hardware development was 'probably as far advanced 18 months ago'.[10]

While Turing waited for engineering developments to start, he and his assistants spent their time pioneering computer programming, preparing a large library of routines for the not-yet-existent computer.[11] Once the hardware was finally working, it was the availability of these ready-made programs that explained the success of NPL's scientific computing service.[12] This took on commissions from government, industry, and the universities, and was the first such service in the world.

In the end, the pilot model of Turing's ACE did not run its first program until May 1950 (Fig. 6.2). By that time, a handful of stored-program computers were working successfully, but Turing's 1-MHz design left them all in the dust.

Open for business in Manchester

Manchester, though, had won the race fair and square in 1948, and Turing moved there shortly after Baby first came to life.[13] He was tired of NPL's ineffectual bureaucrats, and itched to get his hands on an electronic universal Turing machine.

Figure 6.2 The pilot model of Turing's Automatic Computing Engine.

© Crown Copyright and reproduced with permission of the National Physical Laboratory.

When he arrived at the Manchester Computing Machine Laboratory, Baby was working in only a rudimentary way, and Turing rolled up his sleeves to turn it into a fully functioning computer. The original Baby had no input mechanism apart from a bank of manual switches.[14] These were used to insert a program into memory one bit at a time—not much use for real computing. The arrangements that Williams and Kilburn had included for output were equally crude: the user had to try to read patterns of zeros and ones (dots and dashes) as they appeared on what was essentially a TV screen.

Turing used codebreaking technology from Bletchley Park to get the computer working properly, designing an input–output system based on the same punched paper tape that ran through Colossus.[15] He also designed a programming system for the computer, and wrote the world's first programming manual.[16] Thanks to Turing, the first electronic universal computing machine was open for business.

Other electronic universal computers soon followed. In 1949 came Maurice Wilkes's EDSAC, at the University of Cambridge Mathematical Laboratory; Jay Forrester's Whirlwind I, at the Massachusetts Institute of Technology; the Eckert–Mauchly Computer Corporation's BINAC in Philadelphia; and Trevor Pearcey's CSIRAC (pronounced 'sigh-rack') at the University of Sydney in Australia. SEAC, built in Washington, DC for the US Bureau of Standards Eastern Division, by Sam Alexander and Ralph Slutz, ran its first program in 1950, just a few weeks before the pilot model of ACE. Later the same year SEAC was followed by SWAC, built in Los Angeles for the US Bureau of Standards Western Division by Harry Huskey. Von Neumann's 'Princeton computer', engineered by Julian Bigelow at the Princeton Institute of Advanced Study, began working in 1951; informally known as MANIAC, this was the most influential of the early computers. Also in 1951, electronic stored-program computers began arriving in the marketplace. The first were the Ferranti Mark I, based on Baby and produced by the Manchester company Ferranti Ltd; the market-dominating UNIVAC I, produced by Eckert–Mauchly; and LEO, based on EDSAC. In 1953, IBM manufactured the company's first stored-program electronic computer, the 701, based on the Princeton MANIAC. Electronic computing's rough pioneering days were coming to an end.

Turing's universal machine: an in-depth look

A nationwide poll in the UK in 2013 judged Turing's universal machine of 1936 to be the most important British innovation of the last 100 years, ahead of the World Wide Web, the discovery of penicillin, and even the discovery of the structure of DNA.[17] Let's take at look at the workings of Turing's great invention.

The universal computing machine consists of a scanner and a limitless memory-tape. The tape, which is divided into squares, runs back and forth through the scanner. Each square might be blank, or might contain a single symbol, for example '0' or '1'; the scanner can view only one square of tape, and so at most one symbol, at any given time. This tape contains not only the data needed for the computation but also the program.

The Turing machine's basic (or 'atomic') actions are very simple: the scanner contains mechanisms that enable it to *erase* the symbol on the scanned square, to *print* a symbol on the scanned square, and to *move* the tape to the left or right, one square at a time.

The scanner can also alter the position of a dial (or some functionally equivalent device) that is located inside it. The function of the dial is to provide the scanner with a rudimentary short-term memory; for example, the moving scanner can 'remember' that the square it is vacating contains the symbol '1' (say), simply by setting the dial to position '1' as it leaves the square. In Turing-machine jargon, the dial's current position is known as the *current state*.

Turing showed that the universal machine can carry out highly complex tasks by chaining together large numbers of these five elementary actions: *erase, print, move left, move right*, and *change state*. It is a remarkable fact that, despite the spartan austerity of Turing's machines, they are able to compute everything that any computer on the market today can. Indeed, by reading and obeying instructions that the programmer stores on the memory-tape, the universal Turing machine—Turing argued—can perform *every possible* computation. This is called 'Turing's thesis' (see Chapter 41).[18]

In the universal machine, as in every modern computer, programs and data take the same form: numbers stored in memory. Turing showed how to construct a programming code in which all possible sequences of instructions—that is, all possible Turing-machine programs—can be represented by means of numbers, and so ultimately by groups of zeros and ones. This idea of storing programs of coded instructions in the computer's memory was simple, yet profound, and made the modern computer age possible.

In Turing's 1936 article 'On computable numbers', often regarded as marking the birth of theoretical computer science, Turing had brought *machinery* into discussions of the foundations of mathematics. That innovation was one of the features that made his approach to mathematical foundations so novel—even daring. In a biographical memoir prepared for London's Royal Society shortly after Turing's death, Max Newman wrote:[19]

It is difficult to-day to realize how bold an innovation it was to introduce talk about paper tapes and patterns punched in them, into discussions of the foundations of mathematics.

Ironically, however, today's computer science textbooks usually present the universal Turing machine as a purely mathematical entity, an abstract mathematical idea. Purely mathematical notions, such as sets of symbols, and functions from state–symbol pairs, replace Turing's scanner, tape, and punch-holes. (For example, according to one influential modern textbook, a *Turing machine* is defined to be an ordered quadruple $<K, s, \Sigma, \partial>$, where K is a finite set of states, s is a member of K (called the *initial state*), Σ is a set of symbols, and ∂ is a transition function from $K \times \Sigma$, the values of which are state–symbol pairs, including the special symbols L and R, replacing the idea of left and right motion.[20]) So Turing's bold innovation has been purified and rendered into the conventional coin of mathematics. Turing machines are no longer objects located in time and space, and subject to cause and effect. The paper tape, and the punched patterns that cause the machine to act in certain ways, are gone.

A 'Turing-machine realist' such as myself rejects this modern view of the Turing machine. Turing-machine realists regard the mathematics of the preceding paragraph merely as a useful formal *representation* of a Turing machine. But, just as a mathematical representation of digestion should not be confused with the process of digestion itself, so too the mathematical representation of a Turing machine must not be confused with the thing that is represented—namely, an idealized physical *machine*.[21]

It seems transparently clear that Turing himself was a Turing-machine realist. There is a positively industrial flavour to his account of his machines, with its references not only to punched paper tape, but even to 'wheels' and 'levers' within the scanner.[22] Yet the realism of Turing's own account is lost in the modern purified version.

Turing and von Neumann: two founding fathers of the computer age

Probably Turing's greatest peacetime achievement was the proposal of two closely related technological ideas on which modern computing is based: his twin concepts of the *universal computer* and the *stored program*. Yet historians of the computer have often found Turing's contributions hard to place, and sadly many histories of computing written during the did decades since his death did not so much as mention him.

Even today there is still no real consensus on Turing's place in the history of computing. An article published in 2013 by the editor of the Association for Computing Machinery's flagship journal *Communications of the ACM* objected to the claim that Turing invented the stored-program concept.[23] The article's author, my good friend Moshe Vardi, dismissed the claim as 'simply ahistorical'.

Vardi emphasized that it was not Turing but the Hungarian–American mathematician John von Neumann (Fig. 6.3) who, in 1945, 'offered the first explicit exposition of the stored-program computer'. This is true (see Chapter 20), but the point does not support Vardi's charge of historical inaccuracy. Although von Neumann did write the first paper explaining how to convert Turing's ideas into electronic form, the fundamental concept of the stored-program universal computer was nevertheless Turing's.

Von Neumann was actually very clear in attributing credit to Turing, both in private and in public: it is unfortunate that his statements are not more widely known. He explained in 1946 that Turing's 'great positive contribution' was to show that 'one, definite mechanism can be "universal" ', and in a 1949 lecture he emphasized the crucial importance of Turing's research, which lay (he said) in Turing's 1936 demonstration that a single appropriately designed machine 'can, when given suitable instructions, do anything that can be done by automata at all'.[24] Von Neumann's friend and scientific colleague Stanley Frankel observed:[25]

Many people have acclaimed von Neumann as the 'father of the computer' . . . but I am sure that he would never have made that mistake himself.

Figure 6.3 John von Neumann.

Photographer unknown. From the Shelby White and Leon Levy Archives Center, Institute for Advanced Study, Princeton.

Frankel explained that:

he firmly emphasized to me, and to others I am sure, that the fundamental conception is owing to Turing.

I have more to say about Turing's influence on von Neumann in Chapter 20, where I also describe von Neumann's influence on the design of the Manchester Baby.

So von Neumann set out the electronic basis for a practical version of the universal Turing machine, with considerable assistance on the engineering side from his associates Presper Eckert and John Mauchly. In von Neumann's design the picturesque 'scanner' and 'memory-tape' of Turing's 1936 machine were replaced with electronic equipment, and he also replaced Turing's pioneering programming code with what he later described as a 'practical' code for high-speed computing.[26] Von Neumann's design went on to become an industry standard, so although he was not the father of the computer, he was certainly one of the founding fathers of the modern computer age. Yet among his many contributions to the development of the computer, perhaps the greatest of all was simply informing America's electronic engineers about Turing's concept of the stored-program universal computer.

During 1945, hard on the heels of von Neumann's groundbreaking paper reporting his (and Eckert and Mauchly's) design, Turing had designed his own electronic version of his universal machine, ACE.[27] This was radically different from von Neumann's design (Chapters 21 and 22 describe Turing's design in detail). Turing sacrificed everything to speed, launching a 1940s version of what today's computer architects call RISC (reduced instruction set computing). In 1947 he gave a clear statement of the connection, as he saw it, between the universal computing machine of 1936 and the electronic stored-program universal digital computer:[28]

Some years ago I was researching on what might now be described as an investigation of the theoretical possibilities and limitations of digital computing machines. I considered a type of machine which had a central mechanism, and an infinite memory which was contained on an infinite tape . . . [D]igital computing machines . . . are in fact practical versions of the universal machine. There is a certain central pool of electronic equipment, and a large memory, [and] the appropriate instructions for the computing process involved are stored in the memory.

Returning to Moshe Vardi's efforts to refute the claim that Turing invented the stored-program concept, Vardi states (defending von Neumann's corner) that 'we should not confuse a mathematical idea with an engineering design'. So at best Turing deserves the credit for an abstract mathematical idea? Not so fast. Vardi is ignoring the fact that some inventions belong equally to the realms of mathematics and engineering. The universal Turing machine was one such, and this is part of its brilliance.

What Turing described in 1936 was not an abstract mathematical notion, but a solid three-dimensional machine containing (as he said) wheels, levers, and paper tape; and the cardinal problem in the pioneering years of electronic computing was in effect just this: *how best to build practical electronic forms of the universal Turing machine*?

CHAPTER 7

Hilbert and his famous problem

JACK COPELAND

I n 1936 mathematics was changing profoundly, thanks to Turing and his fellow revolutionaries Gödel and Church. Older views about the nature of mathematics, such as
those powerfully advocated by the great mathematician David Hilbert, were being
swept away, and simultaneously the foundations for the modern computer era were being
laid. These three revolutionaries were also catching the first glimpses of an exciting new
world—the hitherto unknown and unimagined mathematical territory that lies beyond the
computable.[1]

The 1930s revolution

In the 1930s a group of iconoclastic mathematicians and logicians launched the field that we
now call theoretical computer science.[2] These pioneers embarked on an investigation to spell
out the meaning and limits of computation. Pre-eminent among them were Alan Turing, Kurt
Gödel, and Alonzo Church.

These three men are pivotal figures in the story of modern science, and it is probably true
to say that, even today, their role in the history of science is underappreciated. The theoretical
work that they carried out in the 1930s laid the foundations for the computer revolution, and
the computer revolution in turn fuelled the rocketing expansion of scientific knowledge that
characterizes modern times. Previously undreamed of number-crunching power was soon
boosting all fields of scientific enquiry, thanks in large part to these seminal investigations. Yet,
at the time, Turing, Gödel, and Church would have thought of themselves as working in a most
abstract field, far flung from practical computing. Their concern was with the very foundations
of mathematics.

Figure 7.1 Kurt Gödel

From the Kurt Gödel Papers, the Shelby White and Leon Levy Archives Center, Institute for Advanced Study, Princeton, NJ, USA, on deposit at Princeton University; reproduced courtesy of Princeton University.

Eternally incomplete

Kurt Gödel (Fig. 7.1), a taciturn 25-year-old mathematician from Vienna University, ushered in a new era in mathematics with his 1931 theorem that arithmetic is *incomplete*.[3] In a sentence, what Gödel showed is that more is true in mathematics than can be formally *proved*.

This sensational result shocked, and even angered, some mathematicians. It was thought that everything that matters *ought* to be provable, because only rigorous proof by transparent and obvious rules brings certainty. But Gödel showed that, no matter how the formal rules of arithmetic are laid down, there must always be *some* mathematical truths—complicated relatives of simpler truths such as $1 + 1 = 2$—that *cannot* be proved by means of the rules. Paradoxically, the only way to eradicate incompleteness appeared to be to select rules that actually contradict one another.[4]

Figure 7.2 David Hilbert

Posted to Wikimedia Commons and licensed under public domain, https://commons.wikimedia.org/wiki/File:Hilbert.jpg

Hilbert's *Entscheidungsproblem*

Gödel's epoch-making result, however, seemed to leave open one of the most fundamental problems of mathematics—and that is where Turing entered the picture.[5] This unresolved problem was known as the *Entscheidungsproblem*, or the 'decision problem'.

The German mathematician David Hilbert (Fig. 7.2) had brought the *Entscheidungsproblem* into the spotlight.[6] Hilbert, who was 50 years Turing's senior, set the agenda for much of twentieth-century mathematics in a lecture that he gave in Paris at the turn of the century, and by the 1930s he was virtually the pope of mathematics.[7] Turing—untidy, irreverent, and scarcely older than an undergraduate—took on the *Entscheidungsproblem* and showed that matters were absolutely not as Hilbert thought.

Turing summarized the *Entscheidungsproblem* like this:[8]

Is there a 'general mechanical process' for telling whether any given formula is provable?

Here 'provable' means that the formula can be derived, step by logical step, using the rules and principles of Hilbert's logico-mathematical calculus—known to mathematics buffs as the *engere Functionenkalkül*.[9] Hilbert thought that there must be a general procedure for telling whether any specified formula is provable, and the challenge—a fundamentally important challenge in his view—was to find it.

Gödel gave a dramatic and easy-to-understand explanation of what Turing established about the *Entscheidungsproblem*. He described an imaginary machine—a computing machine—looking something like a typewriter, except for having a crank-handle at the side.[10] 'Turning the handle' was Gödel's metaphor for what we now call 'executing an algorithm'. The user typed a mathematical or logico-mathematical formula at the keyboard (the formula that the user elected to type could be either true or false), and then the user turned the crank-handle round once. On Hilbert's view of mathematics, this computing machine could, in principle, be designed to behave as follows: when the crank was turned, the machine would sound its bell if the typed formula was a provable one, but if the formula was not provable then the bell would remain silent.

In a word, this machine is able to 'decide'—in the sense of 'figure out' or 'calculate'—whether the typed formula is provable or not (hence the term 'decision problem'). What Turing established is very simply stated: such a machine is impossible. Once you stray beyond the most elementary areas of mathematics (such as Boolean algebra and the 'pure monadic' functional calculus), it is simply not possible to design a finite computing machine that is capable of deciding whether formulae are or are not provable. Hilbert was quite wrong to think that there must be a general mechanical process for telling whether or not any specified formula is provable.

It was in a lecture in the United States in 1939 that Gödel gave this summary, and he rounded off his exposition with the dramatic remark that what this shows is essentially that 'the human mind will never be able to be replaced by a machine'.[11] Had Turing been there, he might have objected that the issue about the mind is not quite so simple—but that's another story.[12]

Turing's universal computing machine

It was precisely in order to prove the impossibility of this decision-machine that Turing dreamed up his universal computing machine—his ingeniously simple digital processor with an indefinitely extendable memory (see Chapter 6).

By using programs stored in memory, Turing argued, the universal machine can carry out every possible computation (see Chapter 41).[13] Scarcely more than a decade later, on the back of wartime technology, Turing's invention moved out of the realm of purely theoretical constructs to become real hardware. As we now know, the electronic universal machine went on to transform both science and society, but when Turing invented it he was not thinking about electronic computers at all. Strangely enough, he was intent on proving the existence of mathematical realms that, in a certain sense, lie *beyond* the range of any computer, no matter how powerful.

Neither Gödel nor Church had come close to inventing the universal machine.[14] Gödel's own greatest gift to computer science was the branch of mathematics known as *recursive function theory*, and like Turing's universal machine, this was also a product of the curiosity-driven exploration of the deepest foundations of mathematics. Recursive function theory is essentially the theory of computer programs, yet when Gödel pioneered it (along with Thoralf Skolem, Stephen Kleene, and Hilbert himself) he was not thinking about what we now call computing at all. He was working on the proof of his incompleteness theorem, and experiencing the first hazy glimpses of an unknown mathematical world—the strange and exciting landscape that lies beyond the computable.

As well as recursive function theory, Gödel also contributed another fundamental idea to computing in the same mathematical paper: published in 1931, and containing proofs of his first and second incompleteness theorems, this paper is one of the classics of twentieth-century mathematics.[15] Gödel's idea was that logical and arithmetical statements can be represented by *numbers*. Turing, von Neumann, and others developed this idea much further, and it is now ubiquitous in computing, with everyone's laptop containing binary numbers that represent program instructions, logical and arithmetical statements, and data of all types. Gödel himself, though, took little or no interest in the development of the electronic computer.[16] Unlike Turing and von Neumann, he was the sort of mathematician who did not care to descend from the mountain-tops of abstraction.

Digital super-optimism

Returning to the question of the computer's limits: would you yourself agree with the claim that computers can perform any and every well-defined mathematical task—if not in practice, then at any rate in *principle*?

In practice, some tasks would involve such vast amounts of processing that the fastest computers would take zillions of years to complete them, and the numbers involved would get so large as to swamp any realistic amount of memory. Let's use the term 'digital super-optimism' for the view that computers can in *principle* do anything and everything mathematical—if only there is enough memory and processing power on tap.

Even *infinite* tasks can be programmed, so long as it is assumed that the computer running the program has access to unlimited memory (as does the universal Turing machine). One example of an easily programmed infinite task is to calculate x^2, first for $x = 1$, then $x = 2$, then $x = 3$, and so on, ever onwards through the integers. There is no end to this program's calculations. Whatever gigantic number x you might consider—a billion times the number of nanoseconds in a billion years, for instance—the program will eventually reach that number and calculate its square.

Turing showed, however, that digital super-optimism is *false*.[17] It is definitely not the case that all well-defined mathematical tasks can be done by computer—not even in principle. Some tasks simply cannot be performed by computing machines, no matter how good the programmers or how powerful the hardware.

Turing was the first to describe examples of tasks like this, and in discovering them he blazed a trail through the new landscape glimpsed by Gödel—the unknown world of uncomputability.

Exploring the unknown world

Some of Turing's examples are simpler than others, and the simplest one is surprisingly straightforward. The task is this: to work out, about any given computer program in some specific programming language (C++, for instance), whether or not running the program will involve outputting a given keyboard character, say '#'.

It might seem a relatively easy task for someone who understands the programming language to scan through the lines of a program and figure out whether any instruction will result in the computer's outputting the character '#', but Turing showed that no computer can perform this task. The reason, essentially, is that the scope of the task covers *all* computer programs, including the program of the very computer attempting the task.

Another example of an uncomputable task is provided by Turing's result about the decision-machine: no computer can decide whether or not arbitrarily selected formulae are provable in Hilbert's *engere Functionenkalkül*. Chapters 37 and 41 give more information on Turing's negative result about the decision-machine.

Between them, Gödel and Turing had delivered multiple blows to Hilbert's account of the nature and foundations of mathematics, from which it never recovered. Hilbert's overarching concern had been to 'regain for mathematics the old reputation for incontestable truth', as he put it;[18] but caring less about incontestability and ancient reputation, Gödel and Turing looked towards new mathematical realms that Hilbert had not even imagined.

Hilbert on axioms

Axioms are mathematical propositions so basic that they stand in no need of proof. Hilbert described axioms as principles that are 'as simple, intuitive, and comprehensible as possible'.[19] Examples are the so-called Peano axioms, named after the Italian mathematician Giuseppe Peano. The Peano axioms are about the 'natural' numbers: 0, 1, 2, 3, 4, etc. Two examples of Peano axioms are:

(1) if the immediate successor of natural number x = the immediate successor of natural number y, then $x = y$; and
(2) 0 is not the immediate successor of any natural number.

The immediate successor of a natural number is simply the number immediately following it in the sequence 0, 1, 2, 3, ... For example, the immediate successor of 7 is 8, and the immediate successor of 3 + 4 is also 8. Most people are able to see intuitively that both these axioms are true.

Axioms are the starting point of all mathematical proofs: by a process of rigorous logical deduction, mathematicians prove *theorems* from axioms. From the total of five Peano axioms, a vast quantity of other truths about natural numbers can be deduced: for example, that multiplying a natural number by 0 results in 0; and that if any natural number has a selected mathematical property, then there is a *least* number with that property (the 'least-number principle'). It would be nice to be able to say that *all* truths about natural numbers can be deduced from the Peano axioms, but Gödel proved that this is not so: the Peano axioms are incomplete, in precisely the sense that not *all* truths about the numbers 0, 1, 2, 3, . . . can be proved from them. This came as an unpleasant surprise to Hilbert.

To be precise, Gödel proved this incompleteness result on the assumption that the Peano axioms are *consistent* with one another, as certainly *seems* obvious once you see all the axioms together. Consistency, or freedom from contradiction, is traditionally regarded as fundamentally important in mathematics, and Hilbert expended a lot of mathematical energy on the very difficult problem of how to prove that arithmetic is consistent: 'we can never be *certain*', he said, 'of the consistency of our axioms if we do not have a special proof of it'.[20] On the other hand, once we do have a consistency proof, 'then we can say that mathematical statements are in fact incontestable and ultimate truths'.[21]

So Hilbert desired a consistent and complete set of axioms for arithmetic; but, in what has become known as his *first incompleteness theorem*, Gödel proved that Hilbert could not possibly have what he desired. As if this were not bad enough, Gödel then dealt a second crushing blow to Hilbert's attempt to provide what he called 'a secure foundation for mathematics'.[22] In his *second incompleteness theorem*, Gödel proved that it is impossible to find a proof of the consistency of arithmetic satisfying Hilbert's stringent conditions. The proof that Hilbert thought mathematicians had to find—because it would fill in the missing step in his case that mathematics consists of 'incontestable truth'—turned out to be an impossibility.[23]

Turing exposed yet another of the flaws in Hilbert's thought. This concerned the process of adding *new* axioms to a set of axioms that has already been laid down—whether to the Peano axioms or to any other set of axioms (e.g. for geometry or for analysis). Suppose a mathematician wanted to add some extra axioms to the Peano axioms—what additional axioms could legitimately be laid down? Hilbert's answer was simple and powerful: a mathematician could lay down any new axioms that he or she saw fit, just so long as the new axioms could be proved consistent with the ones that had already been adopted. Turing pointed out that (for technical reasons) Gödel's second incompleteness theorem did not rule out this form of consistency proof, and he tackled this aspect of Hilbert's thought himself.[24]

In 1928, just a few years before Turing struck, Hilbert had put matters like this:[25]

The development of mathematical science accordingly takes place in two ways that constantly alternate: (i) the derivation of new 'provable' formulae from the axioms by means of formal inference; and, (ii) the adjunction of new axioms together with a proof of their consistency.

It is this view of mathematics that explains why Hilbert regarded the decision problem as being of such supreme importance. For, he announced, also in 1928:[26]

Questions of *consistency* would also be able to be solved by means of the decision process.

For this reason Hilbert called the decision problem 'the main problem of mathematical logic'.[27] No doubt these stirring words influenced the young Turing in his own decision to take on this particular problem. He would also have read Hilbert's 1928 statement that:[28]

in the case of the *engere Functionenkalkül* the discovery of a general decision process forms an as yet unsolved and difficult problem.

Hilbert emphasized that a 'general solution of the decision problem . . . does not yet exist', but his 1928 account exuded optimism, and he wrote buoyantly of moving 'closer to the solution of the decision problem'.[29]

Just a few years later, in 1936, Turing managed to prove that no such decision process was possible. So Hilbert's proposal to resolve questions of consistency by means of such a process lay in tatters. Turing put it in his usual terse way:[30]

I shall show that there is no general method which tells whether a given formula is provable in **K** [the *engere Functionenkalkül*], or, what comes to the same, whether the system consisting of **K** with [a formula] adjoined as an extra axiom is consistent.

All mathematicians make mistakes, but it is a mark of Hilbert's greatness that the mistakes he made were so fundamental, so profound, that exposing them changed mathematics irrevocably.

Turing visits Church

In the autumn of 1936 Turing left England for the United States where he would write a PhD thesis at Princeton University. His supervisor at Princeton was Alonzo Church, the third of the three central figures in the 1930s revolution (Fig. 7.3). Church was a few years older than Turing and a Young Turk in the Princeton mathematics department. At the end of the 1920s he had spent six months studying with Hilbert's group in Germany. Church's leading contribution to the 1930s' revolution—he called it the λ-*calculus*: the Greek letter λ is pronounced 'lambda'— later became part of the basic toolset of computer programming.

Before Turing's departure from England, he and Church had (almost simultaneously) proposed two different mathematical definitions of the idea of computation. Gödel was not at all impressed with Church's definition, bluntly telling Church that his approach was 'thoroughly unsatisfactory'.[31] On the other hand Gödel found Turing's definition 'most satisfactory'.[32] Even the modest young Turing admitted that his definition was 'possibly more convincing' than Church's.[33] It was clear from the start that the relationship between Turing and Church was not

Figure 7.3 Alonzo Church

From the Alonzo Church Papers (C0948); Manuscripts Division, Department of Rare Books and Special Collections, Princeton University Library; reproduced courtesy of Princeton University

going to fit into the usual student–supervisor pattern. Unfortunately, no spark seems to have developed between the two, and ultimately Church was probably a disappointment to Turing.

For his thesis topic Turing chose another of Hilbert's pontifical claims about the nature of mathematics—and once again he argued that mathematical reality is not as Hilbert thought.[34] This time the disagreement concerned mathematical intuition.

Intuiting the truth

Most people can see by intuition that elementary geometrical propositions are true—for instance, that a straight line and a circle can intersect no more than twice. There is no need to go through a train of logical steps to convince yourself that this proposition is true. Hilbert gave $2 + 3 = 3 + 2$ and $3 > 2$ as examples of arithmetical propositions that are 'immediately intuitable'.[35] The hallmark of intuition is this simple 'seeing', in contrast to following a conscious train of logical deductions (if this then that, and if that then such-and-such). 'The activity of the intuition', Turing said, 'consists in making spontaneous judgments which are not the result of conscious trains of reasoning'.[36] The more skilled the mathematician is, the greater his or her ability to apprehend truths by intuition.

Turing and Hilbert both emphasized the importance of intuition. Hilbert declared:[37]

we are convinced that certain intuitive concepts and insights are necessary for scientific knowledge, and logic alone is not sufficient.

Nevertheless Hilbert wanted both to tame and to deflate intuition. He said that mathematical intuition, far from supplying knowledge of an abstract mathematical realm as others had thought—a realm containing abstract objects such as numbers, many-dimensional figures and (Turing might have added) endless paper tapes—has instead a much more down-to-earth role, supplying knowledge concerning 'concrete signs' and sequences of these.[38] Hilbert's examples of concrete signs included '1 + 1', '1 + 1 + 1' and '$a = b$'; and he used the sign '2' to abbreviate the pattern '1 + 1', '3' to abbreviate '1 + 1 + 1', and so on.[39] His concrete signs are nothing other than the pencil or ink marks that a human computer could write when calculating. Hilbert insisted, provocatively and with great originality, that 'the objects of number theory' are 'the signs themselves'; and he urged that this standpoint suffices for 'the construction of the whole of mathematics'.[40]

With these simple but profound statements Hilbert deflated mathematical intuition: numbers are simulated by pencil marks, and intuition consists in perceiving truths about the concrete signs and the patterns in which the signs stand. The structure of these patterns 'is immediately clear and recognizable' to us, Hilbert said.[41] For example, we can immediately grasp that appending '+ 1' to '1 + 1' produces '1 + 1 + 1': or, in other words, that adding '1' to '2' gives '3'.

Hilbert also wanted to *limit* the role of intuition as far as is possible. Intuition often serves to give us certainty that *axioms* are true (although even so, he recognized that sometimes mathematicians disagree about axioms, and said that there might on occasion be 'a certain amount of arbitrariness in the choice of axioms').[42] Other mathematical propositions are *proved* from the finite list of axioms by means of 'formulable rules that are completely definite'.[43] These are rules that any human computer can apply, such as that B follows from A and $A \rightarrow B$ (A implies B) taken together. The rules are, like the axioms, finite in number, and they too are seen to be correct by intuition.[44] Once the axioms and rules are in place, however, intuition is no longer needed: from there on, everything is just a matter of *following* the rules.

In his 1938 doctoral dissertation Turing said, referring to Hilbert and his followers:[45]

In pre-Gödel times it was thought by some that . . . all the intuitive judgments of mathematics could be replaced by a finite number of these rules [and axioms].

Turing argued that, on the contrary, mathematics is just too unruly for intuition's required role to be limited in the way Hilbert wished. Summing up his argument, he explained: 'We have been trying to see how far it is possible to eliminate intuition'; and his conclusion was that in fact the need for intuition turns out to be boundless. Hilbert might have responded that Turing's notion of an untamed and vaultingly powerful faculty of intuition belonged to what he called 'mysterious arts'.[46] Turing could have replied that Hilbert's thinking was simply too narrow, in consequence of his old-fashioned quest for 'complete certitude' in mathematics.

Conclusion

Hilbert's views on the nature of mathematics, although innovative at the time, were the product of a more uncomplicated mathematical era, now long gone: Gödel and Turing changed mathematics forever. Nevertheless, Hilbert's 'proof theory'—probably his proudest invention—is still very much alive and well today, and is a core topic in every university curriculum on mathematical logic.

Hilbert certainly felt the force of Turing's attack on his thinking about the computational nature of mathematics. The following bold statement, made in 1928 in the first edition of his famous book *Grundzüge der Theoretischen Logik* (*Principles of Mathematical Logic*, written with Wilhelm Ackermann) was omitted from the much-revised second edition, published in 1938:[47]

it is to be expected that a systematic, so to speak computational treatment of the logical formulae is possible . . .

CHAPTER 8

Turing and the origins of digital computers

BRIAN RANDELL

I n this chapter I describe my initial attempts at investigating, during the early 1970s, what Alan Turing did during the Second World War. My investigations grew out of a study of the work of Charles Babbage's earliest successors—in particular, the Irish pioneer Percy Ludgate—a study that led me to plan an overall historical account of the origins of the digital computer. The investigation resulted in my learning about a highly secret programmable electronic computer developed in Britain during the Second World War. I revealed that this computer was named Colossus, and had been built in 1943 for Bletchley Park, the UK government's wartime codebreaking establishment. However, my attempt to get the details of the machine declassified were unsuccessful, and I came to the conclusion that it might be a long time before anything more would become public about Bletchley Park and Colossus.[1]

Introduction

Around 1970, while I was seeking information about the work of Charles Babbage and Ada Lovelace to use in my inaugural lecture at Newcastle University, I stumbled across the work of Percy Ludgate. In a paper he wrote about Babbage's 'automatic calculating engines',[2] Ludgate mentioned that he had also worked on the design of an Analytical Engine, indicating that he had described this in an earlier paper in the *Proceedings of the Royal Dublin Society*.[3] From a copy of that paper I learned that an apparently completely forgotten Irish inventor had taken up and developed Babbage's ideas for what would now be called a program-controlled mechanical computer. Previously I had subscribed to the general belief that over a century had passed before anyone had followed up Babbage's pioneering 1837 work on Analytical Engines.[4] This discovery led me to undertake an intensive investigation of Ludgate, the results of which I published in the *Computer Journal*.[5]

With the help of a number of Irish librarians and archivists I managed to find out quite a few details about the tragically short life of this Irish accountant, and even to make contact with one of his relatives. Unfortunately, I found nothing more about his design for a paper-tape-controlled analytical machine beyond what was given in his 1909 paper.

My investigations into the background to Ludgate's work left me with a considerable amount of information on pre-computer technology and on other little-known successors to Babbage. So I came up with the plan of collecting and publishing in book form a set of selected original papers and reports on the many fascinating inventions and projects that eventually culminated in the development of the 'modern' computer.

I took Charles Babbage's work as my starting point, and decided on a cut-off date of 1949, when the first practical stored-program electronic computer became operational. So I planned to include material on ENIAC, EDVAC, and the Manchester 'Baby' machine, ending with the Cambridge EDSAC, but decided to leave coverage of all the many subsequent machines to other would-be computer historians.

I circulated a list of my planned set of documents to a number of colleagues for comment, and one of the responses I received queried the absence of Alan Turing. My excuse was that, to the best of my knowledge, Turing's work on computers at the National Physical Laboratory (NPL) had post-dated the successful efforts at Manchester and Cambridge, and that his pre-war work on computability, in which he described what we now call a Turing machine, was purely theoretical and so fell outside the chosen scope of my collection.

I first became interested in computers in 1956, in my last year at Imperial College. There were few books on computers at that time, but one was Bowden's *Faster than Thought*.[6] My copy of this book was probably my first source of knowledge about Babbage and Turing, and indeed about the work of the various early UK computer projects, though soon after-wards I learned much more about Babbage and his collaboration with Lady Lovelace from the excellent Dover paperback, *Charles Babbage and his Calculating Engines*.[7] In Bowden I read:

The basic concepts and abstract principles of computation by a machine were formulated by Dr. A. M. Turing, F.R.S. in a paper read before the London Mathematical Society in 1936, but work on such machines in Britain was delayed by the war. In 1945, however, an examination of the problems was made at the National Physical Laboratory by Mr. J. R. Womersley, then Superintendent of the Mathematics Division of the Laboratory. He was joined by Dr. Turing and a small staff of specialists . . .

Piqued by the query about the omission of Turing from my envisaged collection, I set out to try to find out more about Turing's work during the period 1936–45. I obtained a copy of the 1959 biography by his mother, Sara Turing,[8] to find that the only indication it contained of what her son had done during the Second World War was the following:

. . . immediately on the declaration of war he was taken on as a Temporary Civil Servant in the Foreign Office, in the Department of Communications . . . At first even his whereabouts were kept secret, but later it was divulged that he was working at Bletchley Park, Bletchley. No hint was ever given of the nature of his secret work, nor has it ever been revealed.

By this time I had learned that his wartime work was related to codebreaking, though neither I nor any of my colleagues were familiar with the name Bletchley Park. On rechecking my copy of David Kahn's magnificent tome, *The Codebreakers*,[9] I found his statement that Bletchley Park was what the Foreign Office 'euphemistically called its Department of Communications'—that is, it was the centre of Britain's wartime codebreaking efforts. However, Kahn gave little information about what was done at Bletchley Park, and made no mention of Turing.

Around this time I came across the following statement by Lord Halsbury:[10]

One of the most important events in the evolution of the modern computer was a meeting of two minds which cross-fertilised one another at a critical epoch in the technological development which they exploited. I refer of course to the meeting of the late Doctors Turing and von Neumann during the war, and all that came thereof . . .

I wrote to Lord Halsbury, who in 1949 was Managing Director of the National Research Development Corporation, the UK government body that had provided financial support to several of the early UK computer projects. Unfortunately he could not recollect the source of his information, his response (quoted in Randell[11]) to my query being:

I am afraid I cannot tell you more about the meeting between Turing and von Neumann except that they met and sparked one another off. Each had, as it were, half the picture in his head and the two halves came together during the course of their meeting. I believe both were working on the mathematics of the atomic bomb project.

Enquiries of those of Turing's colleagues who were still at the NPL proved fruitless, but Donald Davies, who was then Superintendent of the Division of Computing Science at the NPL, arranged for me to visit Sara Turing. She was very helpful and furnished me with several further leads, but was not really able to add much to the very brief and unspecific comments in her book.

Various other leads proved fruitless, and my enthusiasm for the search was beginning to wane. I eventually had the opportunity to inspect a copy of Turing's report giving detailed plans for the Automatic Computing Engine (ACE).[12] This proved to post-date, and even contain a reference to, von Neumann's draft of a report on the EDVAC,[13] so I did not examine it as carefully as I later realized I should have done. However, I did note that Turing's report alluded to the fact that he had obtained much experience of electronic circuits.

Secret wartime computers

My investigations then took a dramatic turn. I had written to a number of people, seeking to understand more fully whether, and if so how, Turing had contributed to the initial development of practical stored-program computers. One of my enquiries (to Donald Michie) elicited the following response (quoted in Randell[14]):

I believe that Lord Halsbury is right about the von Neumann–Turing meeting . . . The implication of Newman's obituary notice, as you quote it, is quite misleading; but it depends a bit on what one means by: a 'computer'. If we restrict this to mean a stored-program digital machine, then Newman's implication is fair, because no-one conceived this device (apart from Babbage) until Eckert and Mauchly (sometimes attributed to von Neumann). But if one just means high-speed electronic digital computers, then Turing among others was thoroughly familiar during the war with such equipment, which predated ENIAC (itself not a stored-program machine) by a matter of years.

The obituary notice for Turing mentioned by Michie was written by Professor M. H. A. Newman who was associated with the post-war computer developments at Manchester University. It stated that:[15]

At the end of the war many circumstances combined to turn his attention to the new automatic computing machines. They were in principle realisations of the 'universal machine' which he had described in the 1937 paper for the purpose of a logical argument, though their designers did not yet know of Turing's work.

I then found that there had already been several (rather obscure) references in the open literature to the work at Bletchley Park with which Turing was associated, of which the most startling was a paper by I. J. (Jack) Good. This gave a listing of successive generations of general-purpose computers, including:[16]

Cryptanalytic (British): classified, electronic, calculated complicated Boolean functions involving up to about 100 symbols, binary circuitry, electronic clock, plugged and switched programs, punched paper tape for data input, typewriter output, pulse repetition frequency 10^5, about 1000 gas-filled tubes; 1943 (M. H. A. Newman, D. Michie, I. J. Good and M. Flowers. Newman was inspired by his knowledge of Turing's 1936 paper).

(The initials of Tommy Flowers were in fact 'T. H.'.)

Furthermore, Good's paper went on to claim that there was a causal chain leading from Turing's 1936 paper,[17] via the wartime cryptanalytic machine, to the first Manchester computers, although it stated that the main influence was from von Neumann's plans for the IAS machine at Princeton University's Institute for Advanced Study. Moreover, in response to the enquiry that I sent him, Jack Good indicated that the wartime computer he had described was preceded by at least one other computer, although he too was unable to give an explicit confirmation of the reported Turing/von Neumann meeting. In fact, as he said (quoted in Randell[18]):

Turing was very interested in the logic of machines even well before World War II and he was one of the main people involved in the design of a large-scale special purpose electromagnetic computer during the war. If he met von Neumann at that time I think it is certain that he would have discussed this machine with him. There were two very large-scale machines which we designed in Britain, built in 1941–43 . . . The second machine was closer to a modern digital binary general-purpose electronic computer than the first one, but the first also might very well have suggested both to Turing and von Neumann that the time had come to make a general-purpose electronic computer.

Further details of Turing's role, and the wartime codebreaking machines, were provided in a letter I received from Tommy Flowers (quoted in Randell[19]):

In our war-time association, Turing and others provided the requirements for machines which were top secret and have never been declassified. What I can say about them is that they were electronic (which at that time was unique and anticipated the ENIAC), with electromechanical input and output. They were digital machines with wired programs. Wires on tags were used for semi-permanent memories, and thermionic valve bi-stable circuits for temporary memory. For one purpose we did in fact provide for variable programming by means of lever keys which controlled gates which could be connected in series and parallel as required, but of course the scope of the programming was very limited. The value of the work I am sure to engineers like myself and possibly to mathematicians like Alan Turing, was that we acquired a new understanding of and familiarity with logical switching and processing because of the enhanced possibilities brought about by electronic technologies which we ourselves developed. Thus when stored program computers became known to us we were able to go right ahead with their development. It was lack of funds which finally stopped us, not lack of know-how.

Another person whom I had contacted in an effort to check the story of the Turing–von Neumann meeting was Dr S. Frankel, who had known von Neumann while working at Los Alamos. Although unable to help in this matter, he provided further evidence of the influence of Turing's pre-war work (quoted in Randell[20]):

I know that in or about 1943 or '44 von Neumann was well aware of the fundamental importance of Turing's paper of 1936 'On computable numbers . . .' which describes in principle the 'Universal Computer' of which every modern computer (perhaps not ENIAC as first completed but certainly all later ones) is a realization. Von Neumann introduced me to that paper and at his urging I studied it with care. Many people have acclaimed von Neumann as the 'father of the computer' (in a modern sense of the term) but I am sure that he would never have made that mistake himself. He might well be called the mid-wife, perhaps, but he firmly emphasized to me, and to others I am sure, that the fundamental conception is owing to Turing—insofar as not anticipated by Babbage, Lovelace, and others. In my view von Neumann's essential role was in making the world aware of these fundamental concepts introduced by Turing and of the development work carried out in the Moore school and elsewhere.

By now I realized that I was onto a very big story indeed, and that I had been wrong to omit Turing's name from the list of pioneers whose work should be covered in my planned collection of documents on the origins of digital computers.

I prepared a confidential draft account of my investigation, which I sent to each person I had quoted, for their comments and to obtain permission to publish what they had told me, and in the hope that my draft might prompt yet further revelations. This hope was fulfilled when in response Donald Michie amplified his comments considerably. The information that he provided included the following (quoted more fully in Randell[21]):

Turing was not directly involved in the design of the Bletchley electronic machines, although he was in touch with what was going on. He was, however, concerned in the design of electromagnetic devices used for another cryptanalytic purpose; the Post Office engineer responsible for the hardware side of this work was Bill Chandler . . . First machines: The 'Heath Robinson' was designed by Wynn Williams . . . at the Telecommunications Research Establishment at Malvern, and installed in 1942/1943. All machines, whether 'Robinsons' or 'Colossi', were entirely automatic in operation, once started. They could only be stopped manually! Two five-channel paper tape loops, typically of more than 1000 characters length, were driven by pulley-drive (aluminium pulleys) at 2000 characters/sec. A rigid shaft, with two sprocket wheels, engaged the sprocket-holes of the two tapes, keeping the two in alignment. Second crop: The 'Colossi' were commissioned from the Post Office, and the first installation was made in December 1943 (the Mark 1). This was so successful that by great exertions the first of three more orders (for a Mark 2 version) was installed before D-day (June 6th 1944). The project was under the direction of T. H. Flowers, and on Flowers' promotion, A. W. M. Coombs took over the responsibility of coordinating the work. The design was jointly done by Flowers, Coombs, S. W. Broadbent and Chandler . . . There was only one pulley-driven tape, the data tape. Any pre-set patterns which were to be stepped through these data were generated internally from stored component-patterns. These components were stored as ring registers made of thyrotrons and could be set manually by plug-in pins. The data tape was driven at 5000 characters/sec, but (for the Mark 2) by a combination of parallel operations with short-term memory an effective speed of 25,000/sec was obtained . . . The total number of Colossi installed and on order was about a dozen by the end of the war, of which about 10 had actually been installed.

So now the names of these still-secret machines had become known to me, and it had become possible for me to attempt to assess the Colossus machines with respect to the modern digital computer. It seemed clear that their arithmetical (as opposed to logical) capabilities were minimal, involving only counting, rather than general addition or other operations. They did, however, have a certain amount of electronic storage, as well as paper-tape 'backing storage'. Although fully automatic, even to the extent of providing printed output, they were very much special-purpose machines, but within their field of specialization the facilities provided by plug-boards and banks of switches afforded a considerable degree of flexibility, by at least a rudimentary form of programming. There seemed, however, no question of the Colossus machines being stored-program computers, and the exact sequence of development and patterns of influence that led to the first post-war British stored-program computer projects remained very unclear.

At about this stage in my investigation I decided 'nothing ventured, nothing gained' and wrote directly to Mr Edward Heath, the then Prime Minister, urging that the UK government declassify Britain's wartime electronic computer developments. In January 1972 my request was regretfully denied. His reply to me was for some time the only unclassified official document I knew of that in effect admitted that Britain had built an electronic computer during the Second World War!

I responded, requesting that 'for the sake of future attempts to assign proper credit to British scientists and engineers for their part in the invention of the computer, an official history be prepared for release when the wartime computers are eventually declassified'. In August 1972 I received an assurance that this would be done, in a letter that ended:

The Prime Minister is most grateful to you for your suggestion; he hopes that the preparation of the record will go a long way towards ensuring that the considerable work of the British scientists and engineers involved will not, in the future, remain without due recognition.

The classified official history that the Prime Minister had commissioned following my request was, it turns out, compiled by one of the engineers involved with Colossus, Don Horwood.[22] Tony Sale recently described Horwood's report as having been 'absolutely essential' to him when he set out in 1993 to recreate the Colossus.[23]

The stored-program concept

One part of my investigation had concerned just how Turing's pre-war work, and the secret wartime work for and at Bletchley Park, related to the development of the 'stored-program concept', the final major intellectual step involved in the invention of the modern computer. There has been much controversy over the credit due for this development, some of which at least is due to lack of agreement as to just what the concept involves.

The most obvious aspect of the concept is the idea of holding the program that is controlling the operation of a computer, not in some external read-only medium such as punched cards, tape, or pluggable cables, but rather in one memory, together with the information that is being manipulated by the computer under the control of this program. This idea is an essentially practical one, in response to a problem that had not existed with card-controlled or tape-controlled mechanical or electromechanical devices, whose calculation speeds were reasonably well matched to the speed with which the cards or tape that controlled them could be read. The advent of electronics, and the first attempts at building programmable electronic calculating devices such as Colossus and ENIAC, exposed a need for some means of representing programs that could cope with the required program sizes, whose access speed matched that of the (fully electronic) operations they were controlling, and which allowed adequately fast means of replacing a program whose task had been finished with the next program to be executed. Although in particular the plugs, cables, and switches used for programming Colossus and ENIAC were well matched to the calculation speed of their electronics, switching from one program to the next was a slow manual business. This was particularly the case with ENIAC, whose programs were much more complex than those that controlled Colossus.

There was a further practical advantage, over and above the speed of switching between programs, of holding the program and the data in the same memory (once the means of building adequately sized memories had been developed). The various different types of application envisaged required differing relative amounts of storage for instruction and (both variable and constant) data, so having separate memories made for difficult decisions regarding memory provisions. Using a single memory removed the need for such decisions.

These practical considerations, coupled with the crucial realization that a set of pluggable cables and switches were 'just' a means of information representation, seem to be the basic reasoning that lay behind the plans of the Moore School team for a successor machine to ENIAC— namely, EDVAC. It is these plans that are the commonly accepted origin of the stored-program computer concept—at least the aspect related to holding the program in, and executing it from, the computer memory, even if not that of regarding the program as information that could itself be produced or manipulated by a computer.

The suggestion that Babbage had conceived the idea of storing and manipulating program instructions (the second aspect of the stored-program concept, to use modern terminology) rests in part on one brief passage in Lady Lovelace's notes on his Analytical Engine:[24]

Figures, the symbols of numerical magnitude, are frequently also the symbols of operations, as when they are the indices of powers . . . of course, if there are numbers meaning operations upon *n* columns, these may combine amongst each other, and will often be required to do so, just as numbers meaning quantities combine with each other in any variety. It might have been arranged that all numbers meaning operations should have appeared on some separate portion of the engine from that which presents numerical quantities; but the present mode is in some cases more simple, and offers in reality quite as much distinctness when understood.

This rather obscure passage is somewhat contradicted by other statements in Lady Lovelace's notes. In fact, perhaps more significantly, a brief passage in one of Babbage's sketchbooks dated 9 July 1836 comments on the possibility of using the Analytical Engine to compute and punch out modified program cards that could later be used to control a further computation, a relatively clear indication that he viewed programs as just a further form of manipulable information.

In my view, Lady Lovelace's 1843 account of programming (which even includes an allusion to what we would now call indexed addressing) also shows a remarkable understanding of the concept of a program, although the question of the extent to which she, rather than Babbage, was responsible for the contents of her notes on this topic is not at all clear.

In fact, apart from these vague statements by Charles Babbage and Lady Lovelace, and of course the very clear implications of Turing's 1936 paper (with its infinite tape 'memory'), the earliest suggestion I knew of that instructions be treated as information to be stored in the main computer memory was contained in von Neumann's famous EDVAC report.[25] However, I then found a 1945 report by J. Presper Eckert[26] containing the claims that in early 1944, prior to von Neumann's association with the EDVAC project, they had designed a 'magnetic calculating machine' in which the program would 'be stored in exactly the same sort of memory device as that used for numbers'. Regrettably, there was no consensus regarding the relative contributions of Eckert, Mauchly, von Neumann, and Goldstine to the design of EDVAC—a controversy that I did not wish to enter into.

At this stage, the initial major goals of my investigation, in particular that of establishing whether Turing had played a direct role in the development of the concept of the stored-program computer, had not been achieved, though I did feel that my investigations had helped to clear up some of the more important misconceptions and misattributions.

But, unfortunately, my investigation in 1972 of Turing's post-war work at the NPL did not match the thoroughness with which Carpenter and Doran later analysed Turing's 1945 design for ACE.[27] Rather, in 1972, I read too much into the fact that Turing's report[28] on his design slightly post-dated (and indeed cited) the EDVAC report,[29] whereas Carpenter and Doran provided a detailed comparison of the fully developed stored-program facilities that Turing proposed for the ACE, against what they showed were the rather rudimentary ones described in the EDVAC report. (In contrast to Turing's ACE report, the EDVAC report explicitly disallowed the modification of stored-program instructions, although this restriction was later lifted.) In retrospect it is clear that I really should have studied the EDVAC and ACE reports more thoroughly, and perhaps assigned more credit to Turing regarding the origins of the stored-program concept beyond that due to his 1936 paper. Clearly, I should later have included at least some of his unfortunately little-known 1945 ACE report in my collection of selected papers on the origins of computers. It was to be many years before I provided a much fuller analysis of the stored-program concept and of Turing's role in its development.[30]

In fact, the question of who first had the idea (and understood its fundamental importance) of having an extensive addressable internal memory used for both instructions and numerical qualities, and also the ability to program the generation and modification of stored instructions, continues to be a matter of debate. What is indisputable, however, is that the various papers and reports emanating from the EDVAC group from 1945 onwards were a source of inspiration to computer designers in many different countries, and played a vital part in the rapid development of the modern computer.

Concluding remarks

Although my initial goals had not been achieved I had—perhaps more importantly—managed to accumulate some evidence that in 1943, a couple of years before ENIAC (hitherto generally accepted as the world's first electronic digital computer) became operational, a group of people directed by Max Newman and Tommy Flowers, and with which Alan Turing was associated, had built a working special-purpose electronic digital computer, the Colossus.

I had established that this computer was developed at the Post Office's Dollis Hill Research Station and installed at Bletchley Park. The Colossus and its successors were, in at least a limited sense, 'program controlled'. Moreover, there were believable claims that Turing's classic pre-war paper on computability, a paper that is usually regarded as being of 'merely' theoretical importance, was a direct influence on the designers of the British machine and also on von Neumann, at a time when he was becoming involved in American computer developments.

Having obtained permission from all my informants to use the information that they had provided to me, Donald Michie and I were keen that a summary of my investigation be placed in the public domain.[31] The vehicle we chose was his 1972 Machine Intelligence Workshop, the proceedings of which were published each year by Edinburgh University Press. (At the workshop, after my lecture, I overheard two worried Edinburgh University Press staff expressing concerns as to whether it would be safe to publish my account—they comforted themselves that if there were any comeback it would be the official head of the Press, the Duke of Edinburgh, who would be sued!)

Afterwards, I managed to persuade Donald Michie to contribute a two-page summary of my findings, and thus at last some coverage of Turing, to my collection of historical computer documents—the first edition of which was published in 1973 as *The Origins of Digital Computers: Selected Papers*.[32]

With this my investigation of Turing and Colossus ended, since I had come to the conclusion that it might be a long time before anything more would become public about Bletchley Park and Colossus. But within a couple of years much more information about the wartime developments at Bletchley Park started to be made public, with the result that I was able to restart my investigation and find out a great deal more about the Colossus Project. These developments are described in Chapter 17.

PART III

Codebreaker

CHAPTER 9

At Bletchley Park

JACK COPELAND

This chapter summarizes Turing's principal achievements at Bletchley Park and assesses his impact on the course of the Second World War.

War party

On the first day of the war, at the beginning of September 1939, Turing took up residence at Bletchley Park, the ugly Victorian mansion in Buckinghamshire that served as the wartime HQ of Britain's military codebreakers (Fig. 9.1). There Turing was a key player in the battle to decrypt the coded messages generated by Enigma, the German forces' typewriter-like cipher machine.[1]

Germany's army, air force, and navy transmitted many thousands of coded messages each day during the Second World War. These ranged from top-level signals, such as detailed situation reports prepared by generals at the battlefronts and orders signed by Hitler himself, down to the important minutiae of war such as weather reports and inventories of the contents of supply ships. Thanks to Turing and his fellow codebreakers, much of this information ended up in Allied hands—sometimes within an hour or two of its being transmitted. The faster the messages could be broken, the fresher the intelligence that they contained, and on at least one occasion the English translation of an intercepted Enigma message was being read at the British Admiralty less than 15 minutes after the Germans had transmitted it.[2]

Bomber

Turing pitted machine against machine. Building on pre-war work by the legendary Polish codebreaker Marian Rejewski, Turing invented the Enigma-cracking 'bombes' that quickly turned Bletchley Park from a country house accommodating a small group of thirty or so codebreakers into a vast codebreaking factory.[3]

There were approximately 200 bombes at Bletchley Park and its surrounding outstations by the end of the war.[4] As early as 1943 Turing's machines were cracking a staggering total of

Figure 9.1 The Mansion, Bletchley Park.

Reproduced with permission of the Bletchley Park Trust.

84,000 Enigma messages each month—two messages every minute.[5] Chapter 12 describes the bombes and explains how they worked.

The U-boat peril

Turing also undertook, single-handedly at first, a 20-month struggle to crack the especially secure form of Enigma used by the North Atlantic U-boats. With his group he first broke into the current messages transmitted between the submarines and their bases during June 1941, the very month when Winston Churchill's advisors were warning him that the wholesale sinkings in the North Atlantic would soon tip Britain into defeat by starvation.[6] Churchill, Britain's bulldog-like wartime leader, later confessed: 'The only thing that ever really frightened me during the war was the U-boat peril'.[7] Turing's work had at last made it possible to defuse that peril. The U-boats' messages revealed their positions, and so the merchant convoys could simply be rerouted away from the submarines—a simple but utterly effective measure.

Breaking Tunny

Turing also searched for a way to break into the torrent of messages suddenly flooding from a new, and much more sophisticated, German cipher machine (described in Chapter 14). British intelligence code-named the new machine 'Tunny'.

The Tunny teleprinter communications network, a harbinger of today's mobile phone networks, spanned Europe and North Africa. It was used for the highest-level traffic, connecting Hitler and the army high-command in Berlin to the front-line generals. Turing hacked into the Tunny system, building on work done by another solitary and taciturn Bletchley Park genius, Bill Tutte. Turing's breakthrough in 1942 yielded the first systematic method for cracking Tunny messages: this was known simply as 'Turingery' (also described in Chapter 14).

Turingery was the third of Turing's three great contributions to the war, along with designing the bombe and unravelling U-boat Enigma. Using Turingery, Bletchley Park was able to read the lengthy typed exchanges between Berlin and the generals commanding the armies at the battlefronts—conversations that laid German strategy bare. 'Turingery was our one and only weapon against Tunny during 1942–3', explained 93-year-old Captain Jerry Roberts, once section leader in the 'Testery', one of Bletchley Park's two Tunny-breaking units, named after its head Ralph Tester (see Chapter 16).

Turingery was also the seed for the sophisticated Tunny-cracking algorithms incorporated in Tommy Flowers' Colossus, the first large-scale electronic computer (described in Chapter 14). With the installation of the Colossi—there were nine by the end of the war—Bletchley Park became the world's first electronic computing facility.

Sea Lion: the invasion that never was

If Hitler's Operation Sea Lion (*Seelöwe*)—his planned invasion of Great Britain in 1940—had actually been launched, troop carriers would have poured across the English Channel from France, accompanied by fleets of supply barges loaded with tanks, artillery, and heavy machine guns.[8] In July 1940, Bletchley Park cracked an Enigma message revealing that the invasion was imminent.[9]

During the massive attack by sea and air, thousands of gliders crammed with heavily armed crack German soldiers would have descended onto British soil. Paratroops would also have rained down, with swarms of dive-bombers disabling airfields and holding back a British ground response. Once the invaders had secured a foothold—a patch of territory containing suitable harbours and airfields—Hitler's formidable forces would have advanced ruthlessly in every direction until they occupied all Britain's key cities, or so the Führer planned.

In the event, however, the Sea Lion invasion was postponed and then abandoned. But Britain's fate had hung by a thread. If the Royal Air Force had not proved so resilient during the summer of 1940, if the German leader's attention had not been wandering in the direction of Russia, if Turing's second bombe, whimsically named 'Agnus Dei' (the Lamb of God), had not been breaking the Luftwaffe's top-secret Enigma communications . . . then it might all have turned out very differently.[10] When the Imperial Japanese Air Force attacked Pearl Harbor in 1941, Roosevelt might have faced a Europe completely dominated by Emperor Hirohito's ally, Hitler.

Cyberwar on Rommel

The tide began to turn against the German military in 1942, with the neutralization of the North Atlantic U-boat threat and the humiliating rout of Field Marshal Rommel's panzer army at El Alamein in North Africa.

British successes in the U-boat war—where Turing was a key player—freed up the supply routes from North America to Britain, while Hitler's disaster at El Alamein denied him his chance of taking the Suez Canal and capturing the precious Middle Eastern oilfields. Debilitating shortages of fuel plagued the German military for the rest of the war.

During the build-up to the fierce fighting at El Alamein, the Bletchley Park codebreakers were reading Rommel's secret messages.[11] He gave them the welcome news that his tanks had insufficient fuel to fight effectively. The codebreakers themselves had played a leading role in bringing this situation about—for weeks their decrypts had been unmasking the cargoes of the German and Italian ships carrying Rommel's supplies across the Mediterranean. This intelligence enabled the RAF to pick and choose the best targets. Thousands of tons of fuel went blazing into the sea.

Ten days into the Battle of El Alamein, a broken message from Rommel to Berlin confessed that the 'gradual annihilation of the army must be faced'. Hitler's response, eagerly read at Bletchley Park, told Rommel not to yield a single step—he must, Hitler ordered, take 'the road leading to death or victory'.[12]

What if Enigma and Tunny had not been broken?

How different would the world have been today if Turing had not been able to break Germany's codes?

Writing 'counterfactual' history is always speculative, never cut and dried—because, if some key matters had gone differently, the overall outcome of a war, or battle, or election, might have been very different, or might nevertheless have been just the same. If the CIA had killed Osama Bin Laden in 2000, 9/11 might still have happened—perhaps because, following Bin Laden's death, one of his lieutenants would have stepped forward to take control of Al Qaeda and implement Bin Laden's plans.

Suppose, contrary to fact, that Turing and the Bletchley Park codebreakers had not managed to crack the communications of the German and Italian navies in the Mediterranean, and had failed to break the cloaked messages of the North Atlantic U-boats. What would the result have been? The U-boats would certainly have continued to prey with their merciless efficiency on the convoys of merchantmen, bringing precious food, fuel, munitions, and manpower from America to Britain. Untold quantities of sorely needed materials would have plummeted to the bottom of the ocean. Without the RAF's depredations on his own seaborne supplies, Rommel might even have defeated the British at El Alamein, and gone on to capture Middle Eastern oil for Germany. Yet even so, the war in Europe might still have ended at more or less the same time that it did—the spring of 1945—because of other counterfactual events.

For example, in 1945 America or Britain might have dropped an atomic bomb on Berlin. Even without a European atomic bomb, and even if, thanks to tighter Axis cipher security, Bletchley Park had not been able to break the key Enigma, Tunny, and Hagelin cipher systems, the Allies might nevertheless still have prevailed. The German defeat might have been virtually inevitable once Hitler took on the vast Soviet Union. Returning to the actual course of events, Bletchley Park decrypts certainly played a crucial part on the Russian front too—most especially the intelligence wrung by Turingery from top-level Tunny messages between Berlin and the front-line generals.[13]

Can historians quantify Turing's impact on the course of the war? They certainly cannot claim with any great confidence that what Turing did shortened the war, as the atom bomb example illustrates only too well.[14] But in the twenty-first century we can at the very least give a reasonably complete account of what Turing actually achieved during his years at Bletchley Park—a picture largely denied to twentieth-century historians due to the blanketing official secrecy.

Turing's anti-Enigma bombes, together with the Turingery-inspired code-cracking algorithms that ran on Bletchley Park's anti-Tunny Colossus computers, supplied the Allies with an unprecedented inside view of the enemy's military thinking. In particular, Bletchley Park had an unrivalled window on German counter-preparations for the looming D-Day invasion of France, which was launched in June 1944 from the beaches of Normandy.

If Turing had not broken U-boat Enigma, this invasion of mainland Europe, ushering in the final stage of the war, could have been delayed by months, even years. This was because the gigantic build-up of the necessary troops and munitions in the southern English ports facing France could not even have begun while the Atlantic sinkings continued unabated. Without a break into the submarines' messages, the invasion would have had to wait until the Allied navies hunted down the U-boats by conventional means.

Incalculable consequences

Any delay in the invasion would have been strongly in Hitler's favour, since it would have given him more time to prepare for the coming attack from across the Channel—more time to transfer troops and tanks from the Eastern Front to France, and more time to fortify the French coast and the River Rhine, the most crucial of the natural barriers lying between the invasion beaches and the German heartland. Also, more V1 drones and more rocket-propelled V2 missiles would have rolled off the production lines, to rain down on southern England and wreak havoc at the ports and airfields needed to support the invading troops.

History records that the Allied armies took roughly a year to fight their way from the Normandy beaches to Berlin. In a counterfactual scenario, in which Hitler had had more time to consolidate his preparations, this struggle might have taken much longer—twice as long, maybe. That translates into a very large number of lives. At a conservative estimate, each year of fighting in Europe brought on average about 7 million deaths.

Returning to the atom bomb example and to the difficulties of counterfactual history, the killing might still have ended in May 1945, even in a scenario that saw Tunny and U-boat Enigma remaining unbroken throughout the war. Nevertheless, this colossal number of lives—7 million had the war had continued for another year, 21 million if, owing to the Atlantic U-boats and a strengthened Fortress Europe, the war had toiled on for as long as another 3 years—do most certainly convey a sense of the magnitude of Turing's contribution.

Turing, the digital warrior, stands alongside Churchill, Eisenhower, and a short glory-list of other wartime principals as a leading figure in the Allied victory over Hitler. There should be a statue of him in Central London among Britain's other war heroes.

CHAPTER 10

The Enigma machine

JOEL GREENBERG

S hortly after the end of the First World War, the German Navy learned that its encrypted communications had been read throughout the hostilities by both Britain and Russia. The German military realized that its approach to cipher security required a fundamental overhaul, and from 1926 different branches of the military began to adopt the encryption machine known as Enigma. By the start of the Second World War a series of modifications to military Enigma had made the machine yet more secure, and Enigma was at the centre of a remarkably effective military communications system. It would take some of the best minds in Britain—and before that, in Poland—to crack German military Enigma.[1]

Introduction

The exact origins of the encryption machine that played such an important role in the Second World War are not entirely clear.[2] In the early 1920s patent applications for a wheel-based cipher machine were filed by a Dutch inventor, Hugo Koch, as well as by a German engineer, Arthur Scherbius.

In 1923, a company called Chiffrienmaschinen AG exhibited a heavy and bulky encryption machine at the International Postal Congress in Bern, Switzerland. This machine had a standard typewriter keyboard for input, and its design followed Scherbius's original patent closely. Scherbius had named his machine 'Enigma', and this 'Model A' was the first of a long line of models to emerge. Models B, C, and D soon followed, and by 1927 Model D was selling widely for commercial use. A number of governments purchased Enigma machines in order to study them, and Edward Travis—the deputy head of Britain's signals intelligence unit, the Government Code and Cypher School—bought one on behalf of the British government in the mid-1920s.

In 1925, the German Navy decided to put Enigma into use the following year, despite having rejected one of Scherbius's previous encryption mechanisms in 1918. Meanwhile, the German Army began to redesign Enigma, with the intention of strengthening its security. By 1928, Model G was in use,[3] and in June 1930 Model I (Eins) became the standard version, deployed first by the army, then the navy in October 1934, and the air force in August 1935.

The version deployed by the navy, called the M1, was functionally compatible with Model I, but there were some minor differences. For example, the rings around the three wheels were inscribed with letters (A–Z), rather than numbers (01–26), and the M1 had a 4-volt power socket, making it suitable for use on board a ship. Approximately 611 M1 machines were built. In 1938, the M1 was followed by the M2, of which 890 units were delivered. Then in 1940, the M2 was replaced by the M3, some 800 of which were built. All three machines, M1, M2, and M3, had the same internal marking 'Ch. 11g'. 'Ch.' stood for Chiffrienmaschinen AG, and the marking identified all three models as variants of the same design, which was just one of the family of Enigma designs manufactured by the company.[4] This chapter focuses mainly on the naval machines—as did Turing.

Enigma's components

In a nutshell, the Enigma carried out varying letter-for-letter substitutions on plain German text, producing what was called the 'ciphertext' of the German message. The ciphertext looked like a random jumble of letters—although in fact it was far from being truly random.

Enigma was an 'offline' device, in the sense that it was used only to encrypt and decrypt the message; the actual transmission of the message was a separate process, performed with other equipment. The Enigma was ideal for mobile communications. Weighing about 25 pounds, it sat in a wooden box that could be easily carried using a fitted leather strap. The machine ran from either the mains supply or a battery.

Enigma had a keyboard with the twenty-six letters of the alphabet laid out in three rows (Fig. 10.1). The top row, from left to right, read QWERTZUIO, as was standard in Germany (the top row of the British and American keyboard read QWERTYUIOP, as on a modern 'Qwerty' keyboard). The Enigma keyboard had no punctuation or number keys, so if an operator wanted to enter the number 1, for example, he would have to type *eins*, the German word for 1. The three German umlauted letters ä, ö, and ü had to be entered as 'ae', 'oe', and 'ue'.

Adjacent to the keyboard were twenty-six small circles of glass, each stencilled with a letter of the alphabet. (These were probably made of acrylic 'safety glass', known as Plexiglas in Germany and Perspex in the UK.) The circles were again laid out in three rows of letters, in the same order as the keyboard. This complete component was known as the 'lampboard'. Underneath each circle of glass was a small bulb, much like one from a torch of the pre-LED era. Every time a key was pressed, one of the bulbs lit up. The stencilled letter that the bulb illuminated was the encryption of the letter on the key that had been pressed. For example, if the O key was pressed and the bulb under Q lit up, then Q was the encryption of O at that point in the message (see Fig. 12.4).

In the middle of the machine were three wheels. These were the heart of the encryption mechanism. They rotated on a removable shaft that passed through a hole in the centre of each wheel (Fig. 10.2). Every time a key was pressed, some or all of the wheels rotated. Due to their relative rates of rotation, the left, middle, and right wheels were referred to as the slow, middle, and fast wheels, respectively.

The Enigma was designed in such a way that as many as about 17,000 key presses could be required in order to return all three wheels to their initial starting positions, although if several of the three special navy wheels—described below—were in use, this figure could drop to as few as 4056.[5] In practice, however, even this smaller number of key presses would never occur in

Figure 10.1 Enigma M3.

Reproduced with permission of the Bletchley Park Trust.

the course of a message, since these were limited to 250 characters, or 320 in the case of Naval Enigma.

Each wheel had twenty-six electrical contacts on each of its two faces. Fixed wiring inside the wheel led from the contacts on one face to the contacts on the other face. The wiring was different for each wheel. This meant that if the wheels were taken out of the machine and put back in a different order, the machine's wiring was different.

On the right-hand side of each wheel (when viewing the machine from the front) was a toothed cogwheel. On the left-hand side was a metal ring or rim, with the letters A–Z on it (or sometimes the numbers 01–26: the army and air force machine usually had numbers while the naval machines usually had letters). This ring could be fixed in any one of twenty-six positions by means of a clip. The ring had a 'turnover' notch on its extreme left-hand edge (or two notches in some cases). The function of the ring and of the notches will be explained later.

To the right of the three wheels (again when looking at the machine from the front) was the so-called 'entry plate', with twenty-six electrical contacts (depicted in Fig. 12.4). To the left of the three rotating wheels was a fixed wheel called the 'reflector', with twenty-six pin contacts wired together in pairs. Reflectors with different wirings were introduced at various times during the war, but the most commonly used was called Reflector 'B' at Bletchley Park. The function of the reflector is described below.

Figure 10.2 Inside the Enigma M3.

Reproduced with permission of the Bletchley Park Trust.

The wiring inside the wheels of the Enigma M3 was different from that in earlier models. The M3 also had a 'plugboard' at the front. This consisted of twenty-six sockets, one for each letter of the alphabet, again laid out in the same order as the letters on the keyboard, in the same three rows. Cables with plugs at each end were placed in the sockets, and the plugging could be varied in accordance with the day's instructions for setting up the machine. Each cable connected two selected letters together. The purpose of the plugboard was to switch the typed letter for a different one before it entered the wheels (see Fig. 12.4). The letter then exiting from the wheels was switched again by the plugboard before reaching the lampboard. From November 1939, most German Enigma networks connected ten pairs of letters together at the plugboard. Previously, fewer pairs had been connected, making messages easier to break.

When the wheels were put into the machine and the lid was shut, the serrated edge of each wheel protruded through a slot in the lid (see Fig. 10.3). Adjacent to each slot was a small viewing window that displayed the topmost letter on the ring. Each wheel could be turned freely by hand to any one of its twenty-six positions by rotating its protruding edge.

It was central to the encryption process that at least one wheel would move forward one position each time a key was pressed at the keyboard. The right-hand wheel moved at every

Figure 10.3 A window adjacent to the edge of each wheel shows the wheel's current position.

keystroke, while the movement of the middle and left-hand wheels was determined by the position of the ring. As a consequence, the electrical circuit through the machine was dynamic, changing every time a key was pressed. Because of this, typing the same letter again and again at the keyboard would produce a stream of different letters at the lampboard.

Once the wheels were clamped together in the machine, between the entry plate and the reflector, there were twenty-six parallel electrical circuits running between them. When a key was pressed at the keyboard, a current passed from the key through, in turn, the plugboard, entry plate, right-hand wheel, middle wheel, left-hand wheel, and reflector, and then (due to the wiring of the reflector) back again through the left-hand wheel, middle wheel, right-hand wheel, entry plate, and finally the plugboard, eventually reaching the lampboard (again see Fig. 12.4). As the current passed through the machine from point to point, the letter continually changed identity. By the time the bulb lit up on the lampboard, the letter could have switched its identity up to nine times.

By 1939 Naval Enigma operators had a box of eight wheels, from which they chose three. Initially, the M3 had been supplied with five cipher wheels, which were compatible with the wheels of the Enigma Model I. This meant that the navy's Enigma machines could exchange messages with army and air force Enigmas. In 1939, however, three extra wheels, used exclusively by the navy, were added to the M3's original five (the new wheels were named VI, VII, and VIII). German U-boats used the M3 until the North Atlantic U-boat fleet was equipped with the four-wheeled M4 in February 1942.[6]

The M4, backwards compatible with the M3, was a modified three-wheel machine. The width of the reflector was halved and the remaining space taken up by a fourth wheel. (There were two different versions of this extra wheel, known at Bletchley Park as 'Beta' and 'Gamma'.)

The new wheel—which did not move during encryption, and could not be exchanged with the other Naval Enigma wheels—introduced an additional stage into the encryption process.

The M3, broken by Turing during his epic struggle with the U-boats, is discussed in more detail later in this chapter.

Sending a message

To prepare the machine—for example the M3—to encrypt a message, three wheels were taken from the box and placed in the machine in the specified order. The rings of the wheels were then adjusted to their specified positions. Next the operator connected up ten pairs of letters on the plugboard (or fewer than ten in the early weeks of the war). His final adjustment was to turn the wheels to a pre-determined starting position.

The number of possible electrical configurations that can be made in this way is surprisingly large. It is calculated by multiplying the number of plugboard configurations (150.7 million million) by the number of ways of selecting the wheels (336), and then multiplying the result by the number of possible starting positions of the wheels (17,576).

Once the machine was fully set up, encrypting a message by Enigma was a relatively simple process. The operator typed the message—in plain German—at the keyboard, and as he typed, the letters of the encrypted form of the message lit up, one by one, at the lampboard. Decryption was equally straightforward. Once the encrypted version of the message had been received, and the recipient's Enigma had been set up, the recipient simply typed out the encrypted version at the keyboard, and the letters of the plain German text lit up at the lampboard (see Fig. 12.5). All the complexity lay in ensuring that the sender and recipient set up their Enigma machines identically. A widely used procedure for doing this involved what Bletchley Park called the 'daily key'. This procedure, as employed with the principal M3 Naval Enigma ciphers, is described in the next section.

Once encrypted, messages were normally broadcast via radio. A complicated programme of changing frequencies was used during transmissions to and from the U-boats. Naval messages were usually transmitted in four-letter groups, as in the example set out here (although a few naval networks followed the army and air force practice of using five-letter groups). Two additional four-letter groups were placed at the beginning and again at the end of the message. These were called the 'indicator' groups, and their function is described later in the chapter.

An important German naval signal was intercepted on 27 May 1941. The intercepted message would have looked like this (using, for illustration, the indicator groups constructed later in the chapter)[7]:

MMÄ 1416/27/989   38
IJTV USYX DERH RFRS OQRV DTYH QWBV HILS CXHR OPOD
GTQL DDHI KFTG EDZS WXQS EDFR HGYG EDZZ UYQV DTYY
EDGH KIRM SYBK PANX JSTP QXDT ERGP JMSX VFWI FTPZ
ADHK WDLE QPAL ALDH XNDH RYFH IJTV USYX
1231   7640

MMÄ was the call sign identifying the transmitting station, 1416 was the time at which the transmission began, 27 was the day of the month, 989 was the message serial number, and 38 was the total number of four-letter groups (a check that the complete message had been received). IJTV USYX was the indicator (repeated at the end of the message). The final line, added by the British interceptors, gave the time of interception, 1231, and the transmission frequency, 7640 kilocycles (kHz).

The message, from the admiral commanding U-boats to all U-boats in the Bay of Biscay, said:

BISMARCK MUST NOW BE ASSUMED TO HAVE SUNK. U-BOATS TO SEARCH FOR SURVIVORS IN SQUARE BE6150 AND TO NORTH WEST OF THIS POSITION.

The number of intercepted Naval Enigma messages increased steadily throughout the war, from about 300 messages a day in 1940 to 1500–2000 messages a day during 1944–45. The highest number of intercepted naval messages recorded on a single day was 2133, on 12 March 1945.

The daily key

The daily key (or daily setting) told the operator how to set up his machine for the day. The naval daily key had four components:[8]

- 'Plugboard pairs': usually ten letters were connected to ten other letters at the plugboard. The remaining six unplugged letters were known as 'self-steckered', meaning 'self-plugged'. Each of the self-plugged letters remained unchanged when current passed through the plugboard.
- 'Wheel order': three wheels selected from eight were placed in the machine in a specific order. There were $8 \times 7 \times 6$ (i.e. 336) possible permutations of the wheel order (more in the M4).
- 'Ground setting': this was a group of three letters that specified the position of the wheels the operator was to use at the start of encrypting the 'message setting'. The message setting, described in what follows, is itself a trio of letters, such as BDK, denoting the position of the wheels at the start of encrypting the actual message text. In either case, the operator turns the wheels by hand and when the three prescribed letters show in the viewing windows, the wheels are in their starting position.
- 'Ring setting': this was the position of the rotatable alphabetic ring around each wheel. With a separate setting for each wheel's ring, there were $26 \times 26 \times 26$ (i.e. 17,576) possible configurations of the rings. The purpose of the ring setting was to disguise the starting position of the wheels. Even if one knew the message setting, it was not possible to infer the actual starting position of the wheels from it unless the ring setting was also known. This is because the letters making up the message setting were not fixed relative to the wheels but inscribed on the movable ring.

The complete daily key was not in fact changed every day. On adjacent days, called 'paired days', the wheel order and ring setting would remain unchanged; in a 30-day period, the wheel order and ring setting would generally change only fifteen times. In 31-day months, there was normally one 3-day period with the same wheel order and ring setting. On some Naval Enigma ciphers, a complete change of daily key was made as infrequently as every ten days.

The German 'key makers' restricted themselves quite unnecessarily in a number of ways. Hut 6 discovered the rules used by the compilers of army and air force keys, calling these 'rules of keys'.[9] The naval key makers had unnecessary 'wheel order rules': for example, the wheel order generally contained a wheel VI, VII, or VIII. The wheel order rules could on occasion reduce the number of possible wheel orders from 336 to as few as ten or twenty, so making the codebreakers' work much easier (although these rules remained undiscovered until 1944).[10]

However, the greatest weakness of the German Navy's daily key system was the fixed ground setting. Turing's method of 'Banburismus' (described in Chapters 13 and 38) depended entirely on the fact that the ground setting was not changed during the day. Had the Germans used a variable ground setting, it would seriously have impeded the decryption of messages.

The navy had a number of different communication networks, each with its own daily key. The principal network for U-boats and surface ships in home waters, including large parts of the Atlantic, was *Heimische Gewässer*, known at Bletchley Park as 'Dolphin'. Up until September 1942 Dolphin, and its companion Triton (used at that time only on the M3), were the only general Naval Enigma keys being broken. The ins and outs of Dolphin will now be explained.

The message setting

After the Enigma machine had been set up using the daily key, and the operator had carried out some further preliminaries, described below, he would then turn the three wheels to the starting position for the first message (e.g. BDK). Each wheel was turned until the letter in question appeared in its viewing window.

The German Air Force and Army allowed their operators to choose these three letters. Unfortunately, operators often chose rather obvious triples of letters, such as the letters on keys sitting diagonally on the Enigma keyboard (e.g. QAW and WSX) making life simpler for the codebreaker. The German Navy, more security conscious than the army and air force, tightened security by issuing Dolphin operators with a book containing lists of potential message settings.

This book was called the *Kenngruppenbuch* or, at Bletchley Park, the 'K-book'. The K-book contained all 17,576 possible trigrams (i.e. three-letter groups, such as PQR). The same K-book remained continuously in use from 1941 until the end of the war. In addition, once an operator had selected a trigram from the K-book, he employed 'bigram tables' to disguise it before transmitting it. A bigram is a two-letter pair (such as PQ).

Bigram tables were paper tables pairing bigrams with bigrams, for all 676 (i.e. 26^2) bigrams. A table might read AA = PY, AB = ZR, AC = NV, etc. AA = PY told the operator to disguise occurrences of AA by putting PY instead (and vice versa). Nine complete tables constituted a set, and which table of the set was in use on any given day was shown in a special calendar issued to operators. (New sets of bigram tables were introduced in July 1940, June 1941, November 1941, March 1943, and July 1944.)

The procedure for creating and using a message setting was as follows:[11]

1 The operator chose a trigram at random from the K-book, say ARQ. He then set his wheels to the ground setting (the three-letter group fixed for the day and part of the daily key), say JNY; this was done by turning the wheels, letter by letter, until JNY appeared in the viewing windows. He then encrypted ARQ by typing ARQ at the keyboard, giving (say) LVN at the lampboard. LVN was the message setting, the position to which he now set his wheels in order to begin encrypting the message itself (by typing it at the keyboard).

2 The operator needed to send the trigram ARQ to the message's recipient, so that the recipient could decrypt it at the ground setting and discover the message setting. So the

sender proceeded to disguise ARQ before transmitting it. To do so, he chose another trigram from the K-book, YVT say. He then wrote down

```
_ Y V T
A R Q _
```

and filled in the two blanks _ with any two letters he cared to choose, giving (say):

```
W Y V T
A R Q N
```

Next, he consulted the day's bigram table and replaced the vertical pairs of letters, WA, YR, VQ, TN, by their equivalents given in the table—say, IJ, TV, US, YX. Finally, he placed IJTV USYX at the beginning of the encrypted message (see the previous example of an intercepted message). IJTV USYX was called the message's 'indicator'. For good measure, the operator added the indicator once again at the end of the encrypted message.

3 The message's official recipient looked up IJ, TV, US, YX in his copy of the bigram table and replaced the letters by their equivalents (WA, YR, VQ, TN), giving:

```
W Y V T
A R Q N
```

He then set his wheels to the ground setting (JNY in this example) and decrypted ARQ to get LVN, the message setting. Finally, he set his wheels to LVN and decrypted the message by (as usual) typing the message's ciphertext at the keyboard; the German plaintext would light up, letter by letter, at the lampboard.

Technicalities: reciprocity and wheel 'turnovers'

The Enigma machine was designed to be 'reciprocal'. This means that if (say) the machine encrypts O as Q, then typing Q at the same position of the wheels produces O at the lampboard (see Fig. 12.5). It is because Enigma was reciprocal that the two processes of encryption and decryption were essentially the same. Typing the plaintext produces the ciphertext, and typing the ciphertext at the same machine settings produces the plaintext again.

The reciprocal nature of the machine was a consequence of the design of the reflector and the plugboard. There was, however, a cost to the way the reflector worked. Because of the reflector's design, it was impossible for a letter to be encrypted as itself—a crucial weakness of the machine. For a letter to encrypt as itself, the electrical current would have to pass along a wire in both directions at the same time, which is a physical impossibility. It seems that the Germans were prepared to accept this weakness, presumably because they viewed it as a tolerable price to pay for a reciprocal design. A machine whose processes of encryption and decryption were identical was so much simpler to operate. However, the machine's designers completely failed to appreciate how useful this weakness would be to the codebreakers (see Chapter 12).

The function of the wheel 'turnovers' was to ensure that the machine's wiring varied with every key press, or in other words to make sure that at least one wheel moved every time a letter was typed at the keyboard. The right-hand wheel always turned over one step each time a key was pressed at the keyboard, irrespective of the motion of the other wheels. The middle wheel turned over a step whenever the right-hand wheel was in what was called a 'turnover position', and the left-hand wheel turned forward whenever the middle wheel was in a turnover position.

Three turnover pawls controlled the motion of wheels. These were pivoted bars (situated at the rear of the machine), with each wheel's pawl engaging the toothed cogwheel mounted on the side of the wheel. A wheel was in a turnover position whenever one of the turnover notches on the wheel's ring engaged with its pawl. The action of the pawl, the cogwheel, and the turnover notch combined to move the wheel forward a step.

Wheels I to V were at their turnover positions when the letters positioned at the top of the wheel-rings (and visible through the viewing windows) were as follows:

Wheel I: Q   Wheel II: E   Wheel III: V   Wheel IV: J   Wheel V: Z

Wheels VI, VII, and VIII were a little different. Whereas each of wheels I to V had a single notch, at a different place on each wheel, wheels VI, VII, and VIII had two notches each, at the same places on each wheel. These three wheels were in their turnover positions when either M or Z was positioned at the top of the ring.

Extra encryption

In so-called super-encryption a message (or indicator, or some other item) is encrypted multiple times. Some especially important—or personal—naval messages were super-encrypted using the *Offizier* system. First, an officer encrypted the message using an Enigma that he had set up in accordance with the special 'officers' daily key'. Then the resulting ciphertext was encrypted for a second time using the ordinary daily key. At Bletchley Park, these super-encrypted messages were called 'offiziers'. Naturally, offiziers were especially difficult to break.

The officers' daily key had the same wheel order and ring setting as the ordinary daily key, but a different set of plugboard pairs was used, and there were only twenty-six permitted wheel-starting positions. These were denoted by the twenty-six letters of the alphabet, and were valid for a month. Each general key had its associated officers' key. In the case of Dolphin, the officers' key was known at Bletchley Park as 'Oyster'.

Another example of super-encryption was the navy's use of special codebooks for shortening signals, as a precaution against shore-based high-frequency direction finding. The shorter the radio transmission, the less chance interceptors had of determining its precise direction. These special codebooks were employed, for example, to shorten reports of convoy sightings (using the short signal book) and weather reports (using the weather short signal book). What would nowadays be called a 'compression procedure' was applied.

Conclusions

The British wartime leader Winston Churchill said after the war:[12]

The Battle of the Atlantic was the dominating factor all through the war. Never for one moment could we forget that everything happening elsewhere, on land, at sea or in the air depended ultimately on its outcome.

It took the brilliance of Alan Turing, the leadership of Hugh Alexander, and the razor-sharp minds of the other codebreakers working in Hut 8 to overcome Naval Enigma and give the advantage to the Allies in the Battle of the Atlantic.

There was an irony to the navy's using a codebook to disguise the message setting, since Enigma was deployed in the first place to overcome the inherent weaknesses of the German naval codebooks of the First World War. Another irony is that the Germans could easily have made small alterations to the Enigma machine that would have made breaking the messages much more difficult. For example, simply placing the turnover notches at the *same* place on all the wheels would have undercut many of Hut 8's hand-methods for working out the daily settings. As with Germany's even more formidable 'Tunny' cipher machine (described in Chapter 14), some of the complications included by the designers—presumably to make the machine harder to break—in fact made the codebreakers' work easier.

Fundamentally, Germany's cryptographic procedures in the Second World War were dangerously flawed—as shown by Bletchley Park's devastating successes. Yet generally speaking the weakness was not in the Enigma machine itself. As Gordon Welchman said about Enigma:[13]

The machine as it was would have been impregnable if it had been used properly.

CHAPTER 11

Breaking machines with a pencil

MAVIS BATEY*

D illy Knox, the renowned First World War codebreaker, was the first to investigate the workings of the Enigma machine after it came on the market in 1925, and he developed hand methods for breaking Enigma. What he called 'serendipity' was truly a mixture of careful observation and inspired guesswork. This chapter describes the importance of the pre-war introduction to Enigma that Turing received from Knox. Turing worked with Knox during the pre-war months, and when war was declared he joined Knox's Enigma Research Section at Bletchley Park.

Introduction

Once a stately home, Bletchley Park had become the war station of the Secret Intelligence Service (SIS), of which the Government Code and Cypher School (GC&CS) was part. Its head, Admiral Sir Hugh Sinclair, was responsible for both espionage (Humint) and the new signals intelligence (Sigint), but the latter soon became his priority.

Winston Churchill was the first minister to realize the intelligence potential of breaking the enemy's codes, and in November 1914 he had set up 'Room 40' right beside his Admiralty premises. By Bletchley Park's standards, Room 40 was a small-scale codebreaking unit focusing mainly on naval and diplomatic messages. When France and Germany also set up cryptographic bureaux they staffed them with servicemen, but Churchill insisted on recruiting scholars with minds of their own—the so-called 'professor types'. It was an excellent decision. Under the influence of Sir Alfred Ewing, an expert in wireless telegraphy and professor of engineering at Cambridge University, Ewing's own college, King's, became a happy hunting ground

---

*This chapter was revised by Jack Copeland and Ralph Erskine after Mavis Batey's death in November 2013.

Figure 11.1 Dilly Knox, sketched by Gilbert Spencer.

Reproduced with permission of Mavis Batey.

for 'professor types' during both world wars—including Dillwyn (Dilly) Knox (Fig. 11.1) in the first and Alan Turing in the second.

Until the time of Turing's arrival, mostly classicists and linguists were recruited. Knox himself had an international reputation for unravelling charred fragments of Greek papyri. Shortly after Enigma first came on the market in 1925, offering security to banks and businesses for their telegrams and cables, the GC&CS obtained two of the new machines, and some time later Knox studied one of these closely. By the beginning of 1939 he was still the only person at the GC&CS to have worked successfully on Enigma messages. A rewired version of the commercial Enigma machine, lacking the plugboard, had been used for military purposes during the Spanish Civil War (1936–39) and Knox broke the machine's wirings, using methods he devised while experimenting with the (differently wired) commercial machine held by the GC&CS.

When Turing was sent to join him in the pre-war months,[1] Knox was probably working on one of the several German Enigma 'cribs' that the French or Polish cipher bureau had given to the GC&CS.[2] 'Crib' was the term used for a stretch of plaintext corresponding to part of a coded message: cribs would later play a huge part in Bletchley Park's successes against Enigma (see Chapter 12). But not even Turing and Knox could unravel the system from these early cribs. They would not make much headway until August 1939, after Knox met with the Polish codebreakers at their headquarters near Warsaw. The Poles had been breaking German military Enigma messages since the early 1930s, and they decided to hand over their secrets just before their country was invaded. Knox seized on the Poles' information and, shortly before Turing joined the Enigma attack full time in September 1939, Knox's assistant Peter Twinn used a monster crib—supplied, ironically, by the Germans themselves—to deduce the wiring of two wheels of the German plugboard Enigma machine. Twinn, a mathematician who had been recruited to Dilly's permanent staff from Brasenose College, Oxford, in February 1939, was very familiar with Knox's methods. The next section (The race to break Enigma) tells the story of the monster crib that Twinn used to crack the wheels.

It was later in the autumn of 1939 that Turing began his attack on the German Naval Enigma, which nobody was tackling. He worked in the stable-yard cottage that Dilly had chosen in order to be away from the administrators in the Mansion (Fig. 11.2). Dilly, like Turing, was a loner. The cottage was very cramped downstairs, and Turing elected to go into the loft above. Since the only access was a ladder in the wall, two of the girls rigged up a pulley and basket for sending up coffee and sandwiches. Typically, Turing did not wish to come down and socialize.

I arrived as a German linguist from University College, London, in April 1940. By this time the Hut system was in full swing. Turing's first bombe was just going into action against Naval Enigma, and he was now head of Hut 8, the Naval Enigma Section. Dilly was in the Cottage, working on still unbroken versions of the Enigma machine. I was taken over to the Cottage and introduced to him. He looked up, amid wreaths of pipe smoke, and said to me 'Hello, we are breaking machines—have you got a pencil?'.

Figure 11.2 Cottage 3 with Turing's stable lads' loft.

Reproduced with permission of Peter Fox.

Dilly encouraged Turing to think about *mechanizing* the breaking of Enigma messages, and this would be Turing's great achievement. They both knew, however, that Turing's 'bombes' were able to break the daily Enigma settings only because the wiring of the Enigma machine had already been discovered by paper-and-pencil methods—thus Dilly's greeting to me and the title of this chapter.

The race to break Enigma

Enigma was first taken seriously as a potential German war weapon by the Poles, and by Section D of the Service de Renseignements (French military intelligence) whose function was to acquire foreign cryptographic material—or, more precisely, taken seriously by a lone Frenchman, Gustave Bertrand, the head of Section D. Bertrand, who was concerned about the belligerent noises coming from France's powerful neighbour, was the first to make use of cryptographic espionage, in 1932. By devious means, and for a considerable sum of money, he managed to acquire photographs of Enigma manuals and other documents, including invaluable key lists for September and October 1932. Hans-Thilo Schmidt, a traitor in the German war ministry, took the photographs and sold them to Bertrand. Neither Bertrand's own government nor the British government showed any interest in this ultra-secret material, but the Polish government certainly did.

At this time Britain was more concerned about communist subversion than with the rise of fascism, and all our codebreakers were working flat out on Bolshevik codes. The Poles took Hitler's threats more seriously, and moreover could intercept the Germans' low-power radio transmissions, which we were unable to do at that time. The Polish cipher bureau recruited three mathematicians from Poznán University and one of them, Marian Rejewski, succeeded in solving the Enigma machine (Fig. 11.3). He had observed a pattern in the doubly enciphered indicator groups at the beginning of the messages, and he used mathematical methods, together with Schmidt's photographs of the key lists, to solve the machine's wiring.

Although Dilly had had such success during the Spanish Civil War, he lagged well behind Rejewski when it came to German military Enigma. When breaking codes, enemy errors are manna from heaven, and Dilly worked on operators' errors to break Enigma messages during the Spanish war. In the case of German Enigma, he received help from no less a source than the German cipher office itself. In 1930, when the Germans had introduced their plugboard machine, they issued a manual showing operators how to set up their machines, using the daily key.[3] This manual gave an authentic ninety-letter example of plaintext with the resulting ciphertext, along with the machine settings that produced it. It was an almighty blunder on the Germans' part. This authentic example was instantly withdrawn once the blunder was discovered, and a fictitious example was used in place of this monumental crib. Usually cribs were the result of patient guesswork by the codebreakers, and were often speculative and quite short—but this long crib was certain, since the German cipher office had supplied it.

The GC&CS received this manual from the French, probably in 1938, and Dilly attacked German military Enigma using the monster crib. He had no success, however, because there was still a fundamental problem to be overcome. When Dilly had first examined the commercial Enigma machine, he saw that the way the letters of the keyboard were wired to the entry plate of the machine (see Fig. 12.4) was simply in the order of the letters on the German QWERTZU keyboard. All his methods were based on this discovery, and fortunately the same wiring held

Figure 11.3 Marian Rejewski.

From Wladyslaw Kozaczuk, *Enigma* (Arms and Armour Press, 1984).

good throughout the Spanish Civil War. But the wiring was different in the Germans' own military version of the machine, and Knox could not discover it.

Rejewski found the answer by inspired guesswork. Dilly's first agitated question to him at the Warsaw conference was 'What is the QWERTZU?'. 'QWERTZU' was Dilly's way of referring to the entry-plate's wiring. Rejewski answered immediately, now that he was able to reveal his secrets. The keyboard was wired to the entry plate in straightforward alphabetical order, ABCDE...!

The Poles had learned from Bertrand about Dilly's Enigma successes, and they had specifically requested that he should be among the party attending the Warsaw conference. Dilly was in bed with influenza at the time, and had just had his first cancer operation, but his family remembered him getting out of bed ashen grey, determined to make the journey to Warsaw. It was on 27 July 1939 that he met Rejewski and the other Polish mathematicians, Henryk Zygalski and Jerzy Różycki, in a clearing in the Pyry forest near Warsaw. Rejewski recalled in a 1978 interview that Dilly was an excellent cryptographer, a 'specialist of a different kind', and that there was little that he, Rejewski, could tell him. 'Knox grasped everything very quickly, almost as quick as lightning', he said.[4]

As soon as Dilly returned from Warsaw, he sent the three Polish mathematicians a note, thanking them sincerely for their 'co-operation and patience', together with three silk scarves, each showing a horse winning the Derby, a very gracious acknowledgment. He also sent a set of 'rods', his small lettered strips of cardboard that reproduced the action of the Enigma's wheels. These rods were the basis of Dilly's own method, and he had been almost in sight of the winning post himself. As soon as Dilly gave Rejewski's missing piece of information about ABCDE . . . to Peter Twinn, Twinn unmasked two of the Enigma machine's wheels in only two hours, using the monster crib and Dilly's rod method.

Difficult to anchor down

Dilly invited Turing to join him at his home, Courns Wood, so that he could tell him all about the Warsaw conference. Closeted in his study, he held an inquest on his failure to achieve what Rejewski had done. After Turing left Courns Wood, Dilly wrote his official report on the Pyry conference, dated 4 August 1939, and marked 'Most Secret'.[5] In it he reported that Rejewski had indicated to him that the Polish solution was achieved by mathematics. Dilly was anxious to know whether this was true, and whether it was his lack of mathematics that had held things up. His report said that Turing had helped to convince him that this was not so.

Gordon Welchman, a first-class Cambridge mathematician, joined the Cottage party at the same time as Turing. He soon shone, and became the head of Hut 6 in January 1940. Welchman was a prime mover in turning Dilly's cottage industry into a factory production line—rather to Dilly's horror. According to Welchman, Dilly was 'neither an organisation man nor a technical man' but 'was essentially an idea-struck man'.[6] Turing was not an organization man either, but he was a technical man, and (like Knox) was a fountain of brilliant ideas. A 'Most Confidential' memo that Dilly wrote about his newly arrived staff gives a glimpse of how he saw Turing:[7]

Turing is very difficult to anchor down. He is very clever but quite irresponsible and throws out suggestions of all sorts of merit. I have just, but only just, authority and ability to keep his ideas in some sort of order and discipline. But he is very nice about it all.

Much to Dilly's disappointment, the French government would not allow the Poles to leave the Château de Vignolles, the country house near Paris where a joint Polish–French cipher bureau had been set up following the Poles' escape from the advancing German forces. While still in Poland, the Polish codebreakers had invented what were called 'Zygalski sheets'— perforated cardboard sheets showing repeated positions of letters in the doubly enciphered indicators—but they never had the resources to manufacture enough of these. Turing couriered the last package of a consignment of British-made Zygalski sheets to Château de Vignolles, and found the Poles anxious to know more from him about Knox's methods. They probably did not already know about what Dilly called 'cillies', where a slack operator used the finishing position of the wheels for the start of the second part of a two-part message. Cillies helped to reduce quite dramatically the number of Enigma settings that the codebreakers had to test. Rejewski admitted that there was also 'another clue for which we had the British to thank'.[8] This was Dilly's ingenious discovery that a seemingly unimportant weather code in fact gave away the Enigma's plug connections for the day, because the Germans were foolishly using the letter pairings that specified the day's Enigma plug connections to encrypt the weather messages. As Rejewski said, this was 'a cardinal error on the part of the Germans'.

Turing cannot have mentioned much about his own work when speaking to the Poles, since Rejewski explained later that they treated him as 'a younger colleague who had specialised in mathematical logic and was just starting on cryptology'.[9]

Turing, Dilly, and Naval Enigma

Turing himself had no dealings with the making of the Zygalski sheets, nor with using them to discover the wheel order and ring setting. This is probably why the Poles thought him a beginner—whereas he was at the time busily engaged on German Naval Enigma, on which the Poles had done little. It was a friendly few days' stay, and Turing did bring back one vital piece of information, which explained why Bletchley Park was having no success in applying the Polish methods. The Poles had inadvertently given us incorrect information about the turnovers on two wheels (see Chapter 10 for an explanation of wheel turnovers).

Soon after Dilly had met Zygalski at the Warsaw conference he penned a note to Alastair Denniston, his companion in Warsaw and the operational head of the GC&CS. The note survives today, handwritten on notepaper bearing the printed heading 'Hotel Bristol, Warszawa'. Knox ended his note dramatically:[10]

It cannot be too strongly emphasised that all successes have depended on a factor (the machine-coding of indicators) which may at any moment be cancelled.

It was this danger of the Polish methods suddenly ceasing to work that Dilly impressed upon Turing when they met at Dilly's home after Pyry. The need for a more robust method was Turing's guiding star in his subsequent work on Enigma. Dilly was right: the indicator system changed on 1 May 1940 and overnight the Polish methods became inapplicable.[11]

Back from France, Turing closeted himself in his stable-yard loft to grapple with German Naval Enigma. Although the wheel wirings were the same as those that the other German military services were using, the problem was that the message settings of naval messages were super-encrypted (see Chapter 10) and Turing needed to deduce how this super-encryption worked. Dilly was aware of this difficulty when he worked on German Naval Enigma in 1937, but he had soon abandoned the naval code to work on German Army and Air Force Enigma.

The responsibility for cracking German Naval Enigma was now Turing's alone, although Dilly gave him every encouragement: Dilly himself would soon be working on Italian Naval Enigma. In the meantime, Dilly persuaded his good friend Admiral John Godfrey, Director of Naval Intelligence, to allow him to be present at prisoner interrogations: in a paper written after the First World War Dilly had argued that this practice should have been adopted. In November 1939 Dilly managed to extract some crucial information about German Naval Enigma from a *Kriegsmarine* prisoner of war, who explained that operators were now instructed to spell out numbers ('Eins', 'Zwei', 'Drei', etc.) instead of using specific letter keys to represent numbers (the Enigma keyboard had no number keys). In fact, the Germans used both methods for a while, which Turing found very useful.

In the 1940 treatise that Turing wrote on Enigma, known at Bletchley Park simply as 'Prof's Book', he described Dilly's 'pen and pencil methods' in order to illustrate the use of cribs in general.[12] He made extensive use of probability in 'Prof's Book', but probability was anathema to

Dilly. As I soon learned, it was *serendipity* that counted—although this was serendipity requiring infinitely patient work from Dilly's assistants. Another difference between the two men is evident in the relentless seriousness of 'Prof's Book'. Describing Dilly's rodding methods, Turing highlighted two kinds of 'clicks' (useful repeats in cipher messages) that in his logician's way he called 'direct' and 'cross'. In Dilly's section, we used the more whimsical Knox nomenclature of 'beetle' in place of 'direct' and 'starfish' in place of 'cross'.

Trouble at sea

The need to break German Naval Enigma became ever more urgent, as the U-boats were sinking a horrifying proportion of the merchant shipping crossing the North Atlantic. If only the Royal Navy could capture the bigram tables used to super-encrypt naval message settings, this would give Turing a shortcut. (Chapter 10 explains what the bigram tables were.) Admiral Godfrey plotted cunning schemes with Dilly and Frank Birch, the head of Hut 4 Naval Section. Dilly suggested obtaining the daily key by asking for it in a bogus signal. A much more ruthless scheme was put forward by Godfrey's personal assistant and liaison officer at Bletchley Park, Ian Fleming, who later wrote the James Bond novels. Fleming devised a plot to crash a captured German bomber in the English Channel, as close as possible to a suitable German vessel. It was assumed that the German sailors would hasten to rescue the crew (in their disguise as German airmen), and once on board the British would heartlessly overpower their German rescuers and steal whatever Enigma materials could be found.

Turing normally took little interest in the intelligence side of things, but this time he was anxiously following the plot. However, RAF reconnaissance flights failed to locate any small and unaccompanied German vessel with Enigma on board, and eventually Fleming's 'Operation Ruthless' was postponed indefinitely. Frank Birch wrote:[13]

Turing and Twinn came to me like undertakers cheated of a nice corpse.

Turing and Twinn had to wait until March 1941 when, in a ship-to-ship fight near the Norwegian Lofoten Islands, the Royal Navy's destroyer HMS *Somali* captured crucial Enigma documents and several Enigma wheels from an armed German trawler *Krebs*.[14]

The recorded date for breaking into German Naval Enigma was the end of March 1941, by which time Turing and Dilly had parted company. But March 1941 brought a proud moment for Dilly, too. In the early hours of 29 March, when the last gun was fired in the great sea battle at Matapan (near Crete), Bletchley Park received this message from Admiral Godfrey:[15]

Tell Dilly we have had a great victory in the Mediterranean and it is entirely due to him and his girls.

Dilly had been put onto Italian Naval Enigma as soon as Mussolini came into the war, in June 1940, and he soon found that the Italians were using the Spanish Civil War 'K' model with additional wheels. The first message to be broken showed that the Italians now had the helpful habit of spelling out full stops as XALTX, and the full stop at the end of the message was filled up with XXX to complete the four-letter groups, giving a possible eight-letter crib for rodding. So it was that a mere full stop contributed so much to the British victory over the Italian fleet at the Battle of Matapan, described by Churchill as 'the greatest fight since Trafalgar'.

The first brief message that we broke said 'Today is the day minus three', with a command to acknowledge. Unbelievably this had three full stops, including the bonus eight-letter crib at the end. It alerted us to the fact that the Italian fleet was going into action, and so we had to work flat out for three days and nights to discover where and what they were going to attack. It turned out to be a large convoy setting sail from Alexandria and bound for Greece. After the Battle of Matapan was won, Admiral Sir Andrew Cunningham, the British Commander-in-Chief Mediterranean, came to the Cottage for a celebratory drink. He was much amused that Dilly, punning on XALTX, called the celebration our 'exaltations'.

Bombe attack

Turing's gigantic bombe was an awesome sight, with its many rotating drums whizzing around. Dilly, a Lewis Carroll addict, thought of Turing as his 'bombe-ish boy', after Carroll's 'Come to my arms, my beamish boy', and as dealing the final blow in 'slaying the Jabberwock'.

The bombe was inspired by, but certainly not based on, the simpler Polish 'bomba' (see Chapter 12). Dilly had told Turing about the bomba when they met at his home on his return from Warsaw. Dilly emphasized the folly of the Poles in basing all their techniques on the doubly enciphered indicators. He made it clear that, while he himself intended to do everything he could to provide the Zygalski sheets that the Poles needed, Turing should go all out in devising a method for breaking messages using 'cribs', rather than focusing exclusively on the message's indicator, as the Poles had done. The bombe-ish boy followed Dilly's advice: this was indeed a blessing, since it was while the first of the bombes was cranking into action that the Germans abandoned doubly enciphered indicators.

Dilly managed to acquire an empty former plum store, located opposite the Cottage, as a small workshop for trying out experimental gadgets. Turing was in his element in the plum store, as his bombe-ish ideas took shape. When Turing told Dilly that the bombe would indeed test cribs, running at high speed through possible machine settings, Dilly was delighted to hear it. (See Chapter 12 for a full explanation of Turing's bombe.) On 1 November 1939 a meeting was held in the Cottage, with Welchman and Twinn also present, to set down the requirements for the bombe, so that a design could be sent through to the group of engineers who would build it at the British Tabulating Machine Company's factory in Letchworth. The first bombe, Victory, was ready for action against Naval Enigma in the spring of 1940. No doubt Dilly organized a suitable celebration.

All was not going well for Dilly, however, and around this time, early in the spring of 1940, he wrote a six-page letter of resignation to Stewart Menzies, head of SIS, complaining about Denniston's proposals for a production-line system to break Enigma messages. Hut 6, dealing with army and air force traffic, was now up and running, thanks to Welchman, a true organization man. Welchman had already planned all this in the Cottage during October 1939, setting out what was needed—even down to where the electric power points would be placed in the skirting boards, for optimal use of the lamps that they used for inspecting Zygalski sheets on glass-topped tables. Welchman had also specified where the power points should be put for Turing's bombe, even though at that time the bombe was little more than wishful thinking. Dilly's personal beef about the accommodation was that Cottage 3, home to his Research Section, was deemed a security risk and shut down. This was because Turing's loft was open to the adjoining loft in Cottage 2: Cottage 2 was occupied by one of the maintenance staff and his family, who could not be allowed to know that German codes were being broken.

Since there was no specific provision for Dilly in the new plans, he made his own arrangements, moving into the plum store once Turing's machinery was shifted to Hut 1 (itself recently vacated by Station X's wireless operators). A Jabberwocky-esque poem by Dilly began 'Come to the Store my bombe-ish boy', as the plum store was now his only hideout. Dilly's resignation letter finished with an urgent complaint about the provision made for Turing in Denniston's distasteful flow charts. Noting that, according to Denniston's plan, 'Mr. Turing, if successful in finding methods for a solution of German Naval Traffic, should work "under Mr. Birch" ', Dilly continued:[16]

The very suggestion . . . is so absurd and unworkable . . . that I could no longer remain in your service to work with its proposer.

Frank Birch was Dilly's long-standing best friend, but he knew next to nothing about Enigma. Birch thought that Turing, although brilliant, was disorganized, untidy, could not copy things correctly, and dithered between 'theory and cribbing'.[17] Birch had not even grasped the fact that Turing had actually abandoned his theoretical approach, which involved the mathematics of groups, for Dilly's practical cribbing.

It all worked out well in the end. While Birch, an expert on naval matters, was responsible for analysing intelligence in Hut 4 and providing cribs for Turing, he had no direct control over Turing and Hut 8. Dilly was not the only one to be concerned about Turing's position since, although Turing had to be titular head of Hut 8, it really needed a natural organizer like Welchman to run it. Hugh Alexander, an excellent manager and cryptanalyst who was first recruited to Hut 6 in February 1940, was the ideal solution to this problem. He transferred to Hut 8 and Naval Enigma in March 1941, quickly becoming acting head, so leaving Turing free to operate in the way he wished among staff he could feel at ease with.

After the fall of France, in 1940, Welchman went to Cambridge to scoop up more mathematicians once the results of the Mathematical Tripos examinations were in. He recruited Keith Batey (later my husband) for his own Hut 6, and having previously supervised Joan Clarke he earmarked her for Turing in Hut 8. He told both recruits that mathematicians were not actually needed for the codebreaking work, but that they tended to be good at it. Joan later recalled that as a new arrival she was 'collected' by Turing and put on to testing bombe 'stops' in the timber-framed Hut 1.[18] This was very laborious work with the first bombe, Victory, since there were excessively many stops (as explained in Chapter 12).

All was now well with Dilly's position. His former Cottage 3 and the adjoining Cottage 2 were knocked into one, and stairs were put up to Turing's old loft, so providing room for the staff joining Dilly's new Enigma Research Section. Dilly decided that he wanted only female staff, and (even before Joan Clarke's appointment) he had recruited Margaret Rock, a mathematician from London University. The rest of his recruits had language skills. Although linguists were an obvious choice, we also had a speech therapist and someone from drama school. We were known—even in Whitehall—as 'Dilly's girls'.

Railway Enigma and 'Prof's Book'

Turing and Dilly worked together again briefly in June 1940. The Germans had introduced a rewired version of commercial Enigma for use by the *Deutsche Reichsbahn*, the German

Railways authority.[19] Turing and Twinn took this on. Colonel John Tiltman had broken into Railway Enigma, helped by what he called the 'very stereotyped pro-forma nature' of the German messages.[20] We knew that messages of a stereotyped nature were easy prey to crib-based attacks. Turing came over to look at Dilly's crib charts. Italy had just come into the war, and we had found an Italian crib chart from the Spanish Civil War—the railway machine and the machine used by the Italians were closely related.

Soon after this, Turing decided to compile his newcomers' manual, 'Prof's Book'. Chapter 1 set out the workings of the Enigma machine, while Chapters 2 and 3 described Dilly's rodding methods and his 'saga' system for finding wheel wirings. Turing recalled the pre-war work on the monster crib supplied by the German cipher office, observing that at least a ninety-letter crib was necessary for success with the saga method.[21] Chapter 4 covered the 'Unsteckered Enigma', including Railway Enigma and Italian military Enigma, and Turing mentioned the Italian crib chart that we had found—although he copied it out wrongly! It is because Turing gave examples from Railway Enigma that the Knoxian methods he was using have sometimes been said to be his own—but he himself never made such claims. His own work, on Naval Enigma and the bombe, appeared in the final chapters.

Parts of 'Prof's Book' were released in 1996 by the US National Security Agency, as was Patrick Mahon's 'The history of Hut Eight, 1939–1945', early chapters of which described Turing's historic attack on Naval Enigma. As a result, Turing's monumental cryptographic achievements are now understood and appreciated. Dilly, on the other hand, left few records, and for security reasons his section's history remained classified until 2011.

Conclusion

At the end of 1941, when terminally ill, Dilly broke the most difficult multi-turnover Enigma machine, used by the Abwehr, the equivalent of Britain's MI5 and MI6.[22] We liaised with him at his home for the remaining months of his life. At this time a new section called Intelligence Services Knox (ISK) was set up, headed by Twinn, and Keith Batey also joined us. At GCHQ (Government Communications Headquarters, the modern equivalent of Bletchley Park) the official report describing the work of ISK is known simply as 'BBR&T', after its authors, Keith Batey, myself, Margaret Rock, and Peter Twinn, and it describes how ten differently wired Enigma machines were broken using Dilly's methods.[23]

In summary, the young Turing's early association with veteran codebreaker Dilly Knox was a key factor in Bletchley Park's great successes against Enigma.

CHAPTER 12

Bombes

JACK COPELAND (WITH JEAN VALENTINE
AND CATHERINE CAUGHEY)

Turing's Enigma-cracking 'bombe' was one of his major contributions to winning the war. His revolutionary machine gave the Allies open access to Germany's secret military communications. This chapter explains what the bombe was for and how it worked. Jean Valentine and Catherine Caughey, both Bletchley Park 'Wrens', describe their first encounters with Turing's awe-inspiring bombes. They speak of the morale-sapping secrecy that surrounded the work of bombe operators. Jean Valentine also describes a visit to the secret factory where the bombes were made.[1]

From bomba to bombe

Poland's codebreakers were reading German military Enigma from early 1933. Their intellectual leader was the legendary Marian Rejewski, one of the twentieth century's greatest codebreakers.

As explained in Chapter 11, Rejewski and his colleagues invited a party of their British opposite numbers to Poland a few weeks before Hitler's army poured across the frontier. An ultra-secret meeting took place at the Biuro Szyfrów (the cipher bureau), a cluster of buildings hidden in the forest at Pyry, near Warsaw. The visitors included Dilly Knox, at that time Britain's most experienced warrior against Enigma (see Chapter 11), and Commander Alastair Denniston, head of British military and civil codebreaking.

'The Poles called for us at 7 a.m.', recalled Denniston, 'and we were driven out to a clearing in a forest about 20 kilometres from Warsaw'.[2] The Poles had decided to reveal everything before it was too late. 'At that meeting we told everything we knew and showed everything we had', Rejewski said.[3] After the meeting, Knox sang for joy—although his immediate reaction, as he had sat listening to Rejewski's codebreaking triumphs, had been undisguised fury that the Poles were so far ahead of him.[4]

In 1938 Rejewski and his colleagues had built a small machine they called a 'bomba'.[5] Their reasons for choosing this distinctive name, whose literal meaning is 'bomb' or 'bombshell', went unrecorded at the time, and theories later abounded. At Bletchley Park Turing told Joan Clarke, to whom he was briefly engaged, that the bomba 'got this nickname because it made a ticking

Figure 12.1 A bombe at the Eastcote outstation: notice the swastikas that someone has doodled along the top of the bombe cabinet and the Keep Feet Off sign at the bottom right.

The Turing Archive for the History of Computing.

noise, like an anarchist's time bomb' (as she recollected).[6] On the other hand, Jean Valentine (Fig. 12.12) remembers the bombes' operators being told that the name had arisen because, at a certain point in the bomba's operations, it would release a metal component that fell to the ground 'like a bomb'. American military personnel stationed at Bletchley Park and its outstations were told the same story, reporting back to Washington that the Polish machine was so named because 'When a possible solution was reached a part would fall off the machine onto the floor with a loud noise'.[7]

It is certainly possible that a falling weight disengaged the bomba's drive mechanism—an early computer printer designed by the Victorian computer pioneer Charles Babbage, part of his Difference Engine, involved the same idea. However, a sketch of the bomba by Rejewski shows no mechanism involving falling weights, and perhaps indicates that the bomba's stopping mechanism was *magnetic* in nature.[8] Another explanation of the name is that Rejewski was eating a bomba—a type of ice-cream dessert, *bombe* in French—when the idea for the machine struck him.[9] Military historian Michael Foot reported being told by Rejewski himself that the name had originated in this way.[10]

Rejewski's bomba consisted of six replicas of the Enigma. His machine worked well, and the prototype bomba spawned several bomby (the plural of 'bomba'). By November 1938 the Biuro Szyfrów had half-a-dozen bomby slogging away at the German Enigma traffic.[11] Denniston and Knox saw the bomby first-hand: 'we were taken down to an underground room full of electric equipment and introduced to the "bombs"', Denniston remembered.[12]

The bomba depended on a weakness, unnoticed by the Germans, in their Enigma operating procedures. There was a flaw in the method that the sending operator used to tell the receiving operator which positions the wheels had been twisted to before the message was enciphered. As explained in Chapter 11, Knox realized at Pyry that the Germans might eliminate this flaw at any moment—which they soon did, in May 1940, disastrously for the Polish methods. The usefulness of the bomby had in fact steadily diminished up until that point. Major setbacks came in December 1938, when the German operators were given two more wheels to choose from, and in January 1939 when the number of letters that were scrambled by the plugboard was increased from five to eight.[13] The change that the Germans made to their operating procedures in May 1940 was the final nail in the coffin. But fortunately, in the wake of the Pyry meeting, Turing had set out to create an improved version of the bomba, called at first the 'superbombe'.[14] His design incorporated not only the Polish method of attack, but also another more general method intended to remain viable if the loophole that the Poles were using should close.[15]

Victory

Engineer Harold 'Doc' Keen took on the job of transforming Turing's logical design into hardware. Keen's men began building Turing's bombe in October 1939, and the first, named simply 'Victory', was installed at Bletchley Park in the spring of 1940, just a few weeks before the Germans closed the Rejewski loophole.[16] The robustly built Victory spent no more than 42 hours out of action during its first 14 months of service.[17] Because space was in short supply, Turing's bulky new machine was housed in the sick bay, Hut 1.[18] It cost approximately £6500—about one-tenth of the price tag of a Lancaster bomber, and around £250,000 in today's money.[19] In the light of what the bombes would achieve, they were among the most cost-effective pieces of equipment of the war.

During 1940, Victory and the second bombe, named 'Agnus Dei' (Lamb of God) between them broke more than 98% of the messages they attacked.[20] When Victory and 'Aggie' were moved out of the sick bay, around March 1941, their home became Hut 11, constructed in the Mansion's pleasure-garden maze.[21] Hut 11 was called the Bombe Hut until February 1942, when a larger structure, Hut 11A, was erected next to it in order to house the growing number of bombes.[22]

The arrival of Victory marked the start of Bletchley Park's industrialization of codebreaking. By the end of 1941, Keen's factory in Letchworth was turning out new bombes by the dozen. It was cyberwar on a previously unknown scale. Workers arrived at Bletchley's outstations in industrial numbers to operate the bombes. The workers, all female, were members of the Women's Royal Naval Service and known colloquially as 'Wrens'. Eventually more than 1500 women were operating the bombes.[23] Turing called them 'slaves'.[24] The bombes ran 24/7 and the Wrens worked in three shifts. They slept, often in multi-tiered bunks, in outlying dormitories with peeling paint, inadequate heating, and terrible food.

Huts 6 and 8

The bombes were used by Huts 6 and 8. These two small buildings housed many of the front-line Enigma cryptanalysts, Hut 6 dealing with Army and Air Force Enigma, and Hut 8 with Naval Enigma. Like the Wrens, the cryptanalysts worked round the clock in three shifts. Some were civilians and some wore military uniforms, but military discipline never ruled the cryptanalysts. 'We acknowledged only the discipline we imposed on ourselves' recalled Peter Hilton, who was a mere 18 years of age when he joined Hut 8 (see Chapter 3).[25] Turing was head of Hut 8 and Gordon Welchman, another Cambridge mathematics don, headed Hut 6. Like Turing, Welchman took up residence at Bletchley Park on 4 September 1939, the day following Britain's declaration of war on Germany.[26] Welchman's book *The Hut Six Story*, published in 1982, was the first detailed account of how Bletchley Park broke Enigma. The authorities threatened him with prison for revealing classified information.

In Hut 8 Turing was assisted by another mathematician, Peter Twinn, who first joined the Enigma fight in February 1939, as Dilly Knox's assistant (see Chapter 11).[27] Turing's own name was not added to Denniston's 'emergency list' until March 1939—his list of 'men of the Professor type' who would be called up in the event of war—and Joan Clarke recounted that Turing was introduced to the Enigma problem 'in the summer of 1939'.[28] As mathematicians, Twinn and Turing were rare creatures in the pre-war world of British codebreaking. The Poles had understood right from the start that Enigma was fundamentally a mathematical problem, but to the British way of thinking there were, Twinn explained, 'doubts about the wisdom of recruiting a mathematician as they were regarded as strange fellows, notoriously unpractical'.[29] Such abstract thinkers, it was felt, lacked 'appreciation of the real world'.

Other leading lights in Huts 6 and 8 were Stuart Milner-Barry, who played chess for England and was chess correspondent for *The Times,* and Hugh Alexander, the British chess champion. Before the war Alexander had been a director of the department store chain John Lewis. A skilled and inspirational manager, he eventually took over from Turing as head of Hut 8. Turing loathed administration, and had always left most of it to Alexander.[30] Alexander had an earthy sense of humour, and colleagues enjoyed quoting his despairing remark 'We'll have to wait, thumbing our twiddles'.[31]

Joan Clarke, one of the few female cryptanalysts, joined Hut 8 in June 1940 and immediately started work with the prototype bombe, Victory. At Cambridge she had been one of Welchman's students. Clarke recollected that when she entered Hut 8's small inner sanctum, where at that time the codebreakers had their desks, one of them said to her 'Welcome to the sahibs' room'.[32] This was a reference—almost certainly not ironic—to the strictly male enclaves found across the British Empire. As a female codebreaker her rate of pay was even lower than the Wrens', scarcely more than £2 a week.[33] Yet it was Clarke who achieved some of the very first successful breaks with Victory.[34]

The bombes were maintained by a team of Royal Air Force sergeants, 250-strong by the end of the war, and commanded from the early months by Sergeant Elwyn Jones, who was later promoted to squadron leader.[35] Like the Wren operators and the codebreakers, these engineers also worked in three shifts.[36] In the early days their workroom in Hut 1 was also their dining room and sleeping room.[37] The RAF eventually introduced a special trade category for these unique crypto-engineers, and their official title became the obfuscating 'instrument repairer (tabulating mechanic)'.[38]

The outstations

By the end of the war the Letchworth factory had manufactured more than 200 bombes.[39] Most were situated in two large 'outstations' in the London suburbs of Eastcote and Stanmore. Smaller outstations had been established near Bletchley Park earlier in the war—in converted stables in the little village of Adstock, and in a 'hut' in the then-picturesque township of Wavendon. More bombes were installed at Gayhurst Manor, a beautiful Elizabethan mansion located about 8 miles from Bletchley.[40] The energetic Wrens who operated the bombes even produced their own newspaper, the *PAGES Gazette*: 'PAGES' stood for 'Park, Adstock, Gayhurst, Eastcote, and Stanmore', the Wavendon outstation having closed in early 1944.

Special telephone and teletype links connected the outstations to Bletchley Park. In 1943 the first few American-built bombes went into action at Op. 20 G, the US Navy codebreaking unit in Washington, DC. There were approximately 122 American bombes at Op. 20 G by the end of the war.[41] Thanks to excellent transatlantic cable communications, the Washington bombes were absorbed relatively seamlessly into Bletchley's operations. Bletchley Park was, Alexander said, 'able to use the Op. 20 G bombes almost as conveniently as if they had been at one of our outstations 20 or 30 miles away'.[42]

First encounter

Catherine Caughey (Fig. 12.2), one of the Wrens, relates in this section how she arrived at what she calls the 'holy of holies', a vast building at Eastcote ready to house more than seventy bombes in twelve large 'bays' (Fig. 12.3).[43] Eastcote at that time was recently opened, and when Caughey arrived only its first fifteen or so bombes had been installed.[44] Each of Turing's

Figure 12.2 Catherine Caughey.

The Turing Archive for the History of Computing.

Figure 12.3 Bombes at the Eastcote outstation and their Wren operators. There were usually two Wrens per bombe.

Crown copyright and reproduced with permission of the Director of GCHQ.

bombes had a name, ranging from Ajax to Fünf (a mystery character in Tommy Handley's *It's That Man Again*, a popular radio show of the time).[45] The Eastcote bombes were named after places, and the Wrens slaved over machines called Paris and Warsaw, Chungking, and Marathon. Other names are visible in Fig. 12.3.

I joined the Women's Royal Naval Service as a volunteer, not a conscript. I was offered a job in HMS *Pembroke V*, in category SDX ('Special Duties X'). The *Pembroke V* wasn't a ship at all, but a land-based unit, I had no idea where. The officer explained it would be 'hush-hush work'. So, on 1 January 1944 I got into the back of a closed van with two other new Wrens. We pulled up in front of some Nissen huts, in Eastcote it transpired.

They showed us into one of the huts and told us it was our 'cabin'. All Wrens were supposed to use navy terminology. An anthracite stove burning in the middle of the hut kept out the intense cold, although the fumes were most unpleasant. We slept in bunks. I met several girls I already knew, including a cousin of mine. The atmosphere was friendly, but everything about our work was still a mystery.

On my second day at Eastcote I was taken to 'Block B', the holy of holies, and told that I would be decrypting German messages on machines called bombes. Then I was lectured fiercely about security. I can't remember if it was at this point that we each had to sign the Official Secrets Act. As well as the lectures, we were shown films making the point that careless talk costs lives. Eastcote was the final station on one of the London Underground's northern lines, and large posters on the Underground emphasized the same message to all and sundry.

Then, disappointingly, we became not codebreakers but charladies. We cleaned floors and windows, even painting some of them. The worst job was the galley, where the cooks delighted in making us do the smelliest and most revolting tasks. I don't know why we had this frustrating delay before we could begin our important work. Perhaps the security people were investigating our backgrounds. Anyway, after about 10 days we were suddenly taken back to Block B and were shown how to operate a bombe.

Meet the bombe

Housed in a large metal cabinet and weighing about a ton, the bombe was over 7 feet long, nearly 3 feet wide, and stood 6 feet 6 inches tall (see Fig. 12.1).[46] Turing's colleague Patrick Mahon, who later became head of Hut 8, described the bombe with a tinge of awe:[47]

The bombe was a highly complicated electrical apparatus, involving some 10 miles of wire and about 1 million soldered connections. Its intricate and delicate apparatus had to be kept in perfect condition or the right answer was likely to be missed . . . From one side, a bombe appears to consist of 9 rows of revolving drums; from the other, of coils of coloured wire, reminiscent of a Fairisle sweater.

Victory, the prototype bombe, contained thirty replicas of the Enigma machine; later models contained thirty-six replica Enigmas. The codebreakers could connect these replica Enigmas together in whatever configuration was best for attacking a given message. Each of the 'drums' (visible in Fig. 12.1) mimicked a single Enigma wheel, and each trio of drums replicated a single Enigma machine. The three special drums on the right, in the middle section, were called the 'indicator drums': these displayed output. Victory's first prey consisted of naval messages transmitted on 26 and 27 April 1940. This first successful break was hard won: after a 'series of misadventures and a fortnight's work the machine triumphantly produced the answer', Alexander said.[48]

As explained in Chapter 10, before the codebreakers could decrypt a message they needed to know

- how the Enigma machine's plugboard was connected up
- which three wheels were inside the Enigma that day
- what position the wheels were in when the operator had started to encipher the message (see Fig. 12.4).

The bombe searched at high speed, its drums spinning, and then it would suddenly come to a stop, with three letters showing on the indicator drums—BOV, for example. These were the machine's guess at the position of the wheels at the start of the message.[49] A panel on the right-hand side of the bombe (Fig. 12.3) recorded the machine's guesses at plugboard connections—not necessarily all the connections, but (with luck) enough of them to enable the codebreakers to decrypt the particular message under attack. Later bombes, called 'Jumbos', were able to print their output on paper by means of an electric typewriter.[50]

Once the bombe stopped, its guesses were processed by a human codebreaker and tested on a replica Enigma machine. If the message decoded, all was well—and if it didn't, the bombe operator restarted the bombe and the search continued.

As soon as the 'bombe runs' for one message were finished, the Wrens would reconfigure the machine ready for the next message. Working at the rear of the machine, they replugged the bombe's complex inter-drum connections, in accordance with a 'menu' drawn up by the codebreaker in charge (Fig 12.6).[51] 'The back of the machine almost defies description,' said bombe operator Diana Payne, 'a mass of dangling plugs on rows of letters and numbers' (Fig. 12.7).[52]

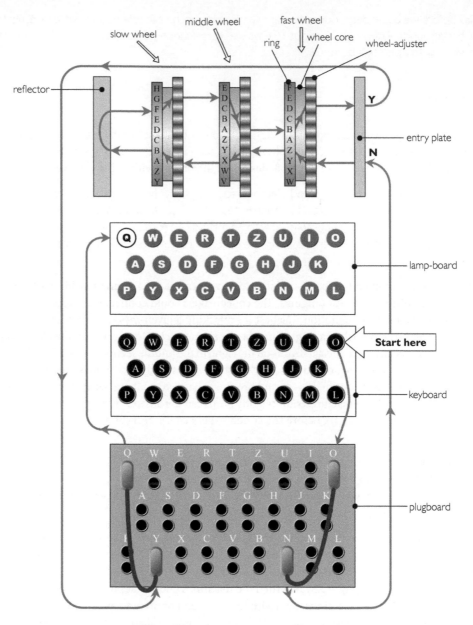

Figure 12.4 How the Enigma machine encrypts.

The bombes were special-purpose electromechanical computers. What they did was to search at superhuman speed through different configurations of the Enigma's wheels, looking for a pattern of keyboard-to-lampboard connections that would turn the coded letters into plain German. As Mahon joked, 'The bombe was rather like the traditional German soldier, highly efficient but totally unintelligent'.[53] Some would say, however, that there was indeed a glimmer of intelligence in the bombe, even though at root the machine depended completely

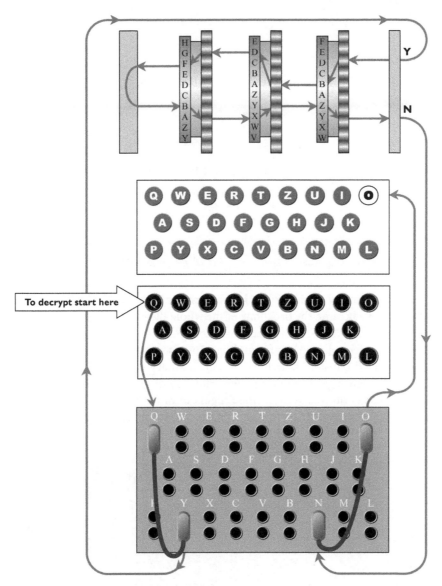

Figure 12.5 How the Enigma machine decrypts.

on human brains: the whole process revolved around a 'crib' that was carefully chosen by a human codebreaker. A crib was a snatch of plain German that the codebreaker believed to occur in the message.

Turing's powerful crib-based method for breaking messages (described in detail in the section 'How it worked') was unaffected by the change to the German system in May 1940—unlike Rejewski's method, which did not involve cribs. Moreover, Rejewski's bomba had simply ignored the Enigma's plugboard.[54] This was possible in those early days, when the Germans used the plugboard to transpose as few as five pairs of letters. Turing's ingenious method for solving

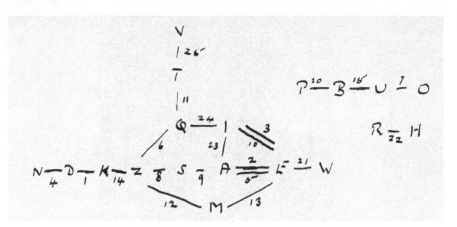

Figure 12.6 An early bombe menu, drawn by Turing. The menu is constructed from the crib depicted in Fig. 12.8.

From 'Prof's Book', p. 99 (see Note 15); digitized and enhanced by *The Turing Archive for the History of Computing*.

the plugboard was the key to the British bombe's power.[55] Soon, the techniques that Turing pioneered in Victory became the backbone of the whole Bletchley attack on the Wehrmacht's Enigma messages.

'Cribbing', the codebreakers' term for finding cribs, was an art in itself. Some codebreakers worked as specialized 'cribsters', searching out phrases that cropped up frequently in German messages, usually stereotyped military jargon. Hut 8's cribsters worked in the 'Crib Room'. One commonly used crib was WEWA, the Enigma term for a weather station (abbreviating the German *Wetter Warte*). Another was 'Continuation of previous message' (*Fortsetzung*,

Figure 12.7 Rear panel of a bombe, photographed in Hut 11A. Here the operator 'plugged up' the menu.

Crown copyright and reproduced with permission of the Director of GCHQ.

abbreviated to FORT): since messages were often sent in several parts, FORT was a very handy crib. The common group of letters EINS (meaning 'one') was an ever-reliable crib—there was about a 90% chance that any given message would contain EINS somewhere within it.[56] Bombes needed longer cribs than human codebreakers; for example 'FORT EINS EINS VIER NEUN' ('Continuation 1149').[57] Cribs as short as fifteen letters were workable on the bombes, but cribs of thirty to forty letters or more were desirable.[58]

More exotic cribs resulted from the sheer naivety of many of Enigma's users. 'German operators were simple souls with childish habits', Denniston said drily.[59] The weather stations regularly sent messages containing routine phrases like 'WETTER FUER DIE NACHT' ('weather for the night') and 'ZUSTAND OST WAERTIGER KANAL' ('situation eastern Channel'). One naval station even transmitted the confirmation 'FEUER BRANNTEN WIE BEFOHLEN' ('Beacons lit as ordered') every single evening. This was 'a very excellent crib', said Mahon, himself a gifted cribster.[60] German Air Force and Army cribs included 'RAF plane over airport', 'Quiet night. Nothing special to report', and 'Wine barrels on hand'.[61]

The cribs used to achieve Victory's first breaks were captured by the Royal Navy from an armed German trawler, *Schiff 26*, which was intercepted and boarded while carrying a cargo of munitions to the Norwegian port of Narvik.[62] Subsequent attacks on German vessels were a rich source of cribs, as well as other Enigma-related materials.

Figure 12.8 shows a crib of 25 letters. The codebreakers could pinpoint the location of a cribbed phrase within a message by using the basic fact that the Enigma machine never encoded a letter as itself. This was the effect of a component called the 'reflector'. Because of the presence of the reflector, the letter that came out of the wheels was bound to be different from the one that went in (see Fig. 12.4). So the cribster would slide a suspected fragment of plaintext (such as ZUSTANDOSTWAERTIGERKANAL) along the ciphertext, looking for a location at which no letters matched up. Positions where a letter in the crib stood alongside the very same letter in the ciphertext were called 'crashes': a crash showed that the crib was in the wrong place. When a crib was more than about thirty letters long, it was more likely than not that it would 'crash out' everywhere except at its correct location in the message.

Figure 12.9 summarizes the action of the bombe. The machine searched at high speed to find

- which wheels were in the Enigma
- the plugboard wiring pattern
- the wheels' starting position at the beginning of the message (such as XYZ).

The correct combination of these elements would produce the right cipher letters from the crib. In an ideal world the bombe could do this by enciphering the crib at *every* possible combination of these elements, looking for the combination that produced (for example)

1	2	3	4	5	6	7	8	9	10	11	12	13	14	15	16	17	18	19	20	21	22	23	24	25
D	A	E	D	A	Q	O	Z	S	I	Q	M	M	K	B	I	L	G	M	P	W	H	A	I	V
K	E	I	N	E	Z	U	S	A	E	T	Z	E	Z	U	M	V	O	R	B	E	R	I	Q	T

Figure 12.8 A crib of twenty-five letters. The crib (bottom line) is positioned against the corresponding ciphertext (middle line).[63] '*Keine Zusaetze zum Vorberiqt*' (the crib) means 'No additions to preliminary report'. ('Vorberiqt' is a shortening of 'Vorbericht': Enigma operators regularly replaced CH by Q.) The numbers along the top indicate the successive steps of the encryption process.

WETTERFURDIENACHT

Settings x, y, z

.... HPZEMUGBNATPRUKAI ....

Figure 12.9 The bombe's function. The bombe searched for a pattern of wiring at the plugboard and wheel start positions that encrypted the crib correctly.

DAEDAQOZSIQMMKBILGMPWHAIV from KEINEZUSAETZEZUMVORBERIQT. In the real world, however, it would take the bombe impossibly long to search through the astronomical number of possible combinations. The challenge facing Turing was to find a faster method.

How it worked

The key to understanding Turing's method is his observation that if a crib involves a number of 'loops', these can be exploited to find the plugboard wiring and the starting position of the wheels, as well as which three wheels are in the Enigma.[64] (Turing himself used the term 'closed chain' for what I call a loop.)

The crib in Fig. 12.8 contains several loops. One occurs at steps 2 and 5: at step 2 E leads to A (that is, E enciphers as A), and at step 5 A leads back down again to E (so A deciphers as E). This short loop is depicted in Fig. 12.10. Figure 12.11 depicts a longer loop occurring at steps 10, 23, and 5: E is linked with I, I is linked with A, and A is linked with E. (Notice that 'linked with' is reversible: if X is linked with Y, then Y is linked with X. This is because, in Enigma, if X encodes as Y at some step, then Y would encode as X at the same step.) Steps 13, 12, 6, 24, and 10 contain a monster loop: E is linked with M at step 13, M is linked with Z at step 12, Z is linked with Q at step 6, Q is linked with I at step 24, and I is linked with E at step 10. Turing called E the 'central letter' of these three different loops.[65]

Turing's great insight was that loops give away information. He put it like this, using the German word '*Stecker*' for the plugboard connections: loops correspond to 'characteristics of the crib which are *independent* of the Stecker'.[66] He realized that a crib's loops can be milked for information by using a short chain of replica Enigmas linked nose to tail. Let's focus on the longest of the three loops just mentioned, E → M → Z → Q → I → E. To milk this five-step loop, we connect five replica Enigmas together in such a way that the output from the first Enigma—i.e.

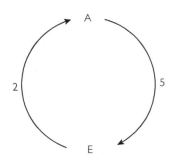

Figure 12.10 A loop involving two letters.

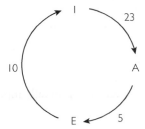

Figure 12.11 A loop involving three letters.

the letter that lights up at its lampboard—is fed into the second Enigma as input, just as though this letter had been typed at the second Enigma's keyboard. Similarly, the output from the second Enigma becomes the input into the third, and so on for the fourth and fifth Enigmas.

In practice, the keyboards and lampboards of the second, third, and fourth Enigmas in the chain can be dispensed with, being replaced by simple wiring. The lampboard of the first Enigma and the keyboard of the fifth can also be replaced by wiring. We are left with a single keyboard, attached to the first Enigma, which serves the purpose of inputting letters into the chain, and a single lampboard, on the fifth Enigma, which records the letter produced at the end of the chain. (In the bombe itself, even these two items were dispensed with, but imagining them to be present will help in visualizing the process for exploiting the loops.) We also strip out all five plugboards: the replica Enigmas used in the bombe had no plugboards and the input letter went straight into the wheels.

By turning the wheels of each machine in the chain we hope to drive these five Enigmas into a loop, so that the same letter that goes in comes out again. Because we want this loop to match our target loop in the crib, the wheels of the first machine should be one step further on than the wheels of the second machine (since the loop begins at step 13 and then jumps to step 12). Similarly, the wheels of the second machine should be six steps further on than the wheels of the third machine, the wheels of the fourth machine should be eighteen steps ahead of those of the third machine, and the wheels of the fifth machine should be fourteen steps behind those of the fourth machine. Programming the bombe involved setting up the machinery so that, as the wheels (drums) of the first replica Enigma rotated, they remained one step ahead of the wheels (drums) of the second—and so on for the other replica Enigmas in the chain.

Since the loop begins with E and ends with E, a simple way of looking at the search process is this: we continually input E into the chain of Enigmas (by repeatedly pressing E at the keyboard), and we keep turning the wheels of the five machines—while maintaining the relative positions of the wheels, as just explained—until we manage to put the Enigmas into a loop so that E lights up at the lampboard at the end of the chain. Once we find the loop, then the position of the wheels of the first machine in the chain must be the position that the wheels of the sender's machine were in at the thirteenth step of the encryption, and from there it is easy to backtrack and discover the position of the sender's wheels at the start of the message.

Although this simple way of looking at things is a good starting point, it is rather *too* simple—not least because it ignores the fact that at the thirteenth step of the encryption (the beginning of the loop) it is *not* the letter E that goes into the wheels of the sender's Enigma, but rather whichever letter E happens to be connected up with at the plugboard of the sender's machine (see Fig. 12.4).

The letter to which E is connected is called E's 'plugboard-mate' (Turing's term was E's 'Stecker value'). It is E's plugboard-mate (not E itself) that we need to input into our chain of replica Enigmas, and we want to drive the Enigmas into a loop so that E's plugboard-mate (and not E) emerges at the other end of the chain. Since we do not know what E's plugboard-mate is—this is one of the pieces of information that we are trying to discover—we proceed by first examining the possibility that E's plugboard-mate is A: we continually input A into the chain and keep turning the wheels, trying to put the Enigmas into a loop so that A emerges at the other end of the chain.

If we fail to create a loop leading from A to A, even after turning the wheels through all their possible positions, then E's plugboard-mate cannot have been A. So now we try B, proceeding as before, and so on. Suppose that our luck is out and we don't find the loop until our final try, when we input Z into the chain and manage to get Z out again. The existence of this loop, the one and only loop to give us the same letter at the start and the finish, establishes that E's plugboard-mate was Z.

Milking the loops

Note that the identity of E's plugboard-mate is not all that we have established. As explained earlier, we now know the position of the sender's wheels at the start of the message. And there is more. Once we have located the loop, the letter that exits the wheels of the first replica Enigma must be M's plugboard-mate, since at the first step of the loop, E is enciphered as M. Other Enigmas in the chain similarly yield up information about the plugboard—for example, the letter exiting the wheels of the third replica Enigma must be Q's plugboard-mate, since Z was enciphered as Q at the third step of the loop. So, as required, Turing's method has supplied not only the starting position of the sender's wheels, but also a quantity of information about the plugboard.

So far, so good. There is a difficulty, however. In practice, many different wheel positions will drive the chain of Enigmas into a loop, and there will be many candidates for E's plugboard-mate. It is rather too easy to find ways of making the input letter light up at the end of the chain.

To counter this embarrassment of riches, Turing used other loops in the crib. As Fig. 12.6 shows, the '*Keine Zusaetze*' crib of Fig. 12.8 contains more than half-a-dozen loops with central letter E. A separate chain of Enigmas was set up within the bombe for each loop; and the same letter, A for example, was continuously input into every chain. The bombe's electrical motors would turn the wheels of all the Enigmas in the chains, searching for a way of making A light up at the end of every chain, and if a way of doing so was found, the bombe would automatically stop. The more loops there were in play, the fewer ways there would be of getting the input letter to light up at the end of each chain, and so the fewer stops there would be.

Given enough loops, this procedure of Turing's usually produced a manageably small number of stops. Each time that the bombe stopped, information was automatically extracted from the chains of Enigmas and displayed by the machine for the Wren operators to read. The Wrens passed the readings on by telephone. Bombe operator Jean Valentine remembers that she had no idea who she was telephoning from Hut 11A or where the readings were going. 'Secrecy was the order of the day', she explains. Long after the war, she discovered that she was phoning the information to Hut 6, only 10 yards away. If a naval message were being attacked, the readings would be telephoned to Hut 8.

In Hut 6 or Hut 8 the bombe's hypotheses were examined and tested on a replica Enigma. The codebreaker typed ciphertext at the keyboard and watched hopefully for fragments of plain German in the output. If fragments of plaintext appeared, the bombe had succeeded, but if no plaintext was uncovered then the information from this stop was discarded and another telephone call told the bombe operators to continue the search. Hut 8 had a room dedicated to testing the bombe's guesses, the 'Machine Room'.[67] The menus that the bombe operators received also came from the Machine Room, although sometimes the menus were drawn up in the Crib Room itself, depending on the difficulty of the job.[68] The sheets containing the menus travelled by a pneumatic tube that connected the Machine Room to the Bombe Hut.

Later, as the numbers of bombes and outstations grew, a control room was set up to allocate the menus from Huts 6 and 8 to specific bombes. The controllers, recruited from the Wrens, routed jobs using a 'wall chart of all machines giving their peculiarities and capabilities'.[69]

A complication glossed over so far is the question of *which* three wheels the sender had used to encrypt the message. At the start of his shift, the sender selected three wheels from his wooden box of wheels, in accordance with his instructions for the day, and placed these into the machine in the order specified in the instructions. German Army and Air Force Enigmas came with a box of five wheels, offering $5 \times 4 \times 3 = 60$ different choices of three, while the box accompanying German Naval Enigmas contained eight wheels, giving a total of $8 \times 7 \times 6 = 336$ choices. Each wheel had a different wiring pattern inside it, so, if Turing's method was to deliver correct answers, it had to use chains of Enigmas containing the same wheels in the same order as the sender's machine.

One way of dealing with this problem was to let the bombe test a number of different possible sets of wheels simultaneously. If, say, a dozen of the bombe's thirty-six replica Enigmas were required to deal with the loops in the crib, then the remaining twenty-four could be used to run the same search using two other possible sets of wheels. But because testing *all* the possible sets of wheels, in all their possible orders, required too much precious bombe time, a hand method called 'Banburismus' was applied before the job was put on the bombes. Banburismus, another of Turing's inventions, was designed to eliminate as many sets of wheels as possible. Banburismus is described in detail in Chapters 13 and 38. It was used only against naval messages.

Another complication not mentioned so far is that, from time to time, the Enigma's middle and left-hand wheels would 'turn over' a notch, as explained in Chapter 10. The search procedure described so far assumes (possibly incorrectly) that no 'turnovers' occurred during the enciphering of KEINEZUSAETZEZUMVORBERIQT. In practice, additional bombe runs would be made to test out the various possibilities for the occurrence of a turnover. Apart from the fact that more bombe time was used up, turnovers presented no great difficulty.

Simultaneous scanning and the diagonal board

There is an obvious way to speed up the bombe's search procedure. The procedure as described so far takes unnecessarily long since it requires the drums to move through all their positions each time a new input letter is tested.

The search can be conducted much more quickly if, instead of making a complete run through all the drum positions every time a new input letter is tested, all twenty-six letters are tested simultaneously at each position of the drums. This was called 'simultaneous scanning'. As Turing put it, in simultaneous scanning 'all 26 possible Stecker values of the central letter [are] tested simultaneously without any parts of the machine moving'.[70] Running a job on Victory without simultaneous scanning normally took about a week, and it was quickly realized that, the official history recorded, 'some radical change was still required for the work to be of any assistance in the War effort'.[71]

Turing was working on a partly electronic method for implementing simultaneous scanning when Welchman came up with an idea for an additional piece of hardware called the 'diagonal board'. The wiring of the diagonal board (which was not electronic) reflected the fact that the plugboard connections were always reciprocal, in the sense that if the plugboard replaced letter X by letter Y, then it also replaced the letter Y by X (where X and Y are any letters). Turing told Joan Clarke that Welchman originally proposed the diagonal board 'simply to provide entry' into additional chains.[72] It was Turing who realized that the diagonal board could be used to implement simultaneous scanning. 'I remember Turing jumping up with the remark that "the diagonal board will give us simultaneous scanning" and rushing across to Hut 6 to tell Welchman', Clarke said.[73] Turing's inspired contribution, Clarke went on to explain, was 'the realization that a wrong stecker assumption for the input letter would imply all wrong steckers, if one allowed an unlimited number of re-entries into the chain'.[74] Because of these implications, testing a single (wrong) letter had the effect of testing all letters simultaneously.

The wiring of the diagonal board also enabled the bombe to exploit deductions about the plugboard-mates of intermediate letters in a crib loop.[75] This added considerable extra power to the process of elimination, power that reduced the stops to a manageable number even when cribs containing only one or two loops were used. Without the diagonal board the bombe needed cribs containing three or more loops, because cribs with fewer loops would usually produce excessively many stops.[76] Welchman's invention of the diagonal board was in fact crucial. It came, Turing said, 'at a time when it was clear that very much shorter cribs would have to be worked than could be managed'—that is, managed by a bombe without the diagonal board.[77]

The diagonal board became a standard component of all bombes from August 1940. Simultaneous scanning produced a stupendous twenty-six-fold speed-up, with the bombe taking only a fraction of a second to test all twenty-six letters, before the fastest-moving drums stepped on to their next position.[78] What took a week on Victory could be done in about 6½ hours with simultaneous scanning. It was not until simultaneous scanning was introduced that the bombe became a thoroughly practical weapon against Enigma.

What if the diagonal board had never been invented? The earlier partly electronic scheme for achieving simultaneous scanning 'would probably have worked', Turing said, after 'a few more months experimenting'.[79] But this delay would not have been the only cost. Without the diagonal board the bombe would in general have been limited to attacking messages containing at least three loops. To break the whole day's traffic on a given Enigma network, the codebreakers would have needed at least one message sent on that day to contain three or four loops. Assuming that this message succumbed to the bombe's analysis, the codebreakers could then use the resulting information to crack the rest of the day's traffic on that network. Even if a network produced messages with the required number of loops only every two or three days on average, this might still have been enough for the codebreakers to keep on top of the traffic, although coverage would have been patchier.

When attempting to assess how useful the bombes would have been if the diagonal board had never been invented, the key question is: how often were messages with the necessary number of loops intercepted? Messages containing three or more loops were certainly common enough in 1944 to be dealt with in a training manual issued to American bombe operators at Eastcote;

this manual gave examples of menus containing three or more loops.[80] The codebreakers are known to have recorded precise information about the number of intercepted messages that contained three or more loops, but if this information has survived, it seems that it is yet to be declassified.[81] In its absence, we cannot reach a final judgement on the question of how successful the bombes would have been without the diagonal board. It seems fair to say, however, that bombes with simultaneous scanning but no diagonal board would still have played a vital part in the Allied struggle, although they would necessarily have provided more limited access to the Enigma traffic than was actually the case.

Between them, Rejewski, Turing, Keen, and Welchman produced a new breed of machine; and their bombe forged a new and enduring connection between warfare and high-speed automated information processing.

A Wren's eye view

This chapter concludes with Jean Valentine's recollections of her bombe training at Eastcote in 1943, her arrival at Bletchley Park, and a visit to the Bombe factory at Letchworth.

I got instructions to report to a training depot for new Wrens, a place in Scotland called Tullichewan Castle. With about forty or fifty other people I was taught to salute and to march—skills that I seldom used again, because Bletchley Park was the most unmilitary place you could ever come across. At the end of our training, we were called in one at a time and told where our futures lay. There were about three or four Wren officers sitting behind a table, and one small hard chair facing them.

I was instructed to sit on the hard chair, which I did. They looked me up and down and said 'Stand up, Valentine!'. I stood up, stood to attention, eyes forward. The officers shuffled some papers around and chatted among themselves, leaving me standing there thinking 'What's all this about?'. Finally they said 'Sit down!'. So I sat down again. 'We're going to ask you to do something

Figure 12.12 Jean Valentine.

Reproduced with permission of Jean (Valentine) Rooke.

that we don't know about ourselves', they said. 'We haven't been told, so we can't pass it on to you—but we have been told to look for girls like you. So tomorrow you are going to London—where you *will* be told. Dismissed.'

The next day I got into an express train for the first time in my life and travelled from my native Scotland to London. There I saw Underground trains—I didn't even know they existed. I reported to a depot in Earls Court, where I waited a week. One day some of us went to the Houses of Parliament and I heard Mr Churchill speak. He stood up to start his speech, but was a bit incoherent, and somebody said 'Sit down, Winnie'. So he sat down, and that was the end of that. Eventually we received orders to report to Eastcote. They told us that we were the first five 'watch-keepers' to be sent there.

Our new quarters at Eastcote were filthy, because the builders had been camping out in them, waiting to leave—which was why we had had to wait in Earls Court. I was shown into a room and told to clean it up, then given a spanner and told to erect two-tier bunks. Next day, we were taken to the working part of the establishment. We walked into a room and my heart sank when I saw a huge machine, about 8 feet long and nearly 7 feet tall.

I wondered how on earth I was going to reach this gigantic machine's highest parts. That's what the 'Stand up, Sit down' business had been all about—to see if I would be tall enough to operate the machine. I discovered later that there was a height minimum, but in my case the panel at Tullichewan seemed to have waived it. I stood diminutively beside the machine and the people there said 'Oh dear'. But a large block of wood was produced, and standing on that I could reach everything well enough.

With 108 revolving drums at the front of the bombe, and a mass of wires and plugs and whatnot at the back, it was quite a frightening machine, especially for someone who didn't know very much at all about machinery of any kind. But actually, once you learned to use the bombe, it really didn't present much of a problem.

We weren't told about Enigma, but we were told that we were helping to break into encrypted German messages. So we knew we were doing something important.

After my training I was sent to Bletchley Park. This was in the autumn of 1943. We drove there in an estate car, about ten of us, and walked the final hundred yards or so to Hut 11A, at that time the Bombe Room. There were five or six bombes inside. We relieved the women who were on duty and spent the next 8 hours working at the bombes, with only a half-hour break in the middle in which to wander up to the Mansion for a quick meal.

I never saw the rest of Bletchley Park, nor even the rest of the Mansion—I saw only those few yards between the Mansion and the hut where I worked. The organization was completely compartmentalized. You didn't go to anybody else's hut, nor they to yours. When you took your meal in the Mansion it was a help-yourself affair (something else I'd never come across before) and you sat down randomly at a table with two or three other people. These might have been fellow Wrens, or civilians, or even officers—there was no hierarchy at all. You talked about the weather, or the films showing in the local cinemas, never about work. To be at Bletchley Park during the war was an isolating experience.

On one occasion we were invited to give up a precious day off in order to visit the factory of the British Tabulating Machine Company in Letchworth, where the bombes were manufactured. We said 'Well, why?', and were told that morale was so low in the factory that the workers needed something to cheer them up. 'But don't tell them what they're doing', we were cautioned. 'Don't tell them what you're doing, don't tell them anything, just go and . . . cheer them up'. A few of us piled into a station wagon and off we went to Letchworth. The British Tabulating Machine

Company had expanded its operations into the factory next door to their own, where previously Spirella corsets had been manufactured. These were ridiculous rigid garments for ladies that allowed very little movement at all. But now the place was fulfilling a useful purpose, turning out bombes.

We were introduced very briefly to Doc Keen, who in fact wasn't a doctor at all. When he was an apprentice, he had carried his tools in what looked like an old-fashioned doctor's bag, so the other apprentices called him 'Doc'. It stuck right through his life. He was the brain behind engineering the bombe. Turing designed it, with some help from Welchman, but Doc Keen was the man who worked out how to actually build it.

No wonder morale at the bombe factory was low. They were even more compartmentalized than we were. I walked into one little room and found two or three women sitting at tables; they were repeatedly counting out piles of nineteen wires. The wires looked a bit like piano wire. I would have needed a psychiatrist by the time I had done that for a few days. These women had no idea what they were doing, and we could not tell them, even though I realized what the wires were. Each of the bombe's rotating drums had a circle of letters on the front of it, A, B, C, . . . ; and behind each of these letters were four tiny little wire brushes, each consisting of nineteen filaments. That's what these poor girls were doing, counting out the nineteen filaments. Can you imagine anything more soul-destroying?

I was myself very happy at Bletchley Park, content in the job of bombe operator. But one day a notice appeared on a board, saying 'The following are required to go overseas'. I didn't want to go one little bit. But I left Bletchley Park in April 1944, and soon I was in Ceylon, working on Japanese codes.

CHAPTER 13

Introducing Banburismus

EDWARD SIMPSON

O nce the bombes got going, they made a massive electromechanical attack on the Enigma's daily-changing key. But with limited numbers of bombes, the demand on them had to be minimized. For the Naval Enigma, Hut 8 used a cryptanalytic process called 'Banburismus' to reduce the amount of processing that the bombes had to do. Banburismus was largely a manual process—although with a vital contribution from the card-sorting machines in the Hollerith section—and employed a handful of the best cryptanalysts, with a large supporting team of WRNS ('Wrens') and civilian 'girls'. The startling recent discovery of Banburies in the roof of Hut 6 at Bletchley Park adds a new twist to the story.

Background

There is much high drama in the story of breaking and reading Enigma:

- the secret meeting of British and French cryptanalysts with their Polish counterparts outside Warsaw in late July 1939 (see Chapter 11)
- Colonel Stewart Menzies (later to be 'C', the head of the Secret Intelligence Service) waiting at London's Victoria Station in mid-August 1939, in evening dress and with the Légion d'honneur rosette in his buttonhole, to receive a Polish-made replica Enigma machine from the French Intelligence's Gustave Bertrand
- the Royal Navy's Anthony Fasson and Colin Grazier of *HMS Petard* kick-restarting the reading of U-boat Enigma in October 1942 when Hut 8 had been shut out of it for ten months, their gallantry commemorated by posthumous George Crosses for securing Enigma materials from the sinking U-559 at the cost of their lives
- some two hundred purpose-built bombes clicking away endlessly at Bletchley Park and its outstations (see Chapter 12).

At first sight, the quaintly named 'Banburismus' component of breaking Enigma offered no high drama. Both the mathematics it was based on and the technology of its application dated back some two hundred years to the eighteenth century. Yet the handful of cryptanalysts and supporting 'girls' who employed Banburismus—building on Alan Turing's genius and carried

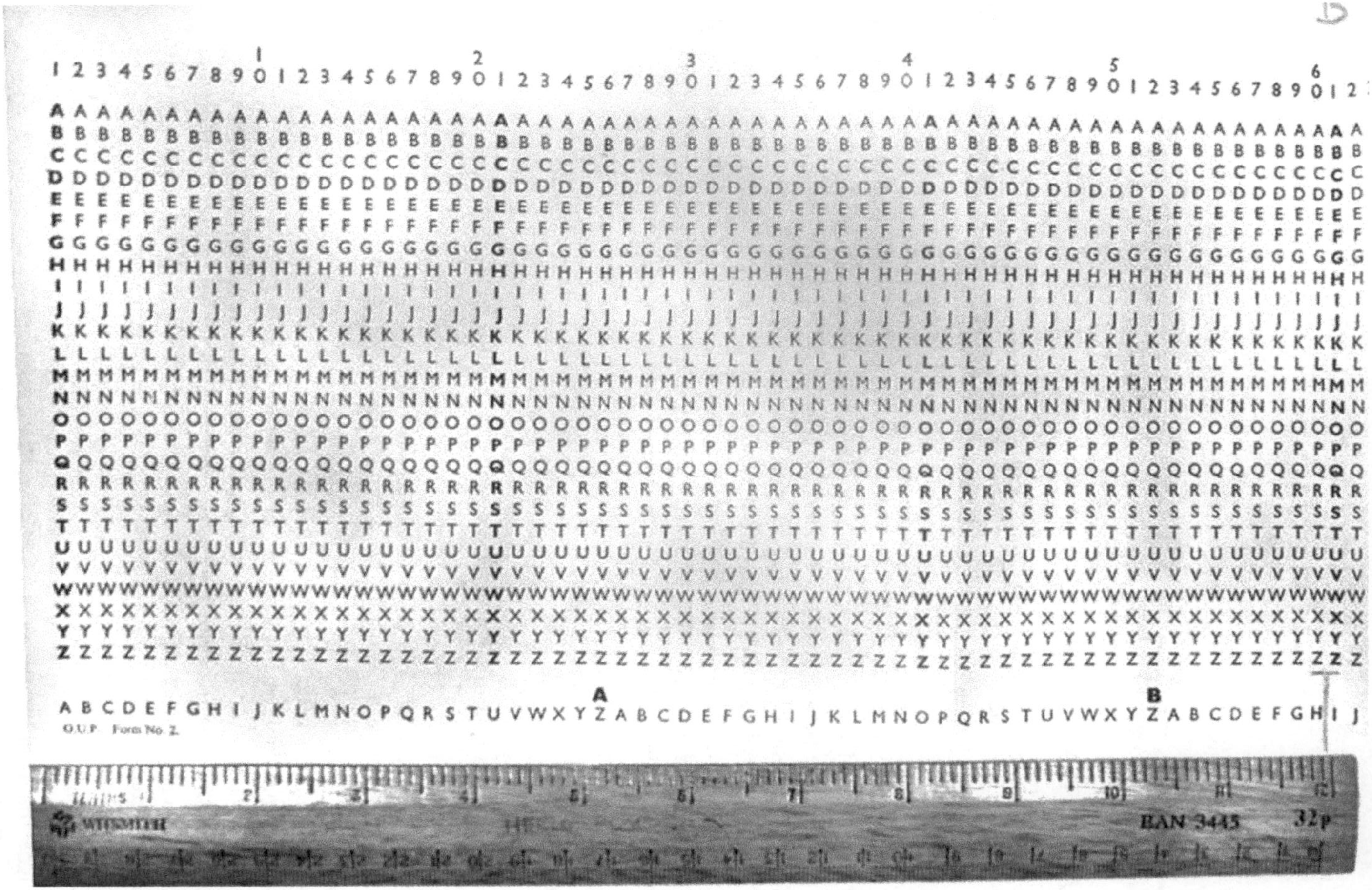

Figure 13.1 A Banbury.

National Archives ref. HW 40/264. Crown copyright and reproduced with permission of the National Archives Image Library, Kew; with thanks to Ralph Erskine and John Gallehawk.

forward by Hugh Alexander's leadership and ingenuity in method—multiplied in manifold ways the quantity of naval intelligence that those bombes could produce.

We are fortunate that Banburismus has been systematically described (although with some gaps) in the two histories of Hut 8 written around 1945 by Hugh Alexander and Patrick Mahon, successive heads of the Hut after Alan Turing.[1,2] This chapter makes continual reference to these histories, and there are more opaque references to Banburismus in some of Alan Turing's own writing.

What Banburismus did

A burglar facing a five-letter combination safe needs only the patience to try all the combinations in turn, in the confident knowledge that one of them will open the door. The theory is simple: the difficulty lies in the scale. The burglar has 26^5, or some 12 million, combinations to try. The cryptanalysts in Hut 8 faced a mind-boggling 890 million million million possible daily keys, which is 75 million million times as many as the burglar. (The arithmetic of this is given in Chapter 38.) Only one of the daily keys was right, and the next day it would be a different one.

The part that the bombe played in this was explained in Chapter 12. The speed at which the bombe worked was mind-boggling: it could test the 26^3 (17,576) ring settings (see Chapter 10) for one wheel order in 15 minutes. Allowing for time to set up, 60 such runs could be made in 24 hours—but to test all 336 possible wheel orders in turn, a total of nearly 6 million (17,576 × 336) ring settings, would still take five and a half days. The war on the Atlantic U-boats could not wait, and Hut 6 also needed its share of the limited number of available bombes for army and air force work. The Banburismus procedure that Alan Turing devised came to the rescue.

To fill the three slots in the machine, the German naval encipherers had selected three out of their eight wheels and placed them in a certain order. Half-a-dozen of the best minds in Bletchley Park on duty at any one time used Banburismus to try to discover which of the eight wheels were in the right-hand and middle slots, thus very significantly reducing the number of tests that the bombes needed to run in order to produce a result.

Suppose, for example, that Banburismus had precisely identified the middle wheel and established that the right-hand wheel was some one of the three special navy wheels (see Chapter 10). Then for each of the three possibilities for the right-hand wheel there would be six out of the eight wheels remaining from which to choose the left-hand wheel: 18 possible combinations. Thus only 18 tests would now be required, instead of the original potential of 336. This chapter and Chapter 38 explain how it was done.[3]

Defining Banburismus

Bletchley Park's 1944 cryptographic dictionary had these (far from self-explanatory) definitions:[4]

BANBURISMUS

1. The use of Banburies to set messages (especially Enigma messages) in depth with each other.
2. Action or process of identifying right-hand and middle wheels of an Enigma machine by relating distances or intervals between message settings . . . to the possible intervals between the enciphered settings . . . .

But it could not be used in all circumstances. Work on Army and Air Force Enigma could not use it because those German encipherers were free to choose their own ground setting when enciphering a message's starting position, instead of having one imposed on all from the centre. In Naval Enigma it could be used only against the three-wheel (and not the four-wheel) version, only when the bigram tables were known, either by reconstruction or thanks to a 'pinch', and even then only if at least some 300 messages on the same version of the cipher were available. A 'pinch' was the capture of actual German enciphering documents (or other Enigma material) from a vessel or a fighting unit in the field. Banburismus was mainly employed against Dolphin, the German Navy's 'home waters' cipher.

Setting 'in depth'

Let us consider the first part of the definition. Depth was the cryptanalyst's staple diet—not just for Enigma but for a wide variety of ciphers. Where two or more messages (whether in the form of letters or numbers or encoded phrases) had been *enciphered in exactly the same way* (whether by manual or electrical substitution or by other methods, like the addition of additives or whatever) they were said to be 'in depth'. Because their encipherments had been identical, certain features of the structure of the underlying language or code (such as the frequencies of its commonest elements) would reappear in the enciphered texts. Very often it was this opportunity to study two or more messages enciphered in exactly the same way that enabled the cryptanalyst to begin to penetrate the cipher. A simple illustrative example of depth is given in Chapter 38.

In Enigma, a depth occurred when two messages were enciphered with the same plugboard settings, the same wheels in the machine (in the same order), and starting positions for the wheels that caused some parts of two messages to overlap and be identically enciphered. Depths in connection with a different cipher, 'Tunny', are described in Chapters 14 and 16.

Mahon's history offers a detailed source of information on how depths were used in breaking Enigma. Mahon joined Hut 8 in October 1941: his history prior to that date relies on conversations he had with Turing. Mahon tells us that Turing devised the Banburismus approach before the end of 1939, on the same night that he worked out the naval indicator system. Turing commented at the time that:[5]

I was not sure that it would work in practice, and was not in fact sure until some days had actually broken.

Enigma's enciphering system, with its several components and many details, changed daily. Turing's shorthand 'days had actually broken' stood for the solving of a day's details which made it possible to decipher all that day's messages. In fact, it was not until November 1940 that Hugh Foss, an old hand from the 1930s who had temporarily joined Hut 8, successfully broke a day using Banburismus. Thanks to several pinches in the months that followed, Banburismus passed from the research stage into operational use.

Turing's invention stemmed from combining two observations. First: when two Enigma messages were in depth, a pair of letters that matched when the two clear texts were written one above the other, properly aligned, would match again in the enciphered texts, because they had been enciphered in exactly the same way. Secondly: letters in language occur unevenly. Writers and students of writing have often turned the spiky distribution of language to their advantage. Edgar Allan Poe observed in *The Gold Bug*, his most popular short story, published in 1843, that:

It may well be doubted whether human ingenuity can construct an enigma of the kind that human ingenuity may not, by proper application, resolve.

Hut 6 and Hut 8 might well have taken encouragement from that. Poe then explained the step-by-step breaking of a substitution cipher, starting from the frequency count of letters in English language.[6]

A century later, just as Turing was devising Banburismus, Udny Yule at St John's College, Cambridge, was statistically contrasting the structure of Macaulay's prose vocabulary with Bunyan's and defining the statistical 'characteristic' for each author. Yule used his 'characteristic' to test whether Thomas à Kempis had written the anonymous Christian spiritual book *De Imitatione Christi*, composed in Latin in the fifteenth century. He published his analysis in 1944, adding weight to its attribution to that author.[7] Yule's statistical analysis of the works of famous authors was a close cousin of Turing's statistical approach to the scoring of depths.

Turing combined these two observations to conclude that, when two messages believed to overlap were speculatively placed in alignment with each other, repeats of letters between them would be more frequent when the messages were aligned correctly (that is, were in depth) than when they were not. Correct alignments could accordingly be distinguished from false ones by counting such repeats.

But first the cryptanalysts had to find pairs of messages that seemed likely to overlap. The preamble to an Enigma message told its recipient how to set the day's three wheels at their correct starting positions, by means of a three-letter indicator such as 'ASL'. Obviously two messages with identical indicators would give immediate access to a depth, but such strokes of fortune were rare. In practice two different pathways, described later, were followed and each of the resulting pairs was tested by counting repeats and judged against a threshold of acceptability (as explained in the sections 'Banburies' and 'Weighing the evidence' and in Chapter 38).

For the first pathway, the day's messages were sorted to extract those pairs that had their first two indicating letters in common—and therefore the starting positions of two of their wheels in common. The starting position of the third wheel was unknown, but the likelihood of two such messages overlapping was sufficient to justify testing every alignment of one against the other. Even though the number of pairs of messages was huge and the testing of them was tediously repetitive, this was within the scope of hand methods.

Next, pairs of messages with only the first indicating letter in common were selected. Such pairs were so numerous that testing them all at all alignments was out of the question. A further selection had to be made within them, and this could be undertaken only by machines. This was the second pathway: the section 'Tetras' explains it further. The remaining pairs, with no indicating letters in the same positions in common, lay beyond even the machines' routine capabilities.

The versatility of punched holes

The testing of speculative alignments had its roots in the eighteenth century, in the exploitation of the potential of holes punched in cards. In 1725 Basile Bouchon, a silk worker in Lyon and the son of an organ maker, noticed how the makers of musical boxes used perforated paper rolls to mark where to insert the tiny pegs into the cylinders: the paper roll represented the musical notes to be played and identified where the pegs should be placed to play them. In modern terms, it contained all the necessary information. Bouchon saw that the same could be done to govern the weaving of patterns by his silk looms. In 1728 his assistant Jean-Baptiste Falcon replaced the paper roll by a set of punched cards connected by ribbons in an endless loop. The idea really took off in 1805 when it was incorporated into the highly successful Jacquard weaving loom.

Charles Babbage began work on his Difference Engine (Fig. 24.1) in 1823 but was unable to complete it. He watched a Jacquard loom at work when travelling in Europe, and this contributed to his plan for a more sophisticated Analytical Engine (Fig. 24.3) which he started designing in 1834. The Jacquard loom also came to the notice of Ada Lovelace (Fig. 24.4). She was the daughter of Lord Byron, but her forceful and intellectual mother took her away at an early age from her father, who then abandoned them for Greece and an early death. As described in Chapter 24, Ada's mother took her to Babbage's popular mathematical evenings, and in 1833 Ada wrote:[8]

This machine [that is, the Jacquard loom] reminds me of Babbage and of his Gem of all Mechanisms.

Lovelace (Ada's married name) and Babbage developed an improbable but prolific mathematical collaboration. In her 'Notes' in 1843 on Menabrea's *'Notions sur la machine analytique de M. Charles Babbage'* Lovelace wrote:[9]

The distinctive characteristic of the Analytical Engine, and that which has rendered it possible to endow mechanism with such extensive faculties as bid fair to make this engine the executive right-hand of abstract algebra, is the introduction into it of the principle which Jacquard devised for regulating, by means of punched cards, the most complicated patterns in the fabrication of brocaded stuffs. It is in this that the distinction between the two engines lies. Nothing of the sort exists in the Difference Engine. We may say most aptly that the Analytical Engine weaves algebraical patterns just as the Jacquard loom weaves flowers and leaves . . . the machine might compose elaborate and scientific pieces of music of any degree of complexity or extent.

In modern terms, Lovelace moved forward from Babbage's concept of a calculator to the concept of a computer.

Some forty years after Lovelace wrote this passage, Herman Hollerith took up the idea that data handling should be carried out by machines based on the Jacquard loom; and he won a competition to process the returns of the 1890 US census with his punched-card machines. His name remained attached to the technology even after the business became IBM and its variants, and 'Hollerith machinery' monopolized the mechanized handling of data into the 1940s when computers such as Bletchley Park's Colossus and its successors elbowed it aside. The massive Hollerith section at Bletchley Park may well have been the technology's zenith in the UK, both in terms of size and of technical skill. Located originally in Hut 7, and from 1942 in Block C, the section was known colloquially as 'the Freebornery' after its head, Frederic Freeborn, seconded from the British Tabulating Machine Company in Letchworth.

Banburies

Hut 8's Banburismus also depended on punched holes, and so can be traced back to Basile Bouchon.

At first the placing of messages correctly in depth by counting the repeats between them was probably done by writing pairs of messages—selected as having their first two indicating letters in common, as already mentioned—on two strips of paper and sliding these against each other. This simple approach was quickly elaborated and a technique was developed that served with distinction until September 1943. The technique used long sheets of paper on which vertical alphabets were printed: variously with 60, 120, 180, or 260 columns. With rows and columns both spaced at 5 mm intervals, each letter occupied a 5 mm square. The longest sheet, the 260-column version, was therefore about 1.3 m long (plus margins). When two of these sheets were at their extreme end of comparison, they would extend to nearly 2.6 metres.

The sheets, printed in Banbury, were whimsically called 'Banburies'. Using them might have been called 'Banburism', but the German form *Banburismus* was preferred: this was the humour of J. B. Morton ('Beachcomber') whose articles in the *Daily Express* had a devoted readership for 50 years before, after, and (more importantly) during the war. One of his recurring characters was 'Dr Strabismus (whom God preserve) of Utrecht'.

The Banburies carried the reference 'OUP Form No 2', so the Banbury printer was presumably a subcontractor of the Oxford University Press. Bletchley Park ordered them through Paymaster Commander Edward Hok, the head of the printing and code production unit at Mansfield College, Oxford; for example, they ordered 7000 of various lengths on 13 January 1942. The National Archives at Kew have a pristine Banbury, shown in Fig. 13.1. This seemed as close as we were going to get to Banburies but there was a surprise to come.

Banburies in the roof

In 2013 a bizarre turn of events catapulted some ancient Banburies into the modern world. During the restoration of Hut 6 at Bletchley Park, a workman found some scrunched-up papers stuffed into the roof. The plausible surmise was that someone had used waste paper to fill a hole and keep out the draught. Fortunately the papers were not discarded, and they turned out to be cryptanalysts' workings, including two used Banburies (Figs. 13.2 and 13.3). No one had seen a used Banbury since Bletchley Park packed up at the end of the war.

Another two of the papers had helpfully written on them 'Yellow' and (respectively) 'April 14' and '15/4/40'. 'Yellow' was Hut 6's shorthand designation for the particular Enigma network used by both the German Army and Air Force during the invasion of Denmark and Norway, launched on 9 April 1940. Hut 6 broke Yellow on 14 April. The relics in the roof are a direct connection back to that historic day.

Both Banburies are incomplete. They differ in several respects from the (presumably) later version held by the National Archives. The roof relics have lettering that is more ornate, no 'OUP Form No 2' and no recurring alphabet along the base. The holes punched in them by their unknown users are vividly clear, each hole centred with impressive accuracy on one letter in each of successive columns. It was not always so. Bletchley Park complained to Commander Hok on 30 December 1941 that:

The printing of the last batch of OUP Forms No 2 . . . is somewhat irregular and the variation in the amount of space between columns is at times sufficiently great for the forms to give the wrong answer.

Figure 13.2 The Banburies found in the roof: Banbury One.

Reproduced with permission of the Bletchley Park Trust. © Bletchley Park Trust.

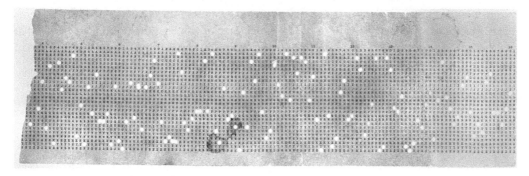

Figure 13.3 The Banburies found in the roof: Banbury Two.

Human fallibility is illustrated too. Holes have been punched on both G and W in column 83, on both B and T in column 112, and on none at all in column 113 of Banbury One: there should have been precisely one hole in each column. Another of the working papers has written on it:

Wheel orders 321 & 123 appear to have punches wrongly punched.

A second hand has ringed this in red and added:

To all [word illegible] very concerning.

Questions abound. For one thing, how could such a blatant breach of security have occurred? The rule was absolute that every scrap of working paper must be locked away when a room was even briefly unoccupied. We know the two points in the Hut 6 roof where the papers were found, but not which staff were occupying the rooms beneath in 1940. Joan Clarke recalled one proper means of disposal in Hut 8:[10]

Chores performed in the slack periods included removing pins and tearing up the Banbury sheets for solved days, and other workings, as required for re-pulping.

Weather reports do not in fact endorse the conjectured purpose of stopping a draught. There had been intense cold during the preceding winter when Hut 6 was being constructed: the coldest January since 1838 at nearby Oxford. Mid-February continued very cold with snow, and this lasted into early March. April was on the whole dull and wet, warm in the last week, the time of the dated papers; and May's temperatures substantially exceeded the average. But no more plausible explanation can be offered.

And why in Hut 6? For technical reasons arising from the cipher systems, Banburismus was neither needed nor used against the Army and Air Force Enigma broken in Hut 6, but only against the Naval Enigma dealt with in Hut 8. Here we do have a clue. Mahon wrote in his history that:[11]

Work was fizzling out when Norway was invaded and the cryptographic forces of Hut 8 were transferred en bloc to assist with Army and Air Force cyphers.

Further light is shed by occasional references in the other papers found in the roof. There is frequent use of the verb 'shove' in such contexts as 'shoved and tested' and 'shove anew'. These, along with the already mentioned 'punches not punched' and other textual details, are terms used in Hut 8, but Hut 6's operations did not involve 'punching' and 'shoving'. It thus seems likely that many of the other working papers as well as the Banburies themselves were related to Naval and not to Army and Air Force Enigma, and that the Hut 8 cryptanalysts lent to Hut 6 in April 1940 had continued to do some Naval Enigma work as well.

Using the Banburies

Alexander's history describes how Banburies were used:[12]

If the message was say 182 letters long and the text began IXBNR . . . the girl would take a 200 long Banbury and would mark in red the letters in successive columns. Having done this she would take the Banbury to a hand punch and punch out the marked letters.

The 'hand punch' must have had some degree of sophistication (perhaps guide rails?) to produce the accuracy of centring of the holes. And later:

Suppose we had [message indicators] ASL and ASJ for instance. Then Banburies ASL and ASJ would be taken out and put down on a dark background with, say, ASL underneath and ASJ on top. The 'guide hole' of ASJ would be put over the 1, 2, 3 . . . 25 of ASL successively . . . . The repeats at each position were immediately seen because a repeat implied a hole in each Banbury at the same place and the dark background could then be seen standing out against the white of the sheets. The scores at each position were recorded . . . .

Figures elsewhere in Alexander's history illustrate the scale of this operation. In Banburismus's heyday, and in round figures, 400 messages a day averaging 150 letters each required some 60,000 holes to be punched in a day—that is, 42 each minute, or two every three seconds, around the clock. Of the 80,000 pairs derived from the 400 messages, about 1 in 676 (that is, 120 pairs) would have their first two indicator letters in common and be sent to be compared, each comparison at 50 different alignments. Thus the comparisons had to be made—and repeats counted and recorded—6000 times in a day: four times every minute around the clock.

Early in 1941 Eileen 'Copper' Plowman joined Bletchley Park as a Foreign Office civilian, aged 18 and quite homesick. She recalls:[13]

My first job in Hut 8 was looking for matches in the Banburies. We worked in a group, not in silence but not particularly chattily either: it was quite sombre. The Banburies came to us with the holes already punched. We could sit to slide and count the shorter ones, but had to stand up to manage the longer ones. This was definitely the most boring job that I did at Bletchley Park. It was tedious, my eyes hurt and it was difficult for us to see the point of what we were doing.

Christine Ogilvie-Forbes also joined as a Foreign Office civilian and was posted to Hut 8 in July 1941. She describes working on Banburies:[14]

You took a Banbury, a piece of paper 10″ × 24″ or longer for longer messages, printed with vertical alphabets. Next you took a coded message, wrote its number and three-letter setting (gleaned from the first two and last two groups). Then you pushed a ruler along the Banbury, pencil-marking a letter in each column from the message, no gaps. Another table had a little hole with a light below. Sliding the paper along, you punched a confetti-sized hole on each pencil mark.

You took two messages with the first two wheel-settings matching, for instance CCE and CCK or ALP and ALA. One Banbury on top of the other, you would slide along 25 places one way, 25 the other. Depending on the number of times each letter appears in German, lots of holes co-incide, particularly when German code passes over German code—one must be correct.

Noted the number of holes in each position, hoping for a group. A group of five, six or more had you jumping up and down. The results were passed on to the Crib Room, to be matched with clear German and make the menu for the Bombe.

The German Navy in the Mediterranean had much less traffic. Same machine, same system, but it ran from mid-day to mid-day. One person could cope with it all. I rather fancied myself at the job and was called 'Med Queen' on my shift. It was a lovely rush every second day on day shift to get a 'throw-on' menu by 4 p.m.

Eileen Plowman's and Christine Ogilvie-Forbes' 'slide' must have been a synonym for the cryptanalysts' 'shove'.

Banbury One from the Hut 6 roof (Fig. 13.2) illustrates these accounts. Along the top are written, not at all carefully for posterity, 'TGZ' above columns 1–6 and 'B/078' above columns 9–13: these are the three-letter setting—or indicator—and number recalled by Christine Ogilvie-Forbes. Over columns 41–52 we have 'with TGC 11^{xx}/159'. Alexander's history enables us to interpret this as 'when this message and another with indicator TGC were compared, with an overlap of 159 letters, 11 repeats were found of which four came as two bigrams' (two-letter sequences). Chapter 38 pursues the significance of these figures.

After careful preservation, a selection of the papers from the roof—including of course the two Banburies—went on display at Bletchley Park in March 2015.

Tetras

In Banbury One the indicators (TGZ and TGC) have their first two letters in common. However the second of the two pathways mentioned used pairs of messages that had only their first indicator letters in common (and therefore only one wheel position in common). Such pairs were so much more numerous that a stringent criterion for which pairs to test was needed. Fortunately, the structure of language again came to the rescue.

In practice, the letters repeated between messages often came not as single spies but, if not in battalions, in bigrams, trigrams, and up to enneagrams (sequences of nine letters) and beyond. The tetragram, a sequence of four letters, proved particularly useful, and was colloquially a 'tetra'.

The observation that some letters occur more frequently than others in language becomes more marked when we look at such sequences of letters occurring together. A two-letter sequence such as 'en' occurs more frequently in English than the combination of 'e' and 'n' counted separately, but for some bigrams (say, 'mf' or 'bv') the opposite applies. As before, this spiky distribution of bigrams carries through into the repeats between identically enciphered messages. Thus a repeating bigram is more powerful evidence of the alignment's being true than two single repeats would be. A three-letter repeat is better still—and so it goes on.

Alexander's history stated that a four-letter repeat was 100 times as likely to occur in German naval language as in a random series of letters. With a six-letter repeat the likelihood was 15,000 times greater.

Experience showed that with these much more numerous pairs of messages, those to be examined had to be confined to pairs with tetragram repeats or better, in order to keep the task manageable. The 'traffic', the colloquial term for the totality of enciphered messages received, was sent in batches to Freeborn's Hollerith section, where this input was searched for tetragram or better repeats between pairs of messages.

Enigma's day generally ended, and the new one began, at midnight (the German Naval Enigma in the Mediterranean was an exception) and the Freebornery's punching was done around the clock. Each day's 'Tetra catalogue', listing all the tetragram or better repeats found, was delivered to Hut 8 by instalments throughout the day. Iris Brown joined Hut 7 in February 1942, at age 17; later the assistant to Freeborn's secretary, she was initially a 'message girl'. She recalls:[15]

In that winter I was taking original Tetra messages back to Hut 8 many times a day.

Alexander acknowledged the importance of the Freebornery's search for repeated groups:[16]

Without this work we should have been helpless and we owe much to the efficient and speedy service we received.

Weighing the evidence

So from the two parallel operations of sliding Banburies and finding tetragram repeats, both following Basile Bouchon in exploiting holes punched in cards, Hut 8 had an array of pairs of messages which might, from the evidence of those repeated letters, be aligned in depth. An arithmetic was needed for deciding how much evidence was 'enough'—that is, sufficient to decide that a particular speculative alignment could be relied upon before using it in the complex reconstruction process that was to follow. Donald Michie, echoing Max Newman, called the scoring system that Turing devised for this purpose his 'greatest intellectual contribution during the war'.[17]

Neither of the histories explains the derivation of this arithmetic: it was in fact based on Bayes' theorem and the calculation of the mathematical probabilities of different numbers of repeats, as Chapter 38 explains. But with the Enigma details changing daily and the Battle of the Atlantic in the balance, there was no time for calculating individual probabilities. Instead, Hut 8 devised a streamlined process of scores that required only to be added or subtracted by non-mathematical (but clever) 'girls'. This process borrowed the ready-to-hand decibel unit from acoustics and re-named it the 'deciban', and eventually worked in units half that size, the half-deciban or hdB. Chapter 38 fills in the main details of this most elegant procedure.

Then the really hard bit

So far this chapter has considered only the first part of the cryptographic dictionary's definition of 'Banburismus'. The second part of the definition is opaque, because the process it describes is incapable of simple explanation. It is valid (although trite) to liken explaining it to describing chess to someone who has not even played draughts. The two histories chose instead to exemplify it, each with a worked example that was complex and required several pages. These are beyond the scope of this book. They depended on a mixture of practised ingenuity, trial and error, and successive approximation.

This is where the most skilled cryptanalysts came into their own and used Banburismus to achieve the purpose of partially identifying the Enigma's wheels and other elements for the day. Alexander, always vigilant for the most effective use of staff, observed that:

It paid to have first-class people plus routine clerical assistance; it was very difficult to find useful work for people of intermediate ability.

He was equally vigilant for their well-being. Seventeen-year-old Hilary Law was a Foreign Office civilian in Hut 8 from autumn 1943; she recalls:[18]

Hugh Alexander . . . a very thoughtful man . . . on a visit to the USA he brought us back nylons and lipsticks . . . he once took us to see the Wrens at work on the Bombes.

In the autumn of 1944 Alexander left Hut 8 and became head of the Japanese Naval Section NS IIJ. When he visited the staff at the Colombo outpost 'he proved a very friendly character, and actually played in one of our cricket matches'.[19]

Alexander's 'first-class people', referred to above, were some eight or ten cryptanalysts, all civilians, usually two on each of the three shifts. Turing was not among them: once the system he had devised was up and running, he moved to research and development elsewhere. Alexander himself was the first amongst them, by repute as well as alphabetically, a top Banburist and superb manager of cryptanalysts too. Joan Clarke, the only woman among these cryptanalysts, was described by him as 'one of the best Banburists'. Hilary Law recalls that:

We always regarded her as a true 'blue stocking' who always had an ethereal look about her—on another planet from us mortals.

And from Christine Ogilvie-Forbes:[20]

We found Joan Clarke very amenable in the Registration Room. She was engaged to Prof (A. Turing) and would walk two steps behind him. Prof avoided eye contact, whether over a piece of paper or a cup of tea. They were always very good at telling us our results—naturally it made us work harder. Particularly the *Scharnhorst* and *Gneisenau*. And telling us on the 12–9 night shift on D-Day that the invasion was about to start.

The *Scharnhorst* and *Gneisenau* were important German battleships, respectively sunk in December 1943 and irreparably damaged in February 1942, with assistance from Bletchley Park.

However, it was symptomatic of Bletchley Park's downgrading of able women that the director, Commander Travis, advised Joan Clarke that she might have to transfer to the Wrens to attain the promotion she deserved. She declined.[21]

The cryptanalysts who used Banburismus recalled the experience with great affection. Mahon called it 'one of our two most pleasurable pastimes' and 'a delightful intellectual game'.[22] Jack Good wrote that it was 'enjoyable, not easy enough to be trivial, but not difficult enough to cause a nervous breakdown'.[23] But the recollections of the many more numerous women who had tediously punched cards in Hut 8 and in the Freebornery, and had strained their eyes counting where the holes in Banburies coincided, may have been less affectionate. Their contributions to the achievement of Banburismus must not be undervalued.

As more and more bombes came into service, it became simpler and quicker to run all wheel orders, and so the need to pick out plausible alignments diminished. Mahon records that at the beginning of July 1943 a new policy 'not to wait for the Banburismus to come up' before starting the bombe runs was introduced, and three weeks later that 'we have had no Banburismus to do lately'. Finally abandoned in mid-September 1943, it had earned an honourable retirement.

CHAPTER 14

Tunny: Hitler's biggest fish

JACK COPELAND

After 1942, the Tunny cipher machine took over increasingly from Enigma for encrypting Berlin's highest-level army communications. Hitler used Tunny to communicate with his generals at the front lines. Turing tackled Tunny in the summer of 1942, and altered the course of the war by inventing the first systematic method for breaking into the torrent of priceless Tunny messages.

A new supercode

The story of Enigma's defeat by the Bletchley Park codebreakers astonished the world. Less well known is the story—even more astounding—of the codebreakers' success against a later, state-of-the-art German cipher machine (Fig. 14.1).[1] This new machine began its work encrypting German Army messages in 1941, nearly two years into the war. At Bletchley Park it was code-named simply 'Tunny'. Broken Tunny messages contained intelligence that changed the course of the war and saved an incalculable number of lives.

How Bletchley Park broke Tunny remained a closely guarded secret for more than 50 years. In June 2000 the British government finally declassified the hitherto ultra-secret 500-page official history of the Tunny operation. Titled 'General report on Tunny', this history was written in 1945 at the end of the war by three of the Tunny codebreakers, Donald Michie, Jack Good, and Geoffrey Timms.[2] Finally the secrecy ended: the 'General report' laid bare the whole incredible story of the assault on Tunny.

Far more advanced than Enigma, Tunny marked a new era in crypto-technology. The Enigma machine dated from the early 1920s—its manufacturer first placed it on the market in 1923—and even though the German Army and Navy made extensive modifications, Enigma was certainly no longer state-of-the-art equipment by the time the war broke out in 1939. From 1942, Hitler and the German Army High Command in Berlin relied increasingly on the Tunny machine to protect their ultra-secret communications with the front-line generals who commanded the war in the eastern and western theatres. Germany's compromised Tunny radio network carried the highest grade of intelligence, giving Bletchley Park the opportunity

Figure 14.1 The Tunny machine in its case (top) and with its twelve encoding wheels exposed.

Reproduced with permission of *The Turing Archive for the History of Computing* (upper image).
Reproduced with permission of the Imperial War Museum (lower image: Hu 56940B).

to eavesdrop on lengthy back-and-forth communications between the grand architects of Germany's battle plans. Tunny leaked detailed information about German strategy, tactical planning, and military strengths and weaknesses.

The Berlin engineering company C. Lorenz AG manufactured the Tunny machine. The first model bore the designation SZ40, 'SZ' standing for '*Schlüsselzusatz*' ('cipher attachment'). A later version, the SZ42A, was introduced in February 1943, followed by the SZ42B in June 1944 ('40' and '42' refer to years, as in Windows 98). Tunny was one of a family of three new German cipher machines, manufactured by different companies. Bletchley Park gave all three the general cover name 'Fish'. The Fish family's other members were codenamed Sturgeon and Thrasher; Thrasher had only limited use, while Sturgeon provided nothing more than a trickle of intelligence compared with the deluge produced by Tunny.

'Listening stations' in the south of England first intercepted Tunny radio messages in June 1941. After a year-long struggle with the new supercode, Tunny messages were being broken in quantity by the middle of 1942. Naturally, Turing played a key role in the attack on Tunny.

The Tunny system

The Tunny machine, measuring 19″ × 15½″ × 17″ high, was a 'cipher attachment': attached to a teleprinter or teletypewriter, it automatically encrypted the outgoing stream of pulses produced by the teletype equipment. It also automatically decrypted incoming messages, enabling the teletype equipment to print out the unenciphered message.

At the sending end of a Tunny link the operator typed plain German (the 'plaintext' or 'clear text' of the message) at the teletype keyboard, and at the receiving end the plaintext was printed out automatically, usually onto a paper strip as in old-fashioned telegrams. With the Tunny machine in 'auto' mode, many long messages could be sent at high speed, one after another. When the transmitting machine was operating in auto mode, the plaintext spooled into the teletype equipment on pre-punched paper tape.

Morse-based Enigma was clumsy by comparison. With Tunny, the German operators at the sending and receiving ends of the radio link did not even see the ciphertext (the encrypted form of the message). With Enigma, on the other hand, the ciphertext appeared at the machine's lampboard, letter by letter. As the keyboard operator typed the plaintext at the machine's keyboard, an assistant laboriously noted down the ciphertext letters as they appeared one by one at the Enigma's lampboard; the radio operator then transmitted the ciphertext in Morse code. Tunny was faster, could handle much longer messages, and required two people (rather than six) to send and receive a message. Morse was not needed in the Tunny system: the Tunny machine's output, encrypted teleprinter code, went directly to air.

International teleprinter code—which was at the time in widespread use around the globe, and remains so today—assigns a pattern of five zeros and ones to each letter and keyboard character. In this system '1' represents a pulse and '0' represents the absence of a pulse. Using Bletchley Park's now antiquated convention of representing a pulse by a cross and the absence of a pulse by a dot: the letter C, for example, is •**xxx**• (no-pulse, pulse, pulse, pulse, no-pulse). Today we write 0 instead of • and 1 instead of **x**, so C is 01110. To give some other examples: O is •••**xx**, L is •**x**••**x**, U is **xxx**••, and S is **x**•**x**••. Figure 14.2 shows a wartime memory aid that lists the pattern of dots and crosses associated with each keyboard character.

CONVENTIONAL NAME	1	2	3	4	5
/	•	•	•	•	•
9	•	•	x	•	•
H	•	•	x	•	x
T	•	•	•	•	x
O	•	•	•	x	x
M	•	•	x	x	x
N	•	•	x	x	•
3	•	•	•	x	•
R	•	x	•	x	•
C	•	x	x	x	•
V	•	x	x	x	x
G	•	x	•	x	x
L	•	x	•	•	x
P	•	x	x	•	x
I	•	x	x	•	x
4	•	x	•	•	•
A	x	x	•	•	•
U	x	x	x	•	•
Q	x	x	x	•	x
W	x	x	•	•	x
5 or +	x	x	•	x	x
8 or −	x	x	x	x	x
K	x	x	x	x	•
J	x	x	•	x	•
D	x	•	•	x	•
F	x	•	x	x	•
X	x	•	x	x	x
B	x	•	•	x	x
Z	x	•	•	•	x
Y	x	•	x	•	x
S	x	•	x	•	•
E	x	•	•	•	•

Figure 14.2 A war-time memory aide showing the teleprint code for each of the keyboard characters used in the Tunny system. In modern notation the dot would be written '0' and the cross '1'.

The Turing Archive for the History of Computing.

Teleprinter code produces a rhythmic, warbling sound when transmitted at high speed by radio. The first interceptors on British soil to listen to the German transmissions described what they heard as a new kind of music.

The first Tunny radio link, between Berlin and Greece, went into operation on an experimental basis in June 1941. Then, in October 1942, this experimental link closed down, and for a short time it was thought the Germans had abandoned Tunny. But later that month Tunny reappeared in a modified form on a link between Berlin and Salonika, and also on a new link between Königsberg and south Russia: this was the start of a rapid expansion of the Tunny network. By the time of the Allied invasion of France in 1944, when the Tunny system was at its peak, twenty-six different links were known to Bletchley Park. The codebreakers gave each link a fishy name: Berlin–Paris was 'Jellyfish', Berlin–Rome was 'Bream', Berlin–Copenhagen was 'Turbot', and so on.

The two central exchanges for Tunny traffic were at Strausberg (near Berlin) for the western links, and Königsberg for the eastern links into Russia. In July 1944 the Königsberg exchange closed down and a new hub was established for the eastern links at Golssen, about 20 miles from the Wehrmacht's underground command headquarters south of Berlin. During the final stages of the war the Tunny network became increasingly disorganized, and by the time of the German surrender the remaining central exchange had been transported from Berlin to Salzburg in Austria.

There were also fixed exchanges at some other large centres, such as Paris. The distant ends of the links, usually close to the fighting, were mobile. Each mobile Tunny unit consisted of two trucks. One carried the radio equipment, which needed to be kept well away from the teletype equipment because of interference. The other truck carried two Tunny machines, one for sending and one for receiving. Sometimes a landline was used instead of radio and the truck carrying the Tunnies was connected up directly to the telephone system. Only Tunny traffic sent by radio could be intercepted in Britain.

How Tunny encrypts

As with Enigma, the heart of the Tunny machine was a system of wheels. Some or all of the wheels moved each time the operator typed a character at the teleprinter keyboard (or, in the case of an 'auto' transmission, each time a new letter was read in from the pre-punched tape). There were twelve wheels in all. They stood side by side in a single row, like plates in a dish rack. The rim of each wheel was marked with numbers (01, 02, 03, etc.), and these were visible to the operator through a small window.

From October 1942 the following operating procedure was used. Before starting to send a message, the operator would use his thumb or fingers to turn the twelve wheels to a given combination, such as 02 14 21 16 03 36 21 16 43 21 50 26: he obtained this combination from a codebook. This codebook, known as a 'QEP' book, contained a hundred or more combinations. At Bletchley Park, the selected combination was called the 'setting' for that particular message. The wheels were supposed to be turned to a new setting at the start of each new message, although because of operator error this did not always happen. The operator at the receiving end, who had the same QEP book, set the wheels of his Tunny machine to the same combination. This enabled the machine to decrypt the message automatically as it was received.

The Tunny machine encrypted each letter of the message by *adding* another letter to it, so masking the identity of the original letter. Letters were added by means of adding the individual dots and crosses that composed them. The rules the Tunny's designers selected for dot-and-cross addition are simple: dot plus dot is dot; cross plus cross is dot; dot plus cross is cross; and cross plus dot is also cross:

$$\bullet + \bullet = \bullet \qquad x + x = \bullet \qquad \bullet + x = x \qquad x + \bullet = x$$

In short, adding two instances of the same thing produces a dot, whereas adding a mixed pair produces a cross. (Computer techies will recognize Tunny addition as Boolean XOR.)

The Tunny machine's internal mechanisms produced a stream of letters that Bletchley Park called the 'key-stream', or simply the 'key'. Each letter of the ciphertext was produced by adding a letter from the key-stream to a letter of the plaintext. For example, if the first letter of the plaintext was H, and the first letter of the key-stream was W, then the first letter of the ciphertext would be the result of adding W to H. Under the rules of dot-and-cross addition, adding W (**xx••x**) to H (**••x•x**) produces U (**xxx••**):

H		W		U
•	+	x	=	x
•	+	x	=	x
x	+	•	=	x
•	+	•	=	•
x	+	x	=	•

The German engineers selected these particular rules for dot-and-cross addition because they wanted to arrange things so that adding one letter to another and then *adding it again a second time* leaves you where you started. In symbols, $(y + x) + x = y$, for every pair of keyboard characters x and y. For example, adding W to H produces U (as we have just seen), and then adding W to U leads back to H:

U		W		H
x	+	x	=	•
x	+	x	=	•
x	+	•	=	x
•	+	•	=	•
•	+	x	=	x

This explains how the receiver's Tunny decrypted the ciphertext. The ciphertext was the result of adding key to the plaintext, so if the receiver's machine added exactly the same letters of key to the ciphertext, the encryption was wiped away and the plaintext was exposed.

For example, suppose that the plaintext was the single word HITLER. The stream of key added to the plaintext by the sender's Tunny might be WZTI/N. In the sender's machine, these characters were added serially to the letters in HITLER:

H + W   I + Z   T + T   L + I   E + /   R + N.

As you can check from Fig. 14.2, this produces UQ/HEI, which was then transmitted over the radio link. The receiver's Tunny added the same letters of key to the encrypted message:

U + W   Q + Z   / + T   H + I   E + /   I + N.

This uncovers the letters HITLER.

The twelve wheels

The Tunny machine produced the key-stream by adding together two other letter streams, called by Bletchley Park the 'psi-stream' and the 'chi-stream', from the Greek letters psi (ψ) and chi (χ). The twelve wheels produced the psi-stream and the chi-stream; in fact, the wheels were divided into three groups, five 'psi-wheels', five 'chi-wheels', and two 'motor wheels' (Fig. 14.3).

Each wheel had different numbers of cams (sometimes called 'pins') arranged evenly around its circumference (the numbers varying from twenty-three to sixty-one). The function of the cam was to push a switch as it passed it, so that a stream of electrical pulses was generated as

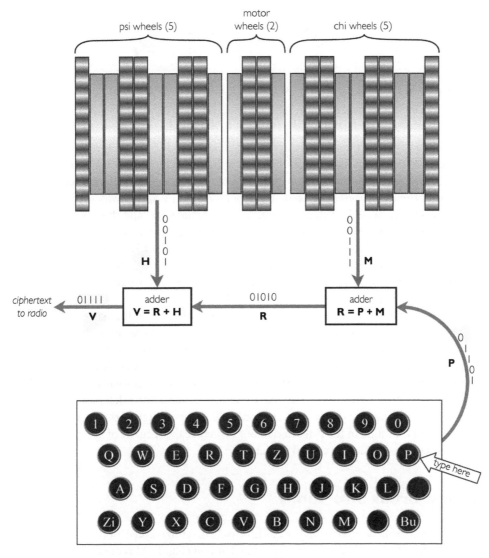

Figure 14.3 How the Tunny machine encrypts a letter.

the wheel rotated. The operator could adjust the cams, sliding any that he selected sideways, so that the selected cams became inoperative and no longer pushed the switch when they passed it (Fig. 14.4). As the wheel turns, it now no longer produces a uniform stream of pulses, but a pattern of pulses and non-pulses (crosses and dots). Once the operator had adjusted the cams in this way, their arrangement around the wheel, with some operative and some inoperative, was called the 'wheel pattern'.

It was the pattern of the cams around the wheels that produced the chi-stream and the psi-stream. Whenever a key was pressed at the keyboard (or a letter read in from the tape in 'auto' mode), it caused the five chi-wheels to turn in unison, just far enough for one cam on each wheel to pass its switch. Depending on whether or not that cam was operative, a cross or a dot was produced.

Operative   Inoperative

Figure 14.4 A wheel cam in the operative (left) and inoperative position.

Suppose, for example, that the cam at the first chi-wheel's switch produced a dot (no pulse) and the cam on the second likewise produced no pulse (a dot) at its switch, but the cams on the third and fourth both produced a cross (pulse), and the cam on the fifth produced a dot: then the letter that the chi-wheels produced at this point in their rotation was ••xx• (N). The five psi-wheels also contributed a letter (or other keyboard character) and this was in effect added to N to produce a character of the key-stream. Under the rules of Tunny addition, it makes no difference whether the letter produced by the chi-wheels and the letter produced by the psi-wheels are regarded as being added *to each other*, with the resulting letter then being added to the plaintext letter, or whether the letters produced by the wheels are regarded as being added *successively* to the plaintext letter as shown in Fig. 14.3.

There was an important complication in the motion of some of the wheels. Although the chi-wheels always moved round by one cam *every* time a key was pressed at the keyboard (or a letter arrived from the tape, or from the radio receiver), the psi-wheels moved irregularly. The psi-wheels might all move round one cam in unison with the chi-wheels, or they might all stand still, missing an opportunity to move. This irregular motion of the psi-wheels was described as 'staggering' at Bletchley Park. The motions of the two motor wheels determined whether the psi-wheels moved with the chi-wheels, or stood still. Tunny's designers presumably believed that this irregular motion of the psi-wheels would enhance the security of the machine—but in fact it turned out to be Tunny's crucial weakness.

The first break into Tunny

When messages from the unknown new machine were first intercepted, nothing of their content could be read. But in August 1941 Colonel John Tiltman (Fig. 14.5) scored a tremendous success, managing to break a message about 4000 characters long. One of codebreaking's legends, Tiltman had made a series of major breakthroughs against Japanese military ciphers before tackling Tunny. He had very little to go on. Tunny was believed to encipher teleprinter messages. It also seemed likely that Tunny was an 'additive' cipher machine. An additive machine adds key to the plaintext to form the ciphertext, as already described. Symbolically, $P + K = Z$, where P is the plaintext, K is the key, and Z is the ciphertext.

Sometimes the sending Tunny operator would foolishly use the same setting for two messages: at Bletchley Park this was called a *depth*. Depths were often the result of something going wrong during the encryption and transmission of a message—radio interference perhaps, or a jammed or torn paper tape. So the sending operator would start again from the beginning of the message. If, instead of selecting new starting positions for the wheels, he stupidly used the same ones, a depth resulted. However, if the message was repeated identically on the second transmission, the depth would be of no help to the codebreakers—they simply ended up with two copies of the same ciphertext. But if the sending operator introduced typing errors during the second attempt, or abbreviations, or other variations, then the depth would consist of two not-quite-identical messages, both encrypted by means of exactly the same key—a codebreaker's dream.

It was such a depth that Tiltman decrypted in the late summer of 1941, giving the codebreakers their first entry into Tunny. On the hypothesis that Tunny was additive, he added the two intercepted ciphertexts—call them Z_1 and Z_2. If Tunny were indeed an additive machine, this would have the effect of cancelling out the key, and would produce a sequence of approximately 4000 letters that consisted of the two plaintexts, P_1 and P_2, added together character by character. This is because

$$Z_1 + Z_2 = (P_1 + K) + (P_2 + K) = (P_1 + P_2) + (K + K) = P_1 + P_2.$$

Tiltman managed to prise the two separate plaintexts out of the sequence $P_1 + P_2$. It took him 10 days. He had to guess at words in each message, and Tiltman was a very good guesser. Each time he guessed a word from one message, he added it to the characters at the appropriate place in the $P_1 + P_2$ sequence, and if the guess were correct an intelligible fragment of the second message would pop out. For example, adding the probable word '*geheim*' (secret) at a particular place in the $P_1 + P_2$ sequence revealed the plausible fragment 'eratta'.[3] This short break could then be extended to the left and right. More letters of the second message were obtained by guessing that 'eratta' is part of '*militaerattache*' (military attaché), and if these letters were added to their counterparts in the $P_1 + P_2$ sequence, further letters of the first message were revealed—and so on.

Eventually, Tiltman achieved enough of these local breaks to realize that long stretches of each message were the same, and so he was able to decrypt the whole thing. By adding one of the resulting plaintexts to its ciphertext, he was then able to extract the 4000 or so characters of key that had been used to encrypt the message (since $P + Z = K$).

Tutte and Turing join the attack

However, breaking one message was a far cry from knowing how the Tunny machine worked. The codebreakers still had no idea how the machine produced the key that Tiltman had extracted. In the case of Enigma, Bletchley Park knew the internal workings of the machine before the war even began, thanks to Poland's codebreakers, and then once the fighting started a number of Enigmas were captured on land and sea. Tunny, on the other hand, was an unknown quantity: no machine had been captured, and nor would one be until the war was almost over.

Figure 14.5 The great codebreakers John Tiltman (left) and Bill Tutte.

Left-hand image reproduced with permission of Barbara Eachus and Government Communications Headquarters, Cheltenham. Right-hand image with permission of Jack Copeland.

Nevertheless, in January 1942, a young member of Bletchley Park's Research Section, William Tutte (Fig. 14.5), was able to deduce the internal workings of the Tunny machine simply by studying the key that Tiltman had retrieved. It was one of the most astonishing pieces of cryptanalysis of the war.

Tutte deduced that the machine produced key by means of adding two separate streams of letters. He inferred further that each of these two streams was produced by a different set of five wheels (five because there are five bits in the teleprint code for each character), and he made up the names 'chi-wheels' and 'psi-wheels'. The chi-wheels, he deduced, moved regularly, whereas the psi-wheels moved in an irregular way under the control of two further wheels, to which he gave the name 'motor wheels'. Tutte's secret deductions would save untold lives.

There are two separate steps in breaking a new cipher system. First comes what is called 'breaking the machine': the codebreakers must discover enough of the design of the cipher machine, and this Tutte had now done. Second, a fast and reliable method must be found for breaking the intercepted messages on a day-to-day basis. This method must work quickly enough to allow the codebreakers to decipher the messages in a timely fashion, before the intelligence they contain goes out of date.

Turing devised the first method used against the daily Tunny message traffic. It was known simply as 'Turingery'.[4] Turingery depended on a technique that Turing invented, a process of 'sideways' addition called 'delta-ing' (after the Greek letter delta, Δ). To delta the four characters

ABCD the codebreaker used the rules of Tunny addition to add A to B, B to C, and C to D. One might think that adding the letters of an encrypted message together in this way would scramble the message still further, but Turing showed that delta-ing in fact *reveals* information that is otherwise hidden. Delta-ing is a crypto-technique that remains applicable today. It was the basis not just of Turingery but of every Tunny-breaking algorithm used in Bletchley Park's soon-to-be-developed electronic computers. The entire computer-based attack on Tunny flowed from this basic insight of Turing's.

Turingery itself, though, was a paper-and-pencil method that was carried out by hand. From July 1942 until June 1943 Turingery was the codebreakers' only weapon against Tunny, and during this period they broke an astonishing total of approximately one-and-a-half million letters of ciphertext. But no matter how fast the hand-breakers worked, they could not keep pace with the increasing volume of Tunny messages. The codebreakers needed a machine to help them—and because of Tunny's complexity, a machine much faster than Turing's bombe was required.

Codebreakers using Turingery had to make use of their *insight*—Turingery depended on what you 'felt in your bones', Tutte said. Deprecatingly, he described Turingery as 'more artistic than mathematical'.[5] Tutte desired a Tunny-breaking method that required no use of what you felt in your bones—the sort of method that a machine could carry out. He found it in November 1942, using an ingenious extension of delta-ing; it was his second great contribution to the attack on Tunny.[6] But there was a snag. Tutte's method demanded a huge amount of calculation. If the method were to be carried out by hand, it might take as long as several hundred years to decrypt one sizeable message.[7]

Machine against machine

At this point Turing's teacher from Cambridge, Max Newman, entered the battle against Tunny (Fig. 14.6). Back in 1935, in a lecture on mathematical logic, Newman had launched Turing on the research that led to the universal Turing machine, and in the following year he had assisted Turing with preparing his groundbreaking paper 'On computable numbers' for publication (see Chapter 40).[8] At the end of August 1942 Newman left Cambridge for Bletchley Park to join the codebreakers of the Research Section in their struggle against Tunny.[9]

When Tutte explained his theoretical Tunny-breaking method to Newman, Newman immediately suggested using electronic counters to do the necessary calculations at high speed. It was a 'Eureka' moment. Newman was aware that, before the war, electronic counters had been used to count radioactive emissions in Cambridge's Cavendish Laboratory, and in a flash of inspiration he saw that this technology could be applied to the very different problem of breaking Tunny messages. With Turing's assistance, Newman sold his idea to the head of codebreaking operations at Bletchley Park, Edward 'Jumbo' Travis, and was put in charge of building a suitable machine (Fig. 14.7).

The prototype of Newman's machine, installed in 1943, was soon dubbed 'Heath Robinson', after the famous cartoonist who drew absurd devices. Smoke rose from Heath Robinson's innards the first time it was switched on.[10] Newman's wonderful contraption proved the feasibility of carrying out Tutte's method by machine, but Heath Robinson was slow, and prone to inaccuracy and breakdowns.

Figure 14.6 Max Newman, codebreaker and computer pioneer.

Reproduced with permission of William Newman.

Figure 14.7 Robinson, precursor to Colossus. This machine, eventually called 'Old Robinson', replaced the original 'Heath Robinson' (the two were similar in appearance).

From 'General Report on Tunny', National Archive ref. HW 25/5 (Vol. 2). Crown copyright and reproduced with permission of the National Archives Image Library, Kew.

A Colossal computer

Telephone engineer Tommy Flowers (Fig. 6.1) knew that he could build something better than Heath Robinson. In the years before the war he had begun to pioneer large-scale digital electronics. Few outside his small group at the Post Office Research Station at Dollis Hill in North London knew that electronic valves could be used reliably in large numbers.

According to conventional wisdom, valves were just too flaky to be used en masse: each one contained a hot glowing filament, and the delicacy of this filament meant that valves were prone to sudden death. The more valves there were in the equipment, the greater the chance that one or two would fizzle out in the middle of a job. Plagued by continually blowing valves, a big installation would be completely impractical, or so most engineers believed: it was a belief based on experience with radio receivers and the like, which were switched on and off frequently. But Flowers had discovered that so long as valves were left running continuously, they were very reliable. He also discovered ways of boosting the reliability even further, such as using lower than normal electric currents. As early as 1934, he had successfully wired together an experimental installation containing 3000 to 4000 valves. The equipment was for controlling connections between telephone exchanges by means of tones, like today's touch-tones.

Digital electronics was a little-known field in those days. In 1977 Flowers told me that, at the outbreak of war with Germany, he was possibly the only person in Britain who realized that valves could be used for large-scale high-speed digital computing.[11] So he turned out to be the right man in the right place when he was sent to assist the codebreakers at Bletchley Park. He realized that Heath Robinson could not give them the accuracy and speed they needed. Heath Robinson contained no more than a few dozen valves—a puny number as far as Flowers was concerned. He knew that by using valves in large quantities he could store information that Heath Robinson could store only on wear-prone paper tape. Flowers proposed an electronic monster containing approximately 2000 valves.

Newman took advice on Flowers' proposal, consulting other specialists in electronics, and was told that such a large installation of valves would never work reliably. In consequence, Bletchley Park declined to support it. But in his laboratory at Dollis Hill Flowers quietly got on with building the all-electronic machine that he knew the codebreakers needed. He and his team of engineers and wiremen worked day and night for ten months to build Colossus, as the giant computer was called—they worked until their 'eyes dropped out', Flowers recalled. All this was 'without the concurrence of BP' (Bletchley Park), he said, and was done 'in the face of scepticism'. 'BP weren't interested until they saw it [Colossus] working', he recalled with wry amusement. Fortunately his boss, Gordon Radley, had greater faith in Flowers and his ideas than Bletchley Park did, and gave him whatever he needed.

On 18 January 1944, Flowers' lads took Colossus to Bletchley Park on the back of a lorry. It was a significant moment in the history of computing—the delivery of the world's first large-scale electronic computer. The computer's arrival caused quite a stir. 'I don't think they understood very clearly what I was proposing until they actually had the machine', Flowers said. 'They just couldn't believe it!' The codebreakers were astonished by the speed of Colossus—and also by the fact that, unlike Heath Robinson, it would always produce the same result if set the same problem again.

The name 'Colossus' was apt—the computer weighed about a ton (Fig. 14.8). Flowers' gigantic computer began active service against the German messages on 5 February 1944. As Flowers noted laconically in his diary, 'Colossus did its first job. Car broke down on way home'.

Figure 14.8 Colossus. Notice the early computer printer in the foreground. The long punched tape containing the message to be analysed is mounted on aluminium wheels.

From 'General Report on Tunny', National Archive ref. HW 25/5 (Vol. 2). Crown copyright and reproduced with permission of the National Archives Image Library, Kew.

Colossus used Tutte's method to cleanse the ciphertext of the part of the key that the chi-wheels had contributed. The result of stripping away the contribution of the chi-wheels was called the 'de-chi' of the message: this consisted of the plaintext obscured only by the contribution of the psi-wheels. De-chis could almost always be broken by hand, because of distinctive patterns in the component of the key contributed by the psi-wheels. These patterns were due to the irregular movement of the psi wheels, the Tunny machine's greatest weakness.

De-chis were broken in a section called simply the 'Testery', after its head, Major Ralph Tester. Captain Jerry Roberts, one of the Testery's codebreakers, explains this part of the process in Chapter 16.

The first computer centre

By the end of the war Newman had nine Colossi working round the clock in the Newmanry, and another stood in the factory almost ready for delivery. The nine computers were housed in two vast steel-framed buildings. It was the world's first electronic computing facility, with job-queues, teams of operators working in shifts, specialized tape-punching crews, and engineers on hand day and night to keep the machinery running smoothly. Not until the 1960s was anything like it seen again, when the first modern computing centres started to develop around large mainframe computers.

Newman's incredible factory of giant computers did not outlast the war, however. With Germany's capitulation, in May 1945, Tunny went off the air. The Newmanry and Testery were out of work. Germany's teleprinter cipher machines would define a new era in encryption technology—the Tunny machine, and its fundamental encryption principle of $Z = P + K$, would continue to be used around the globe for many years to come. But at war's end, the British government saw no need to retain all the Colossi. In a hush-hush operation, two were transferred to Eastcote, the codebreakers' new headquarters in suburban London. A gigantic grabber lifted the two computers aloft and they disappeared from the Newmanry through a hole in the roof.[12] The rest were dismantled. A lorry loaded with some of the parts drove in secrecy to Manchester (see Chapter 20). 'Suddenly the Colossi were gone', Dorothy Du Boisson recounted, 'broken up on the orders of Churchill, we were told at the time'.[13] (Du Boisson is the Colossus operator with her back to the camera in Fig. 15.2.) 'All that was left were the deep holes in the floor where the machines had stood', she remembered.

So most of Britain's secret stock of electronic computers, the most sophisticated machines in Europe, were knocked to pieces. That this brutal reversal of scientific progress was unknown to the outside world hardly lessens the magnitude of the blow. The Colossi might have become part of public science. Turing, Newman, and Flowers would quickly have adapted them for new applications, and they could have become the heart of a scientific research facility. With eight massive electronic computers in the public arena from mid-1945, the story of modern computing would have begun very differently. Who can say what changes this different start would have brought? With clones of the Bletchley computing centre popping into existence elsewhere, the Internet—and even the personal computer—might have been developed a decade or more earlier. Even before the new millennium began, social networking might have changed the political map of the world.

Keeping the secret

The secret of Tunny and Colossus was a very long time coming out—much longer than with Enigma. Ex-staff of the Newmanry and the Testery scrupulously followed their orders to reveal nothing of what they knew. Catherine Caughey, one of Colossus's operators, even feared going to the dentist, in case she talked while under the anaesthetic.[14]

Caughey's greatest regret was that she could never tell her husband about her extraordinary wartime job, operating the first large-scale electronic computer. Jerry Roberts, too, regretted that his parents died knowing nothing of his war work in the Testery—work of such importance that, in different circumstances, he might reasonably have expected an honour from the British Crown. Helen Currie, who produced the complete German plaintext after Roberts or one of his colleagues had broken 15–30 consecutive letters of a de-chi, spoke of the immense burden of being unable to share her memories with her family. During the 'years of silence', she said, her wartime experiences 'took on a dream-like quality, almost as if I had imagined them'.[15]

Turing's mother, Sara, wrote ruefully of the 'enforced silence concerning his work'.[16] 'No hint was ever given of the nature of his secret work', she said, complaining that the necessary secrecy 'ruined' their communication. Max Newman's son William, himself a leading light in the computer graphics industry, said that his father spoke to him only obliquely about his war work and died 'having told little'.[17]

At the time the Colossi were dismantled, news was leaking out about another electronic computer that was being built in America. The ENIAC (Electronic Numerical Integrator and Computer) was the brainchild of John Mauchly and Presper Eckert, two engineering visionaries who were commissioned by the United States Army to build a high-speed calculator.[18] The Army wanted this for the mammoth job of preparing complex tables that gunners needed to aim artillery. Construction got under way in Philadelphia, at the University of Pennsylvania, and ENIAC eventually went into operation at the end of 1945, almost two years after the first Colossus. Flowers himself saw ENIAC shortly after the end of the war. Colossus, with its elaborate facilities for logical operations, was he said 'much more of a computer than ENIAC'.[19]

While Colossus remained cloaked in secrecy, ENIAC became public knowledge in 1946, and was trumpeted as the first electronic computer. John von Neumann—wholly unaware of Colossus—told the world, in his prominent scientific writings and charismatic public addresses, that ENIAC was 'the first electronic computing machine'.[20] Secrecy about Flowers' achievement still bedevils the history of computing even today. The myth that ENIAC was first became set in stone soon after the war, and for the rest of the twentieth century book after book—not to mention magazines and newspapers—told readers that ENIAC was the first electronic computer. An influential textbook for computer science students gave this woefully inaccurate historical summary: 'The early story has often been told, starting with Babbage [and] up to the birth of electronic machines with ENIAC'.[21]

Flowers was simply left out of the picture. It was monstrously unfair, although inevitable given the secrecy. In later life, Flowers became tinged with bitterness:

When after the war ended I was told that the secret of Colossus was to be kept indefinitely I was naturally disappointed . . . I was in no doubt, once it was a proven success, that Colossus was an historic breakthrough, and that publication would have made my name in scientific and engineering circles—a conviction confirmed by the reception accorded to ENIAC . . . I had to endure all the acclaim given to that enterprise without being able to disclose that I had anticipated it.

Worse still, his views on electronic engineering carried little weight with his colleagues, who had no idea what he had achieved, and he gained a reputation for 'pretentiousness', he said:

Matters would have been different, I am sure, both for myself and for British industry, if Colossus had been revealed even ten years after the war ended.

One day Flowers called Ken Myers into his Dollis Hill office.[22] Myers had worked with him on Colossus from the very beginning, and had been there at Bletchley Park on that historic day in 1944 when the first of the gigantic computers was delivered. Flowers gestured towards his office safe. That was where he kept all his papers and records about Colossus, including his blueprints. He told Myers in his quiet way that instructions had come from on high to destroy all the documents without trace. The two men carried armloads of paper to a ground floor workshop where there was a coal stove for heating. Myers told me, 'I stood there with him and we burned it all'.

Next—the universal Turing machine in hardware

Colossus was far from being a universal machine. As Jack Good explained to me, even long multiplication—unnecessary for Tunny-breaking—was found to be just beyond the computer's scope. Nor did Colossus incorporate Turing's all-important stored-program concept. The computer was programmed by hand, by means of large panels of plugs and switches. The female operators used these to (quite literally) re-wire parts of the computer each time it was required to follow a new set of instructions. This laborious method of programming seems unbearably primitive from today's perspective when we take Turing's glorious stored-program world for granted. ENIAC, too, was programmed by re-routing cables and setting switches. It might take ENIAC's operators as long as three weeks to set up and debug a program.[23]

The rest of the world did not know about Colossus, but its impact on Turing and Newman was colossal indeed: Colossus was the connection between Turing's groundbreaking article of 1936 and his and Newman's post-war projects to build a universal Turing machine in hardware. Turing's opportunity to realize his dream of a stored-program all-purpose electronic computer came in October 1945 when he joined the National Physical Laboratory, a government research establishment situated in the London suburb of Teddington. Newman's opportunity came almost simultaneously, with his appointment in September 1945 to the Fielden Chair of Mathematics at the University of Manchester.

Turing's ambitious Automatic Computing Engine, or ACE, is described in Chapters 21 and 22, and the more modest Manchester 'Baby' computer in Chapter 20. It would have been fitting if Turing and the National Physical Laboratory had won the race to build the first universal Turing machine in hardware. But life is not always fair: as Chapter 20 relates, Newman's group at Manchester romped home.

CHAPTER 15

We were the world's first computer operators

ELEANOR IRELAND

I n 1944 I worked at Bletchley Park and lived at nearby Woburn Abbey. My job was to assist the codebreakers by operating one of the Colossus machines. In this chapter I describe how this came to be and what it was like to live and work at Bletchley Park during the last months of the war.

In 1944 I was working in London, in a philatelist's business. One of my friends joined the Women's Royal Naval Service (WRNS) as a motor transport driver, and I decided to join too. In great trepidation I went to Queen Anne's Gate and volunteered. I was interviewed immediately and very soon was called to a medical. Not long after, I received a letter telling me to report on 2 August 1944 to a WRNS establishment at Tullichewan Castle near Glasgow. I found out much later that it would have been more usual to report to Mill Hill in London, but there had been a spate of bombing and the powers that be did not wish to take any chances with the new intake, so they sent us up to Scotland.

Strangely enough, in the week before I was due to set off on this adventure I met a school friend who was also joining the WRNS and had been asked to report to Tullichewan on the same day. I was pleased to discover that I had a companion to go with. As it turned out, we were to stay together until we were demobbed at the end of December 1946.

We travelled to Glasgow and then out to a small station on the edge of Loch Lomond, where we were picked up and taken to Tullichewan WRNS reception camp, a requisitioned castle standing in a large hillside estate. At the bottom of the hill was the Regulating Office, together with a large number of Nissen huts—the sleeping quarters, a mess, the stores hut, and so forth. Opposite the huts was an enormous parade ground, while at the top of the hill was the castle, used by the officers, and another parade ground with the naval flag.

Every day a bell sounded at 5 a.m. to get us up, after which we were required to do various menial tasks, such as cleaning out the huts, potato peeling, and blancoing the steps of the castle.

Figure 15.1 Eleanor Ireland.

The Turing Archive for the History of Computing

Perhaps this was a test: some people left at this stage. Those of us who remained were kitted out with temporary garb and then, eventually, with uniforms (Fig. 15.1). We were each given a service number. This one never forgets—it seems to be engraved on the soul.

We did hours of squad drill to smarten us up, and were lectured on the Senior Service and its history since the time of Samuel Pepys. There were interviews to find out which category we would like to go into: I cannot remember what I said. We found out later that this was all a terrific blind—they had already decided where we were going. Just before we left Tullichewan we were told we were being posted to a station in the country 50 miles from London. 'Very depressing', we thought.

We travelled down on the night train from Glasgow. It was absolutely packed with service personnel and we arrived at Bletchley completely exhausted. From the station we were taken by transport to Bletchley Park, only a few minutes away. There was a high security fence. The transport stopped at an entrance manned by guards, and we were taken a few at a time into a concrete building where we were issued with our security passes. Without our pass we would be unable to enter the compound, and we were told to protect it with our lives.

Before us was a large Victorian house with a sward of grass in front of it. We learned that it was called the Mansion. A Wren officer escorted us into a low building adjacent to the Mansion, where she gave us a very intimidating lecture about the extreme secrecy of Bletchley Park and every aspect of the work being done there. We were never to divulge any information about our work or about the place where we worked, on penalty of imprisonment or worse. We were never to discuss our work when 'outside', not even with those with whom we worked. We were not to ask anyone outside our own unit what they did. We were not to keep diaries. Our category, we were told, was PV Special Duties X: 'PV' stood for 'Pembroke V' (pronounced 'Pembroke

Five')—our notional ship. We would wear no category badges, and if anyone asked us what we did we were to say we were secretaries. We were told we would never be posted anywhere else, because the work was too secret for us to be released. Everyone had to sign the Official Secrets Act. So effective was this lecture that each time I left the building where I worked, I just dropped a shutter and blanked it all out. There were Foreign Office, naval, army, and air force personnel at the Park, but we never knew what was done in other sections.

We returned to the transport bemused and subdued by all this secrecy, still with no idea where we were going. The transport left the sleepy town of Bletchley and drove nine miles into the country, through woodland, until we came to the village of Woburn. We turned by a church and drove through a very imposing set of gates into beautiful parkland. In front of us stood the magnificent stately home of Woburn Abbey. This was HMS *Pembroke V*.

The transport stopped at the main entrance, where a WRNS petty officer met us. The officer took us into an enormous hall that had been made into a Regulatory Office. There we were issued with station passes for the Abbey and told that every time we went out our passes must be handed in, and picked up again when we came back. The only exception to this was that when we came back at midnight from the evening 'watch' (the naval term for 'shift'), we would find them in our own labelled post boxes—a huge rack of cubby holes on the opposite side of the hall.

After climbing up the grand staircase to the second floor we were allocated temporary accommodation. All the off-duty Wrens were very helpful and showed us everything that we would need. I can still remember being impressed by all the double green baize doors. The rooms were very grand—formerly they were bedrooms used by the Duke of Bedford and his family. The toilets were of Delft china and were raised two steps above the floor. The walls were lined with red silk. The bathrooms were impressively large, with the bath on a 'throne' two steps up, encased in mahogany and very gloomy. One of the first things I heard was that a nun haunted our corridor. A girl called Dawn told this to me with great relish and assured me that her friend had actually seen the ghostly nun.

After we finished our fortnight's initiation at Bletchley we were allocated to watches, A, B, C, or D: fortunately, I was put on C watch together with four friends I had made. At that stage we were moved out of our temporary accommodation and up into a room under the eaves at the front of the house, the servants' quarters, where eight of us shared a 'cabin' called 'Swordfish 50'. The cabin was spartan, just four bunk beds, four chests of drawers, and a built-in cupboard where we kept our cases, food, and so on—until we discovered the resident mice.

Being up under the eaves, the room was very hot in summer and cold in winter. With eight of us packed into it we had to leave the windows open, and when snow drifted in onto the window sill it would stay there for about three weeks: Bedfordshire is supposed to be the coldest county in England.

Our sitting room, or fo'c'sle as it was called in the WRNS, was less cramped, being the largest and grandest room in the house. The walls were completely boarded up. Standing in the room were three electrically heated metal tubes, each four feet long. When we were off duty we sat huddled around these tubes with our greatcoats on. Later on, a pleasant square room next door to our cabin was converted for our use. There was a marble fireplace and, joy of joys, a fire was lit for us in the winter. We were provided with sofas with pretty cretonne covers: this was the nearest thing to comfort that I came across during my career in the WRNS.

Our mess was the original kitchen, situated at the further end of a ground-floor passage paved with stone flags that were worn down with age. We ate off scrubbed tables, and we all kept our own mugs in preference to the metal mugs we were issued with. The food was just about edible. At Bletchley Park we ate in the Mansion with the Foreign Office personnel at first, and the food was good. When there became too many of us, however, they built some huts near our block and we ate there. The food was grim. I found it difficult to eat on night watch and never became used to eating at such hours. Instead of going to the canteen, we would often walk out of the main gate and down a side alley to Bletchley Station. At the end of one of the platforms was a NAAFI hut and we would eat buns and drink a decent cup of tea before walking back. Better than cold liver and prunes: I did not eat another prune for over 30 years.

Most of the buildings at Woburn Abbey had been commandeered by the WRNS. The Foreign Office had the stables at the rear of the buildings—also hush-hush. The Duke of Bedford was fearful that fire might destroy the Abbey, so at night when not on duty we had to take it in turns, two at a time, to patrol the building for two hours. We patrolled in the dark, holding torches. I hated doing this—it was very eerie.

The Duke lived in a house in the grounds and he would come and have a look around every now and then to make sure that everything was all right. All the family pictures and furniture were stored away in another wing of the Abbey. Some of my friends had a lovely cabin on the ground floor, which I recognized when I went back many years later: it is now a dining room, hung with yellow silk and a magnificent collection of Canaletto paintings. The park was magnificent too, with seven lakes and several herds of rare deer. I loved the view from our cabin window.

There were very few officers at the Abbey, and none of them (not even the officer in charge) had any idea of the work that we did. I remember that when we first went there, if we had a Sunday off watch, we were expected to join the church parade and march two miles to the Woburn village church and back. Later on, the First Officer was warned that we were under terrific pressure at work and was told not to stress us unnecessarily. After that I do not remember going to church. Discipline became more relaxed.

The day after we arrived at the Abbey, we were driven into Bletchley Park in an old army transport bus with a soldier at the wheel. The bus stopped at the main gate and we all got out and showed our passes. We were told to wait outside the Mansion. From there we were escorted past a tennis court and past some hideous low concrete buildings—we later learned that they were bomb proof. Journey's end was Block F, another grim concrete building.

Initiation

Max Newman himself met us at the entrance to Block F, introducing himself and welcoming us. We were taken into a long low room with a large blackboard and long tables. Mr Newman took his position in front of the blackboard and we all sat at the tables. He had a very pleasant manner and put us at our ease. He told us that this section had only recently been set up and that we would be working with mathematicians and engineers. He said he had specifically asked for Wrens to staff the section, run the machines, and organize the Registry office.

For a fortnight we went in every day and he lectured us on a new type of binary mathematics that he would write up on the blackboard. We were shown the tapes that were used on the machines we would soon become so familiar with—these tapes were an inch wide and very strong. They had a continuous row of small sprocket holes running along the centre of the tape, and these were used to drive it through the machinery. We had to learn the alphabet punched on the tapes and become adept at reading them: this alphabet was the same as the GPO (General Post Office) teleprinter alphabet. On either side of each sprocket hole there was space for two larger holes to be punched above and three below: the letter 'A', for instance, was two holes above the sprocket hole and nothing below.

We were taken round the section and shown what everyone was doing. We saw the room where the messages came in on teleprinter tape on two separate machines. Most of the messages came from Knockholt in Kent and also Kedleston in Derbyshire, I learned later. We were shown into a long room where tapes were cut and joined, and tapes that had split on the machines were repaired. Then we went to the Registry itself (or 'Ops', as it was called), where all tapes were registered and tabulated and put into a series of cubbyholes. We were also shown the two Colossus computing machines the section had at that time. I was overawed by them, a mass of switches, valves, and whirring tape: I thought they were incredible—quite fantastic. It is now known that the computers were built to take over codebreaking work originally done by means of a hand method invented by Alan Turing himself, and it is also known that although Turing had no hand in designing Colossus, some of the ideas involved in his hand method were fundamental to the algorithms used by Colossus. But, of course, we humble computer operators were told nothing about any of this at the time.

At the end of the fortnight we were tested on our knowledge and, depending on how well we performed, were selected for various tasks—administration, dealing with the tapes as they came in, and so forth. I was delighted to be chosen to operate Colossus, which I considered the plum job (Fig. 15.2). A Wren named Jean Bradridge taught me how to operate the machine. She explained what all the switches were for, and showed me how to peg a wheel pattern on the grid at the back of the machine, using pins that looked like very large and very strong hairpins, copper–nickel plated.

The tape was shut into position in front of the photoelectric cell, which had a small gate for the tape to slide through. Metal wheels supported the tape, and depending on the length of the tape we would use as many wheels as necessary to make the tape completely taut. This

Figure 15.2 Colossus with two operators.

From 'General Report on Tunny', National Archive ref. HW 25/5 (Vol. 2). Crown copyright and reproduced with permission of the National Archives Image Library, Kew.

was a tricky operation, getting the tape to the right tension: it took a little time and had to be done with great care. We were terrified of the tape breaking should the tension be wrong, as breaks meant that valuable time was lost. (I can remember that when I was given a new Wren to instruct, I was worried about leaving her for very long, and would hurry back from my meal-break to make sure nothing awful had happened.) All the 'break-ins' that we put on the machine were timed—they generally took about an hour to run. Every tape was logged on and off in a book—the time that we had received the tape and the time that it was taken off the machine. It was instilled into us that time was of the very essence. We knew that we were working against the clock and that people's lives depended on what we were doing.

We worked in four watches. I was on C watch, working from 9 a.m. to 4 p.m., 4 p.m. to 12 p.m., and 12 p.m. to 9 a.m.—a week of days, a week of evenings, a week of nights, and a week of 'changeovers'. During this fourth week we filled in any gaps in A watch. The changeover week could be very tiring—off at 9 a.m. and on again at 4 p.m., for instance. We had one weekend off every month and an occasional additional weekend. Small buses took us between Woburn and Bletchley Park: these would be lined up on the drive opposite the Mansion. Other destinations included Gayhurst Manor and Wavendon.

A big block was put up to house two more Colossi. I was sent to work on Colossus 3 and my friend Jean Beech was on Colossus 4. These stood in an enormous room, and were about twice as big as Colossus 1. Later, Block H was built to house yet more Colossi.

I operated Colossus 3 alone under the direction of a mathematician codebreaker, or 'crypt-ographer' as they were generally called, who would sit at a long table facing Colossus. Others, such as Jack Good, Donald Michie, and Shaun Wylie, would come in to discuss what was going on and to make suggestions. On the table in front of the cryptographer were sheets of codes and he used a slide rule to make his calculations. He would tell me what he wanted from the machine. I would pin up on the grid, at the back of the machine, whatever pattern he was working on, and would put on the tape that he wished to run. At the front of the Colossus were switches and plugs. We could set switches to make letter counts, and the machine had its own electronic typewriter to record the results. Sometimes I was given a norm, and as each figure came up on the typewriter I did a calculation and wrote down against the figure how much above or below the norm it was. I became very good at mental arithmetic.

What I did not know then, but learned 50 years later, was that Colossus was designed to break the messages sent by means of a machine called 'Tunny', which had been especially ordered by the German High Command to enable them to communicate in complete secrecy. Hitler, Göring, and Goebbels all used the Tunny machine, as did the field marshals and generals. The Germans thought the coded messages sent out by the Tunny machines were completely unbreakable. The Tunny wheel patterns were pegged up at the back of Colossus, and the tape contained an intercepted message. The purpose of Colossus was to find the positions of the codewheels at the beginning of a message.

If anything went wrong with Colossus we would contact the maintenance team. The officer in charge of the team was an extraordinarily clever man, Harry Fensom, while another brilliant engineer working with us was Ken Myers, who after the war worked on the coordination of the traffic lights in London. The magnificent work done by these engineers has had little or no recognition.

At the end of the war, we helped to break up most of the Colossi: this was a sad job. Then we were made to sign the Official Secrets Act again. We all remained completely silent—until now.

CHAPTER 16

The Testery: breaking Hitler's most secret code

JERRY ROBERTS[*]

I joined the Intelligence Corps in autumn 1941. At that time few people were allowed into the Mansion at Bletchley Park, the nerve centre. I was fortunate enough to work in the Mansion and was one of the four founder members of the Testery, set up in October 1941 to break 'Double Playfair' cipher messages. Then in July 1942 the Testery was switched to breaking Tunny traffic.[1]

Alan Turing

Before reminiscing about the breaking of the Tunny code I should like to recall Alan Turing himself. If it had not been for him everything would have been very different, and I am eternally grateful to him that I did not have to bring up my children under the Nazis. We would have entered a dark age of many years—once the Nazis had got you down, they did not let up.

Here is just one example of what life was like under the Nazis. After the war I met a brave Belgian lady called Madame Jeanty. Her family was one of those who kept a safe house for Allied airmen, shot down over Europe and trying to make their way back to Britain to fly again. Helen Jeanty and her husband had a hidey-hole in their house, and had an airman in there one day when the Gestapo came calling, at the usual time of 6 a.m. They searched the house up and down but did not find him, and went away. Everybody was delighted and relieved—claps on the back or whatever the Belgians do. But the Gestapo came back again to find this celebration in progress. Her husband was arrested and taken away and she never saw him again. That sort of thing would have happened time and time again here in Britain if the Nazis had managed to invade.

One reason Britain did not fall to the Nazis is that in 1941 Turing broke U-boat Enigma. The decisive effect he had on the Battle of the Atlantic can be seen from the tonnages sunk. The tonnages lost to sinkings dropped by 77% after Turing broke into U-boat Enigma in June 1941,

*This chapter was revised by Jack Copeland following Jerry Roberts' death in March 2014 at the age of 93.

from approximately 282,000 tonnes of shipping lost per month during the early part of 1941, to 64,000 tonnes per month by November. If Turing had not managed that, it is almost certain that Britain would have been starved into defeat.

I had a glimpse of the Battle of the Atlantic when I met a friend of a neighbour not long ago. He had been in the SAS during the war, a tough unit, and he made a crossing of the Atlantic early in 1941. As he told me about it, his face darkened:

The worst days of my life. We left port with 21 ships and arrived in Britain with only 11. The U-boats had downed 10 ships and left the survivors struggling in the water.

I used to see Turing from time to time. I have a strong mental image of him walking along the corridor in one of Bletchley Park's huts. With his gaze turned downwards, he was a shy and diffident man, flicking the wall with his fingers as he walked, dressed in a mid-dark brown sports jacket and baggy grey trousers. This is not your Achilles figure; he was not a warrior king. But at that juncture, he was the most influential man in Europe bar none, and we owe our freedom to him.

What was Tunny?

Tunny (or 'Fish', or as the Germans called it, the Lorenz SZ40) was a new, specially designed machine, ordered by Hitler and used to encipher top-level messages between the German Army headquarters in Berlin and the top generals, field marshals, and commanders-in-chief of the huge forces on all the battlefronts (see Chapter 14). Tunny began with one link but spread rapidly throughout Europe, since the German Army top brass valued it highly.

In 1940 Hitler's mind was turning to his plans to dominate the whole of Europe. He knew that there was going to be extensive fighting on land and that his generals were going to be pretty busy. Up to this point, the German Army, Navy, and Air Force had used Enigma for enciphering their messages, but Hitler was not satisfied with that: communications between his generals and his headquarters needed something more powerful, more secure. So the Lorenz company, a German electrical outfit, gave him a superb machine—Tunny. With twelve wheels it was very advanced, highly complex, faster, and much more secure than the three- or four-wheeled Enigma.

If you or I had to encipher a message we would probably use simple letter substitution, and that would give us *one* level or layer of encryption. Tunny had not one, not two, but three levels of encryption. Even two was a cryptologist's nightmare: three was out of this world!

Why was Tunny so important?

Tunny carried only the highest grade of intelligence. Top people signed the messages and we used to see their names cropping up time and again in the messages that we broke: Keitel (head of the whole German Army), von Jodl (Chief-of-Staff of the Army, the CEO), Rommel (Commander-in-Chief, Normandy), von Runstedt (the Western Front), Kesselring (Italy), von Manstein (the Russian Front)—and, of course, even messages signed by Adolf Hitler himself. That's not a bad start!

Those were the people at one end. People at the other end included the generals and field marshals in charge of the fighting forces on each of the battlefronts, people like von Weichs, von Kluge, and Model, who headed up the three Russian sectors. We covered the five main battlefronts. There were the three Russian fronts: north, central, and south. Then there was the front in the west, and the front in the Italian Peninsula, important because the Germans moved their troops to Sicily when they were routed out of North Africa by Montgomery.

Army Enigma at that stage dealt with divisions and regiments, whereas Tunny dealt with whole armies and army groups. As you can imagine, then, Tunny continuously produced a very high grade of intelligence.

Our best successes

There are many situations in which our ability to break Tunny played a major role, but there are four absolutely outstanding examples. First and foremost there was D-Day, the Allied landings on the beaches of Normandy, in June 1944. In the run-up to D-Day it was important to know whether the Germans were going to keep their large Panzer divisions further north in the Calais–Boulogne region, which was what Hitler wanted, or whether they were going to move them to Normandy, as the generals wanted. The generals were the professionals and they knew that the Allies needed good landing beaches—but Hitler was the boss. In fact, we helped him to make the wrong decision: one of our clever tricks, in this period before D-Day, was to moor objects along the Kent and Essex coasts that looked from the air like a large number of landing craft, poised for an invasion in the Calais–Boulogne region.

Hitler fell for it. Through Tunny decrypts we knew that his assessment of the situation had prevailed, and that Normandy would therefore be more lightly defended than if the generals had had their way. Moreover our decrypts told us the location of pretty much every Panzer division in the area. You can imagine how much help all this gave Eisenhower and Montgomery, when they and their staff were planning the invasion. If, on the other hand, Hitler had made the opposite decision, and moved all those tanks and men down to Normandy, then there would have been a good chance that the D-Day landings would have failed, and then it would have taken at least two years to mount another similar effort—during which time Hitler could well have made Europe impregnable. Not only that, but his scientists were working on their own atom bomb, and had brought together 25% of the uranium needed. It was an absolutely critical time.

The second outstanding example of how Tunny decrypts turned things our way was the great tank battle near the Russian city of Kursk. The Germans, always short of petroleum, had made an enormous effort to get at the Russian oil fields, but in February 1943 they were badly defeated at Stalingrad. Then, in April 1943, we broke Tunny messages revealing that the Germans were planning another assault, a huge one, at Kursk, southwest of Moscow. We gave the Russians months of advance warning, and they used that time wisely, including putting pressure on their factories to turn out more tanks. Eventually the Battle of Kursk began on 4 July 1943, and the Germans were defeated in the biggest tank fight ever to take place.

The third outstanding example was the German defeat in Italy. Once the German Army, retreating from North Africa, had been driven out of Sicily, it moved to the Italian mainland. Here Kesselring, the German supreme commander in the Mediterranean region, mounted a very stubborn defence. But we broke a large number of Tunny messages to and from Kesselring, and from these decrypts we knew the enemy's strategic decisions and their overall thinking. Kesselring was pushed steadily northwards by the Allies until the German Army was out of the way.

The fourth example came after D-Day, during a period of stalemate in France when the battle lines got bogged down. The Allies had tried but failed to take Caen in Normandy and the Germans were holding out resourcefully. Hitler tasked von Kluge, who was in charge of the German forces in France, with pushing the Allies back into the sea—or at the very least holding the line. He managed to do so for 3 or 4 weeks, and then Patton broke through in the west of France and Montgomery in the north. Between them they managed to surround part of the German army and took 90,000 prisoners. Von Kluge was ordered back to Berlin, but committed suicide rather than face the furious Hitler.

It is often said that Tunny decrypts helped to shorten the European war by at least two years. This was a war in which approximately 7 million people on average died each year—so our decrypts saved millions of lives. Much of this success was down to the work of Bill Tutte and the Testery. Before Tutte's work on Tunny he had been interviewed for a job by Turing, but was rejected; luckily he was interviewed again and taken on by Colonel Tiltman. As it happened, I then worked in the same office as Tutte and still have a clear mental image of him staring into the middle distance for long periods, twiddling his pencil, and covering reams of paper with calculations. I used to wonder if he was actually getting anything done, but my goodness, he was!

Breaking Tunny

Once Tutte had opened the door to Tunny, and some broken messages had revealed its importance, we realized that we must break it on a daily basis. At that point the Testery, which had until then been working on a somewhat old-fashioned crypto-system called Double Playfair, was switched entirely to Tunny. At the time—the middle of 1942—the Testery consisted of a linguist and three senior cryptographers, of whom I was one. Six additional cryptographers were brought in as the Tunny work increased, and by the end of the war the Testery had a total staff of 118. I estimate that we produced some 64,000 decrypted messages, often running into many pages. In terms of sheer numbers, this is mere peanuts compared with Enigma, where more than 100 messages were being broken every hour—day and night. But Tunny messages were nuggets of gold.

One of the most remarkable things about the Testery was how young we all were. Peter Hilton, for example, joined during his first year at Oxford University, when he was just 19 years old (Fig. 3.1). Donald Michie was even younger, literally a schoolboy. I was pushing it a bit, age-wise, because I was 20 when I joined (Fig. 16.1). Tutte was 24 when he broke the Tunny system (Fig. 14.5). As for Turing, at 29 he was positively in his dotage. But don't get the wrong impression: breaking Tunny was never child's play. It was always a challenge to get each new day's wheel patterns for each of the five key links we worked on, since to do this we had to break from thirty to fifty consecutive places of a message's ciphertext. To break into a message in two or three different places is one thing, but to break so many consecutive places is very much more difficult. For the first year we did everything by hand, without any machines to help us, and we broke one-and-a-half million letters of ciphertext. That takes a lot of doing.

I will explain one of the most important methods that we used in the Testery to break Tunny messages. In the earlier days, before the tightening up of German security during 1943, there

Figure 16.1 Jerry Roberts at Bletchley Park.

Reproduced with permission of Mei Roberts.

were plenty of what we called 'depths'—messages that had been enciphered using the same key (see Chapter 14 for an explanation of key). Our intercept stations were often able to detect when a pair of messages were depths, and they passed this information on to us. I would select two longish ciphertexts that the intercept people thought had probably been enciphered with the same key, and I added the two ciphertexts together. Because the messages (if depths) had had exactly the same encryption applied to them, this process of addition had the effect of 'cancelling out' the common key. Although the resulting string of letters looked random, in fact it consisted of the two plaintexts added together. My problem now was to separate out these two plaintexts.

For example, the process of adding the two ciphertexts together might produce this stream of letters:

$$...FJM5XEKLRJJ...$$

(the dots indicate simply that the letters carry on to the left and to the right). I would start by choosing a German word—the trick was to choose a good word—and I would then add this to the stream of letters, beginning at some position in the stream that my instinct told me was promising. If this didn't work out, I would try adding my chosen word at a number of other positions. If I was lucky, sooner or later I found a position where adding the chosen word produced plaintext from the other message of the pair. For example, I might try the common plaintext expression 9ROEM9. This related to Roman numerals. Units in the German Army were commonly referred to by Roman numerals, I, XXII, CXX, and so on. The Tunny operators dealt with these numerals by typing the German equivalent of 'Roman 1', 'Roman 22', and so on. The abbreviation for 'Roman'—ROEM (from 'Roemisch')—therefore cropped up a lot in messages, usually with a '9' stuck on at either end, since Tunny operators used '9' to indicate a space between words.

Adding 9ROEM9 to FJM5XE—six consecutive letters in the stream—might give DE9GES. DE9GES is possible plaintext, so I would push on with a number after ROEM, in this case EINS (one). Adding 9ROEM9EINS9 to the letters in the stream produces good 'clear':

```
  F J M 5 X E K L R J J
+ 9 R O E M 9 E I N S 9
= D E 9 G E S C H I C K
```

Having got this far, I would assume that the plaintext ran 9WURDE9GESCHICKT9 (the German for 'was sent', with three word-spacers) and then see what letters that gave in the other message. If adding 9WURDE9GESCHICKT9 did not produce plausible letters of plaintext, I would discard WURDE and try other likely words. Plugging away like this I would, if lucky, reveal longish stretches of the clear text of the two messages.

Old habits die hard: if the Germans had only known how useful we found those four characters ROEM then they would have stopped using them at once! Even today I can remember all the basic letter additions, such as J + R = E and M + O = 9, since we had to know them absolutely by heart, and be able to recall them instantly, because we were testing all the time—adding this word, adding that word, adding, adding.

We had to keep extending the break in whatever direction we could, left or right, and in whichever of those two messages we could, until we got thirty to fifty letters of clear text for each, and then we could work out the wheel patterns of the day. Once we had the day's wheel patterns, the problem of breaking other messages sent on the same day was simplified a bit: we only had to break fifteen or so continuous letters of a message and then we could 'set' the wheels, which is to say, find the positions that the wheels had started at when the message was encrypted—and so decipher the entire message.

Mechanical aids

That was how things were during the first year of our attack on Tunny: the codebreaking process consisted entirely of what we called 'hand-breaking'. Then another Tunny unit was set up, the Newmanry, where machines were used to speed up one stage of the breaking process, setting the chi-wheels, as explained in Chapter 14. The Testery carried on with setting the psi-wheels and the motor wheels by hand. The Newmanry also depended on us to find the daily patterns of the wheels, which we did by means of the method that I have described. We called this 'breaking the wheels', and the Newmanry was helpless until the Testery had successfully completed this initial stage of the process—without the Testery there would have been no decrypts at all. We also carried out the final stage of the process, the actual decipherment of the messages, which was done on our own machines, operated by a team of twenty-four ATS 'girls' (women in the Auxiliary Territorial Service).

These machines did the opposite of what the German Tunny machines had done. The German machines transformed plaintext into ciphertext. Our machines started with ciphertext and transformed it into plaintext. Once the ATS girls had set the machines up correctly—using information about the wheels that we supplied—they then typed in the complete ciphertext and out would come the plaintext. It was magic!

There were three broad phases to the operational work on Tunny. During the first phase, which lasted for a whole year, from mid-1942 to mid-1943, the Testery worked alone, breaking

Tunny messages entirely by hand. It was during this first phase that the messages giving away the German plans for an assault at Kursk were broken. During the second phase, lasting from mid-1943 to February 1944, we received a limited amount of help from the Newmanry's Heath Robinson machine. The Heath Robinson was quite useful, but a bit slow and unreliable, and it kept on breaking down. So it was a relief when Colossus I appeared in the Newmanry in the middle of February 1944, initiating the third and final phase of the work.

The Newmanry people operated the Colossi and Robinson machines and did a good job. They were totally different from the Testery crowd, largely mathematicians and engineers, whereas the Testery people were hand-breakers and linguists. It was rather like two Native American tribes who live in peace with each other, but do not understand each other's language!

Three heroes

Some years ago now I referred to Alan Turing, Bill Tutte, and Tommy Flowers as the 'three heroes' of Bletchley Park, and I am pleased to see that other people are using that terminology, because heroes they were: Turing, about whom we know plenty by now, Tutte, whose name is recognized by few, and Flowers, also largely unknown, who designed and built Colossus. In my view, this puts any one of our heroes in the class of the Duke of Wellington or the Duke of Marlborough. The effect of what they did was enormous—in fact, even greater than what

Figure 16.2 Jerry Roberts being presented to the Queen at Bletchley Park in 2011. A Tunny machine is behind.

Wellington or Marlborough achieved. Without those three great minds, and the many supporting personnel at Bletchley Park, Europe would be a very different place today. Britain was lucky to have these brilliant men in the right place at the right time to break Tunny.

So were they rewarded? Marlborough was given Blenheim Palace for his military achievements, and Wellington was given No. 1 Piccadilly. Turing was given a bonus of £200 and an OBE. This was more in those days, but doesn't compare with the massive bonuses and gongs given to top civil servants when they retire. They have kept their desks clean for twenty years—not quite the same as saving the country! Tommy Flowers did rather better: he got an innovation award of £1000 for introducing the technology that has totally changed the way we work and play worldwide. He was already an MBE for his other work. Yet Turing and Flowers were lucky in comparison with Bill Tutte: he got nothing, absolutely nothing! No wonder he went off to Canada. The whole thing is a national scandal. Any other country would be proud of these people and would shout their achievements from the rooftops. I don't understand why cryptographers who did such astonishing things get no official credit, nothing. Why aren't there statues of the three heroes in Whitehall, and a memorial on the Thames Embankment for all those cryptographers who contributed so much to the Allied victory over Hitler?

I am lucky to be the last survivor of the nine leading cryptographers who worked on Tunny. I was pleased to be presented to HM the Queen in July 2011, when she and the Duke of Edinburgh visited Bletchley Park (Fig. 16.2). Three months later, the BBC made a film about the Tunny story called *Code-breakers: Bletchley Park's Lost Heroes*, which many of you may have seen. It is not perfect, but it is good, and it tells the story accurately—and it does do justice to Tunny, Bill Tutte, and Tommy Flowers.

After the filming, I drove to the BBC in Cardiff with my wife Mei to see the first cut. There were several errors that Jack Copeland and I put right, and we also asked the producer, Julian Carey, to strengthen even further the coverage of Tutte and Flowers, which he did. To date, television audiences for this film are in excess of 10 million.

In 2012 the film won two BAFTA Awards, and in 2013 was listed as one of the year's three best historical documentaries at the Media Impact Awards in New York City. Word is getting out at last about Bill Tutte and Tommy Flowers.

CHAPTER 17

Ultra revelations

BRIAN RANDELL

In 1974 and 1975 two books (The Ultra Secret and Bodyguard of Lies) were published. These books alerted the general public for the first time to some of the secrets of Bletchley Park's wartime activities, and caused a great sensation. These developments provided me with an excuse to enquire again about the possibility of persuading the British government to declassify the Colossus project. This second account describes how, following a partial such declassification, I received official permission in July 1975 to undertake and publish the results of a detailed investigation into the work of the project. As a consequence, at the 1976 Los Alamos Conference on the History of Computing I was able to describe in some detail, for the first time, how Tommy Flowers led the work at the Post Office Dollis Hill Research Station on the construction of a series of special-purpose electronic computers for Bletchley Park, and to discuss how these fitted into the overall history of the development of the modern electronic computer. The present chapter describes the course of this further investigation.[1]

Introduction

In the spring of 1974 the official ban on any reference to Ultra, a code name for information obtained at Bletchley Park from decrypted German message traffic, was relaxed somewhat, and Frederick Winterbotham's book *The Ultra Secret* was published.[2] Described as the 'story of how, during World War II, the highest form of intelligence, obtained from the "breaking" of the supposedly "unbreakable" German machine cyphers, was "processed" and distributed with complete security to President Roosevelt, Winston Churchill, and all the principal Chiefs of Staff and commanders in the field throughout the war', this book caused a sensation, and brought Bletchley Park, the Enigma cipher machine, and the impact on the war of the breaking of wartime Enigma traffic, to the general public's attention in a big way. The book's single reference to computers came in the statement:

It is no longer a secret that the backroom boys of Bletchley used the new science of electronics to help them . . . I am not of the computer age nor do I attempt to understand them, but early in 1940 I was ushered with great solemnity into the shrine where stood a bronze coloured face, like some Eastern Goddess who was destined to become the oracle of Bletchley.

No mention was made of Alan Turing, or any of the others who I had learned were involved with Bletchley's code-breaking machines.

A further, even more sensational, book *Bodyguard of Lies*[3] then revealed more about how the Germans had been using Enigma cipher machines. It gave some information about the work of first the Polish cryptanalysts, and then of Turing and others at Bletchley Park on a machine called the 'bombe' that was devised for breaking Enigma codes, but made no mention of computers and referred to electronics only in connection with radar and radio. Its main topic was the immense impact of all this work on the Allies' conduct of the war.

Emboldened by what seemed to be a rather significant change in government policy concerning any discussion of Bletchley Park's activities, I made some enquiries as to whether another request to declassify the Colossus project might now have a chance of being treated favourably. I was strongly urged not to write to the Prime Minister again—apparently my earlier request had caused considerable waves on both sides of the Atlantic. Instead, on the advice of David Kahn, I wrote on 4 November 1974 to Sir Leonard Hooper, described by Kahn as being the former head of the Government Communications Headquarters (GCHQ), who was by then an Under Secretary in the Cabinet Office with, I believe, the title of Coordinator for Intelligence and Security. After a brief exchange of correspondence, I received a letter from Sir Leonard dated 22 May 1975 with the welcome news that 'approval had been given for the release of some information about the equipment', and that it was proposed to release some wartime photographs of Colossus to the Public Record Office. I was invited to come to London for discussions at the Cabinet Office.

This visit occurred on 2 July 1975. When I arrived, somewhat nervously, in the Cabinet Office building I was escorted along a corridor, past what I was told was the famous internal doorway that provided a discreet means of gaining access to 10 Downing Street, to a panelled room where I met Sir Leonard Hooper, his personal assistant, and Dr Ralph Benjamin. (I do not recall whether it was then or later that I learned that Dr Benjamin was Chief Scientist at GCHQ.) I was shown the photographs, and we discussed in detail the wording of the explanatory document. (Their concern was that some of the wartime terminology used in their explanation might not be understood by present-day computer people.) We readily agreed on some modest rewording of the one-page document.

I was then told that the government were willing to facilitate my interviewing the people who had led the Colossus project after they had been briefed as to just what topics they were allowed to discuss with me—this was with a view to my being allowed to write a history of the project, provided that I would submit my account for approval prior to publication. Needless to say, I agreed. My assumption was that the fact that I was viewing the Colossus solely from the viewpoint of the history of computing, and that I had little interest in (and less knowledge of) cryptography, contributed to the decision to facilitate my investigation.

The photographs and explanatory document were made available at the Public Record Office (now The National Archives) on 25 October 1975, and I had the pleasure of sending a letter (now on display in the Turing Exhibition at Bletchley Park) to Mrs Turing, saying:

I thought you would like to know that the Government have recently made an official release of information which contains an explicit recognition of the importance of your son's work to the development of the modern computer. They have admitted that there was a special purpose electronic computer developed for the Department of Communications at the Foreign Office in 1943. Their information release states that Charles Babbage's work in the 19th century and your son's work in the 1930s laid the theoretical foundations for the modern computer and that the Foreign Office's special purpose computer was, to their belief, the first such electronic computer to be constructed anywhere. Moreover, they credited your son's work with having had a considerable influence on the design of this machine.

Further, a book has recently been published in the United States entitled Bodyguard of Lies which describes the work of the Allies during the war to deceive the Germans and to prevent them from anticipating the D-day invasion. This book credits your son as being the main person involved in the breaking of one of the most important German codes, the Enigma Code, and thus implies that his work was of vital importance to the outcome of World War II. This latter information in the book Bodyguard of Lies is not, of course, official information but nevertheless it will, I believe, enable many people to obtain a yet fuller understanding of your son's genius. I am very pleased that this is now happening.

I was delighted to receive a very nice reply from Mrs Turing, dated 2 December 1975, saying that she was 'very much gratified' by my letter, and that she regarded it of such importance that she would like to have a few photostat copies so that she could distribute them.

During the period October–December 1975 I interviewed the leading Colossus designers: Tommy Flowers, Bill Chandler, Sidney Broadhurst, and Allen 'Doc' Coombs. I found all four of them to be delightful individuals, immensely impressive and amazingly modest about their achievements. All were unfailingly pleasant and helpful as they tried to recollect happenings at Dollis Hill and Bletchley Park. I had the further pleasure of interviewing Max Newman and Donald Michie, and David Kahn kindly interviewed Jack Good for me at his home in Roanoke, Virginia. I also corresponded, in some cases quite intensively, with all these interviewees and with a considerable number of other people, including several of the Americans who had been stationed at Bletchley Park. Indeed, it was one of these who alerted me to what had been said in the Washington Post in 1967 in a review of Kahn's book The Codebreakers:[4]

Magnificent as the book is, it is like a history of English drama that takes no account of Shakespeare. Kahn notes, correctly, that 30,000 people were involved in cryptanalysis in Britain during World War II . . . In importance, not to mention drama and sheer intellectual brilliance, the British effort pales every other account in Kahn's huge book, or perhaps all of them combined.

Each interview was tape-recorded, and I had the tapes transcribed in full. The people I interviewed and corresponded with were being asked to recall happenings of thirty or so years earlier, and to do so without any opportunity of inspecting original files and documents. Secrecy considerations had been paramount, and had given rise to a rigid compartmentalization of activities. Few had any detailed knowledge of the work of people outside their own small group. Many of them had made conscious efforts to try and forget about their wartime work.

Piecing together all the information I thus obtained, and even establishing a reasonably accurate chronology, was therefore very difficult. I was greatly aided in this task by the advice I had read in Kenneth May's magnificent Bibliography and Research Manual on the History of Mathematics;[5] for example, the techniques that he described for creating and using a set of correlated card indexes greatly helped me in sorting out a major chronological confusion amongst my interviewees concerning the development of the Robinson machines. The indexes that I constructed employed unpunched IBM cards, rather than conventional index cards. They filled a standard box, so I knew that there were approximately 2000 of them.

What became clear from my discussions with the Colossus designers was that their interactions with Turing had mainly occurred on projects that preceded Colossus. My investigation led me to summarize their and other's attitude to him as follows (quoted in Randell[6]):

Turing, clearly, was viewed with considerable awe by most of his colleagues at Bletchley because of his evident intellect and the great originality and importance of his contributions, and by many with considerable discomfort because his personality was so outlandish. Many people found him incomprehensible, perhaps being intimidated by his reputation but more likely being put off by his character and mannerisms. But all of the Post Office engineers who worked with him say that they found him very easy to understand—Broadhurst characterised him as 'a born teacher—he could put any obscure point very well'. Their respect for him was immense, though as Chandler said 'the least said about him as an engineer the better'. This point is echoed by Michie who said 'he was intrigued by devices of every kind, whether abstract or concrete—his friends thought it would be better if he kept to the abstract devices but that didn't deter him'.

I submitted the draft of my paper on Colossus to Dr Benjamin on 12 April 1976. Subsequent correspondence and discussions with him and Mr Horwood led to my incorporating a number of relatively small changes into the paper and its abstract, the main effect of which was to remove any explicit indication that the projects I was describing were related to codebreaking. I was merely allowed to say that:

The nature of the work that was undertaken at Bletchley Park during World War II is still officially secret but statements have been appearing in published works in recent years which strongly suggest that it included an important part of the British Government's cryptologic effort.

However, this, and the fact that I was allowed to retain references to books such as *The Ultra Secret* and *Bodyguard of Lies*, meant that readers would be left in little doubt as to what Turing and his colleagues had been engaged in and the purpose of the Robinson and Colossus machines.

The outing of Colossus

The cleared paper was then submitted to the International Conference on the History of Computing, which was held in Los Alamos in June 1976. (No attempt is made to detail the contents of that 21,000-word paper here!)

Doc Coombs and his wife were planning to be on vacation in the United States at about the time of the conference, so to my delight he suggested that he accompany me to the conference and I arranged for him to participate. It is fair to say that my presentation created a sensation—how could it not, given the material that I had been allowed to gather?

I have recently found that Bob Bemer has reported his impressions of the event:[7]

I was there at a very dramatic moment of the invitational International Research Conference on the History of Computing, in Los Alamos . . .

Among the many that I conversed with was a medium-sized Englishman named Dr. A. W. M. Coombs, who was so excited about something that he was literally bouncing up and down. Not being bashful I asked (and he didn't mind) about the cause of his excitement, and he replied 'You'll know tomorrow morning—you'll know'.

Saturday morning we regathered in the Auditorium of the Physics Division. I sat third row from the front, a couple of seats in from the right, to get a good view of all the famous attendees.

To my left in the same row, three empty seats intervening, was the bouncy Englishman, all smiles and laughter. In front of him, two seats to his left, was Professor Konrad Zuse . . . In the fifth row, again to the left, was Dr. John Mauchly, of ENIAC fame.

On stage came Prof. Brian Randell, asking if anyone had ever wondered what Alan Turing had done during World War II? He then showed slides of a place called Bletchley Park, home base of the British cryptographic services during that period. After a while he showed us a slide of a lune-shaped aperture device he had found in a drawer whilst rummaging around there.[8] Turned out it was part of a 5000-character-per-second (!) paper tape reader. From there he went on to tell the story of Colossus, the world's really first electronic computer . . .

I looked at Mauchly, who had thought up until that moment that he was involved in inventing the world's first electronic computer. I have heard the expression many times about jaws dropping, but I had really never seen it happen before. And Zuse—with a facial expression that could have been anguish. I'll never know whether it was national, in that Germany lost the war in part because he was not permitted to build his electronic computer, or if it was professional, in that he could have taken first honors in the design of the world's most marvelous tool.

But my English friend (who told us all about it later) was the man doing the day-to-day running of Colossus. I saw then why he was so terribly excited. Just imagine the relief of a man who, a third of a century later, could at last answer his children on 'What did you do in the war, Daddy?'.

The conference organizers hurriedly arranged an additional evening session at which Doc Coombs and I fielded a barrage of questions from a packed audience. Coombs's role at this session became that of adding detail to some of the events that my paper described rather guardedly, and mine became (at least in part) that of endeavouring to make sure that his splendidly ebullient character did not lead him to too many indiscretions. Tommy Flowers had warned me beforehand that 'in his natural exuberance [Doc Coombs] is likely to give away too much for the Foreign Office and you should be careful not to provoke him!'

My paper was promptly published and circulated widely as a Newcastle University Computing Laboratory Technical Report[9]—the proceedings of the Los Alamos conference did not appear until four years later.[10] In addition, a summary version of my paper, including all the Colossus photographs, was published in the *New Scientist* in February 1977 after I had also cleared this with the authorities;[11] this version was afterwards included in the third and final edition of my book, *The Origins of Digital Computers*, in place of the earlier two-page account by Michie.

Sometime in early 1976 I became aware that BBC Television were planning a documentary series, *The Secret War*, and that the sixth, and originally last, episode (entitled 'Still Secret') was going to be about Enigma. I met with the producer of this episode, Dominic Flessati, and told him—very guardedly—about Colossus. I showed him the Colossus photographs, at which he became very excited. This meeting was in Bush House, the home of the BBC World Service, and took place in March 1976. For our discussion Flessati took me to a rather gloomy and forbidding wood-panelled studio, reached via corridors that seemed to be full of people speaking in every language but English. The architecture reminded me of several East European government buildings that I had lately been in. This environment—together with my memories of my recent meeting in the Cabinet Office—made me feel the need to be very careful about what I could say to him.

The result of this meeting was that Flessati revised his plans for the sixth episode in *The Secret War* series, so as to cover Colossus as well as Enigma. The BBC brought their formidable research resources to bear on the making of this episode. The Enigma section of the episode gave extensive details of the work of the Polish cryptanalysts who originally broke Enigma, how Enigma worked, and how Bletchley Park made use of a large number of machines, the so-called 'bombes', designed by Alan Turing and Gordon Welchman to break Enigma traffic on an industrial scale. It also took the Colossus story on somewhat further than I had managed. For the Colossus section of 'Still Secret' they interviewed Tommy Flowers, Gordon Welchman, Max Newman, and Jack Good, mainly on camera, and filmed a number of scenes at Dollis Hill and Bletchley Park, as well as showing the official Colossus photographs.

Whereas I'd had to be very guarded in my paper regarding the purpose of Colossus, 'Still Secret' made it abundantly clear that Colossus was used to help break high-level German messages sent in a telegraphic code, via a machine that it said was called a *Geheimschreiber* ('secret writer'). However, the machine that it described, and whose workings it showed, was a teleprinter-based device made by Siemens & Halske. It was a number of years before this inaccurate identification of the target of the Colossus project was corrected, and it became known that Colossus was used to help break teleprinter messages that were enciphered using a separate ciphering device (the SZ40/42 made by Lorenz AG) to which an ordinary teleprinter was connected, rather than an enciphering teleprinter.

One further event occurred in late 1976, while I was on sabbatical at Toronto University, and is worth mentioning. I was invited to visit Professor May and to give a seminar at his Institute for the History and Philosophy of Science. He sensed that I felt concerned about my temerity in undertaking historical research without having had any formal training as a historian. I've always remembered his very comforting reassurance: 'Don't worry—there is as much bad history of science written by historians who don't understand science as by scientists who don't understand history!.'

The aftermath

The television series was very successful when it was broadcast in early 1977.[12] Undoubtedly it, and the accompanying book by the overall editor of the series,[13] did much to bring Bletchley Park, Alan Turing, Enigma, and Colossus to the public's attention, although it was some years before there was a general awareness that Colossus was not used against Enigma, and one still occasionally sees confusion over this point.

My original query, concerning the story of a wartime meeting between Turing and von Neumann at which the seeds of the modern computer were planted, remained unanswered. The present general consensus, with which I tend to agree, dismisses this as a legend, but I should mention that after my account was published one senior US computer scientist, well connected with the relevant authorities there, did hint to me rather strongly that it would be worth my continuing my quest! But nothing ever came of this.

There was one amusing aftermath of my involvement with the BBC television programme. I had been asked by Domenic Flessati to tell him the next time I would be in London after the TV series had been broadcast, so that we could have a celebratory dinner. This I did, and we met on the front steps of Bush House, where he introduced me to Sue Bennett, his researcher for 'Still Secret', in the following terms: 'Miss Bennett, I'd like you to meet Professor Randell, the "Deep Throat" of *The Secret War* series'. I'm rarely left speechless, but this was one of the occasions!

One final happening in 1977 needs to be mentioned—the conferment on Tommy Flowers of an honorary doctorate by Newcastle University, an event that was reported prominently by

The Times the next day.[14] I take great pride in the fact that I played a role in arranging this very belated public recognition for his tremendous achievement.

Concluding remarks

One further Newcastle-related incident is worth reporting. At my invitation Professor Sir Harry Hinsley, a Bletchley Park veteran and senior author of the multi-volume official history, *British Intelligence in the Second World War*,[15] gave a public lecture at Newcastle University soon after the first volume was published in 1979: his lecture was on the subject of the impact of Bletchley Park's activities on the war. One of the questions he received after his lecture was: 'If this work was so significant, why didn't it shorten the war?' His reply was short and to the point: 'It did, by about two years!' It is a tragedy that Turing did not live, and that Flowers and his colleagues had to wait more than thirty years, to receive any public recognition for their role in this great achievement.

In the years since 1977 more information has become available about the work and the impact of Bletchley Park, and of Alan Turing in particular. Turing's brilliant yet ultimately tragic life has been well documented in Andrew Hodges' excellent biography,[16] later made into a superb play, *Breaking the Code*, by Hugh Whitemore. Wonderful museum-quality replicas of the bombe and Colossus have been created and made operational at the National Museum of Computing at Bletchley Park, and during 2012 there were all manner of fascinating events worldwide, marking the centenary of Turing's birth. Hopefully this behind-the-scenes account of how, after decades of silence, the wartime work of Turing and his colleagues at Bletchley Park started to be revealed to the general public is a useful contribution to the growing pool of public knowledge about Alan Turing.

CHAPTER 18

Delilah–encrypting speech

JACK COPELAND

O nce Enigma was solved and the pioneering work on Tunny was done, Turing's battering-ram mind was needed elsewhere. Routine codebreaking irked him and he was at his best when breaking new ground. In 1942 he travelled to America to explore cryptology's next challenge, the encryption of speech.[1]

Welcome to America

Turing left Bletchley Park for the United States in November 1942.[2] He sailed for New York on a passenger liner, during what was one of the most dangerous periods for Atlantic shipping. It must have been a nerve-racking journey. That month alone, the U-boats sank more than a hundred Allied vessels.[3] Turing was the only civilian aboard a floating barracks, packed to bursting point with military personnel. At times there were as many as 600 men crammed into the officers' lounge—Turing said he nearly fainted.

On the ship's arrival in New York, it was decreed that his papers were inadequate, and this placed his entry to the United States in jeopardy. The immigration officials even debated interning him on Ellis Island. 'That will teach my employers to furnish me with better credentials' was Turing's laconic comment.[4] It was a private joke at the British government's expense: since becoming a codebreaker in 1939, his employers were none other than His Majesty's Foreign Office.

America did not exactly welcome Turing with open arms. His principal reason for making the dangerous trip across the Atlantic was to spend time at Manhattan's Bell Telephone Laboratories, where speech encryption work was going on, but the authorities declined to clear him to visit this hive of top-secret projects.[5] General George Marshall, Chief of Staff of the US Army, declared that Bell Labs housed work 'of so secret a nature that Dr. Turing cannot be given access'.[6]

While Winston Churchill's personal representative in Washington, Sir John Dill, struggled to get General Marshall's decision reversed, Turing spent his first two months in America advising Washington's codebreakers—no doubt this was unknown to Marshall, who might otherwise have forbidden Turing's involvement. During this time Turing also acted as consultant to the engineers who were designing an electronic version of his bombe for production in America.

When Turing arrived in Washington, Joe Eachus of the US Navy's codebreaking unit showed him around the city.[7] Turing was fascinated by Washington's letter–number system of street names—M Street, K Street, 9th Street, 24th Street, and so on. It was reminiscent of Enigma. The number streets run north–south and the letter streets east–west, but because twenty-six letters were not enough, the city planners started the letter names over again at the Capitol, with C Street SW running parallel to C Street NW but a few blocks further south. Turing's first reaction was to ask the key question 'What do they do in the numbered streets when they get to 26?' He joked that the numbers should go up as high as 26×26—corresponding to a potential 26 distinguished alphabets of 26 letters each—and grinned when he learnt that actually there are two 1st streets, two 2nd streets, two of everything.

Helping out

Turing had already been involved in US liaison for at least a year by the time he visited Washington in person. In 1941 he had written out a helpful tutorial for the US Navy codebreakers, whose attempts to crack U-boat Enigma were going nowhere; Turing had politely made it clear that their methods were utterly impractical.[8] The United States is often portrayed as responsible for breaking Naval Enigma—for example, in the Hollywood movie *U-571* starring Harvey Keitel and Jon Bon Jovi—but in actuality the truth could hardly be more different.

Another report of Turing's impressions of the US codebreaking effort survives in the archives, prosaically titled 'Visit to National Cash Register Corporation of Dayton, Ohio'.[9] At Dayton, NCR engineer Joseph Desch was heading up a massive bombe-building programme for the US Navy. In his report Turing explained that the US bombes would be used as additional muscle, while the brain work would still be done at Bletchley Park, observing that 'The principle of running British made cribs on American Bombes is now taken for granted'.

Turing took the train over to Ohio to advise Desch and electronics expert Robert Mumma, Desch's right-hand man on the bombe project. Jack Good, Turing's Bletchley Park colleague, described Desch as 'a near genius', but Turing still found a lot to criticize in the Dayton bombe design.[10] 'I suspect that there is some misunderstanding', he said sharply at one point in his report. He also complained:

I find that comparatively little interest is taken in the Enigma over here apart from the production of Bombes.

It is not widely appreciated that Turing contributed significantly to the eventual design of the American bombes. Mumma himself explained the importance of Turing's visit to the Dayton project in a 1995 interview.[11] Turing simply 'told us what we wanted to do and how to do it', Mumma said. He emphasized that Turing 'controlled the design more than anyone else did'.

Secret speech

In January 1943, after a protracted exchange of letters between General Marshall and Sir John Dill, Bell Labs finally opened its fortress-like metal gates to the shabby British traveller.[12] Turing based himself there for the next two months.[13] The imposing thirteen-storey Bell Labs building was on Manhattan's lower West Side, close to the Hudson waterfront and a short stroll from the gay bars, clubs, and cafeterias of Greenwich Village.[14] New York's relatively open gay community must have seemed a million miles from Bletchley Park, where Turing kept his sexual orientation to himself.

Bell Labs was developing a speech encryption system based on a voice synthesizer called the 'vocoder'. Speech encryption was cryptology's new frontier. If speech could be encrypted securely, then top-level secret business could be conducted person to person, by radio or even by telephone: this was a more natural way for military commanders to communicate than by written text. In addition, interactive voice communication was less open to misinterpretation and to the perils of incompleteness. The Bell Labs system, codenamed SIGSALY, was used by America and her allies from 1943 to 1946. Turing may have contributed some finishing touches to the SIGSALY system during his two months at Bell Labs.

The Bell Labs vocoder lives on as a musical instrument; anyone who has listened to the music of T-Pain, Herbie Hancock, Kraftwerk, or the Electric Light Orchestra will have heard its weird and unearthly sound. Today's vocoders are not much larger than a laptop, whereas the original 1943 model occupied three sides of a room and consisted of a number of cabinets each taller than Turing himself.[15] SIGSALY was about the size of Colossus: Turing decided to miniaturize speech encryption.

Thinking small

Once back in England Turing set up an electronics lab in a Nissen hut at Hanslope Park, a Buckinghamshire country house situated a few miles from Bletchley Park.[16] Nowadays Hanslope Park is one of Britain's most secure sites and home to HMGCC (Her Majesty's Government Communications Centre), where, in the Turing tradition, mathematicians, engineers, and programmers supply Britain's intelligence spooks with specialized hardware and software. At Hanslope Turing designed the portable voice encryption system that he named 'Delilah'; it consisted of three small boxes, each roughly the size of a shoebox (Fig. 18.1).

Turing's talent for electronic circuit design blossomed in his Hanslope laboratory. Tommy Flowers also had a hand in this new and highly secret project.[17] He and Turing spoke each other's language. 'Turing had the reputation of being practically incoherent in explanation', Flowers said in an amused way—yet he himself never had any trouble at all understanding Turing.[18] 'The rapport was really quite remarkable', he remembered.

The surviving blueprints for Delilah, each the size of a desk and still, after all these years, bright powder-blue in colour, depict the complex electronic system in detail.[19] Its method of encrypting speech was analogous to the way that the Tunny machine encrypted typed text (see Chapter 14). Tunny added obscuring key to written words, while Delilah added obscuring key to spoken words: in Delilah's case, the key was a stream of random-seeming numbers. The first step in the encryption process was to 'discretize' the speech, turning it into a series of individual numbers: each number corresponded to the voltage of the speech signal at that particular moment in time.[20] Delilah then added key to these numbers, creating the enciphered form of the spoken message. This was then automatically transmitted to another Delilah at the receiving end of the link.

Figure 18.1 Turing's Delilah.

From A. M. Turing and D. Bayley, 'Speech secrecy system "Delilah", Technical description', c. March 1945, National Archives ref. HW 25/36. Crown copyright and reproduced with permission of the National Archives Image Library, Kew.

As with Tunny, the receiver's Delilah had to be synchronized with the sender's, so that both the transmitting and receiving machine produced identical key. In designing Delilah, Turing ingeniously adapted existing cryptographic technology. At the heart of Delilah's mechanism for producing the key was a five-wheeled text-enciphering machine modelled on Enigma.

The receiving machine stripped the key from the enciphered message and the resulting decrypted numbers (specifying voltages) were used to reproduce the original speech. The result was a bit crackly, but was generally quite intelligible—although if the machine made a mistake, there would be 'a sudden crack like a rifle shot', Turing said.[21] It must have been hard on the receiving operator's straining ears. But Turing had succeeded: Delilah was a functioning portable speech-encryption system that was only a fraction of the size of the gigantic SIGSALY.

Mess life

At Hanslope, Turing lived in an old cottage and took his meals in the army mess.[22] As the commanding officer recollected:[23]

In spite of having to live in a mess and with soldiers, Turing soon settled down and became 'one of us' in every sense; always rather quiet but ever ready to discuss his work even with an ignoramus like myself.

After a few months the army sent several recent university graduates to Hanslope and Turing made two firm friendships. Robin Gandy shared Turing's Nissen hut, working on improvements to equipment used in intercepting German messages.[24] At first Gandy thought Turing a bit austere, but later was 'enchanted to find how human he could be, discussing mutual friends, arranging a dinner-party, being a little vain of his clothes and his appearance'.[25] The second new friendship sprouted when Don Bayley came to Hanslope in March 1944 to assist with Delilah.[26] The three of them would take long walks together in the Buckinghamshire countryside.

One day Turing entered his name for the mile race in the regimental sports. Some of the soldiers thought it must be a leg-pull, but when the race was run Turing 'came in a very easy first', the C.O. remembered.[27] It was the beginning of Turing's career as an Olympic standard runner. Soon after the war ended he began to train seriously and was to be seen 'with hair flying', his Cambridge friend Arthur Pigou said, as he flashed past on a '10, 15 or 17 miles solitary "scamper"'.[28]

Victory

Turing's invention was never used in earnest, for Germany fell in May 1945, not long after Delilah was completed. As Russian troops poured into Berlin, members of the Nazi elite barricaded the windows of the Reich Chancellery with heavy crates of never-to-be awarded Iron Cross medals.[29] Turing celebrated victory quietly, ambling through the countryside with Bayley and Gandy.[30] 'Well, the war's over now', Bayley said to Turing as they rested in a clearing in the woods, 'it's peacetime so you can tell us all'. 'Don't be so bloody silly', Turing replied. 'That was the end of that conversation', Bayley recollected 67 years later.

CHAPTER 19

Turing's monument

SIMON GREENISH, JONATHAN BOWEN,
AND JACK COPELAND

Today Bletchley Park is a thriving monument to Turing and his fellow codebreakers. This did not happen easily. After the Second World War, the Bletchley Park site went into gradual decline, with its many temporary wartime buildings left unmaintained. The intense secrecy still surrounding the codebreakers' wartime work meant there was no public awareness of what they had achieved, nor even that they had existed. It was only when the information embargo finally began to lift, decades later, that Bletchley Park's importance became more widely known—but by that time the site was in danger of being razed to the ground to make way for housing estates. This chapter tells how Bletchley Park was rescued from property developers and from financial failure to become a national monument. The true heroes of this story are too numerous to mention by name—the hundreds of committed hard-working people, many of them volunteers, whose collective efforts over time saved Bletchley Park.

Background

Bletchley Park, also known as Station X, is arguably the most important single site associated with the Second World War. One of the best kept of all wartime secrets, it was acquired by the Foreign Office not long before the start of the fighting and at that time comprised some 55 acres of land, together with a large and architecturally odd mansion (see Fig. 9.1) and various associated outbuildings of the type common in large estates of the period. The mansion was built in Victorian and Edwardian times and its grounds were laid out as formal gardens, with a lake and many specimen trees. Curiously, these trees subsequently played a role in saving the site.

During the war there was an almost continuous programme of construction. Numerous typical Ministry-of-Defence-style brick buildings were erected, as well as an assortment of timber huts. The huts included Hut 8, where Turing worked on Naval Enigma (Fig. 19.1). His

Figure 19.1 Hut 8, where Alan Turing worked on Naval Enigma.

Copyright Shaun Armstrong/mubsta.com. Reproduced with permission of Bletchley Park Trust.

introduction of the bombe, in 1940, was the start of Bletchley Park's conversion into a code-breaking factory (see Chapter 12) and by the end of the war there were more than seventy buildings on the site, including the original mansion and its outbuildings. These structures supplied the capacity to house more than 3000 people per shift, as well as the large quantities of machinery and other equipment involved in the codebreaking work.

The codebreakers move out

In April 1946 Britain's military codebreaking HQ was transferred from Bletchley Park to Eastcote in suburban London, where the codebreakers took over what was previously an outstation housing bombes (see Fig. 12.3).[1] They transported as much of their codebreaking machinery as was needed for their peacetime work, including two of the nine Colossi and two British Tunny machines (larger replicas of the German cipher machine, used for decoding broken Tunny messages). At the time of the move, the organization's old name 'Government Code and Cypher School' was formally changed to 'Government Communications Headquarters' (GCHQ). Six years later another large move began, and during 1952–54 GCHQ transferred its operations away from the London area to a large site in Cheltenham, the quiet town where it remains to this day.

Burying the whole Bletchley Park operation in secrecy was inevitable. With fresh conflicts—and the Cold War—just around the corner, GCHQ did not want the world knowing how good it was at breaking codes. So the documents recounting Bletchley Park's wartime story were imprisoned in classified archives, and few had any inkling of the site's historical importance. It was not until the 1970s and early 1980s—with the publication of

Frederick Winterbotham's *The Ultra Secret* in 1974, followed by the BBC TV series *The Secret War* in 1977 and Gordon Welchman's controversial book *The Hut Six Story: Breaking the Enigma Codes* in 1982—that the story began to come out about Bletchley Park's boffins and their behind-the-scenes contribution to the Allied victory. Large amounts of information remained classified, however, and it was only after a number of key documents emerged into the light of day, during the period 1996–2000, that the full nature and scope of the Bletchley Park operation was revealed.

Fighting off the property developers

Bletchley Park's history in the post-war years was relatively undistinguished. Ownership of the site passed eventually to British Telecom (BT), and the Civil Aviation Authority also used the site for training purposes. When BT decided to dispose of Bletchley Park in the 1990s developers showed a keen interest, planning to demolish the existing buildings. Local and central government raised no objections, since the site's historical value was unknown.

But before a sale could go through an enterprising local history society arranged for former wartime staff to visit the mansion en masse.[2] About 400 attended, many expressing anger at the plans to redevelop the site. The Bletchley Park Trust was established in an attempt to prevent the sale—although the Trust had no resources at this time, nor any clear idea of what it was going to do with the 55 acres of land and seventy buildings should it succeed.

In 1992 Milton Keynes Council agreed to make Bletchley Park a conservation area.[3] This, together with tree preservation orders, helped deter commercial developers. BT decided to support the trust, and agreed to pay the site's operating costs until the Trust became financially viable. It was at this time that the Trust first opened the site to the public, although on a very limited basis. Optimism was in the air. Over time, volunteers installed exhibits and displays, all with minimal funding, and everyone involved felt that the site had been saved for posterity.

The site is nearly lost—again

Storm clouds gathered when, after a few years, BT withdrew its financial support: the Trust now had to take on the site's huge operating expenses. There was no viable solution in sight and disaster seemed imminent. The Trust continued to operate as best it could, using its army of volunteers; but it experienced great difficulty in getting potential funders to appreciate Bletchley Park's outstanding historical value. The Trust's vigorous CEO Christine Large returned empty-handed from a fundraising visit to the United States. Nor could the Trust get its message across to the UK Government. It was natural to hope for government or local authority funding: museums typically do not make enough money to cover their costs, so regional museums usually receive significant support from their local authorities while national museums are supported by government grants. But Bletchley Park had nothing of that sort: the Trust lurched from one financial crisis to another.

In fact very little money was raised in the 10 years or so up to 2005; but fortunately the Trust's bank permitted the site to remain open while new solutions were sought. When part of the site was leased to an innovation company things looked up financially; but expenditure still exceeded income by almost £500,000 in 2006. A controversial decision was taken to sell a parcel of land for housing development. This enabled the Trust to clear its debts, but the core problem of balancing expenditure and income remained. Another seemingly intractable problem was how to maintain the buildings themselves: the Trust had no money for this. The site continued to deteriorate alarmingly; some buildings were hazardous to enter and others were coming close to being unsafe. The nation's heritage was crumbling away.

During this stressful time the Trust was plagued by bitter infighting. In 2007, when tensions were at their height, the National Museum of Computing—originally set up by Tony Sale and Margaret Sale in the 1990s—broke away and became a separate entity on the site. Today, the National Museum of Computing is a separate attraction and is housed in Block H, a large concrete building into which the Newmanry with its Colossi had expanded in September 1944.[4] The museum contains Tony Sale's magnificent rebuilds of Colossus and a British Tunny machine,[5] as well as an exhibition about codebreaking and a priceless collection of later computers (many in working order) illustrating the history of computing from Colossus through to more recent times.

Things come right at last

In 2007 the Trust introduced some radical changes: cost savings, new initiatives for income generation, and a generally more aggressive approach to cost control. Finally the Trust's income was running ahead of expenditure. But this only bought time. There remained the gigantic problem of where to find the money to save and restore the buildings, and to develop Bletchley Park into the world-class heritage site that it is today.

In the long run, effective publicity was the key to securing Bletchley Park's future. A first step followed on from the previously mentioned document declassifications of 1996–2000. National and international attention was drawn to Bletchley Park by a clutch of new books: packed with fresh and often revolutionary information from the recently declassified wartime documents, these took the Bletchley Park story to a new level. They included Jack Copeland's *The Essential Turing* (2004), Michael Smith's *Station X: The Codebreakers of Bletchley Park* (1998), Hugh Sebag-Montefiore's *Enigma: The Battle for the Code* (2000), Ralph Erskine's and Michael Smith's *Action This Day* (2001), and Jack Copeland's *Colossus: The Secrets of Bletchley Park's Codebreaking Computers* (2006). *The Essential Turing* published extensive extracts from Turing's wartime treatise on the Enigma, known at Bletchley Park as 'Prof's Book', and also from Patrick Mahon's fascinating 'The History of Hut Eight, 1939–1945', both declassified in 1996; while *Colossus* and *Action This Day* contained eye-opening new accounts by Bletchley Park's leading codebreakers, including Keith Batey, Mavis Batey, Peter Edgerley, Tom Flowers, Jack Good, Peter Hilton, Donald Michie, Rolf Noskwith, Jerry Roberts, Bill Tutte, Derek Taunt, and Shaun Wylie. An important book from the era preceding the 1996–2000 declassifications was Harry Hinsley's and Alan Stripp's 1993 *Codebreakers: The Inside Story of Bletchley Park*.

By 2008 the Trust's in-house publicity machine was in top gear and interest from the news media in the Bletchley Park story was beginning to surge. Radio coverage during the first months of that year included a 45-minute reunion interview by the BBC involving a group of five Bletchley Park veterans, Sarah Baring, Mavis Batey, Ruth Bourne, Asa Briggs, and John

Herivel.[6] On 24 July a letter was published in *The Times* headed 'Saving the heritage of Bletchley Park: We cannot let Bletchley go to rack and ruin'.[7] Ninety-seven computer scientists signed the letter, including Sue Black, who highlighted the plight of Bletchley Park on BBC TV news the same day. The BBC reported:[8]

The centre of the British code-breaking effort during World War II is in urgent need of saving, a group of scientists says. The group, led by Dr Sue Black, say the mansion and the huts surrounding it, where the Enigma code was cracked and one of the first computers was built, are in serious disrepair.

The effect of increased publicity on Bletchley Park's visitor numbers was dramatic: the numbers doubled and then doubled again in 2008–12. These annual visitor numbers are the standard measure of Bletchley Park's health, since they are not only the main determinant of the Trust's operating revenue but also an indicator of public enthusiasm for what Bletchley Park has to offer. In March 2008, when visitor numbers stood at less than 50,000 a year, the Trust's CEO, Simon Greenish, announced that Bletchley Park was teetering on a financial knife-edge. But publicity roared ahead, with the wartime story finally reaching a sizeable proportion of the population. By the end of 2015, annual visitor numbers had climbed to a thumping 286,000.

Getting the message out

Movies helped to publicize the Bletchley Park story. *Enigma*, produced by Mick Jagger of the Rolling Stones and released in 2001, doubled visitor numbers. More recently Hollywood's *The Imitation Game* (2015) has entertained audiences around the world. Although widely criticized for inaccuracy, this movie nevertheless brought home the mammoth importance of Bletchley Park's attack on the German naval ciphers. It evokes the wartime atmosphere at Bletchley Park with its odd mix of military types and civilians: in among the crisp army and navy uniforms the film shows the codebreakers dressed down in pullovers and flannels, and when somebody starts to get a break into a message there are lifelike scenes of codebreakers clustering around excitedly and looking over each other's shoulders. Also realistic are the shots of coaches ferrying workers in and out, and of the wide-open spaces where people could get away and talk, as when Turing (Benedict Cumberbatch) and Joan Clarke (Keira Knightley) settle down on the grass with sandwiches to discuss mathematics.

Bletchley Park movies have tended to focus on Enigma, with no mention of the monumentally important attack on Tunny (see Chapters 14 and 16). In 2008 Jerry Roberts and Jack Copeland (friends and collaborators since 2001) initiated discussions with several movie companies about that aspect of Bletchley Park's story. Their efforts culminated in the BBC's *Code-Breakers: Bletchley Park's Lost Heroes*, directed by Julian Carey: this won two BAFTAs in 2012 and was rated as one of the year's three best historical documentaries at the 2013 Media Impact Awards in New York City. *Code-Breakers* led on to director Denis van Waerebeke's 2015 movie about Turing, *The Man Who Cracked the Nazi Codes*, distributed throughout Europe by Arte TV. Unlike *The Imitation Game*, *The Man Who Cracked the Nazi Codes* gives an accurate portrayal of Turing and his work at Bletchley Park. Other important movies, such as Patrick Sammon's *Codebreaker*, released in 2011, and the BBC's *Bletchley Park: Codebreaking's Forgotten Genius* of 2015 (directed by Russell England and based on Joel Greenberg's book *Gordon Welchman: Bletchley Park's Architect of Ultra Intelligence*) also played their part.

Social media too helped spread the word. In 2009 Stephen Fry, encouraged by Sue Black, tweeted:[9]

#bpark You might want to sign the Save Bletchley Park petition. Read @Dr_Black's reasons why on http://savingbletchleypark.org – BP won us the war!

It is certainly an appealing idea that Bletchley Park was saved by social media, and so was ultimately the beneficiary of what was pioneered under its own roof nearly 70 years previously, namely the modern digital electronic age. But opinions differ over how significant a role social media actually played. In the final analysis, Twitter and other social media appear to have had only a relatively minor impact on visitor numbers. Surveys filled in by visitors to Bletchley Park indicate that less than one per cent of the meteoric rise in visitor numbers during 2008–14 was directly attributable to social media. More significant factors included a visit by Prince Charles and the Duchess of Cornwall in 2008, and two *Antiques Roadshow* TV programmes, filmed at Bletchley Park and broadcast in 2009, each to an audience of 6 million.

The modern era

In July 2011 Queen Elizabeth visited Bletchley Park, declaring:[10]

It is impossible to overstate the deep sense of admiration, gratitude and national debt that we owe to all those men and, especially, women. . . . This was the place of geniuses such as Alan

Figure 19.2 The bombe rebuild at Bletchley Park.

Photograph by Jonathan Bowen.

Turing. But these wonderfully clever mathematicians, language graduates and engineers were complemented by people with different sets of skills, harnessing that brilliance through methodical, unglamorous, hard slog. Thus the secret of Bletchley's success was that it became a home to all the talents.

The Queen was introduced to Jean Valentine (see Fig. 12.11), Jerry Roberts (see Fig. 16.2) and other veterans, and also to John Harper's wonderful recreation of the bombe shown in Fig. 19.2.[11]

By the time of the Queen's visit, the site had been saved from developers and rescued from bankruptcy, and most of its buildings were by then properly maintained; also its significance was recognized nationally and internationally. Attention now focused on using the unique site to dramatize the codebreakers' story for the general public. To this end a number of new permanent exhibitions were installed, including 'Life and Works of Alan Turing' (curated by Gillian Mason and Jack Copeland) and 'Hitler's "Unbreakable" Cipher Machine', telling the story of the Tunny machine and how it succumbed to the ingenuity of Tutte, Turing, Flowers, and others (also curated by Mason and Copeland). Stephen Kettle's statue of Turing is a prominent feature of the 'Life and Works of Alan Turing' exhibition (Fig. 19.3).

The 2012 *Alan Turing Year*, celebrating Turing's centenary (see Chapters 2 and 42) helped to make Bletchley Park widely known internationally. Bletchley Park's own contribution to the Turing year 'The Turing Education Day' sold out weeks in advance: like this book, the Bletchely Park event aimed to explain Turing's scientific ideas to a general audience (the event was organized by Jack Copeland, Claire Urwin, and Huma Shah). In 2014, the first phase of a new restoration and development project was opened by Her Royal Highness the Duchess of Cambridge, thanks to a grant in 2011 of £5 million from the Heritage Lottery Fund, reeled in by a small group that included Kelsey Griffin, Bletchley Park's Director of Operations and Communication—and additional contributions totalling £3 million from the Foreign Office, Google, and other companies, trusts and foundations.[12] Bletchley Park's Director of Development Claire Glazebrook says:

The public also got behind the campaign and gave generously.

The money funded not only state-of-the-art museum facilities and the imaginative new visitor centre in Block C, but also extensive upgrades to the site and its landscaping, all helping to return Bletchley Park towards its wartime appearance (even to the point of cars being excluded from the central site). Two codebreakers' huts neighbouring Turing's already magnificently restored Hut 8—Hut 3 and Hut 6—were put back into their original condition and lovingly repainted in their wartime colours.[13]

In 2015, on the anniversary of VE Day, current CEO Iain Standen unveiled the plan for the next stage of development:[14]

The future of Bletchley Park has already been safeguarded for the nation, but this Masterplan is the next milestone on the road to completing its restoration for all future generations to experience, learn from and enjoy.

Figure 19.3 Bletchley Park's slate statue of Turing with an Enigma machine, by Stephen Kettle, part of the 'Life and Works of Alan Turing' exhibition.[15]

Stephen Kettle (http://www.stephenkettle.co.uk). Photograph by Jonathan Bowen.

Conclusion

Since the war Bletchley Park has fortunately remained largely intact, except for the loss of some of the original smaller huts and also of Block F, which housed the Newmanry and the first Colossus computers. Today Bletchley Park looks much as it did in Turing's time.

The *Guardian* newspaper aptly said:[16]

It's terrific that Bletchley Park has not only been rescued from the decay into which the site had fallen, but brilliantly restored . . . Even at its lowest ebb, Bletchley had a magical aura.

Magical indeed. Google's head of communications Peter Barron speaks for many young people when he says:[17]

a lot of our staff feel that if they had been around during the war they would have wanted to work at Bletchley Park.

PART IV

Computers after the war

CHAPTER 20

Baby

JACK COPELAND

T he modern computer age began on 21 June 1948, when the first electronic universal stored-program computer successfully ran its first program. Built in Manchester, this ancestral computer was the world's first universal Turing machine in hardware. Fittingly, it was called simply 'Baby'. The story of Turing's involvement with Baby and with its successors at Manchester is a tangled one.[1]

The modern computer is born

The world's first electronic stored-program digital computer ran its first program in the summer of 1948 (Fig. 20.1). 'A small electronic digital computing machine has been operating successfully for some weeks in the Royal Society Computing Machine Laboratory', wrote Baby's designers, Freddie Williams and Tom Kilburn, in the letter to the scientific periodical *Nature* that announced their success to the world.[2] Williams, a native of the Manchester area, had spent his war years working on radar in rural Worcestershire. Kilburn, his assistant, was a blunt-speaking Yorkshireman. By the end of the fighting there wasn't much that, between them, they didn't know about the state of the art in electronics. In December 1945 the two friends returned to the north of England to pioneer the modern computer.[3]

Baby was a classic case of a small-scale university pilot project that led to successful commercial development by an external company. The Manchester engineering firm Ferranti built its Ferranti Mark I computer to Williams's and Kilburn's design: this was the earliest commercially available electronic digital computer. The first Ferranti rolled out of the factory in February 1951.[4] UNIVAC I, the earliest computer to go on the market in the United States, came a close second: the first one was delivered a few weeks later, in March 1951.[5]

Williams and Kilburn developed a high-speed memory for Baby that went on to become a mainstay of computing worldwide. It consisted of cathode-ray tubes resembling small television tubes. Data (zeros and ones) were stored as a scatter of dots on each tube's screen: a small focused dot represented '1' and a larger blurry dot represented '0'. The Williams tube memory, as the invention was soon called, was also used in Baby's immediate successors, built at Manchester

Figure 20.1 Baby. Freddie Williams is on the right, Tom Kilburn on the left.

Reproduced with permission of the University of Manchester School of Computer Science.

University and by Ferranti Ltd. Computers using the Williams tube memory included TREAC, built in Worcestershire at the secret establishment where Williams had done his radar work; SEAC, constructed at the US National Bureau of Standards in Washington, DC; SWAC, built at the US National Bureau of Standards Western Division at the University of California, Los Angeles; and the highly influential computer built at Princeton's Institute of Advanced Study (officially nameless, but informally known as MANIAC), as well as the Princeton computer's epoch-making successors, the IBM 701 and 702—and also AVIDAC, ORDVAC, ORACLE, ILLIAC, SILLIAC, and other engagingly named first-generation computers.[6]

It was Turing's Bletchley Park colleague Max Newman (Fig. 14.6) who founded the Manchester Computing Machine Laboratory, Baby's birthplace. When Bletchley Park wrapped up its operations at the end of the war, Newman became Fielden Professor of Mathematics at Manchester. His dream was to create a new Newmanry, this time housing a true universal Turing machine in hardware—a stored-program electronic computer that would revolutionize peacetime mathematics and science, just as Colossus had revolutionized warfare. (For more information on Max Newman, and his work at Bletchley Park and Manchester University, see Chapters 14 and 40, and also my previous publications listed in Note 1.)

Unfortunately, traditional histories of the Manchester Baby have either ignored or under-rated the role that Newman played in the Manchester computer project—and also the role played by Turing himself. In his classic 1975 history of the Manchester computer project, Simon Lavington stated that 'Newman and his mathematicians took no active role in the design of Manchester computers', while Mary Croarken said in her 1993 history that 'neither Newman nor Turing had any influence on Williams's designs for the computer'.[7] In fact, both played very active roles in the computer developments at Manchester, as did their Bletchley Park colleague

Jack Good. Until recently, Turing's influence on Kilburn's early thinking about computer architecture went completely unrecognized. Turing showered Kilburn with original ideas, while Good (although a brilliant and original thinker himself) acted mainly as a go-between, supplying Kilburn with design ideas that originated in the United States. Newman, too, exposed Williams and Kilburn to American thinking.

As I shall explain, the logical design of the 1948 Baby was virtually identical to a 1946 design produced by John von Neumann (Fig. 6.3) and his computer group at Princeton's Institute of Advanced Study. It was as electronic engineers, not as computer architects, that Williams and Kilburn led the world in 1948. In effect Newman and Good acted as messengers between the Princeton group and Kilburn and Williams, transferring information about the Princeton design. Kilburn seems to have remained quite unaware of the true source of the logical ideas he acquired from Newman and Good, while Williams acknowledged that his and Kilburn's thinking might (via Newman) have been indirectly influenced by von Neumann, although he too seemed to have no inkling of the true extent of this influence.[8]

The intellectual traffic was not all one way, however. In the summer of 1948, when the Princeton project was hopelessly bogged down with memory problems, von Neumann's chief engineer, Julian Bigelow, visited Williams's laboratory at Manchester.[9] Bigelow, who extolled Williams's 'inventive genius', realized that the Manchester tube memory was exactly what the Princeton group needed. When their vast computer was finished in 1951—the first of the so-called Princeton class computers—its main memory consisted of forty Williams tubes, compared with just one in Baby. The Manchester tube memory made 'a whole generation of electronic computers possible', said Herman Goldstine, one of the Princeton group's leading lights.[10] The Princeton computer's commercial offspring, the IBM 701, was IBM's first stored-program electronic computer. Its unveiling, complete with the Williams tube memory, marked the dawn of a new era in computer sales.

By a wonderful coincidence, Turing himself had outlined a design for a computer memory closely resembling the Williams tube six months or more before Williams even entered the field. In his report 'Proposed electronic calculator' (the document setting out the design of his ACE), Turing wrote that 'a suitable storage system can be developed without involving any new types of tube, using in fact an ordinary cathode-ray tube with tin-foil over the screen to act as a signal plate'. This was exactly the arrangement later adopted by Williams. Turing continued:[11]

It will be necessary to furbish up the charge pattern from time to time, as it will tend to become dissipated . . . If we were always scanning the pattern in a regular manner as in television this would raise no serious problems. As it is we shall have to provide fairly elaborate switching arrangements to be applied when we wish to take off a particular piece of information. It will be necessary to . . . switch to the point from which the information required is to be taken, do some scanning there, replace the information removed by the scanning, and return to refurbishing from the point left off.

Turing was certainly correct that the key problem was how to 'furbish up' the stored patterns—the zeros and ones—before they leaked away. The solution he proposed in this quotation was essentially the method subsequently adopted by Williams: the way to stop the pattern disappearing—in other words, the way to make the tube 'remember' the data written on its screen—was to scan the face of the tube continually, reading and then rewriting each digit. As Williams later explained it, 'you go and look at a spot and say "what's there?", all right that's there, so you put it back'.[12]

The National Physical Laboratory (NPL) had mailed Williams a copy of Turing's 'Proposed electronic calculator' in October 1946, in the month before Williams lodged a draft application for a patent on the Williams tube.[13] It isn't recorded whether Williams read it at that time, but there is no reason to think otherwise. He need only have opened the envelope and glanced at the table of contents to see the irresistible chapter title 'Alternative forms of storage', the very field he himself was pioneering. Racing to the chapter, he would have found Turing's pithy description of the Williams tube—enough to make his ever-present cigarette pop out of his mouth with surprise. He might well have objected, though, to Turing's airy statement that turning the basic ideas into engineering reality would not involve 'any fundamental difficulty'. There is now no way of telling how much or how little Williams learned from Turing's document. All that can be said for certain is that he quickly dropped his original 'anticipation pulse' method, in favour of the simpler read–rewrite method for refreshing the data that Turing described.

Too little credit

Although Williams and Kilburn had brilliantly translated the logico-mathematical concept of the stored-program computer into electronic hardware, the mathematicians in Newman's department gave them too little credit.[14] Peter Hilton, who joined the Manchester mathematicians in 1948, explained that Williams and Kilburn were regarded as excellent engineers but not as 'ideas men'. Nowadays, though, the tables have turned too far and the triumph at Manchester is usually attributed to the engineers alone. Fortunately, Williams's own words survive to remind us of the true situation. 'Tom Kilburn and I knew nothing about computers', Williams said.[15] 'We'd had enough explained to us to understand what the problem of storage was and what we wanted to store, and that we'd achieved, so the point now had been reached where we'd got to find out about computers', he recalled, continuing: 'Professor Newman and Mr A. M. Turing . . . took us by the hand'.[16]

Kilburn, in later life, was an important source for what became the canonical view of the roles of Turing and Newman (or rather, their lack of role) in the origin of Baby. In his first papers on the Manchester computer, Kilburn gave credit to both Turing and Newman,[17] but in later years he was at pains to assert the independence of his and Williams's work from outside influence, presenting the history of Baby in a way that assigned no role to Turing or to Newman. In an interview with me in 1997, Kilburn emphasized that Newman 'contributed nothing to the first machine' (Baby).[18] Turing's only contributions, Kilburn told me, came after the computer was working, and included preparing what he called a 'completely useless' programming manual.

Yet Turing's logico-mathematical ideas of 1936 had led, via Newman, to Manchester's stored-program computer project. Even in the midst of the attack on Tunny, Newman had been thinking about Turing's universal machine: when Flowers was designing Colossus, Newman showed him Turing's 1936 paper ('On computable numbers') about the universal computing machine, with its key idea of storing coded instructions in memory.[19] By 1944 (as Newman related in a letter to von Neumann) he was looking forward to setting up an electronic computer project as soon as the war was over.[20] The Princeton computer, too, was a physical embodiment of Turing's universal computing machine of 1936: Julian Bigelow told me that von Neumann even gave his Princeton engineers 'On computable numbers' to read.[21] The reason why von Neumann became the 'person who really . . . pushed the whole field ahead', Bigelow explained, was that he understood what was implied by Turing's universal machine idea:[22]

Turing's machine does not sound much like a modern computer today, but nevertheless it was. It was the germinal idea.

Although Turing is not mentioned explicitly in von Neumann's papers developing the design for the Princeton computer, his collaborator and co-author Art Burks told me that their key 1946 design paper (co-authored also by Goldstine) did contain a reference to Turing's 1936 work.[23] There they emphasized that 'formal-logical' work—in particular, Turing's 1936 investigation—had shown 'in abstracto' that stored programs can 'control and cause the execution' of any sequence (no matter how complex) of mechanical operations that is 'conceivable by the problem planner'.[24] More information about Turing's influence on von Neumann can be found in Chapter 6.

So Turing's theoretical ideas of 1936 were the ultimate inspiration for both the Manchester and Princeton computer projects. In 1947, thanks to Newman and Good, an early version of the Princeton design ended up as the logical blueprint for the Manchester Baby, as we shall see. But the Princeton computer worked only because Williams baled out von Neumann's group, tossing them the memory design he had invented for Baby (in return for substantial royalties, of course). It was a wonderfully tangled loop, with Turing at its centre.

Prequel: from Bletchley Park to Manchester

It was, of course, no coincidence that, as soon as the war ended, Turing and Newman both embarked on projects to create a universal Turing machine in hardware. But before the 1970s, when the British government declassified captioned photographs of Colossus,[25] few people had any idea that electronic computers had been operational at Bletchley Park from 1944 (see Chapter 14). It was not until 2000 that the British government finally declassified a complete account of Colossus (as explained in Chapter 17).[26] Having no inkling of Colossus, early historians of computing formed the view that the British pioneers inherited their vision of large-scale electronic computing machinery from the ENIAC group in the United States.[27]

ENIAC (Chapter 14) and its successor EDVAC were certainly the fountainhead of inspiration for some British computer pioneers (Douglas Hartree and Maurice Wilkes, for instance).[28] However, it was Colossus, not ENIAC, that formed the link between Turing's pre-war exposition of the universal machine and Newman's post-war project to build an electronic stored-program computer. Colossus was a pivotal influence on Turing also. Flowers told me that once Turing had seen Colossus, it was just a matter of his waiting for an opportunity to put the idea of his universal computing machine into practice.[29] Other mathematicians and engineers working in Newman's section at Bletchley Park also made the connection between Flowers' racks of electronic equipment and the idea of an all-purpose stored-program computer. Members of the Newmanry were (so Newmanry codebreaker Donald Michie reported) 'fully aware of the prospects for implementing physical embodiments of the UTM [universal Turing machine] using vacuum-tube technology'.[30]

Newman laid plans for his Computing Machine Laboratory as soon as Manchester University told him that they were going to appoint him as professor. His formidable talent as an organizer, honed in the Newmanry, was now brought to bear on the problem of designing and constructing an electronic stored-program computer. He applied to the Royal Society of London for a sizeable grant to develop such a machine (finally approved by the Treasury in May 1946),[31] and in August 1945 he wrote to Bletchley Park, requesting that 'the material of two complete Colossi' be sent to Manchester. Newman said: 'We should like the counter racks and the "bedsteads" (tape-racks) to be in working order but the rest could be dismantled so far as is necessary to make the circuits unrecognisable'.[32] A 6-ton lorry and trailer piled high with goodies from Bletchley Park arrived in Manchester on 7 December 1945.[33] The first work on the Manchester computer made use of one of the bedsteads.[34] 'It reminds me of Adam's rib', said Jack Good.

By the time that Freddie Williams was interviewed for Manchester's recently vacated Edward Stocks Massey Chair of Electro-Technics, in 1946, Newman had already established his Computing Machine Laboratory at the university and was on the lookout for the right engineer to bring into the project. What he needed, he explained in his funding application to the Royal Society, was a 'circuit-designing engineer' who, although 'he would not be expected to provide the main ideas', would 'need a rare combination of wide practical experience in circuit design, with a thorough understanding of the abstract ideas involved'.[35] Newman was on the panel that interviewed Williams,[36] and when Williams began explaining the virtues of the new form of computer memory he was perfecting, the prospect of joining forces must have started to look like a marriage conceived in heaven. It was 'a very fruitful opportunity for collaboration', Williams said.[37] Only 35 years old, he acquired the title 'professor'. When Williams started his new job, Newman's computer lab was little more than an empty room. Williams later poked fun at the room's 'fine sounding' title of Computing Machine Laboratory, recollecting that the 'walls were of brown glazed brick and the door was labelled "Magnetism Room" '.[38] It was, he said with his blunt humour, 'lavatorial'.

At Bletchley Newman had been chief executive of a project with staff numbering over 300. He initiated and oversaw the creation of a dazzling array of machines, all lying at the frontier of the then-current technology.[39] Newman, himself no engineer, achieved these outstanding successes by the skilful use of a simple principle: get the right engineers involved, explain to them what needs to be done, and let them get on with it. As Michie wrote:[40]

Once he [Newman] placed his trust in people he cut them loose to manage according to their own judgement.

Not surprisingly, Newman followed the same method at Manchester: he educated Williams in the fundamentals of the stored-program computer, and then (as Newman wrote) the 'design of the machine was naturally placed in his hands'.[41]

By this time, Turing himself was already providing Williams's assistant Tom Kilburn with a much more extensive education in computer design.

Turing's Adelphi lectures, 1946–47

Tom Kilburn stepped into Turing's world at the end of 1946, when he entered a dingy London lecture room and sat down to listen to Turing explaining how to build a computer.[42] Williams had recently succeeded in storing a single binary digit on the face of a cathode-ray tube, proving that his memory design worked in principle, and so (as Williams said) 'the point now had been reached where we'd got to find out about computers'.[43] They heard that Turing was giving a series of lectures on computer design in London, and it was decided that Kilburn would attend.[44] The lectures ran from December 1946 to February 1947 and were held in a conference room at the Adelphi Hotel in the Strand.[45]

When asked, in later life, where he got his basic knowledge of the computer from, Kilburn usually said, rather irritably, that he couldn't remember.[46] In an interview, he commented vaguely:[47]

Between early 1945 and early 1947, in that period, somehow or other I knew what a digital computer was . . . Where I got this knowledge from I've no idea.

However, Kilburn had referred to 'unpublished work' by Turing in his first report on his computer research, dated December 1947, and he had used a number of Turing's technical terms.[48] These included Turing's 'universal machine' and 'table of instructions', and also various terms that, whether or not they originated with Turing, were distinctive of Turing's approach in the Adelphi lectures: for example 'source', 'destination', 'temporary store', 'staticisor', and 'dynamicisor'. In a subsequent report, written when the computer was working, Kilburn said:[49]

I wish to acknowledge my indebtedness to Prof. M. H. A. Newman, and Mr. A. M. Turing for much helpful discussion of the mathematical requirements of digital computing machines.

Williams summed up Turing's role like this:[50]

[Turing was] instrumental, with Newman, in instructing us in the basic principles of computing machines, not on the engineering side of course, on the mathematical, and we had very close collaboration with both of them.

The Adelphi lectures played a key role in the Manchester project, by introducing Kilburn to the fundamentals of computer design. There is in fact no mystery about where Kilburn got his basic knowledge of the computer from—Turing taught him.

Kilburn's first computer

Kilburn was a good pupil, quickly progressing during the lectures from not knowing the 'first thing about computers'[51] to the point where he could start designing one himself: in fact, Kilburn's initial design for what would eventually be the Manchester computer followed Turing's principles closely. Turing, unlike von Neumann and his group, advocated a *decentralized* computer with no central processing unit (cpu)—no one central place where all the logical and arithmetical operations were carried out. (The term 'decentralized', and its opposite 'centralized', were due to Jack Good, for whom Newman had created a special lectureship in Mathematics and Electronic Computing in his Manchester department.[52]) Kilburn designed a decentralized computer very much along the lines that Turing set out in his lectures: it was quite different from the centralized type of design that von Neumann was proposing.

In Version V of Turing's ACE design, which formed the subject matter of the first five Adelphi lectures, there was a collection of different 'sources' and 'destinations' in place of a central accumulator. The central accumulator is a storage unit that forms the sum of an incoming number and the number already stored: this sum then replaces the previous contents of the accumulator. Early centralized computers did all their calculations by means of transferring numbers to and from the accumulator. In Turing's decentralized Version V, on the other hand, the arithmetic and logical operations were carried out at various 'destinations': these were typically mercury delay lines. One destination performed addition, another subtraction, others logic operations, and so forth. ACE's programs were made up of instructions like 'Transfer the contents of sources 15 and 16 to destination 17'.[53] Turing wrote this simply as '15–17'. The instruction's effect was to apply the operation associated with destination 17 to the two numbers that were transferred. In this case, the operation was addition, and the instruction 15–17 summed the numbers stored in sources 15 and 16. Since the identity of the operation that would be performed was implicit in the destination number 17 (or in the case of some instructions, in the source number), there was no explicit term in the instruction specifying the operation itself—the instruction contained no 'op-code' meaning ADD.

The computer designed by Kilburn had no central accumulator. In his Turing-style decentralized design, each of the machine's elementary arithmetical and logic operations was implemented by a separate destination.[54] Every instruction, Kilburn explained, transferred a number 'from one part of the machine—a "source"—to another—a "destination" '. As in the Turing paradigm, instructions consisted of pairs of numbers, a source number s and the destination number d. These numbers controlled what Kilburn called, following Turing, a 'source tree' and a 'destination tree'. A 'tree' is a system of interconnections and gates. It is the source tree that accesses the number with address s in the store, while the destination tree accesses the destination that will perform the required operation (such as the adder). An instruction's effect was to route the operand(s) from the main memory, via the source tree, to the destination tree, and thence to the adder (or other unit specified in the instruction). All of these ideas, including the concepts of source and destination trees, were explained to Kilburn during the Adelphi lectures. He went on to summarize his design in a block diagram that depicted a very ACE-like computer.[55]

Meanwhile, Newman spent the autumn and winter of 1946 in Princeton,[56] and by the time he returned he had decided to adopt von Neumann's centralized design for his own computer at Manchester. When Turing's NPL colleague Harry Huskey dropped in to visit the Newman–Williams project (as he called it) early in 1947, he learned that the Manchester plan was to 'more or less copy the von Neumann scheme'.[57] Newman had originally intended to copy Turing's ACE design, and had assumed that Flowers and his engineers in London would be involved in building the Manchester computer,[58] but he abandoned this plan. He and Jack Good had travelled down from Manchester to the NPL to learn in detail about Turing's design for ACE, but Newman grew so irritated by Turing's inability to make himself clear that he excused himself and returned to Manchester, leaving Good to hold the fort.[59] Good later explained to Michie that 'Turing plunged into some arbitrary point in the thicket, and insisted on defining and describing every leaf and branch at that particular spot, fanning out from there'.[60]

ACE was not for Newman, and he found von Neumann's design principles simple and straightforward. When Newman discovered that Williams's bright young assistant Kilburn was full of ideas for an ACE-like computer, he may not have been too impressed.

From Princeton with love

Not long after Williams and Kilburn arrived in Manchester, Newman himself gave them a few lectures on how to design a computer. Turing's Adelphi lecture series was probably just about finishing at this time.[61] Naturally, Newman's lectures laid emphasis on the von Neumann centralized design.[62] 'Newman explained the whole business of how a computer works to us', Williams remarked.[63] As he recalled in a letter to Brian Randell:[64]

I remember Newman giving us a few lectures in which he outlined the organisation of a computer in terms of numbers being identified by the address of the house in which they were placed and in terms of numbers being transferred from this address, one at a time, to an accumulator where each entering number was added to what was already there. At any time the number in the accumulator could be transferred back to an assigned address in the store and the accumulator cleared for further use.

At this point, Williams left it to Kilburn to fathom out the details of what they were going to build; apart from being preoccupied with perfecting the new memory, Williams now had his Department of Electro-Technics to run.[65] Kilburn eagerly put his ideas for an ACE-like design behind him, and began to plan out a centralized computer.

Kilburn consulted Good, asking him to suggest an instruction set for a centralized computer.[66] The instruction set is the logical heart of any computer; it details the computer's 'atomic' operations, the elementary building blocks for all its programmed activity. In May 1947 Good suggested a set of twelve basic instructions to Kilburn, as shown in Table 20.1.[67] Good announced this important historical fact to the computing world in 1998, in his acceptance speech for the IEEE Computer Pioneer Award, where he explained that he had 'made a proposal, at Kilburn's request, for the basic mathematical instructions for the Baby machine'.[68]

Kilburn simplified the twelve instructions that Good gave him, reducing them to five. (I established this by comparing Baby's instructions as given by Williams and Kilburn in their

Table 20.1 *Good's Princeton instructions. This table shows the twelve basic instructions that Good recommended to Kilburn in May 1947. Good himself gave only the symbolic form of each instruction, noted here in square brackets.*

1	Transfer the number in storage register x (there were sixty-four storage registers or 'houses') to the accumulator	$[x \rightarrow A]$
2	Add the number in x to the number in the accumulator and store the result in the accumulator	$[A + x \rightarrow A]$
3	Transfer the negative of the number in x to the accumulator	$[-x \rightarrow A]$
4	Subtract the number in x from the number in the accumulator and store the result in the accumulator	$[A - x \rightarrow A]$
5	Transfer the number in the accumulator to the (arithmetic) register R	$[A \rightarrow R]$
6	Transfer the number in R to the accumulator	$[R \rightarrow A]$
7	Transfer the number in the accumulator to x	$[A \rightarrow x]$
8	Transfer the number in x to R	$[x \rightarrow R]$
9	Shift the number in the accumulator one place to the left	$[l]$
10	Shift the number in the accumulator one place to the right	$[r]$
11	Transfer control unconditionally to the instruction in x	$[C \rightarrow x]$
12	Conditional transfer of control, viz, transfer control to the instruction in x if the number in the accumulator is greater than or equal to 0	$[CC \rightarrow x \text{ if } A \geq 0]$

1948 letter to *Nature* with the instructions in Good's May 1947 set.) Kilburn decided that subtraction should be Baby's only atomic arithmetical operation, making Good's first two instructions unnecessary. Williams later explained why addition was not needed:[69]

The facilities we decided to provide were the absolute minimum . . . . [Y]ou can do addition by means of subtraction, because you can subtract something from nothing and get its negative, and then subtract its negative from what you want to add it to, and you get the sum . . . So we had the one basic arithmetic operation, subtraction.

The two shift instructions *l* and *r* were logically redundant, and could be dispensed with in a minimal machine: as Good noted,[70] the left shift is just multiplication by two, and the right shift is division by two. Good's instructions $A \rightarrow R$, $R \rightarrow A$, and $x \rightarrow R$ were also unnecessary, since there was no arithmetic register *R* in Baby.

So Kilburn's reduced instruction set contained Good's instructions numbered 3, 4, 7, and 11. There was also a modified form of Good's instruction 12—namely, skip the next instruction in the store if the number in the accumulator is less than zero—and additionally a sixth instruction, stop the machine. Now that Kilburn had an instruction set, he knew what he and Williams needed to build—as he said to me, 'You can't start building until you have got an instruction code'.[71] The tiny centralized computer they built contained three Williams tubes, forming the store, accumulator, and control that were presupposed by Good's instructions. Baby performed calculations by transferring numbers between the store and the accumulator, and as Table 20.2 shows,[72] Baby's initial instruction set consisted exactly of the five instructions that Kilburn had derived from Good's twelve, plus the additional 'stop' instruction.[73]

Good did not, it seems, mention to Kilburn that he had distilled his twelve instructions from a 1946 design paper by von Neumann, Burks, and Goldstine.[74] Like Newman, Good had studied that paper closely.[75] Of his twelve instructions, only instruction 5—transfer the number in *A* to *R*—was not in the Princeton set. Baby was a very close relative indeed of the more complex computer described in the Burks–Goldstine–von Neumann paper. Both were single-address machines with an accumulator;[76] and moreover the control arrangements for the two computers were effectively the same.[77]

The extent to which the design of Baby followed Princeton thinking becomes evident if Table 20.2 is compared with the following 1947 summary by Harry Huskey of logical aspects of

Table 20.2 *Baby's instruction set given in Williams and Kilburn's 1948 letter to Nature (see Note 2). (The square brackets show my translation into Good's Princeton-style notation of the instructions as Williams and Kilburn presented them.) Instructions 1–4 are all members of Good's May 1947 instruction set, and instruction 5 is a simplified form of Good's CC instruction.*

1.	If *x* is any number in the store, $-x$ can be written into a central accumulator *A*	$[-x \rightarrow A]$
2.	*x* can be subtracted from what is in *A*	$[A - x \rightarrow A]$
3.	The number *A* can be written in an assigned address in the store	$[A \rightarrow x]$
4.	Control can be shifted to an assigned order in the table	$[C \rightarrow x]$
5.	The content of *A* can be tested for whether $x \geq 0$, or $x < 0$; if $x < 0$, the order standing next in the store is passed over	$[CC]$
6.	The machine can be ordered to stop	[No equivalent]

the Princeton design. Apart from Kilburn's idea of using subtraction as the single basic arithmetical operation, Baby's instruction set was virtually identical to what was being proposed at Princeton:[78]

Princeton. In the von Neumann plan the machine consists essentially of a memory and a static accumulator. All transfers from the memory to the accumulator are essentially additions. The orders used are of the following type:

(1) Clear A   This clears the accumulator.

(2) x to A   Adds the number in position x in the memory to the contents of the accumulator.

(3) A to x Transfers the number in the accumulator to position x of the memory.

(4) C to x Transfers control to x; i.e., the next order to be obeyed is in position x in the memory.

(5) CC to x Conditional transfer of control; i.e., control is transferred to x in the memory if the number in A is negative. Otherwise, control obeys consecutive orders in the memory.

Baby's originality certainly did not lie in its logical design, but in its cathode-ray tube memory and its electronic engineering. Baby is regarded as a British triumph, a world first for British computing. Yet one of computing's greatest ironies is that, thanks to Good and his Princeton-based instruction set, the logical design of Baby was imported from America—seemingly without Kilburn or Williams ever realizing that this was so.

Von Neumann and ACE

Von Neumann also had some influence on Turing's design for ACE. Von Neumann's first paper on computer design, now known universally as 'First draft of a report on the EDVAC', was circulated as early as June 1945 and was widely read and used.[79] Turing's great friend Robin Gandy recalled that Turing 'became excited' when he first read von Neumann's paper.[80] He studied it closely—although, with his focus on squeezing as much speed as possible from the hardware, he went on to design a very different, decentralized, computer. Turing expected the readers of his own design paper 'Proposed electronic calculator' (completed later the same year) to be familiar with the 'First draft', and he recommended that his paper be read 'in conjunction with' von Neumann's.

In fact, the design that Turing set out in 'Proposed electronic calculator' was much more concrete than the proposals in the von Neumann paper, where EDVAC was described at a high level of abstraction: von Neumann hardly mentioned electronics at all. Huskey, the engineer whose job it was to draw up the first detailed hardware designs for EDVAC, said that he found von Neumann's paper to be of 'no help'.[81] Turing's design paper, on the other hand, gave detailed specifications of the hardware and also included sample programs in machine code. Turing was, though, perfectly happy to borrow some of the more elementary material in von Neumann's report. For example, his diagram of an adder is essentially the same as von Neumann's diagram (Figure 10 of Turing's 'Proposed electronic calculator' shows an adder identical to the one depicted in Figure 3 of von Neumann's 'First draft'), although in general Turing's logic diagrams—which set out detailed circuits for the arithmetical parts of the calculator and also the logical control—go far beyond anything to be found in the 'First draft'. This minor borrowing is probably what Turing was referring to when he told a newspaper reporter in 1946 that he gave 'credit for the donkey work on the A.C.E. to Americans'.[82]

Yet the similarities between Turing's design and the von Neumann proposals are relatively minor compared with the striking differences. Moreover, the fact that von Neumann exerted some influence on 'Proposed electronic calculator' should not be allowed to mask the extent to which Turing's universal machine of 1936 was itself a fundamental influence on von Neumann.

Making history

By the start of summer 1948, the Manchester Baby was wired together and ready to try out. Painstakingly, Kilburn and Williams entered the first program by hand, literally bit by bit, using a panel of switches to plant each bit in memory. The program's function was to find the highest factor of a given number, a task that many people can carry out easily on the back of an envelope. Eventually the program was stored successfully on the screen of a single Williams tube. It was a mere seventeen instructions long.

Gingerly, the start switch was pressed. 'Immediately the spots on the display tube entered a mad dance', Williams related.[83] This turned out to be a 'dance of death', he said, a hiding to nowhere that was repeated again and again during the following week. 'But one day', he recounted, 'there, shining brightly in the expected place, was the expected answer'. It was Monday 21 June 1948, the first day of the modern computer age—never before had electronic hardware run a stored program. Williams said drily 'We doubled our effort immediately by taking on a second technician'.

At Manchester Turing finally had his hands on a stored-program computer. He was soon using Manchester University's Ferranti Mark I to model biological growth (as described in Chapters 33–35). Others also began to explore the computer's potential, creating a starburst of 'firsts' in the Manchester laboratory. Christopher Strachey, soon to be one of computing's leading luminaries, got an artificial intelligence program up and running as early as 1952: it played draughts (checkers), soliciting its human opponent's next move with a peremptory 'pip pip' sound.[84] If the human player delayed too long, the computer's printer would chatter out impatiently

YOU MUST PLAY AT ONCE OR RESIGN

or even

I REFUSE TO WASTE ANY MORE TIME. GO AND PLAY WITH A HUMAN BEING.

Strachey hijacked the computer's monitor tube—normally used for showing the engineers how things were ticking over in the inner workings—and set it up so as to display a virtual draughts-board (Fig. 20.2, which is from Strachey's notes). This was the first time that a computer screen was used for interactive game playing.

Williams and Kilburn had used a Williams tube to store digital text several years earlier, in the autumn of 1947—another historic moment—and used the term 'picture elements' (nowadays always shortened to 'pixels') to refer to the discrete blips making up the letters.[85] A photograph of their historic experiment showed primitive pixels emblazoned across the face of the tube to read 'C.R.T. STORE' (for 'cathode-ray tube store').[86] What's more, Williams and Kilburn

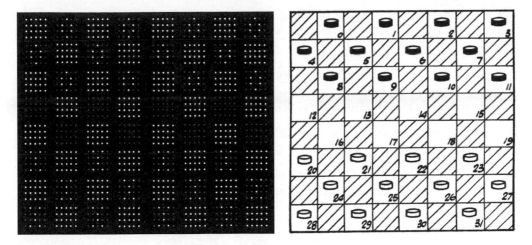

Figure 20.2 The Manchester computer playing draughts (checkers). Strachey's hand-drawn diagram explains the symbols on the screen. The computer is Black.

also created the first *moving* digital images ever to appear on a computer screen, in June 1948: rhythmically changing patterns of pixels on the Baby's screen represented—or pictured—the changing contents of the computer's memory as the program ran. These images, flickering across the tiny glass-and-phosphor screen, were the first step toward today's digital movies and computer-generated animations as well as the images on the screens of our phones and tablets. You can watch a recreation of these images at *tinyurl.com/first-moving-digital-images*.[87]

Turing and Strachey were both interested in the computer's capacity to process words as well as numbers. Strachey went so far as to design a program that wrote love letters. The steamy notes were signed 'M. U. C.' (for 'Manchester University computer'). Strachey used programming tricks that he described as 'almost childishly simple' (and to him they probably were): he made up some recipes for building sentences and stored the recipes in the computer's memory, together with lists of words and expressions that he culled from *Roget's Thesaurus*.[88] The program would follow one or another recipe step by step, including steps like 'Here choose a word from list so-and-so'. The program made its selections using the random number generator that Turing had designed, a kind of electronic roulette wheel (see Chapter 39).[89] The computer's letters were eerily alien:[90]

DARLING SWEETHEART
YOU ARE MY AVID FELLOW FEELING. MY AFFECTION CURIOUSLY CLINGS TO YOUR PASSIONATE WISH. MY LIKING YEARNS FOR YOUR HEART. YOU ARE MY WISTFUL SYMPATHY: MY TENDER LIKING.
YOURS BEAUTIFULLY
M. U. C.

HONEY DEAR
MY SYMPATHETIC AFFECTION BEAUTIFULLY ATTRACTS YOUR AFFECTIONATE ENTHUSIASM. YOU ARE MY LOVING ADORATION: MY BREATHLESS ADORATION. MY FELLOW FEELING BREATHLESSLY HOPES FOR YOUR DEAR EAGERNESS. MY LOVESICK

ADORATION CHERISHES YOUR AVID ARDOUR.
YOURS WISTFULLY
M. U. C.

Turing, rather more practically, typed his own letters at the computer keyboard, making use of the slash symbol to produce punctuation and crude formatting. He printed these out and dropped them into the post. One was to Robin Gandy concerning his PhD viva.[91] It began:

DEAR/ROBIN///////////////SORRY////IT/REALLY/ISNT/POSSIBLE/TO/MAKE/YOUR/ORAL/ANY/
EARLIER////

Trust Turing to be the first person on earth to deal with his correspondence by word-processor. He and Strachey also began to explore the possibilities of the computer as a musical instrument, as explained in Chapter 23.

CHAPTER 21

ACE

MARTIN CAMPBELL-KELLY

I n October 1945 Alan Turing was recruited by the National Physical Laboratory to lead computer development. His design for a computer, the Automatic Computing Engine (ACE), was idiosyncratic but highly effective. The small-scale Pilot ACE, completed in 1950, was the fastest medium-sized computer of its era. By the time that the full-sized ACE was operational in 1958, however, technological advance had rendered it obsolescent.[1]

Introduction

Although the wartime Bletchley Park operation saw the development of the electromechanical codebreaking bombe (specified by Turing) and the electronic Colossus (to which Turing was a bystander), these inventions had no direct impact on the invention of the electronic stored-program computer, which originated in the United States.

The stored-program computer was described in the classic 'First draft of a report on the EDVAC',[2] written by John von Neumann on behalf of the computer group at the Moore School of Electrical Engineering, University of Pennsylvania, in June 1945. The report was the outcome of a series of discussions commencing in the summer of 1944 between von Neumann and the inventors of the ENIAC computer—John Presper Eckert, John W. Mauchly, and others. ENIAC was an electronic computer designed primarily for ballistics calculations: in practice, the machine was limited to the integration of ordinary differential equations and it had several other design shortcomings, including a vast number of electronic tubes (18,000) and a tiny memory of just twenty numbers. It was also very time-consuming to program. The EDVAC design grew out of an attempt to remedy these shortcomings. The most novel concept in the EDVAC, which gave it the description 'stored program', was the decision to store both instructions and numbers in the same memory.

It is worth noting that during 1936 Turing became a research student of Alonzo Church at Princeton University. Turing came to know von Neumann, who was a founding professor of the Institute for Advanced Study (IAS) in Princeton and was fully aware of Turing's 1936 paper 'On computable numbers'. Indeed, von Neumann was sufficiently impressed with it that he invited Turing to become his research assistant at the IAS, but Turing decided to return to

England and subsequently spent the war years at Bletchley Park. Although there is no explicit citation of Turing's work in von Neumann's EDVAC report (unlike the work of the mathematical biologists McCulloch and Pitts), it is likely that von Neumann—and therefore the EDVAC—were influenced by Turing. To go further than this, however, and to assert that Turing was the true begetter of the stored-program computer—as some commentators do—is not supported by the evidence (see Chapter 6 for further discussion of the idea that Turing was 'the inventor' of the computer). But, if Turing did not invent *the* computer, he certainly invented *a* computer, and a successful one at that.

ACE and its context

Immediately after the Second World War, three major British computer developments were set up at Manchester and Cambridge universities and at the National Physical Laboratory (NPL) in Teddington, near London. These projects were headed by Frederic Williams and Tom Kilburn at Manchester, Maurice Wilkes at Cambridge, and Turing at the NPL.

Although the primary aim of all these projects was to build a computer and provide a computing service, their secondary objectives were somewhat different. At Manchester University, Williams and Kilburn were located in the Department of Electrical Engineering and the computer was fundamentally an engineering research project. They developed a novel form of memory based on a cathode-ray tube, and demonstrated it successfully in the 'Baby' computer in June 1948 (see Chapter 20). Although Baby was the world's first stored-program computer it was too small to solve realistic problems. A full-scale computer was developed over the following year (see Chapter 23). A fully engineered version of this machine was manufactured by Ferranti and provided a university computing service starting in early 1951. Ferranti subsequently became a leading British computer manufacturer.

At the Mathematical Laboratory in Cambridge University, Wilkes's objective was to build a computer as soon as possible and to use it for a computing service and research into programming methods. Wilkes's design was straightforward, closely following the EDVAC report, and was electronically conservative; for example, it used relatively slow electronics and a mercury delay-line memory, for which he copied a working design developed by the Admiralty for radar echo cancellation. EDSAC became operational in May 1949 and was offering a university-wide computing service by the beginning of 1950. In 1951 Wilkes, with research students David Wheeler and Stanley Gill, published the classic *Preparation of Programs for an Electronic Digital Computer*, which influenced the development of programming worldwide. A commercial version of EDSAC, Lyons Electronic Office (LEO), was developed by the J. Lyons bakery company.

The NPL established a Mathematics Division immediately after the end of the war in May 1945, with John R. Womersley as its head. (The NPL was founded in 1900 as a government-funded standards setting and physical sciences research centre.) The aim in creating the Mathematics Division was to provide industry and government research organizations with a computing service similar to that existing in, or planned by, the universities. The division consisted of five sections: the differential analyser section, the Hollerith punch-card computing section, statistics, desk machines, and the ACE section.

The aim of the ACE section was to build an electronic computer and to conduct research into numerical methods (the study of the mathematics of approximations and errors in computation). Turing was recruited to lead the section and took up his post in October 1945. At first he was the sole member of his section, but he was joined by James Wilkinson in May 1946. Although Turing left the NPL in September 1947, Wilkinson proved an outstanding successor. He made fundamental contributions to numerical methods, was elected a Fellow of the Royal Society in 1969, and in the following year received the Turing Award of the Association for Computing Machinery, computing's highest honour. Others who joined the ACE section later included Mike Woodger and Brian Munday.

The ACE design

Almost all of the early computer projects in the United States and the United Kingdom were based closely on the EDVAC report of June 1945 and the subsequent reports produced by von Neumann and his colleague Herman Goldstine at the IAS. Although Turing took many ideas from EDVAC, ACE was by far the most original design of its era. Turing submitted his 'Proposal for the development in the Mathematics Division for an Automatic Computing Engine (ACE)' in March 1946.[3] It was remarkably detailed, and even included a cost estimate of £11,200 (which, inevitably, was optimistic by a factor of at least 10). The proposal cited the EDVAC report and used its terminology, memory structure, adding technique, and logic notations.

To appreciate ACE's novelty it is necessary to know a little of how EDVAC operated and its memory technology. The EDVAC report specified that a computer would consist of five functional parts: the control, the arithmetic unit, the memory, and the input and output devices. A key characteristic of EDVAC was that both programs and numbers were stored in the memory. The memory consisted of a sequence of storage locations numbered from zero upwards, each of which could store either an instruction or a number. A program was placed in a sequence of consecutive memory locations. To obey the program, instructions would be fed from the memory into the control unit and then executed in ascending sequence (unless interrupted by a 'branch' instruction). Turing recognized that executing instructions sequentially in this way was inherently inefficient.

A typical delay line consisted of a 5-foot mercury-filled steel tube with acoustic transducers at each end. In order to store data, electronic pulses ('bits') were converted to acoustic energy at one end of the tube; these travelled at the speed of sound down the mercury column and were reconverted to electrical pulses at the other end. The electric pulses would then be re-injected into the input end of the delay line. In this way, a stream of bits would be trapped inside the delay line indefinitely. Sound travelled very much more slowly than electricity, so a sonic pulse would take about 1 millisecond (ms) to travel down the tube. The delay line could store approximately 1000 bits (or, more usefully, 1024 bits). The complete memory would consist of a bank of several mercury delay lines. Turing envisaged between 50 and 500 delay lines, giving a total memory capacity of between 1500 and 150,000 instructions or numbers—or 'words'.

The delay-line memory had a 'waiting time' problem, later called 'latency'. An instruction or number could not be utilized by the computer until it emerged from the delay line. On average, the waiting time was 0.5 ms. The result of this latency was that the speed of an EDVAC-type machine was constrained by the speed of the delay line, almost irrespective of the speed of the arithmetic circuits. EDSAC, for example, which used one-millisecond delay lines, managed just 650 instructions per second, well below its potential.

Turing's insight—which was not described in his ACE report of March 1946, but appeared a few months later—was that, instead of arranging instructions sequentially in the memory, they could be placed so that they emerged from the delay line just at the moment they were needed: this would eliminate the waiting time, making the machine potentially much faster. To achieve this, each instruction would nominate the location of its successor. Numbers would also be strategically placed to minimize waiting time. Although this made programming more difficult, Turing was convinced that the superior performance would justify the added complication: this coding style was later known as 'optimum programming'. The idea of optimum programming was wholly original and is perhaps the most direct connection with Turing's 1936 paper. In both the Turing machine and ACE, each instruction nominated a successor.

From ACE to Pilot ACE

Following acceptance of Turing's ACE report, Womersley explored the options for the machine's construction. These options included construction by the Post Office or collaboration with the Telecommunications Research Establishment (TRE), both of which came to nothing. In November 1946 he approached Wilkes at Cambridge. Wilkes, who had his own funding uncertainties, responded positively by sending Womersley his initial plans for EDSAC. Womersley passed these to Turing for expert comment. Turing's memo—admittedly a private communication—was nonetheless extraordinarily barbed and condescending. He wrote:[4]

I have read Wilkes' proposal for a pilot machine, and agree with him as regards the desirability of the construction of some such machine somewhere. I also agree with him as regards the suitability of the number of delay lines he suggests. The 'code' which he suggests is however very contrary to the line of development here, and much more in the American tradition of solving one's difficulties by means of much equipment rather than by thought. I should imagine that to put his code (which is advertised as 'reduced to the simplest possible form') into effect would require a very much more complex control circuit than is proposed in our full-sized machine.

Womersley was a good deal more diplomatic than Turing in his response to Wilkes—who was not shown Turing's memo—but it was clearly not possible to bridge the gulf between EDSAC and ACE and the two projects went their separate ways. Wilkes, in fact, did not see Turing's memo until 1977, when it was published by Woodger. In 1946 Wilkes had considered that Turing's ideas were wrong-headed, and in 1977 he believed that history was on his side.

During the period from December 1946 to February 1947 Turing gave a series of lectures on computer design at the Adelphi offices of the Ministry of Supply in London (see Chapter 20). Wilkes attended the first lecture, but no more—he thought that Turing's ideas were so far removed from the mainstream that there was no value for him in attending.

Meanwhile, construction of ACE was going nowhere fast. In January 1947, however, an American named Harry Huskey arrived for a 1-year visiting position in the Mathematics Division. He had worked on ENIAC and came to the NPL on the recommendation of Douglas Hartree, Professor of Mathematical Physics at Cambridge University, one of the first British experts to visit the Moore School after the war and a member of the NPL Executive Committee.

Huskey made the suggestion that instead of going outside the NPL to get the machine constructed they should build it themselves. Well seasoned by his ENIAC experience, Huskey advised the construction of a small-scale prototype which he called the 'Test Assembly'. Turing

Figure 21.1 The Pilot ACE in the Mathematics Division of London's National Physical Laboratory.

© Crown Copyright and reproduced with permission of the National Physical Laboratory.

reacted negatively to this suggestion, no doubt supposing that it would delay the building of a full-scale machine. He took no part in it, left the NPL for a sabbatical year at King's College, Cambridge, in September 1947, and never returned. At the instigation of Charles Darwin, director of the NPL, construction was handed over to the Radio Division, with the Mathematics Division acting as consultants. Unfortunately, the Radio Division had no experience in digital electronics, and it was not until the spring of 1948 that they established an electronics section with some expertise. Meanwhile, by the end of 1947, Huskey's year at the NPL was over and the project was stalled.

The impasse was broken when members of the ACE section negotiated with the head of the electronics section, F. M. Colebrook, and an active collaboration was established between the two groups to build what had by then become known as the 'Pilot ACE' (Fig. 21.1). From this point on, progress was rapid and the Pilot ACE sprang into life in May 1950. The machine was equipped with thirty-two delay lines, each holding thirty-two words, giving a total storage capacity of 1024 words—which was comparable to that of the Cambridge and Manchester machines.

Optimum programming

From the moment that it began operation it was clear that the Pilot ACE was more than a small-scale prototype. It was potentially a highly effective computing machine in its own right, with a theoretical maximum speed of 16,000 operations per second.

The ACE section developed an impressive subroutine library, which included common requirements such as trigonometric functions and matrix operations. The subroutine library was notable for the sophistication of its numerical methods, developed under the leadership of Wilkinson. Using subroutines in programs typically reduced their length by a factor of 3, and correspondingly reduced debugging and the chore of optimum coding.

Programming for the Pilot ACE was a two-stage process: first, the program was written out on paper as a sequence of instructions, much the same as for any other computer; second, the program was 'optimized' by the programmer, who assigned every instruction, data item, and variable to a particular storage location in the delay line memory.[5] Optimization was very time consuming: it was rather like a sliding block puzzle, and it took a degree of experience to get the balance right between programmer time spent optimizing and machine time saved.

A potential solution to the optimum programming problem was an automatic optimization program. Although some independently developed American machines were supplied with automatic optimization routines, this was never done for the Pilot ACE.

An attempt to write such a program was made by Christopher Strachey, starting in January 1951. (Strachey was then a schoolteacher who had become interested in computers and obtained an entrée to NPL through a mutual friend of his and Woodger's. He never looked back: see Chapter 23.) Writing an optimization routine was a formidable challenge and Strachey did not get far before he was seduced in July 1951 by the larger capacity of the Ferranti Mark I at Manchester University. By that time Turing had been appointed deputy director of the Computing Machine Laboratory at Manchester; he and Strachey already knew one another slightly as they were both at King's College before the war. Turing suggested that Strachey write a 'trace' program so that the machine simulated itself—this was potentially a useful debugging tool (see Chapter 23). Neither Strachey nor anyone else ever successfully took up the automatic optimization challenge for the Pilot ACE.

Because subroutines were written once, but used many times, they justified a big investment in optimization and they showed the Pilot ACE at its best. This was particularly the case with the 'floating-point' subroutines. Floating-point representation enabled calculations to be performed without regard to the magnitude of the numbers involved, as on a modern electronic calculator. Without floating-point operations, numbers had to be kept within the (typically ten decimal digit) capacity of the machine, which involved tricky scaling operations. Table 21.1 shows the comparative speeds of the floating-point subroutine libraries for the Pilot ACE and the Cambridge and Manchester computers: the Pilot ACE was roughly ten times faster than its competitors. Noting that the Pilot ACE contained 800 electronic tubes, whereas the Ferranti Mark I and EDSAC both contained over 3000, one can argue that the Pilot ACE was perhaps twenty times more cost-effective. It was a *tour de force*.

Table 21.1 *Floating-point operations.*

Operation	Time (milliseconds)		
	Pilot ACE	EDSAC	Mark I
Add/subtract	8	90	60
Multiply	6	105	80
Divide	34	140	150

The Pilot ACE was later augmented with a magnetic drum which put its total storage capacity on a par with the Manchester computer, and enabled it to tackle very large matrix problems.

ACE's legacy

It happened that George Nelson, the managing director of English Electric, was an external member of the NPL Executive Committee from 1946 to 1948. Learning of the ACE project, he saw its potential both for his firm's engineering calculations and as a possible product. A young graduate, George Davis, was appointed as a liaison to transfer the technology between the NPL and English Electric's Nelson Research Laboratory in Kidsgrove, Staffordshire. There was a good working relationship between the two organizations.

In 1952, when the Pilot ACE turned out to be such a success, Nelson decided that English Electric would build a copy; the English Electric machine was, unsurprisingly, called DEUCE, a contrived acronym for 'digital electronic universal computing engine'. By this time the computer industry in the UK was beginning to blossom, with about a dozen firms eventually manufacturing computers. The first in the market was Ferranti, followed over the next few years by several of the electronics and control firms, including EMI, AEI, Metropolitan Vickers, General Electric, and Elliott Brothers. In 1955 the Lyons Company formed a subsidiary, Leo Computers Limited, and by the late 1950s the office machine companies BTM and Powers-Samas had begun computer developments. Thus, despite its early commitment, English Electric soon found itself in a crowded marketplace.

DEUCE had a similar specification to the Pilot ACE, with thirty-two delay lines and a drum store, and it sold for about £60,000. It was a scientific machine unsuitable for commercial data processing, but it sold well to organizations with big computational requirements. Indeed, in terms of raw computational speed it was the best value machine on the market.

DEUCE was particularly popular with aircraft companies. A major driver for aircraft firms to acquire computers was the fall-out from the Comet disasters of 1954. The de Havilland Comet was Britain's (and the world's) first passenger jet airliner, and during 1954 many lives were lost following crashes that were found to be caused by metal fatigue resulting from an instability at speed (so-called 'flutter'). The government mandated that extensive stress calculations must be undertaken for all new aircraft to prove their airworthiness. It was a case of an ill wind, and English Electric was a primary beneficiary.

Although DEUCE, like the Pilot ACE, was fundamentally difficult to program, in 1955 a user-friendly interpretive matrix package known as GIP (for 'general interpretive program') was developed for the Pilot ACE by Brian Munday. This was adapted for DEUCE and greatly simplified the development of the matrix programs needed for stress calculations.

Over the next five years DEUCEs were sold to the Royal Aircraft Establishment (RAE) and to the aerospace manufacturers Short Brothers & Harland, Bristol Aeroplane, Bristol Siddeley Aero-engines, and English Electric Aviation (subsequently the British Aircraft Corporation). The RAE eventually acquired a second machine, and the two were happily known as Gert and Daisy, in homage to two popular comediennes of the time.

Between its launch in 1955 and its final sales five years later, over thirty DEUCEs were sold. As well as the aerospace industry, several went to defence and research organizations with intensive computational needs, including the Atomic Weapons Research Establishment, the Atomic Energy Authority, the National Engineering Laboratory, and the Central Electricity Generating Board.

Another machine was exported to the New South Wales Institute of Technology in 1956. Generally, however, DEUCE was not popular in higher education. Computer centres in universities had two missions: to provide a computing service and to train undergraduates in programming. DEUCE was excellent for the first mission, but was horrendously difficult to program for a novice. As a result, a competing machine, the Ferranti Pegasus, was a much more popular choice for universities. The Pegasus had been specified by Christopher Strachey and, informed by his early experience with the Pilot ACE, he had avoided confronting the user with the need for optimum programming. The Pegasus was a drum-based machine with a similar specification and price to DEUCE, but by a neat architectural innovation Strachey achieved approximately half the performance of DEUCE without the need for optimum programming. The Pegasus gained a well-deserved reputation as a programmer's dream machine.

In economic and industrial terms, the most important legacy of the ACE design was the Bendix G-15 computer, of which over 400 were sold in the United States. In a parallel development to the English Electric DEUCE, the Bendix Aviation Corporation in California wanted a computer for its own use and as a potential product. In order to specify the machine, Bendix retained Harry Huskey as a consultant (he was then employed by Wayne State University in Detroit). The first G-15 was delivered in 1956 and sold for a modest $45,000 (about £16,000). The machine dispensed with the difficult and expensive mercury delay-line technology and instead used a magnetic drum as its primary store. The G-15 drew much from the ACE design, and optimum programming gave it an acceptable speed despite the slow memory technology. For a decade it was one of the workhorses of the US aerospace and engineering industries.

Although there were other direct derivatives of the ACE design (including the one-of-a-kind MOSAIC at the TRE and the Packard-Bell 250 in the United States), none was as important as DEUCE or the Bendix G-15. All of the derivatives of ACE had an exceptional price-performance ratio, thanks to optimum programming.

In the United States optimum programming was used most prominently in the IBM 650 magnetic drum computer, of which 2000 were sold in the second half of the 1950s. IBM's president called it the company's 'Model T' and it established the firm's dominance of the computer industry: it is unlikely, however, that this was directly or indirectly influenced by Turing's work. By the mid-1950s optimum programming was a well-known and frequently re-invented technique used in several machines. Moreover, programming techniques had been developed to automate optimum coding which, along with high-level languages, made the machines as easy to use as more conventional designs. Optimum coding was an idea of its time, however, and it disappeared when true random-access memories came along.

Conclusion

The Pilot ACE, and a DEUCE acquired from English Electric which replaced it in 1956, provided the main computing resource for the NPL throughout the 1950s. Work on the full-scale ACE continued, however, and it finally came to life in late 1958. It was a spectacular and elegant machine—its curved floor plan complemented by the modernist floor-to-ceiling windows of its surroundings (Fig. 21.2). Sadly, by this date the mercury delay-line memory on which ACE was premised was obsolete, and with it the *raison d'être* of the ACE design. Of course, since the NPL had the machine, it was obliged to use it, which it did until 1967.

Figure 21.2 The full-scale ACE in 1958.

© Crown Copyright and reproduced with permission of the National Physical Laboratory.

In a way, the demise of ACE vindicated Wilkes's original objections to Turing's design. Wilkes argued that delay lines were an ephemeral technology, that true random-access memories would eventually come along, and that with a conventional design it would be relatively easy to replace one memory technology with another. Indeed, this is exactly what happened with EDSAC 2, which was designed for a mercury delay-line memory but switched to a random-access core memory midway through the project. This was not possible with ACE. Turing's viewpoint was undoubtedly coloured by the fact that he was blessed with the manipulative skills of a first-class mathematician. He believed that the trade-off between a little extra effort with programming was amply justified by the superior performance of the ACE design, though he underestimated the difficulty ordinary users would have in programming the machine.

It is perhaps fair to say that Turing was right in the short term, and Wilkes was right in the long run. But in the early 1950s the short term counted for a lot.

CHAPTER 22

Turing's Zeitgeist

BRIAN E. CARPENTER AND ROBERT W. DORAN

This chapter reviews the history of Alan Turing's design proposal for an Automatic Computing Engine (ACE) and how he came to write it in 1945, and takes a fresh look at the numerous formative ideas it included. All of these ideas resurfaced in the young computing industry over the following fifteen years. We cannot tell to what extent Turing's unpublished foresights were passed on to other pioneers, or to what extent they were rediscovered independently as their time came. In any case, they all became part of the Zeitgeist of the computing industry.

Introduction

At some universities, such as ours in New Zealand, the main computer in 1975 was a Burroughs B6700, a 'stack' machine. In this kind of machine, data, including items such as the return address for a subroutine, are stored on top of one another so that the last one in becomes the first one out. In effect, each new item on the stack 'buries' the previous one. Apart from the old English Electric KDF9, and the recently introduced Digital Equipment Corporation PDP-11, stack machines were unusual. Where had this idea come from? It just seemed to be part of computing's Zeitgeist, the intellectual climate of the discipline, and it remains so to this day.

Computer history was largely American in the 1970s—the computer was called the von Neumann machine and everybody knew about the early American machines such as ENIAC and EDVAC. Early British computers were viewed as a footnote; the fact that the first stored program in history ran in Manchester was largely overlooked, which is probably why the word 'program' is usually spelt in the American way.[1] There was a tendency to assume that all the main ideas in computing, such as the idea of a stack, had originated in the United States.

At that time, Alan Turing was known as a theoretician and for his work on artificial intelligence. The world didn't know that he was a cryptanalyst, didn't know that he tinkered with electronics, didn't know that he designed a computer, and didn't know that he was gay. He was hardly mentioned in the history of practical computing.

There were clues. The first paper published on the ACE did say 'based on an earlier design by A M Turing'.[2] A trade press article by Rex Malik[3] described Turing as 'a four in the morning system kicker'—that didn't sound like a theoretician. There were growing rumours of secret stuff happening during the Second World War, and there were Brian Randell's 1972 paper and 1973 book.[4] In 1972, the National Physical Laboratory (NPL) issued a reprint of Turing's 1945/46 ACE proposal as a technical report,[5] and the Ultra secret was finally blown in 1974.[6]

The ACE proposal

The present authors read the ACE proposal in 1975, and were enchanted by its style and originality—and its contrast with conventional wisdom about the state of the art in 1945 (such as in John von Neumann's EDVAC report).[7] Subsequently, we analysed the original ACE design in some detail and wrote 'The other Turing machine', published in the *Computer Journal* in 1977.[8] Turing had of course read the EDVAC report, but apart from his very different and lucid writing style—he couldn't match von Neumann's ponderous 'octroyed temporal sequence'—he provided a complete and quite detailed design for the ACE. (Unless otherwise noted, in this chapter we will use the word 'ACE' specifically to refer to the late 1945 design described in Turing's original report.)

The report was not detailed enough to act as an engineering blueprint, and it was not the design that was actually used, either for the Pilot ACE or the full-scale ACE. However, compared with the EDVAC report, much less was left as an exercise for the reader. Many of the ideas in the ACE design were common knowledge by the early 1960s, yet Turing was unknown as a computer designer.

In forty-eight typed pages Turing described the concepts of a stored-program universal computer, a floating-point library, artificial intelligence, details such as a hardware bootstrap loader, and more, down to the level of detailed circuit diagrams and sample programs. Not all of these ideas were his, but as an act of synthesis the proposal was remarkable.

How was a theoretician able to write such a report in 1945? To a large extent, that is a trick question: Turing *wasn't* just a theoretician by that time. Even his most famous theoretical work, 'On computable numbers' (1936), was a thought experiment invoking a memory tape and logic circuits. He had made two serious attempts to build mathematical machines before the Second World War. During the war he designed information-processing machines and witnessed large-scale data processing at Bletchley Park, and had personally built electronic devices at Hanslope Park (see Chapter 18). Even in 1936, the basic components needed to build an electronic universal Turing machine existed: magnetic wire recording dated back to 1898, and the flip–flop (multivibrator) was around in 1919. Electronic AND gates (coincidence circuits) and binary counters came along in 1930. By 1945, Turing, well aware of Colossus, was ideally positioned to design a fully fledged machine.

Credit should be given to the NPL for making this possible. The NPL Mathematics Division was approved in late 1944, supported by the Ministry of Supply, Commander Edward Travis of the Government Code and Cypher School, Douglas Hartree, and Leslie J. Comrie, the New Zealander who founded Scientific Computing Service Ltd. in 1938.[9] The job of the Mathematics Division was to provide and coordinate national facilities for automated computation, including military applications.

The first head of the division was John R. Womersley (Fig. 22.1), better known today for his later work on fluid dynamics. He had worked with a differential analyser and read 'On computable numbers' before the war. He was sent to the United States in early 1945 to learn about

Figure 22.1 John R. Womersley.

Reproduced from D. A. MacDonald, *Blood Flow in Arteries*, 2nd edition, Arnold: London (1974).

ENIAC and the plans for EDVAC after being appointed to NPL but just before taking up the job. Womersley understood the potential of universal automatic computers and was willing to foster unconventional ideas. He showed Turing the EDVAC report and hired him as a one-man section of his division to study the design of an 'Automatic Computing Engine'. By the end of 1945, Turing's report, 'Proposed electronic calculator', also known as 'Proposals for development in the Mathematics Division of an Automatic Computing Engine (ACE)', was finished. It was presented to the NPL Executive Committee in March 1946, supported by Womersley and Hartree. The ACE project (but not the detailed design) was approved by the committee, chaired by the Director of the NPL, Sir Charles Darwin (grandson of *the* Charles Darwin).

Turing's proposal gave an outline of the principles of stored-program computers, binary representation, and floating-point arithmetic; a detailed architecture and instruction set; detailed logic diagrams; electronic circuits for various logic elements; and example programs. It also gave a budget estimate of £11,200 (twenty times Turing's annual salary at NPL), starting a long tradition in the IT industry of hopelessly underestimating costs. This version of ACE was to be a serial machine operating with a 1-MHz clock and 32-bit words. Turing was an early adopter, if not the originator, of the word 'word' in this usage. Completely unlike EDVAC, ACE was to have thirty-two registers in the central processing unit, known as TS1–TS32, where 'TS' meant 'temporary storage', actually a short mercury delay line. (The instruction set was register-to-register, whereas EDVAC was an accumulator machine.) There were only eleven instructions, closely reflecting the hardware design, giving Turing a fair claim to being the first RISC (reduced instruction set computer) designer.[10]

The proposed applications ranged from numerical analysis—as expected by NPL—to counting butchers, solving jigsaws, and playing chess. The last of these was certainly *not* expected, and it was probably a topic that Turing had discussed with Claude Shannon during the war (see Chapter 31). Turing foresaw relocatable code and something very like assembly language, called 'popular' form. He also foresaw a subroutine library, including floating-point routines. Among his examples were two routines named BURY and UNBURY, which implemented a stack for nested subroutine calls.

Formative ideas

In our 1975 paper, we identified the set of formative technical ideas clearly found in the 1945 ACE proposal (Table 22.1); only a few of these are also to be found in von Neumann's EDVAC report. We explain most of these concepts in more detail in this section.

The stored-program concept—that a computer can contain its program in its own memory—derived ultimately from Turing's paper 'On computable numbers'; Konrad Zuse also developed

Table 22.1 *The set of formative technical ideas clearly found in Turing's 1945 ACE proposal.*

Formative idea	Present in the EDVAC report	Present in the ACE proposal
Stored program	✓	✓
Binary implementation using standardized electronic logic elements	✓	✓
Complete notation for combinational and sequential circuits	✓	✓
Memory-control-arithmetic unit-input/output architecture	✓	✓
Conditional branch instructions (although clumsy)	✓	✓
Address mapping (in a simple form)		✓
Instruction counter and instruction register		✓
Multiple fast registers for data and addressing		✓
Microcode (in a simple form); hierarchical architecture		✓
Whole-card input/output operations (similar to direct memory access)		✓
Complete set of arithmetic, logical and rotate instructions		✓
Built in error detection and margin tests		✓
Floating point arithmetic		✓
Hardware bootstrap loader		✓
Subroutine stack		✓
Modular programming; subroutine library		✓
Documentation standards		✓
Programs treated as data; link editor; symbolic addresses		✓
Run time systems (input/output conversions; hints of macro expansion)		✓
Non-numerical applications		✓
Discussion of artificial intelligence		✓

it in Germany, in the form of his *Plankalkül* language, without having read Turing's paper. The next four ideas in Table 22.1 were also to be found in the EDVAC report which Turing had recently read. This might suggest that they derived indirectly from various American sources via von Neumann. However, since binary implementation using standard electronic elements had been used in Colossus by Tommy Flowers, it seems unlikely that Turing derived his knowledge of this from American sources, and of course the idea of binary working was already there in 'On computable numbers'. Colossus also had an elementary conditional branch, as Max Newman emphasized. The circuit notation in the ACE proposal derived (via von Neumann) from the famous American pioneers of computational neurophysiology, Warren S. McCulloch and Walter Pitts,[11] who in turn had cited Turing's 'On computable numbers'.

All the other concepts in Table 22.1 appeared in Turing's report but not in von Neumann's. They were without published antecedents, and one should not overlook the fact that they preceded the influential Moore School lectures in Philadelphia by several months. We now know about Turing's experiences at Bletchley Park, and about his knowledge of Colossus and of the work of Jacquard, Babbage, and Lovelace in the previous century, as well as his discussions with Claude Shannon during the war (see Chapter 31). While it is very likely that some of Turing's ideas were not brand new in the ACE report, it nevertheless remains quite startling to find them all in one place at such an early date. Other pioneers took a few more years to reach a similar point.[12]

What happened to Turing's ideas?

Next we consider in turn the formative ideas in Table 22.1 that were apparently unique to Turing in 1945. All of these resurfaced over the following fifteen years or so. The question is: how much of that was rediscovery, and how much was unacknowledged reuse?

Turing's descriptions of these ideas were not in the open literature. The 1945 ACE report, mimeographed in a limited number of copies, was out of stock by 1948 and vanished from view until 1972. The Pilot ACE was well known in itself, but with little mention of Turing's contributions, although two reports by James Wilkinson on the Pilot ACE were widely circulated.[13] The Cambridge computer design team (see Chapter 21), especially Maurice Wilkes, never admitted to much influence from Turing—although Stanley Gill, for one, worked alternately on the Pilot ACE and the Cambridge EDSAC, but in both cases after the main design choices were fixed. The University of Manchester team mainly followed the EDVAC line (see Chapter 20).

Yet all of the ACE ideas showed up in later designs. We aim to give the flavour of Turing's insights and a feel for why they were so prophetic in 1946. It is difficult to convey the breadth and depth of Turing's originality in the ACE design without going into technical detail and using modern jargon, but we have kept the technicalities to a bare minimum.

Several of Turing's ideas were bound up with the notions of 'words in memory', 'addresses', and 'registers'. Computer memory is divided up into small pieces, known as 'words', and in the case of ACE these consisted of thirty-two bits each, a word size still commonly used today. The position of a given word in memory is simply a number (0 for the first word, 1 for the next word, and so on) and this is called its 'address'. When the content of a word in memory has to be processed by a computer instruction it is often copied into a temporary storage device called a 'register', which is part of the computer's central processing unit. Registers in ACE each held thirty-two bits.

A common technique in all modern computers is that memory addresses are 'mapped' for the convenience of the programmer; in this way the programmer can assume that programs and data are always in the same place, even if they are actually in different parts of the memory during different runs of the program. ACE presaged address mapping with a memory inter-leaving trick (although this was not carried through to the Pilot ACE). Address mapping later resurfaced in Manchester, most famously in the Atlas computer of 1962.

Turing made it clear that a special register was needed to contain the instruction currently being obeyed by the machine, while another one had to contain the address of the following instruction. These two registers, usually called the 'instruction register' and the 'instruction counter', became universal in computer designs, perhaps because there is really no other way to do it—but Turing wrote it down first.

Turing planned multiple registers, whereas EDVAC had very few. Apart from DEUCE, the first production machine with multiple registers was the Ferranti Pegasus in 1956. Multiple fast registers, used to contain data or addresses, became widespread in the 1960s—in the IBM 360, for example. A particular kind of register that Turing foreshadowed in ACE was an 'index regis-ter', which was used as a pointer in an array of data. Index registers (known at Manchester as 'B-lines') were first implemented in 1949, at the Manchester Computing Machine Laboratory.

Many modern computers use what is called a 'register-to-register design', in which instruc-tions use registers as the source and destination for arithmetical or logical operations. This was first described in the 1945 ACE proposal and was implemented in the Pilot ACE and DEUCE (Fig. 22.2). The 1956 Ferranti Pegasus also had this design. The idea famously reappeared in

Figure 22.2 DEUCE, the production model of the ACE.

© Crown Copyright and reproduced with permission of the National Physical Laboratory.

1970 in the Digital Equipment Corporation's PDP-11/20, designed by Gordon Bell, who used DEUCE while he was a Fulbright Scholar in Australia.

ACE had a rather hierarchical design, with a simple basic arrangement and simple basic instructions. However, some instructions could be modified to perform quite complex operations, by setting extra bits on or off in their binary code. Today we recognize this technique as 'microcode', a concept that was also presaged in the MIT Whirlwind of 1947 and which reappeared most famously in Cambridge in EDSAC2 (1956).

Another important modern technique is 'direct memory access', whereby data are transferred to or from an external device automatically without the need for a complex sequence of machine instructions. A feature of the ACE design was the input or output of a whole punched card 'in one go'. This clearly presaged direct memory access, even though this is generally credited to the US National Bureau of Standards DYSEAC (1954), or else to a technique called 'channels', first used in the IBM 709 of 1957.

As was to be expected, ACE had a complete set of arithmetic instructions, but it also had logic instructions (AND, OR, etc.), the latter probably suggested by requirements from cryptanalysis. Logical instructions re-emerged in the Manchester Mark I of 1949 and the IBM 701 of 1952.

Turing recommended built-in error detection and operating margin tests. He presumably knew about the need for these from his Bletchley Park experience, or perhaps directly from Tommy Flowers, the designer of Colossus (although Colossus itself did not include margin tests). Other builders of thermionic valve computers had to learn about this the hard way.

ACE was to have floating-point software. Floating-point arithmetic effectively means that a machine stores a certain number of significant digits (such as 3142) and a decimal multiplier (such as 0.001) separately, instead of storing the single value 3.142. The multiplier is economically stored as an exponent (such as -3 for 10^{-3}). This allows a machine to store and process a very wide range of numbers. The technique had been known conceptually since 1914 and was found in various electromechanical machines, such as the Zuse Z1 (1938), the Harvard Mark II (1944), and the Stibitz Model V (1945). Whether Turing absorbed the idea from elsewhere or reinvented it is not clear. Going forward, in 1954 floating-point electronics appeared in the Manchester MEG, the prototype of the Ferranti Mercury, and in the IBM 704.

When computers were first invented they normally did nothing when first switched on, and a program had to be laboriously inserted by hand using switches on the front panel. Today, computers always start up an elementary program—they 'pull themselves up with their own bootstraps'—and the small built-in program that does this is called a 'bootstrap loader'. Amazingly, ACE was designed in 1945 to have a form of bootstrap loader, although this idea is conventionally credited to the IBM 701 of 1952.

Turing also clearly described what we would now call 'modular programming' and a 'subroutine library'. He recognized that large programs needed to be built up out of smaller ones (called 'modules' or 'subroutines'), and that many of the smaller ones could be kept and re-used later on, thus becoming a 'library'. These ideas were reinvented at least twice, by Grace Hopper in the United States (in 1951–52) and by Maurice Wilkes, David Wheeler, and Stanley Gill in Cambridge (in 1951). Software documentation standards are usually credited to Grace Hopper around 1952, but Turing recognized the need for them in 1945 three years before any stored-program machine was built.

A very important concept in computer science is 'recursion', in which a subroutine calls itself. Technically this is a bit tricky, because the computer has to keep track of where it is. This is done by stacking data items on top of each other and unstacking them in reverse. Turing described exactly that in his ACE proposal, calling the stacking and unstacking processes BURY and UNBURY, as already mentioned. A recursive stack appeared in the LISP programming language in 1958 and in the European Algol language by 1960. It then appeared in hardware in the English Electric KDF9 (1960), the Burroughs B5000 (1961), and the Manchester University and Ferranti Atlas (1962). There were multiple contributors to this aspect of Algol including Mike Woodger, Turing's NPL colleague, who was an author of the original Algol report. (Woodger has told us that the legendary Edsger Dijkstra approved the explicit notion of recursion in Algol.[14])

Turing also anticipated several basic types of software development tool. In modern terms the first of these was the equivalent of a 'link editor', a program that takes several pieces of machine code and combines them into a single program—in 1945 the notion of a program that manipulates another program was truly spectacular. Secondly, he suggested programming using symbolic addresses: instead of writing numerical memory addresses, the programmer could use comprehensible names. Thirdly, he suggested writing programs not in numerical machine code but in a readable ('popular') form of machine code very similar to a modern assembly language. EDSAC had alphabetic instruction mnemonics along these lines by 1949, and EDSAC itself translated these into machine code. 'Autocode' programming systems later appeared in Manchester (1952) and Cambridge. Turing contributed directly to the Manchester software effort, but the first fully viable Autocode solution was due to Tony Brooker (1954).

Turing described a simple run-time system for input–output conversions. Such systems soon became universal, representing another of Turing's remarkable foresights in 1945. He also discussed non-numerical applications, presumably inspired by cryptanalysis (which could not then be discussed in writing due to the Official Secrets Act). He mentioned artificial intelligence too in the ACE proposal. At least Turing got credit for that, along with Claude Shannon, with whom Turing had discussed artificial intelligence during the war.

In 1945, Turing even foresaw the profession of computer programmer.[15] He showed remarkable sociological foresight, and we can do no better than to quote his own words:

Instruction tables will have to be made up by mathematicians with computing experience and perhaps a certain puzzle-solving ability . . . This process of constructing instruction tables should be very fascinating. There need be no real danger of it ever becoming a drudge, for any processes that are quite mechanical may be turned over to the machine itself.

However, not much later, in 1947, he wrote:[16]

One of our difficulties will be the maintenance of an appropriate discipline, so that we do not lose track of what we are doing. We shall need a number of efficient librarian types to keep us in order . . . I have already mentioned that ACE will do the work of about 10,000 [human] computers.

He also foresaw what systems programmers would be like, with their mystique and gibberish:

The masters [programmers] are liable to get replaced because as soon as any technique becomes at all stereotyped it becomes possible to devise a system of instruction tables which will enable the electronic computer to do it for itself. It may happen however that the masters will refuse to do this. They may be unwilling to let their jobs be stolen from them in this way. In that case they would surround the whole of their work with mystery and make excuses, couched in well chosen gibberish, whenever any dangerous suggestions were made.

Conclusion

How much credit should Turing get? It's clear that he was the first to work seriously on a general-purpose computer design in the UK, in late 1945, and that he showed remarkable foresight and inventiveness. Soaking up ideas from many sources, and contributing many of his own, he proceeded to write them all down clearly and coherently in a few pages. The community of pioneers in the UK and the United States was relatively small until well after 1954, when Turing died, and we can assume that word of mouth had a significant effect—and also that ideas were not always properly credited.

In the end, though, it is now impossible to say to what extent Turing's amazing and wide-ranging foresights were passed on directly or indirectly to other pioneers, and to what extent they were simply rediscovered as their time came. Regardless of that, Turing's ideas indeed formed part of the computing Zeitgeist for many years to come.[17]

CHAPTER 23

Computer music

JACK COPELAND AND JASON LONG

O ne of Turing's contributions to the digital age that has largely been overlooked is his groundbreaking work on transforming the computer into a musical instrument.[1] It is an urban myth of the music world that the first computer-generated musical notes were heard in 1957, at the Bell Laboratories in the United States.[2] In fact, computer-generated notes were heard in Turing's Computing Machine Laboratory at Manchester University about nine years previously. This chapter establishes Turing's pioneering role in the history of computer music. We also describe how Christopher Strachey, later Oxford University's first professor of computing, used and extended Turing's note-playing subroutines so as to create some of the earliest computer-generated melodies.

Introduction

A few weeks after Baby ran its first program (see Chapter 20) Turing accepted a job at Manchester University. He improved on Baby's bare-bones facilities, designing an input–output system based on wartime cryptographic equipment (see Chapter 6). His tape reader, which used the same teleprinter tape that ran through Colossus, converted the patterns of holes punched across the tape into electrical pulses and fed these to the computer. The reader incorporated a row of light-sensitive cells that read the holes in the moving tape—the same technology that Colossus had used.

As the months passed, a large-scale computer took shape in the Manchester Computing Machine Laboratory. Turing called it the 'Manchester Electronic Computer Mark I' (Fig. 23.1).[3] A broad division of labour developed that saw Kilburn and Williams working on the hardware and Turing on the software. Williams concentrated his efforts on developing a new form of supplementary memory, a rotating magnetic drum, while Kilburn took the leading role in developing the other hardware. Turing designed the Mark I's programming system, and went on to write the world's first programming manual.[4]

The Mark I was operational in April 1949, although additional development continued as the year progressed.[5] Ferranti, a Manchester engineering firm, contracted to build a marketable

Figure 23.1 Baby grows into what Turing called the 'Mark I'.

Reproduced with permission of the University of Manchester School of Computer Science.

version of the computer, and the basic designs for the new machine were handed over to Ferranti in July 1949.[6] The very first Ferranti computer was installed in Turing's Computing Machine Laboratory in February 1951 (Fig. 23.2), a few weeks before the earliest American-built marketable computer, the UNIVAC I, became available.[7]

Turing referred to the new machine as the 'Manchester Electronic Computer Mark II', while others called it the 'Ferranti Mark I'; in this chapter we follow Turing's nomenclature. His programming manual was written in anticipation of the Mark II's arrival, and was titled *Programmers' Handbook for Manchester Electronic Computer Mark II*, but it was the outcome of his programming design work undertaken on the Mark I.[8] Turing's *Programmers' Handbook* contains what is, so far as is known, the earliest written tutorial on how to program an electronic computer to play musical notes.

How to program musical notes

The Manchester computer had a loudspeaker—called the 'hooter'—that served as an alarm to call the operator when the machine needed attention.[9] With some simple programming, the loudspeaker could be made to emit musical notes.

The computer's 'hoot instruction' worked like this. There was an electronic clock in the computer synchronizing all the operations. This clock beat steadily, like a silent metronome, at a rate of thousands of noiseless ticks per second. Executing the hoot instruction once caused a sound

Figure 23.2 Turing at the console of the Ferranti Mark I computer, which *he* called the 'Mark II'.

Reproduced with permission of the University of Manchester School of Computer Science.

to be emitted at the loudspeaker, but the sound lasted no longer than a tick, a tiny fraction of a second: Turing described the sound as 'something between a tap, a click, and a thump'.[10] Executing the hoot instruction over and over again resulted in this brief sound being produced repeatedly, on every fourth tick: tick tick tick *click*, tick tick tick *click*.[11]

If the clicks are repeated often enough, then the human ear no longer hears discrete clicks but a steady note. Turing realized that if the hoot instruction is repeated not simply over and over again but in different patterns the ear hears different musical notes: for example, if the pattern 'tick tick tick *click*, tick tick tick tick, tick tick tick *click*, tick tick tick tick' is repeated, then the note of C_5 is heard. (The subscript indicates the octave in which the note occurs.) Turing described C_5 as middle C, as musicians sometimes do, especially if playing an instrument with a very high register; although it is more usual to refer to middle C as C_4, an octave below C_5.[12] Repeating the different pattern 'tick tick tick *click*, tick tick tick *click*, tick tick tick tick, tick tick tick *click*, tick tick tick *click*, tick tick tick tick' produces the note F_4, and so on. It was a wonderful discovery.

Turing seems not to have been particularly interested in programming the machine to play conventional pieces of music. The different musical notes were used as indicators of the computer's internal state—one note for 'job finished', others for 'error when transferring data from the magnetic drum', 'digits overflowing in memory', and so on.[13] Running one of Turing's programs must have been a noisy business, with different musical notes and rhythms of clicks enabling the user to 'listen in' (as Turing put it) to what the program was doing. He left it to someone else, though, to program the first complete piece of music.

A royal surprise

One day Christopher Strachey (Fig. 23.3) turned up at the Computing Machine Laboratory. Before the war he had known Turing at King's College, Cambridge. Strachey was soon to emerge as one of Britain's most talented programmers, and he would eventually direct Oxford University's Programming Research Group.

When he first strode into the Manchester Laboratory Strachey was a mathematics and physics teacher at Harrow, a top school. He had felt drawn to digital computers as soon as he heard about them (in about January 1951) and, taking the bull by the horns, had written to Turing in April 1951.[14] Turing replied with a copy of his *Programmers' Handbook* and Strachey studied it assiduously.[15] This was 'famed in those days for its incomprehensibility', Strachey said.[16] An ardent pianist, Strachey appreciated the potential of Turing's terse directions on how to program musical notes. His first visit to the laboratory was in July 1951; Turing decided to drop him in at the deep end and suggested he try writing a program to make the computer check itself.[17] When Strachey left the laboratory, Turing turned to his friend Robin Gandy and said impishly, 'That will keep *him* busy!'.[18]

It did keep him busy, during the school summer holidays of 1951.[19] Strachey was a precocious programmer and when he 'trotted back to Manchester', he recollected, he took twenty or so pages covered in lines of programming code—at that time it was by far the longest program to be attempted.[20] 'Turing came in and gave me a typical high-speed, high-pitched description of how to use the machine', Strachey recounted.[21] Then he was left alone at the computer's console until the following morning.

'I sat in front of this enormous machine', he said, 'with four or five rows of twenty switches and things, in a room that felt like the control room of a battle-ship.'[22] It was the first of a lifetime of all-night programming sessions. He worked on debugging his monster program, which he called 'Checksheet'.[23] The name was a variation on a term Turing had used in his *Programmers' Handbook* for a hand method of checking programs. Turing called his method 'Check Sheets'. The method was 'done on paper with quarter inch squares on which vertical lines are ruled in ink', Turing explained in his *Programmers' Handbook*.[24]

As well as spending the night struggling to debug Checksheet, Strachey prepared a surprise. He managed to debug and get running another program that he'd brought with him.

Figure 23.3 Christopher Strachey sunbathing in the garden of his cottage 'The Mud House' in 1973, two years before his untimely death.

To the astonishment of onlookers, the computer raucously hooted out the British national anthem, *God Save the King*.[25] A budding programmer could hardly have thought of a better way to get attention. A few weeks later Max Newman heard the computer grinding out *God Save the King* and quickly wrote a letter to Strachey suggesting he might like a programming job in the lab.[26]

Manchester's musical computer also caught the attention of the popular press, with headlines like 'Electronic brain can sing now'.[27] The accompanying article explained that 'the world's most powerful brain' had been 'given a coded version of the score', from which it 'constructed the necessary waveform'. The BBC sent a recording team, together with a *Children's Hour* radio presenter known as Auntie, to capture a performance by the computer.[28] As well as *God Save the King*, the BBC recorded a version of Glenn Miller's *In the Mood*, a reedy and wooden performance of the famous hit. There was also an endearing, if rather brash, rendition of the nursery rhyme *Baa Baa Black Sheep*. The Mark II, still full of glitches, managed to crash in the middle of its Glenn Miller party piece: 'The machine's obviously not in the mood', Auntie gushed.

The unedited BBC recording of the session conveys a sense of people interacting with something entirely new. 'The machine *resented* that', Auntie observed at one point. The idea of a thinking machine, an electronic brain, was in the air at Manchester. Turing merrily fanned the flames. He provocatively told a reporter from *The Times* that he saw no reason why the computer should not 'enter any one of the fields normally covered by the human intellect, and eventually compete on equal terms'.[29]

Max Newman, now Professor of Mathematics at Manchester and founder of the Computing Machine Laboratory, lectured on the new computer music to 250 professional musicians who attended the annual conference of the Incorporated Society of Musicians in 1952, and his lecture was reported in the national press.[30] After explaining that to make the Manchester computer play melodies, 'All you have to do is to send an instruction to the hooter with the frequency of the note you want it to play', Newman described the discovery that the computer could be programmed to *compose* tunes for itself. So far these were, he admitted, 'very bad tunes'. (Quite possibly the program used Turing's random number generator, a standard hardware component of the Ferranti computers.) According to the *Manchester Guardian*:[31]

The next step, said Professor Newman, would be to make a machine which could compose *good* tunes, but so far no method of bridging the gap had been devised.

The article continued:

Professor Newman ended with this note of comfort for the assembled musicians: 'All this appears much more alarming and dangerous than it really is. When you see how it is done and how far it is from genuine composition, composers will realise they need not start taking steps to protect themselves against competition from machines.'

Extract from Turing's programming manual

Turing's brief tutorial in his *Programmers' Handbook* on how to program musical notes was typically compressed and demanding—yet, equally typically, his terse account told readers everything they needed to know in order to start writing note-playing programs. Turing called the hoot instruction '/V' (pronounced 'slash vee'). The complete tutorial occupied little more than half a page:

<u>The hooter</u>. When an instruction with function symbol /V is obeyed an impulse is applied to the diaphragm of a loudspeaker. By doing this repeatedly and rhythmically a steady note, rich in harmonics, can be produced. This is used to enable the operator to be called to attend to the machine in some way. The simplest case is where the whole of a job is completed and it is required to clear the electronic stores and start something different. All that is then required is to repeat a cycle of instructions including a hoot, e.g.

$$
\begin{array}{ll}
\text{FS} & \text{NS/V} \\
\text{CS} & \text{FS/P}
\end{array}
$$

In this case every second instruction will put a pulse into the speaker. These pulses will occur at intervals of 8 beats i.e. 1.92 ms giving a frequency of 521 cycles (about middle C). Or one could use the loop of three instructions

$$
\begin{array}{ll}
\text{O@} & \text{/V} \\
\text{G@} & \text{P@/V} \\
\text{M@} & \text{O@/P}
\end{array}
$$

which gives a slightly louder hoot a fifth lower in frequency.[32] Single pulses applied to the loudspeaker are distinctly audible as something between a tap, a click, and a thump. This fact can be turned to good account. By putting hoot instructions into programmes at suitable points one is enabled to 'listen in' to the progress of the routine. Some indication of what is going on is given by the rhythm of the clicks that are heard.[33]

Explaining Turings tutorial

We will decompress this truly historic tutorial by Turing, and after explaining his notationally formidable subroutines will develop a more perspicuous notation that helps bring out the connection between the looping subroutines and the musical notes that they produce. We will then describe the loops that were used to play the notes in the BBC recording, and we will also describe a fascinating discovery we made about the recording itself. This discovery allowed us, eventually, to pass back in time and listen to the computer as it had sounded on that day in 1951 when the BBC visited the Computing Machine Laboratory. You too can listen by following the link at the end of the chapter.

In Turing's two specimen loops, one consisting of two instructions and the other of three, he has used international teleprinter code to abbreviate the strings of binary digits (bits) that constitute the instructions at the level of 'machine code'. Teleprinter code associates a string of five bits with each keyboard character—for example, A is 11000 and B is 10011. Teleprinter code was well known to engineers in that era, and was very familiar to Turing from his wartime work on the Tunny teleprinter code used by Hitler and his generals (see Chapters 14 and 16). To Turing, teleprinter code must have seemed a natural choice for abbreviating the Manchester computer's bitcode. This system's main defect, that the abbreviations give no intuitive sense at all of what is being abbreviated, is one reason why his *Programmers' Handbook* was such heavy going.

In teleprinter code '/' is 00000 and 'V' is 01111: thus /V is the teleprinter code abbreviation of the Mark II's ten-digit hoot instruction 0000001111. Turing's /P is also an instruction: instructions always began with either '/' or 'T' (00001). The other symbols in Turing's sample subroutines ('NS', 'P@', 'FS', 'CS', 'O@', 'G@', and 'M@') are memory addresses: each pair of symbols abbreviates a ten-digit address in the computer's Williams tube memory.

Instruction /P (unconditional transfer of control) tells the machine to obey next the instruction stored at a location specified via the address immediately to the left of the '/'. In effect the second line of the first loop sends the machine back to the first line, and the final line of the second loop again sends the machine back to the first line.[34] The computer continues to loop until an instruction from elsewhere in the program terminates the loop after n repetitions.[35] The programmer selects the number n, thus determining how long the note is held, as required by the rhythm of the melody.

Our analysis of the BBC's recording of the Mark II playing *God Save the King, Baa Baa Black Sheep*, and *In the Mood*, showed that the durations of the played notes varied between 80 and 1100 milliseconds (ms). The analysis also revealed that very short pauses were programmed between consecutive notes, by means of 'silent' loops—short loops containing no hoot instruction. These inter-note pauses helped to define the beginning of each note, and were essential if a sequence of several notes of the same pitch was played: without gaps between the individual notes, a single long note would be heard.

The occurrence of NS to the left of /V in Turing's first subroutine can be ignored for the present purposes, and so can the P@ to the left of /V in the second subroutine: these terms create special effects to do with the computer's visual display, and play no role in the production of musical notes. (The effect produced by including the address NS in line 1 of the first subroutine is that the information stored at NS momentarily brightens on the monitor display as the machine hoots:[36] this provides a visual prompt to assist the operator. Similarly, the effect of P@ in the second instruction of the three-line subroutine is to cause the information stored at P@ to brighten on the display as the hooter sounds.) The two note-playing subroutines can be written more cleanly without these special effects:

```
FS      /V
CS      FS/P

O@      /V
G@      /V
M@      O@/P
```

Taking clarity a step further, we might replace the teleprinter-coded addresses with simple line-numbers:

```
1       /V
2       1/P

1       /V
2       /V
3       1/P
```

As mentioned earlier, the /V instruction took four ticks to complete—four *beats* in the Manchester jargon, with the actual hoot occurring on the fourth beat[37]—and /P also took four beats to complete. (As Williams and Kilburn put it, the basic rhythm of the Manchester computer was 'four beats to the bar'.[38]) Thus, running once through Turing's two-line subroutine produced: 'tick tick tick *click*, tick tick tick tick', and looping repeatedly through the subroutine produced the first of the two sequences of ticks and clicks discussed earlier. Similarly, running through the three-line subroutine once gave: 'tick tick tick *click*, tick tick tick *click*, tick tick tick tick', and looping repeatedly gave the second sequence mentioned earlier.

The precise duration of a single beat was 0.24 milliseconds (ms). The first subroutine produced one click every eight beats (that's every 1.92 ms): thus, the frequency with which clicks are produced, as the machine loops repeatedly through this subroutine, is $1 \div 1.92 = 0.52183$ clicks per millisecond, or 521.83 clicks per second. In standard units, the frequency of the clicks is said to be 521.83 hertz (Hz): this is close to C_5, whose assigned frequency in the standard 'equal-tempered' scale is 523.25 Hz. The equal-tempered scale is the standard scale for keyboard instruments, with adjacent keys playing notes heard as equidistant from one another.[39]

Table 23.1 shows the equal-tempered frequencies of all the notes occurring in the fragments of the scores of *God Save the King, In the Mood*, and *Baa Baa Black Sheep* that were performed in the BBC recording. The computer was by no means always able to hit the equal-tempered frequencies, though, and Table 23.2 shows the actual frequencies that the computer produced.

By dispensing with any reference to memory addresses, and abstracting from which particular instructions are employed, Turing's note-playing subroutines can be represented very transparently by means of what we call 'note-loops'. The note-loop corresponding to Turing's C_5 routine is:

$$\text{START} - - - \mathbf{H} - - - - \text{REPEAT}$$

Each dash represents a single beat, with the hoot **H** occurring on the fourth beat of the first bar. More economically, the note-loop can be represented as <3**H**, 4>.

Table 23.1 *The equal-tempered frequencies of the notes from those parts of the scores of* **God Save the King, In the Mood,** *and* **Baa Baa Black Sheep** *that were performed in the 1951 BBC recording.*

Note	Frequency (Hz)
$F\sharp_2$	92.5
G_2	98
A_2	110
B_2	123.4
C_3	130.8
$C\sharp_3$	138.6
D_3	146.8
E_3	164.8
$F\sharp_3$	185
G_3	196
A_3	220

Table 23.2 *The frequencies that the Mark II was able to play by means of loops containing four-beat instructions or mixtures of four- and five-beat instructions, down to the lowest frequency in the BBC recording. By increasing the number of beats in a loop still further, the machine will play ever lower notes, until at approximately 20 Hz the human ear begins to perceive a series of individual clicks rather than a note. We call the notes that the computer is able to play the 'playable notes'.*

Beats	Frequency (Hz)
8	520.8
12	347.2
13	320.5
16	260.4
17	245.1
18	231.5
20	208.3
21	198.4
22	189.4
23	181.2
24	173.6
25	166.7
26	160.3
27	154.3
28	148.9
29	143.7
30	138.9
31	134.4
32	130.2
33	126.3
34	122.6
35	119.1
36	115.7
37	112.6
38	109.7
39	106.8
40	104.2
41	101.6
42	99.2
43	96.6

Analysing the notes

Subroutines for playing lower notes require the addition of further instructions, since this has the effect of adding extra blocks of four beats between the hoots, so lowering the frequency. Conveniently, several of Turing's instructions served to waste time: their execution took up four beats but they did nothing. An example is /L: the instruction did nothing unless a 'dummy stop' switch had been set manually at the control console before the program started (in which case /L caused the machine to pause). Any of these four-beat 'dummy' instructions can be used for creating lower-frequency notes. For example, the note-loop <3H, 4, 4> produces a frequency of 347.22 Hz, which is approximately F_4 (349.23 Hz), a fifth lower than C_5.

The note loop <3H, 4, 4> produces the same note as Turing's second example of a loop, which in our notation is <3H, 3H, 4>. Adding the second pulse of sound at the same frequency does not alter the note, but (as Turing said) has the effect of making the note louder. We call note-loops that play the same frequency *equivalent*.

Two further examples of note-loops are <3H, 4, 4, 4, 4, 4, 4, 4>, producing a frequency of 130.21 Hz, fairly close to C_3 (130.81 Hz), and <3H, 4, 4, 4, 4, 4, 4>, producing a frequency of 148.81 Hz, lying between D_3 (146.83 Hz) and D_3 sharp or $D\sharp_3$ (155.56 Hz). Both of these note-loops produce a quiet sound: the same notes are played louder if extra hoots are added to form equivalent note-loops, such as <3H, 3H, 3H, 3H, 4, 4, 4, 4> and <3H, 3H, 3H, 3H, 4, 4, 4>, respectively.

We call a note-loop containing only one hoot the *primary* form, and equivalent note-loops containing more than one hoot are called *padded* forms of the loop. Padded note-loops typically produce notes with a different timbre or tone colour from the note produced by the loop's primary form. Timbre is manifested by differences in the shape of wave-forms of the same frequency; for example, if a violin and a flute play exactly the same note at exactly the same volume, the sounds are nevertheless instantly recognizable as different because of their different timbres.

We built a programmable simulator to play the Mark II's note-loops, and we used it to investigate the effects of padding note-loops. An Atmel ATmega168 microcontroller was used to create a functional simulation of the Mark II as a note-playing device. We connected a small loudspeaker directly to one of the digital output pins. Microcontroller programs using pulses and delays reproduced the beat structure of the Mark II and emulated the effects of the Mark II's music routines.

We found that primary note-loops produce relatively thin-sounding notes while their padded equivalents produce somewhat louder and fuller-sounding notes. Over-padding is possible, however. The simulator revealed that including too many hoots adds a high overtone, especially with lower notes containing more beats. Since an uninterrupted sequence of hoot instructions generates the Mark II's highest achievable note of 1041.67 Hz (somewhere in the vicinity of C_6), the result of over-padding a note-loop is that the ear tends to hear not only the intended note but also this maximum note as a high overtone.

The BBC recording indicates that the programmer probably used padding. If only unpadded loops had been used, lower notes would have been quieter than higher notes, since in a lower note there are longer gaps between the hoots. This is not observed in the recording, and in fact some lower notes are louder than some higher notes. Because of the poor quality of the recorded material, however, the present analysis did not reveal the number of hoots used in each individual note-loop.

Although the normal rhythm of the Manchester computer was four beats to the bar, some instructions took five beats to execute. Incorporating a suitable five-beat instruction in note-loops (e.g. Turing's instruction TN) extends the number of playable notes. For instance, adding ten extra beats to either the primary or the padded form of the 148.81 Hz note-loop displayed previously results in a loop that plays 109.65 Hz, which is very close to A_2 (110 Hz); the primary form is <3H, 4, 4, 4, 4, 4, 4, 5, 5>. The following loop plays the low note $F\sharp_2$: <3H, 4, 4, 4, 4, 5, 5, 5, 5, 5>. This loop produces 92.59 Hz, fractionally higher than the note's equal-tempered frequency of 92.5 Hz.

In what follows, note-loops are sometimes written in an abbreviated form; for example, <3H, 4×7> replaces <3H, 4, 4, 4, 4, 4, 4>, and <3H, 4×4, 5×5> replaces <3H, 4, 4, 4, 4, 5, 5, 5, 5, 5>. Table 23.2 shows the full range of frequencies (down to 96.9 Hz) that the Mark II could produce by means of note-loops containing four- or five-beat instructions.

The hoot-stop

As Turing explained in his tutorial, a fundamental use for <3H, 4> was the so-called 'hoot-stop'. If the two lines of code displayed in the tutorial were placed at the end of a routine (or 'program', as we would say today), then once the routine finished running, the computer would sound C_5 continuously until the operator intervened. The intervention might take the form of pressing the 'KEC' key at the control console—the 'clear everything' key—in order to clear out the completed routine, in preparation for running the next one.[40] Without the convenient hoot-stop facility, the operator was obliged to remain at the console, watching the indicators, in order to tell whether the routine had stopped running.

A very different solution to effectively the same problem was found in the case of BINAC, an early US computer. Herman Lukoff, one of BINAC's engineers, explained that the technician whose job it was to monitor BINAC through the night, Jack Silver, had to spend all his time 'looking at the flashing lights; it was the only way of knowing that the computer was working'. One night Silver switched on a radio to alleviate the monotony:[41]

To Jack's surprise, all kinds of weird noises emanated from the loudspeaker, instead of soothing music. He soon realized that the churning BINAC generated these noises because as soon as it halted, the noises stopped. . . . He put the computer-generated tones to good use. Jack found that by turning the volume up he was able to walk around the building and yet be immediately aware of any computer stoppage.

BINAC, built in Philadelphia by Presper Eckert, John Mauchly, and their engineers at the Eckert–Mauchly Computer Corporation, was the stored-program successor to the pioneering Eckert–Mauchly ENIAC (see Chapter 14). Eckert and Mauchly went on to build UNIVAC, one of the earliest electronic digital computers to enter the marketplace.

The first systematic use of the Manchester Mark I's programmable hooter seems to have been to provide the hoot-stop facility.

When were the first notes played?

In a section of his *Programmers' Handbook* devoted exclusively to the Mark I machine, Turing made it clear that the programmable hooter pre-dated the Mark II machine.[42] /V was Mark II notation; in the Mark I era it was the instruction K (11110) that caused the loudspeaker to sound. The Mark I, which was closed down in the summer of 1950,[43] was a slower machine than the Mark II; the duration of a beat was 0.45 ms compared with 0.24 ms in the Mark II. This considerably reduced the number of playable notes: lengthening the beat to 0.45 ms causes the frequency of <3H, 4>, the highest-frequency loop, to drop from 523.25 to 277.78 Hz, which is approximately $C\sharp_4$.

It is not known precisely when a programmable hooter was first added to the computer. Geoff Tootill's laboratory notebook is one of the few surviving documents relating to the transition of Baby into the Mark I. In a notebook entry dated 27 October 1948, Tootill listed the K instruction 11110 among the machine's 32 instructions, but indicated that it was unassigned at that time.[44] Given Turing's focus on the programming side, and the emphasis that he placed on the use of the hoot instruction and pause instructions (which he called 'dummy stops') for 'testing', i.e. debugging, new routines, it seems likely that the hooter was incorporated earlier rather than later.[45] The computer was running complex routines by April 1949, in particular a routine that searched for Mersenne primes (primes of the form 2^n-1),[46] and Turing's debugging toolkit of hoots and dummy stops was probably introduced earlier than this. It also seems likely that the use of loops to increase the volume of the hooter's native clicks occurred at about the same time as the K instruction was introduced. The loud note produced by the loop would have been more useful than the quiet click given by a single instruction. As Dietrich Prinz, a regular user of the Mark I, said: 'By programming a simple loop containing this instruction . . . an *audible* "hoot" is emitted'.[47]

A table in Tootill's notebook dated 28 November 1948, showing the machine's instructions at that time, listed three different dummy stops, N, F, and C. The section of Turing's *Programmers' Handbook* dealing with the Mark I explained that, for checking routines, these dummy stops were operated in conjunction with the hoot instruction K. By the time of the 28 November table, the K instruction had been assigned: Tootill listed its function as 'Stop'. However, his table also contains another instruction labelled 'Stop' (00010). Since the machine had no need of two ordinary stop instructions, it seems very likely that K was being used for hoot-stop at this time. When execution of the program that was being run reached the point where the hoot-stop had been inserted, execution would pause and the hooter would play the note $C\sharp_4$ (middle C sharp) continuously until the operator intervened. We conclude that the Mark I was playing at least one note in about November 1948.

Other early computer music

The Manchester Mark I was not the only zeroth-generation electronic stored-program computer to play music. Trevor Pearcey's Sydney-built CSIRAC (pronounced 'sigh-rack') had a repertoire that included *Colonel Bogey*, *Auld Lang Syne*, and *The Girl with the Flaxen Hair*, as well as brief extracts from Handel and Chopin. Some of the music routines survived on punched paper tape, but seemingly no audio recordings were preserved. Australian composer Paul Doornbusch has re-created some of the music, using reconstructed CSIRAC hardware and the surviving programs.[48] CSIRAC, still complete and almost in working order, is now in Melbourne Museum.

Doornbusch's recordings and the BBC's Manchester recording show that the programmers of both computers ran into the problem of 'unplayable notes'—notes that could not be replicated (or even closely approximated) by means of an available note-loop. An example is the note D_3, which occurs five times in the Manchester rendition of *God Save the King*. This note's equal-tempered frequency is 146.8 Hz, but the closest that the Mark II can approach it is the significantly different note of 148.81 Hz, discussed earlier. To judge from the Doornbusch recordings, $F\sharp_2$, G_2, $C\sharp_3$, $F\sharp_3$, D_4, E_4, F_4, G_4, and A_4 were particularly troublesome for CSIRAC.

The Australian and British solutions to the problem of unplayable notes were distinctively different. At Manchester they opted to use the nearest playable frequency and tolerated the melody being less in tune, selecting frequencies with a view to their overall relationships, rather than trying to hit the equal-tempered frequencies as closely as possible. CSIRAC's programmers, on the other hand, attempted to mimic the unplayable frequency by rapidly moving back and forth between two playable frequencies that bracketed the note in question. The result was a melody in which tuning-related problems were replaced by timbre-related problems, with the Australian technique producing notes that sound grainy and unnatural.

An embryonic CSIRAC first ran a test program in about November 1949.[49] The computer seems to have been partially operational from late 1950, and in regular operation from about mid-1951. The date when CSIRAC first played musical notes is unrecorded; presumably this was in late 1950 or in 1951, at least 2 years later than the Manchester computer. CSIRAC is known to have belted out tunes at the first Australian Conference on Automatic Computing Machines, held at Sydney University in August 1951.[50] A 2008 BBC News article, based on Australian sources, stated that CSIRAC was the first computer to play music.[51] The only evidence offered for this claim was that CSIRAC's performance at the Sydney Conference allegedly preceded the date of the previously described BBC recording of the Manchester computer. But in fact the exact date of the BBC recording is unknown, and in any case the Manchester computer's first performance of *God Save the King*—whose precise date is also unknown—would have occurred some days or weeks, or even months, before the BBC recording was made. Unfortunately, *The Oxford Handbook of Computer Music* also states, without evidence, that CSIRAC was 'the first computer to play music'.[52]

American hoots

CSIRAC, however, was certainly not the first computer to play music. BINAC was playing music before CSIRAC even ran its first test program. BINAC was completed in August 1949 (although it ran a fifty-line test program in April of that year).[53] As Lukoff explained, a party was held to celebrate the machine's completion:[54]

It was held right at the BINAC test area one August evening. In addition to hors d'oeuvres and cocktails, the BINAC crew arranged a spectacular computer show. Someone had discovered that, by programming the right number of cycles, a predictable tone could be produced. So BINAC was outfitted with a loudspeaker attached to the high speed data bus and tunes were played for the first time by program control. The audience was delighted and it never occurred to anyone that the use of a complex digital computer to generate simple tones was ridiculous.... The crowning achievement of the evening came after a long, laborious arithmetic computation; the machine laid an egg! The engineers had programmed the machine to release a hard-boiled egg from its innards.

As far as can be ascertained, therefore, the first melodies to be played by a computer were heard at the Eckert–Mauchly Computer Corporation in the summer of 1949, and it is very likely that individual musical notes were heard earlier at Turing's Computing Machine Laboratory, probably in November 1948. Effectively, the pioneering developments on either side of the Atlantic were contemporaneous, with Australia entering the field a year or two later.

Listening to the BBC recording

The BBC's website offers an edited digitized version of the original BBC recording of the Manchester Mark II, and there is a full-length version of the same digitization of the recording on the Manchester University website.[55] Upon pressing 'play', the listener is greeted by a thick wall of noise—a combination of hissing, humming, and rhythmically repeating crackles from the original acetate disc. Then a tone not unlike a cello cuts through this cacophony to give a mechanical-sounding rendition of the first two phrases of the national anthem. The melody, although familiar enough, is somewhat out of tune, with some notes more distinctly out than others. Moreover, some notes are loud relative to their neighbours (most likely the result of padding). At the end of the second phrase, the performance is suddenly cut short by a glitch and nervous laughter.

The engineers restart the routine and this time the machine energetically plays its way through the entire first verse. Then, with scarcely a pause, it follows up with an unbroken performance of the first line of *Baa Baa Black Sheep*. For its third number the Mark II attempts *In The Mood*, but once again falls victim to an unknown error, causing it to sing out a high pitched beep. The recording team give the computer one more chance to make its way through *In The Mood*, and it proceeds admirably until the final line, when it yet again breaks down. Altogether, the entire recording lasts about three minutes.

Two different acetate discs were cut during the recording session. One was taken away by the BBC and was presumably used in a broadcast. It is unlikely that this disc survives, but a second disc was given to Manchester engineer Frank Cooper as a souvenir. It contained another recording, made at Cooper's request once the main recording session was over.[56] By that time, Cooper recollected, 'the computer was getting a bit sick and didn't want to play for very long'. Eventually he donated this 12-inch single-sided acetate disc to the Computer Conservation Society, and subsequently the National Sound Archive (part of the British Library) made a digital preservation copy of the recording.[57]

Table 23.3 gives the primary forms of the note-loops that were used to play *God Save the King* in the BBC recording; and Table 23.4 gives the primary forms of the loops for the remaining notes in the other two melodies.

There are unsettled questions about the authorship of the routines that played the melodies recorded by the BBC. Cooper related that, in the wake of the computer's virtuoso performance of the national anthem, 'everybody got interested—engineers started writing music programs, programmers were writing music programs'.[58] Nothing about the BBC recording settles the question of authorship: even the routine that played the National Anthem in the recording may have been a retouched version of Strachey's original. However, it can at least be said that the programmer(s) of the routines for *Baa Baa Black Sheep* and *In The Mood* used the same key signature as the programmer of *God Save the King*, and also used the very same primary loops as those selected for *God Save the King*: new loops were introduced only for notes that do not occur in the national anthem. This was so even though some alternative primary loops were available, and in fact it is arguable that some of these choices would have produced frequencies

Table 23.3 *The primary note-loops used to play* **God Save the King.**

Note	Beats	Primary note-loop
$F\sharp_2$	42	<3H, 4×7, 5×2>
G_2	40	<3H, 4×9>
A_2	36	<3H, 4×3, 5×4>
B_2	32	<3H, 4×7>
C_3	30	<3H, 4×4, 5×2>
D_3	26	<3H, 4×3, 5×2>
E_3	24	<3H, 4×5>

Table 23.4 *The primary note-loops for additional notes in the melodies* **Baa Baa Black Sheep** *and* **In the Mood.**

Note	Beats	Primary note-loop
$C\sharp_3$	28	<3H, 4×6>
$F\sharp_3$	21	<3H, 4×3, 5>
G_3	20	<3H, 4×4>
A_3	17	<3H, 4×2, 5>

that sounded more in tune. In sum, it is possible that all three of the routines involved in playing the recorded melodies were coded by someone other than Strachey—although naturally all were inspired by his pioneering routine for *God Save the King*.

Restoring the recording

Our frequency analysis of the BBC recording revealed that the recorded music was in fact playing at an incorrect speed. This was most likely a result of the turntable in the BBC recording van running too fast while the acetate disc was being cut. (Achieving speed constancy was always a problem with the BBC's standard mobile recording equipment, whose mechanical cutter gouged out a groove in the rotating disc.[59]) So when the disc was played back, at the standard speed of 78 r.p.m.,[60] the frequencies were systematically shifted. It is those shifted frequencies, rather than the frequencies that the computer actually played, that we found in the National Sound Archive's digital copy of the recording.

Thus it became apparent to us that what we had thought was a tolerably accurate record of how the computer sounded in fact was not. The effect of the frequency shifts was so severe that the sounds in the recording often bore only a very loose relationship to the sounds that the computer actually produced. So distant was the recording from the original that many of the recorded frequencies were actually ones that it was *impossible* for the Mark II to play.

Naturally we wanted to uncover the true sound of Turing's computer. These 'impossible notes' in the recording proved to be the key to doing so: using the differences in frequency between the impossible notes and the notes that would actually have been played, we were able to calculate how much the recording had to be speeded up in order to reproduce the original sound of the computer.[61] It was a beautiful moment when we first heard the true sound of Turing's computer.

To put our findings to some practical archival use, we restored the National Sound Archive recording: this is now part of the National Sound Archive's collection. We increased the speed of the recording, and we filtered out extraneous noise and also removed the effects of a troublesome wobble in the speed of the recording. The wobble, most likely introduced by the disc-cutting process, was another source of frequency distortion, and even caused notes to bend slightly through their duration.

Nobody had heard the true sound of the computer since the early Ferrantis were scrapped more than half a century ago. A German researcher David Link attempted to re-create the sound by programming his marvellous emulation of the Mark II.[62] But an emulation is far from being the real thing, and without the original physical components, including of course the hooter, an emulation cannot recapture the actual sound. But now, thanks to an improbable meeting—in New Zealand—of the 1951 recording and modern analytical techniques, we really can listen to Turing's Mark II. Our restoration is available at *www.AlanTuring.net/historic_music_restored.mp3*.

CHAPTER 24

Turing, Lovelace, and Babbage

DORON SWADE

'The principles on which all modern computing machines are based were enunciated more than a hundred years ago by a Cambridge mathematician named Charles Babbage.' So declared Vivian Bowden—in charge of sales of the Ferranti Mark I computer—in 1953.1 This chapter is about historical origins. It identifies core ideas in Turing's work on computing, embodied in the realisation of the modern computer. These ideas are traced back to their emergence in the 19th century where they are explicit in the work of Babbage and Ada Lovelace. Mechanical process, algorithms, computation as systematic method, and the relationship between halting and solvability are part of an unexpected congruence between the pre-history of electronic computing and the modern age. The chapter concludes with a consideration of whether Turing was aware of these origins and, if so, the extent—if any—to which he may have been influenced by them.

Introduction

Computing is widely seen as a gift of the modern age. The huge growth in computing coincided with, and was fuelled by, developments in electronics, a phenomenon decidedly of our own times. Alan Turing's earliest work on automatic computation coincided with the dawn of the electronic age, the late 1930s, and his name is an inseparable part of the narrative of the pioneering era of automatic computing that unfolded.

Identifying computing with the electronic age has had the effect of eradicating pre-history. It is as though the modern era with its rampant achievements stands alone and separate from the computational devices and aids that pre-date it. In the 18th century *lex continui in natura* proclaimed that nature had no discontinuities, and we tend to view historical causation in the same way. Discontinuities in history are uncomfortable: they offend against gradualism, or at least against the idea of the irreducible interconnectedness of events.

The central assertion of this chapter is that core ideas evidenced in modern computing, ideas with which Turing is closely associated, emerged explicitly in the 19th century, a hundred years earlier than is commonly credited. While this assertion may come as no great surprise to many, the detailed extent to which it applies is perhaps less fully appreciated than the evidence warrants. Ideas articulated a century apart show a startling congruence, and this agitates an inevitable question: to what extent, if any, were the pioneers of the modern age of computing aware of what had gone before?[2]

In the context of Turing's work this suite of ideas is unmistakably modern, yet each was explicitly articulated in the writings of Charles Babbage (1791–1871) and Ada Lovelace (1815–52). The suite includes:

- algorithm and stepwise procedure
- 'mechanical process' and the idea of computation as systematic method
- computing as symbol manipulation according to rules
- halting as a criterion of solvability
- machine intelligence
- abstract formalism—attempts to express complex relations in a compact language of signs and symbols.

The start of automatic computation

Automatic computation received its first major impetus from Charles Babbage. The designs for his vast mechanical calculating engines mark the genesis of the movement to realize viable computing machines. Babbage bursts out of contemporary practice to startle us with concepts and machines that represent a quantum leap in logical conception and physical scale in relation to what had gone before. Babbage was a towering and controversial figure in scientific and intellectual life in Regency and Victorian England. Supported by his father, a wealthy London banker, he went up to Cambridge in 1810, aged 18, to study mathematics, an early and enduring passion. Already a moderately accomplished mathematician he found his college tutors staid and indifferent to new Continental theories, and he pursued a curriculum largely of his own devising. He formed a lifelong friendship with John Herschel, later a celebrated astronomer and lauded ambassador for science, and enjoyed the company of a wide circle of friends. He played chess, took part in all-night sixpenny whist sessions, and bunked lectures and chapel to go sailing on the river with his chums.

He was instrumental in the founding of the Analytical Society, a group whose mission was to reform Cambridge mathematics. His political views were radical, supporting as he did Napoleonic France, with which England was still at war. In a preliminary part of his final examinations he attempted to defend a proposition regarded by the college moderators as blasphemous, and he suffered their censure. Despite this apparent boldness and independence of thought, he recalls that he was 'tormented by great shyness'.

Babbage married in 1814 and settled in London with his wife, Georgiana. He embraced London scientific life, published several mathematical papers, and went on to author six books and some ninety papers, the scope of which collectively attests to a formidable range of interests. In maturity he was a respected and imposing scientific figure enjoying both fame and notoriety. He was proud, combative, and principled beyond ordinary reason—launching scathing ill-judged public attacks on the Royal Society for its supposedly negligent governance of scientific life. He appeared to prefer protest to persuasion and behaved as though being right entitled him to be rude. His first biographer described him as 'the irascible genius' and this depiction has endured.[3]

For all his wide interests the centre of his life was a near obsession with his calculating engines, and it is for these that he is most widely known. Their design was his great reward, his ill-fated attempts to build them was his downfall. He famously never completed an engine in its entirety and, not without bitterness, saw himself as having failed.

Babbage's epiphany is captured in a well-known vignette in which, in London in 1821, he and John Herschel were checking astronomical tables calculated by human computers. Babbage, dismayed at the number of errors, recalled that he exclaimed:[4]

I wish to God these calculations had been executed by steam.

The appeal to steam can be read as a metaphor for the infallibility of machinery, as well as for the idea of mechanized production as a means of solving the problem of supply. The 'unerring certainty of mechanical agency' would ensure error-free tables as and when needed.[5] Babbage immediately launched himself on his great venture—the design and construction of machines for mechanized computation—a mission that occupied him for much of the rest of his life.

By the spring of 1822 Babbage had made a small working model of a *Difference Engine* powered by a falling weight. His Difference Engine, so called because of the mathematical principle on which it was based (the method of finite differences), would, in intention at least, eliminate all sources of human error at a stroke. Calculation, transcription, typesetting, and verification—manual processes used in the production of mathematical tables and therefore susceptible to error—would now be unfailingly correct, ensured as they were by the certainties of mechanical action.

Up to that point Babbage's main interest and experience had been in mathematics, and his published output to this time consisted entirely of mathematical papers, thirteen in all between 1813 and 1821. So in 1822, when experimenting with his new model, we have a mathematician aged 30 running the first practical automatic computing machine and reflecting for the first time on the implications for machine computation. With his first experiments fresh in his mind he articulated these early reflections in five papers written in the six months between June and December 1822. Their content is as revealing as it is remarkable, although it has been largely overlooked.

The first model completed in the spring of 1822, known as 'Difference Engine 0', has never been found. But a demonstration piece, completed in 1832, representing one-seventh of the calculating section of Difference Engine No. 1, has all the essential features of its lost predecessor and is used here to illustrate Babbage's earliest recorded reflections on machine computation (Fig. 24.1a). The 'beautiful fragment', as the piece was referred to by Babbage's son,[6] is perhaps the most celebrated icon in the pre-history of automatic computing. It was the first successful automatic calculating device, and the significance of it being automatic cannot be overstated.

Number values were represented by the rotation of geared wheels (figure wheels) engraved with the decimal numbers 0–9. These were arranged in columns with the least significant digit at the bottom. Initial values, from a pre-calculated table, were entered by rotating individual figure wheels by hand to the required digit value. The engine was then operated by cranking a handle above the top plate to and fro (see Fig. 24.1b). Each cycle of the engine produced the next value of the mathematical expression in the table, and this could be read from the last column on the right. In this way the engine tabulated a class of mathematical expressions (called polynomials), using repeated addition of the values on the columns, and the results appeared in turn on the last column.

In contemplating his first working model, Babbage was struck by the prospects of solving equations by machine computation. An equation, say $y = x^2 - 1$, defines the relationship

Figure 24.1 Difference Engine No. 1 demonstration piece, 1832 (top). View from above showing crank handle (Bottom).

Photographs by Doron Swade.

between y (the dependent variable) and x (the independent variable) and the equation defines the value of y for each value of x. By convention the 'solution' (also called the 'root') of the equation is the value (or values) of x for which y is zero. In our example there are two roots: +1 and −1. In Babbage's engine the value of y appears on the last column and each cycle of the engine increases the value of x by a fixed increment, usually 1, and produces the corresponding new value of y. When the value of x coincides with a root, the result on the last column is all zeros. So finding the roots of an equation reduces to detecting the all-zero state in the results column and counting the number of times the handle was cranked, which gives the value of x. If two consecutive results straddle zero (i.e., if the result passes through zero without stopping) a root will be signalled by a sign change which can also be detected automatically.

At first the machine automatically rang a bell on the mechanical detection of the all-zero state to signal to the superintendent to stop the machine. Later designs incorporated mechanisms for halting the machine automatically. If there were multiple roots then the operator would keep cranking, so cycling the engine to find the remaining roots. If there were no solutions then the machine continued *ad infinitum*. Babbage wrote explicitly of the machine halting when it found a root: halting, as a criterion of solvability, is explicit in these earliest reflections.

Pre-echoes of Turing's 1936 breakthrough paper 'On computable numbers' are unmistakable. Whether or not a machine would properly complete its calculation was later called the 'halting problem' which, though not explicitly referred to in this way by Turing, is implied in his 1936 paper and is intimately associated with him. For those interpreting the behaviour of Turing's 'circular machine', halting became a central logical determinant as to whether or not machines could decide if a certain class of problems was soluble. Babbage himself did not claim any special *theoretical* significance for the halting criterion. For him it was a practical matter: if there were a solution, the machine would halt; if there were no solutions, the machine would grind on indefinitely.

Machine computation offered the prospect of solutions to mathematical problems which had until then had resisted solution by conventional formal analysis. There were equations, for example, for which there were no known theoretical solutions. By systematically cycling the machine to produce each next value of the expression, solutions could be found by detecting the all-zero state even though the roots that produced this result could not (yet) be found by mathematicians. Having a computational rule could achieve results that had evaded formal analysis.

In addition to instances of solving the hitherto unsolved, machine computation offered a more mundane prospect. There are many important 'series' in mathematics. These are sequences of values defined by a general formula. For example, if you wished to know the value of the 350th term in the series there was no simple way of finding this unless you had a general formula for the nth term. In some instances no such expression was known. By cycling the engine through 350 values the required value could be found, something that was impractical, or at least prohibitively tedious, to do by hand.

For Babbage, his calculating engines represented a new technology for mathematics, and machine computation as systematic method offered new prospects for solving certain mathematical problems.

The Difference Engine embodied mathematical rule in mechanism, and Babbage's 1822 model, driven by a falling weight, was the first physical machine to execute a computational algorithm without human intervention. The 1832 demonstration piece was driven not by a falling weight but by a manual crank at the top of the machine. In both machines the operator did not need to understand the mechanism, nor the mathematical principle on which it was based, to achieve useful results. The steps of the computational algorithm were no longer directed by human intelligence, but by the internal rules embodied in the mechanism and automatically executed. By exerting physical energy the operator could achieve results that, up to that point in time, could be achieved only by mental effort—by thinking. The idea that the machine was

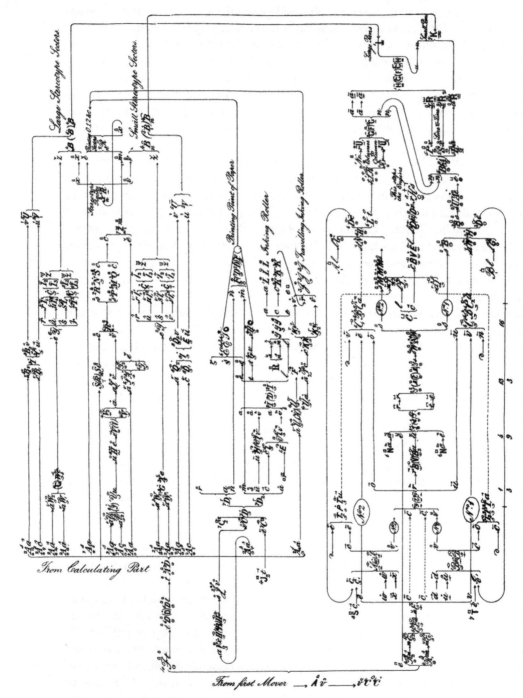

Figure 24.2 From Bromley, Allan, Ed. *Babbage's Calculating Engines: A Collection of Papers by Henry Prevost Babbage. Volume 2.* Los Angeles: Tomash, 1982.

From Bromley, Allan, Ed. *Babbage's Calculating Engines: A Collection of Papers by Henry Prevost Babbage. Volume 2.* Los Angeles: Tomash, 1982 Photograph by Doron Swade.

'thinking' was not lost on Babbage or his contemporaries. Harry Wilmot Buxton, a junior colleague of Babbage wrote:[7]

the marvellous pulp and fibre of a brain had been substituted by brass and iron, he [Babbage] had taught wheelwork to think, or at least to do the office of thought.

In 1833 Lady Annabella Byron, wife of the poet Byron and mother of Ada Lovelace wrote:[8]

We both went to see the *thinking* machine (for so it seems).

The transference from mind to machine is clear in these descriptions.

The 'mechanical notation'

Babbage quickly found that the intricate mechanisms and long trains of action in the Difference Engine were impossible to hold in his mind at once. To alleviate the difficulty he developed his 'mechanical notation', a language of signs and symbols of his own devising that he used to formalize the description of his machines and their mechanisms (Fig. 24.2).

The notation is not a calculus. It is more in the nature of a symbolic description that specifies the intended motion of each part and its relation to all other parts: whether a part is fixed (a framing piece, for example) or free to move, whether motion is circular or linear, continuous or intermittent, driver or driven, and to which other parts it is connected. Each part was assigned a letter of our familiar Latin alphabet and various typefaces were used (italicized for moving parts and upright for fixed framing pieces, for example). The physical form of parts and their motion were indicated by up to six superscripts and subscripts appended to each identifying letter. One of the superscripts (the 'sign of form') indicated the kind of part—gear wheel, arm lever, spring, pinion, etc. Another index (the 'sign of motion') indicated the nature of movement—reciprocating, linear or circular, or combinations of these. Annotations of this kind are liberally distributed throughout the technical drawings.

The notation has three main forms:[9]

1 a tabular form usually included on the drawing of the mechanisms (the drawings were called 'forms')—the table is a compilation or index of the individual notations that appear on the drawing and was used as an easy way to locate particular parts;
2 timing diagrams (called 'cycles') that describe the phasing of motions in relation to each other—that is, the orchestration of motion;
3 a flow diagram form (called 'trains') that depicts the chain of influence of parts on one another.

Babbage was inordinately proud of the Mechanical notation, which he regarded as amongst his finest inventions. He saw it as a universal language with application beyond science and engineering. Two examples he gave of its extended use were the circulation of blood in birds and the deployment of armies in battle. It represents a first level of abstraction and he used it extensively as a design aid to describe the function of the mechanisms.

The fate of the notation, baroque in its intricacy and idiosyncratically novel, has been largely one of obscurity. The notation pre-echoes what we would now call a 'hardware description language' (HDL). Such languages rose to prominence again in the early 1970s to manage complexity in computer circuit and system design, particularly in solid-state chip design where

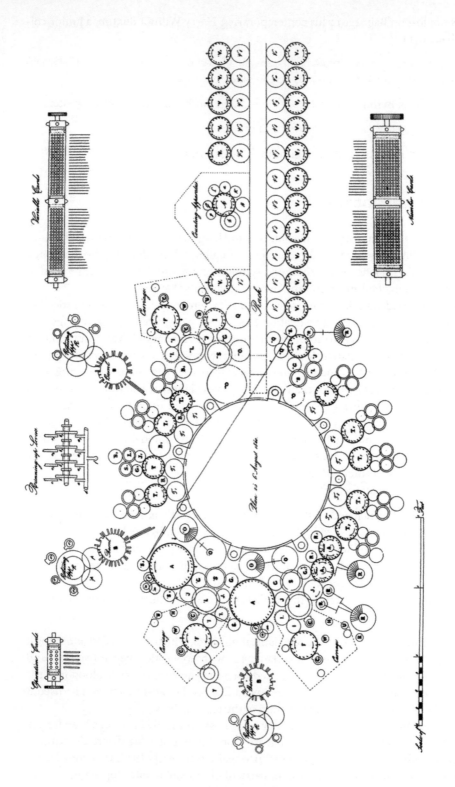

Figure 24.3 Analytical Engine, plan 25 (part), 1840.

Reproduced with permission of the Science Museum, London.

higher-order representation was used to manage vast and complex detail. Babbage had adopted the same route driven by the same imperative—a symbolic language to manage otherwise unmanageable complexity at the component level—the same solution to the challenges of complexity 150 years apart.

In these earliest reflections on the potential of automatic computing machines we can see a clear articulation of a set of core ideas that were evidenced in Turing's work a century later. The idea of 'mechanical process' is central. As related by Max Newman, this idea may well have been the jumping-off point for Turing's 'On computable numbers', following a lecture given by Newman in 1935 that was attended by Turing (see Chapter 40). Here Newman recalled using the idea of 'a purely mechanical machine' to describe a process in mathematical logic:[10] 'mechanical', in the Turing–Newman sense, denoted a systematic procedure that is 'unthinking' or 'unintelligent'. The development of Turing's universal machine involved the stripping out of all but operational (mechanical) features of calculation. A startling feature of the Turing machine was the reduction of calculation, for which the *sine qua non* had until then been human intelligence, to a series of unthinking tasks. For Babbage, in his world of cogs, levers, cams, ratchets, and machine cycles, 'mechanical process' was something literal rather than theoretical, but unmistakeably unthinking in Turing's sense.

Mechanical process, algorithm as a stepwise procedure, computation as a systematic method, halting, machine intelligence, and abstract formalism were intrinsic to both men's work and can be seen as a part of a remarkable congruence between the central features of both Babbage's and Turing's ideas. But there is one critical congruence without which the picture is incomplete.

The Analytical Engine: from calculation to computing

Difference Engines, for all the ingenuity of their design, and all the stimulus that they provided to debates about the prospects and limits of machine calculation, are not computers as we would now understand them. They are strictly calculators that crunch numbers in the only way they know how—by repeated addition, according to the method of differences. They execute a single specific fixed algorithm on whatever initial values they are given. Useful as tabulating polynomials is (they have wide application in science and engineering), in computational terms the Difference Engines have no generality, not even as a four-function arithmetical calculator.

A leap to generality is found in Babbage's *Analytical Engine*, conceived during 1834, shortly after his decade-long project to construct a full-sized and well-engineered Difference Engine had fatally foundered. There are detailed and extensive designs for the Analytical Engine that Babbage worked on intensively for over a decade until 1847, and then intermittently until he died in 1871 (Fig. 24.3). There is no definitive single version of the Analytical Engine, but rather three distinct phases in which different features were developed and refined. Like the first Difference Engine, the Analytical Engine remained unbuilt in its entirety.[11]

The designs for the Analytical Engine describe an automatic programmable general-purpose computing machine, capable of executing any sequence of arithmetical operations under program control. At a user level it was programmable, using punched cards which Babbage introduced in 1836. These were made of paste-board and stitched together in a fan-fold arrangement, where the hole-pattern determined the instruction to be executed.

There were several classes of punched card: 'operation cards' for instructions, 'number cards' for data, 'variable cards' to indicate where in the memory to place or retrieve data, 'combinatorial cards' to indicate how many times a particular sequence of operations should be repeated, and so on. The machine had an internal repertoire of automatically executable functions—direct division, multiplication, subtraction, and addition.

The range of features was dazzling: there was a separate 'store' (memory) and a 'mill' (processor). At machine level, 'microprograms' automatically executed each of the internal repertoire of instructions: these microprograms were coded onto 'barrels' with moveable studs like those on a barrel organ, but substantially larger. The machine was capable of iterative looping, conditional branching, parallel processing, anticipating carriage, transferring data on a multidigit parallel bus or data highway, and output in various forms including printing, stereotyping, graph plotting, and punched cards.[12] Serial operation using a fetch–execute cycle, the separation of store and mill, and the input–output arrangements of the Analytical Engine are signature features of the so-called 'von Neumann architecture', as described by John von Neumann in his classic paper in 1945, and which has dominated computer design since.[13]

Ada Lovelace

The machine conceived and designed by Babbage was an engine for doing mathematics, the driving preoccupation of its inventor—finding the value of any mathematical expression, tabulating functions, solving partial differential equations, enumerating series, and so on. In 1836 Babbage hinted at the prospects of a general-purpose algebraic machine able to manipulate symbols 'without any reference to the value of the letters', but he described the idea as 'very indistinct' and did not elaborate.[14] Symbolic algebra was an abiding aspiration and he returned to this later with some success. The context throughout was nonetheless mathematical.

The transition from calculation to computing, or from number to symbol, was articulated more clearly by Ada Lovelace, Lord Byron's daughter from his short-lived and troubled marriage to Annabella Milbanke. Lovelace (Fig. 24.4) and Babbage met in 1833 when she was 17

Figure 24.4 Ada Lovelace.

Alfred Edward Chalon. Posted to Wikimedia Commons and licensed under public domain, https://en.wikipedia.org/wiki/Ada_Lovelace#/media/File:Ada_Lovelace_portrait.jpg.

and he in his early 40s. Babbage became, and remained, a close family friend until her death in 1852 at the age of 36. Young Ada Byron described Babbage at their first meeting as being 'full of animation—and talked about his wonderful machine'. Lovelace, who was tutored in mathematics by Augustus De Morgan, published her 'Sketch of the Analytical Engine' in 1843.[15] The Sketch consists of Lovelace's translation from the French of the article on the Analytical Engine by Luigi Menabrea, who was present at a convention in Turin in 1840 where he heard Babbage speak about his engines, the only known occasion that he lectured in open forum on his machines. Lovelace appended to her translation extensive notes of her own written in close collaboration with Babbage.[16] To call it a 'Sketch' (translated from the French 'notions') is misleading if this implies that its content is slight: the piece remains the most substantial contemporary account of the capabilities and potential of the engine.

In her notes Lovelace described the mathematical capabilities of the engine, and moreover emphasized the implications of machine computation outside mathematics:[17]

[The Analytical Engine] might act upon other things besides *number* . . . Supposing, for instance, that the fundamental relations of pitched sounds in the science of harmony and of musical composition were susceptible of such expression and adaptations, the engine might compose elaborate and scientific pieces of music of any degree of complexity or extent.

Lovelace's example takes a crucial step in signposting the representational power of number. If meaning is assigned to number then results, arrived at by operating on number according to rules, can say things about the world when they are mapped back onto the world using the meanings assigned to them. The act of abstraction is in the assignment by us of such meanings. Lovelace wrote that the machine could 'arrange and combine its numerical quantities exactly as if they were *letters* or any other *general* symbols'.[18]

The key transitional idea is that number could represent entity other than quantity and that the potential of computing lay in the power of machines to manipulate, according to rules, representations of the world contained in symbols. Nowhere, at least in his published work, does Babbage speak in this way.

Babbage was much taken with Lovelace's ideas, to the extent that he was reluctant at one point to part with her manuscript. He wrote:[19]

. . . the more I read your notes the more surprised I am at them and regret not having earlier explored so rich a vein of the noblest metal.

If Babbage was true to character in this tribute then he was not dissembling, being patronising, or falsely flattering. Yet if he was influenced by Lovelace's ideas, he showed no sign of it. Towards the end of his life he set out, finally, to write a general description of the Analytical Engine. Each of his three separate attempts, none of which was finished, opened with a statement clarifying the purpose of his engine. The third and last of these, dated 8 November 1869, some two years before his death, opens:[20]

The object of the Anal. Eng. Is two fold
  1st The complete manipulation of number
  2nd The complete manipulation of Algebraic Symbols

After half a century of deliberation and inspired design, Babbage in his most mature reflections did not see the scope of the engine extending beyond algebra. And if he saw algebra having a representational reach outside mathematics, he made no reference to it. The three separate opening statements declaring the purpose of the machine differ in specifics, but in each his ambitions for the engine are confined to mathematics with no reference as to how, through representation, computers could make statements about the world. Lovelace's emphasis on symbolic representation, and her mention of the use of arbitrary rules invented by the user[21] pre-echoes features of universality in Turing's work.[22]

Both Lovelace and Turing speculated about machine intelligence and the relationship between computing and the brain. Lovelace argued for the amenability of mental process to logical or mathematical description when, in 1844, three months after her Sketch was published she wrote:[23]

I have my hopes . . . of one day getting *cerebral* phenomena such that I can put them into mathematical equations; in short a law, or laws, for the mutual actions of the molecules of *brain* I (equivalent to the *law of gravitation* for *the planetary and sidereal* world).

. . . none of the physiologists have yet got on the right tack . . . It does not appear to me that cerebral matter need be more unmanageable to mathematicians than *sidereal* and *planetary* matters and movements, if they would but inspect it from the right point of view. I hope to bequeath to the generations . . . a *Calculus of the Nervous System*.

While working on the ACE in the 1940s Turing wrote:[24]

In working on the ACE I am more interested in the possibility of producing models of the action of the brain than in the practical applications to computing.

Lovelace wrote that the 'Analytical Engine has no pretentions whatever to *originate* anything. It can do whatever *we know how to order it* to perform'.[25] In his paper 'Computing machinery and intelligence', published in 1950, Turing called the view that machines were incapable of originating anything 'Lady Lovelace's objection' and he used this as a counterpoint to argue for more generous limits to the potential for intelligent machines.[26]

The apparent paradox of rule-based creativity was one of many issues provoked by the first essay into machine intelligence prompted by Babbage's calculating engines and later engaged with by Turing with challenging originality. Many of the self-same issues that arose in the early 19th century continue to confound and preoccupy us nearly two centuries later, and the vigour of these debates remains unabated as developments in neurosciences and machine intelligence leap-frog each other.

The design of the Analytical Engine was sufficiently advanced by 1837 for Babbage to write 'programs', though neither Lovelace nor Babbage used that term. Between 1837 and 1840 Babbage wrote twenty-four such programs for a variety of problems. His programs were mainly illustrations of how the engine could do things that could already be done by hand—solutions for simultaneous equations of various kinds, and three examples of series calculated using recurrence relations requiring the iteration of the same set of operations.

Babbage's writing was largely technocentric, consisting as it did of how the engine worked and what it did. Lovelace, on the other hand, was concerned with the significance of the machine, its wider implications, and its potential. Her program for the calculation of Bernoulli numbers, for which she is famed, while possessing the same format as Babbage's earlier programs, is significantly more ambitious than Babbage's early examples. It does not appear that Babbage returned to programming after 1840 to provide solutions to new problems, though the capabilities of the engine grew in the years that followed.

If we wished to tag the respective roles of Babbage, Lovelace, and Turing, it would be fair to say that Babbage's interests were essentially hardware-led, Lovelace's interests application-led, and Turing's, at least initially, theory-led, devising as he did a formal description of a computing engine in logically rigorous terms.

Did Babbage and/or Lovelace influence Turing?

Ideas explicitly articulated in the 19th century pre-echo with startling familiarity those of the 20th century, and this raises the inevitable question of influence. Did pioneers of the electronic computer age reinvent core principles largely in ignorance of what had gone before, or was there direct influence of some kind?

Babbage's engines were a false dawn. With Lovelace's early death automatic computation lost a formidable advocate, and when Babbage died some twenty years later the movement lost its most energetic practitioner. There were a few short-lived followers, but no heirs. Developmental continuity was broken, and except for a few isolated twitches, interest was revived in earnest only in the 1940s with the start of the electronic era.[27] Leslie Comrie referred to the gap between Babbage and electronic computing as a 'dark age of computing machinery that lasted 100 years'.[28] There was no chain of intermediaries to the modern era through which to track influence: the baton was dropped and lay where it fell for a hundred years.

Studies show that most of the pioneers of the modern age knew of Babbage.[29] A few did not; and at least one appears to have claimed to know more than he did.[30] The legend of Babbage did not die, and what he had attempted was common knowledge to the small cadre of those involved with automatic computation. So just what was known of Babbage and Lovelace's works by those who knew of them at all?

We know that Turing discussed Babbage with Donald Bayley at Hanslope Park after Bayley was posted there in 1944, but too little is known about the content of these discussions to provide us with any conclusive view of the extent of Turing's knowledge.[31] Similarly, there were accounts of 'lively mealtime discussion'[32] about Babbage at the Government Code and Cypher School at Bletchley Park, which would not have pre-dated September 1939 when Turing reported there at the start of the war. 'Babbage was a household word and very much a topic of conversation . . . very, very early on' attested J. H. Wilkinson, Turing's assistant at the National Physical Laboratory during the design of the ACE computer in the 1940s.[33]

What sources might Turing have had access to? There were several published descriptions of Babbage's his work in the public domain that went beyond general accounts of his work of the kind found in encyclopaedias, or indeed in his own autobiographical outing, *Passages from the Life of a Philosopher*, published in 1864. One was Lovelace's Sketch,[34] which included specific mention of the machine halting automatically at the detection of the all-zero state or a sign change, and of course the idea of symbolic representation. A second was a large collection of articles, book extracts, and a selection of design drawings collated by Babbage's son, Henry, and published in 1889.[35] A third was Dionysius Lardner's article, published in 1834, which gave an extensive account of Babbage's Difference Engine No. 1, and in which there was explicit mention of the machine being cycled to produce successive values of a formula, the detection of the roots, and a bell ringing automatically to signal a solution.[36] Lardner's article was both cited and referred to in Lovelace's Sketch. There was the comprehensive technical archive of Babbage's designs and notebooks held by the Science Museum in London.[37] No more than a handful of the

7000 manuscript sheets had been published, and any significant knowledge would necessarily have involved physical examination *in situ*. The detail of this archival material was not studied in earnest until the late 1960s, and scarcely anything was published on the specifics of the designs until the 1970s.[38] These 19th-century sources were accessible in libraries and archives.

In Turing's own time there was a paper on Babbage's Difference Engine by Dudley Buxton,[39] grandson of a junior colleague of Babbage into whose hands Babbage had entrusted a major cache of manuscripts.[40] The paper, presented at the Science Museum in December 1933 and published by the Newcomen Society, describes the Difference Engine in fair detail, touches on the transition to the Analytical Engine and outlines techniques for direct multiplication and division. Several practitioners allied to Turing's subsequent interests participated in the discussion following Buxton's paper. Included were D. Baxandall, Keeper of the Mathematics Section at the Science Museum, L. J. Comrie, Superintendent of the Nautical Almanac Office who implemented on commercially available calculators Babbage's method of tabulation by differences, and A. J. Thompson from the General Register Office.[41] Buxton's paper signalled that Babbage still had a visible profile in the period of Turing's famous 1936 paper. Turing does not mention Babbage's archival drawings or the other published sources.

In August 1951 Turing, with four Cambridge friends, visited the Science Museum in South Kensington where the science exhibits for the Festival of Britain were displayed.[42] On public view would have been Babbage's small experimental model of a part of the Analytical Engine under construction at the time of his death in 1871 and the 'beautiful fragment' of Difference Engine No. 1, but it is not known whether he saw them.[43] In any event, neither of these physical relics illuminated the design principles or their mechanical operation.

In the context of his reference to 'Lady Lovelace's objection' Turing comments that Lovelace's 'memoir' was the source of 'our most detailed information of Babbage's Analytical Engine'. This was the earliest known reference he made.[44] In quoting Lovelace's 'objection' Turing says that Lovelace's statement was quoted by Douglas Hartree. This opens up the possibility that Turing's source was not Lovelace's Sketch itself, in all its extensive detail, but Hartree's brief overview.[45]

It is reasonable to conclude that Turing did not consult 19th-century sources by H. P. Babbage, by Lardner, or the technical plans by Babbage himself, and that his knowledge of Lovelace and the Analytical Engine was most likely no more than Hartree's summary account. If this is so then, given that Hartree's book was not published until 1949, Babbage's and Lovelace's ideas could not have influenced Turing's 1936 paper. Robin Gandy supports this view. He commented that Menabrea's account and Lovelace's notes appear not to have been known even to some of those familiar with Babbage:[46]

Thus there is no reason to suppose that the mathematicians involved in the discoveries of 1936 (including Hilbert, Bernays, Gödel, and Herbrand) were familiar with Babbage's theoretical ideas: certainly none of them refer to his work.

In overall terms there does not appear to be a direct overt line of influence between Babbage–Lovelace in the 19th century and the pioneers of electronic digital computing in the 20th century. Most leading figures knew of Babbage and the legend of his mission but none appear to have had any detailed knowledge of his designs.

In the light of a strong congruence of ideas between two pioneering eras separated in time we are left with the suggestion, both reassuring and confining, that core ideas, articulated a century apart, embody something fundamental about the nature of computing.

PART V

Artificial intelligence and the mind

CHAPTER 25

Intelligent machinery

JACK COPELAND

T his chapter explains why Turing is regarded as founding father of the field of artificial intelligence (AI), and analyses his famous method for testing whether a computer is capable of thought.

The first manifesto of AI

In the weeks before his 1948 move from the National Physical Laboratory to Manchester, Turing wrote what was, with hindsight, the first manifesto of artificial intelligence (AI). His provocative title was simply *Intelligent Machinery*.[1]

While the rest of the world was just beginning to wake up to the idea that computers were the new way to do high-speed arithmetic, Turing was talking very seriously about 'programming a computer to behave like a brain'.[2] Among other shatteringly original proposals, *Intelligent Machinery* contained a short outline of what we now refer to as 'genetic' algorithms—algorithms based on the survival-of-the-fittest principle of Darwinian evolution—as well as describing the striking idea of building a computer out of artificial human nerve cells, an approach now called 'connectionism'. Turing's early connectionist architecture is outlined in Chapter 29.

AI and the bombe

Strangely enough, Turing's 1940 anti-Enigma bombe was the first step on the road to modern AI.[3] As Chapter 12 explains, the bombe worked by searching at high speed for the correct settings of the Enigma machine—and once it had found the right settings, the random-looking letters of the encrypted message turned into plain German. The bombe was a spectacularly successful example of the mechanization of thought processes: Turing's extraordinary machine performed a job, codebreaking, that requires intelligence when human beings do it.

The fundamental idea behind the bombe, and one of Turing's key discoveries at Bletchley Park, was what modern AI researchers call 'heuristic search'.[4] The use of heuristics—shortcuts or rules of thumb that cut down the amount of searching required to find the answer—is still a fundamental technique in AI today. The difficulty Turing confronted in designing the bombe was that the Enigma machine had far too many possible settings for the bombe just to search blindly through them until it happened to stumble on the right answer—the war might have been over before it produced a result. Turing's brilliant idea was to use heuristics to narrow, and so to speed up, the search.

Turing's idea of using crib-loops to narrow the search was the principal heuristic employed in the bombe (as Chapter 12 explains). Another heuristic that Turing mentioned in his 1940 write-up of the bombe was known as 'Herivelismus' (so named after its inventor, John Herivel).[5] Herivelismus involved a simple yet powerful idea. It could be assumed that when an operator enciphered his first message of the day, the starting positions of his Enigma machine's three encoding wheels would be in the vicinity of the wheels' 'base' position for the day. This base position was determined by one of the variable settings of the Enigma machine, which the operators would change every day (by referring to a calendar that specified the day's setting). This setting was known as the *Ringstellung*, or ring position, and the *Ringstellung* fixed this daily base position of the three wheels. The German operator was supposed to twist the three wheels away from their base position before enciphering the first message, but operators—in a hurry, under fire, or perhaps just lazy—often did not turn the wheels far from their base position before starting to encipher the message. So when the bombe was searching for the positions of the three encoding wheels at the beginning of what was believed to be an operator's first message, only settings in the neighbourhood of the base position were considered.

This procedure narrowed the search considerably and saved a lot of time. The assumption that the wheels were in the vicinity of the base position when operators began encoding their first message was not always true, but it was true often enough. The Herivelismus heuristic broke a lot of messages.

Spreading the word

Thanks to the bombe, Turing glimpsed the possibility of achieving machine intelligence of a general nature by means of guided search. This idea fascinated him for the rest of his life. He was soon talking enthusiastically to his fellow codebreakers about using this new concept of guided search to mechanize the thought processes involved in playing chess (see Chapter 31).[6] He also wanted to mechanize the process of learning itself. At Bletchley Park, he actually circulated a typescript on machine intelligence—now lost, this was undoubtedly the earliest paper in the field of AI.[7]

Once the war was over, Turing began making his radical ideas public. In February 1947, in an ornately grand lecture theatre in Burlington House, a Palladian mansion near London's Piccadilly, Turing gave what was, so far as we know, the first public lecture ever to mention computer intelligence.[8]

Turing offered his audience a breathtaking glimpse of a new field, predicting the advent of machines that act intelligently, learn from experience, and routinely beat average human opponents at chess. Speaking more than a year before the Manchester Baby ran the first computer program, his far-seeing predictions must have baffled many in his audience. In the lecture Turing even anticipated some aspects of the Internet, saying:[9]

It would be quite possible to arrange to control a distant computer by means of a telephone line.

In 1948, with mathematician David Champernowne, Turing wrote the world's first AI program, a heuristic chess-playing program named 'Turochamp'—as described in Chapter 31.

Child machines

Another of the fundamental concepts that Turing introduced into AI was the 'child machine' (see Chapter 30).[10] This is a computing machine whose makers have endowed it with the systems needed in order for it to learn as a human child would. 'Presumably the child-brain is something like a note-book as one buys it from the stationers', Turing said: 'Rather little mechanism, and lots of blank sheets'.[11] 'Our hope', he explained, 'is that there is so little mechanism in the child-brain that something like it can be easily programmed'.[12]

Turing expected that once the child machine had been 'subjected to an appropriate course of education one would obtain the adult brain', and that eventually a stage would be reached when the machine is like the pupil who 'learnt much from his master, but had added much more by his own work'.[13] 'When this happens I feel that one is obliged to regard the machine as showing intelligence', Turing said.

Turing proposed equipping a robot 'with the best sense organs that money can buy' and then teaching it in a way resembling the 'normal teaching of a child'.[14] As early as 1948, Turing was proposing the use of 'television cameras, microphones, loudspeakers, wheels and "handling servo-mechanisms" ' in building a robot.[15] The robot would, Turing said, need to 'roam the countryside' in order to learn things for itself—joking that, even so, 'the creature would still have no contact with food, sex, sport and many other things of interest to the human being'.[16] He worried, tongue in cheek, that the child machine could not be sent to school 'without the other children making excessive fun of it'.[17] Turing's colleagues at the National Physical Laboratory mocked his far-sighted ideas. 'Turing is going to infest the countryside', they laughed, 'with a robot which will live on twigs and scrap iron'.[18]

It was Turing's ex-Bletchley colleague and friend Donald Michie who re-introduced Turing's child-machine concept into modern AI. Michie burst onto the AI scene at the start of the 1960s, and was a powerful advocate for Turing's ideas about machine learning. The AI bug first bit him at Bletchley Park, thanks to Turing: after a heavy week of codebreaking, the two would meet in a village pub on Friday evenings to discuss how to reproduce human thought processes in a universal Turing machine (see Chapter 31). From chess Turing moved on to the idea of automating the whole process of learning. When he introduced the child-machine concept, Michie was gripped: 'By the end of war I wanted to spend my life in that field', he said.[19]

Later, during the 1960s and 1970s, Michie seeded departments of machine intelligence at several UK universities, and exported Turing's child-machine concept to AI labs in North America and around the world. In the 1970s, Michie and his AI group at Edinburgh University built a child-machine robot they named Freddy (for 'Friendly Robot for Education, Discussion and Entertainment, the Retrieval of Information, and the Collation of Knowledge' or FREDERICK). Freddy, like the imaginary robot described by Turing in 1948, had a television camera for an eye and handling servo-mechanisms to guide its pincer-like appendage (Fig. 25.1). Michie's group taught Freddy to recognize numerous everyday objects—a hammer, cup, and ball for example—and to assemble simple objects like toy cars from a pile of parts.

In those early days of computing, Freddy required several minutes of CPU time to recognize a teacup. Turing had predicted, on the basis of an estimate of the total number of neurons in the human brain, that not until computers possessed a memory capacity of around a gigabyte would human-speed AI start to become feasible,[20] and it was about 15 years after Freddy was switched off for the last time that the Cray-2 supercomputer offered a gigabyte of RAM.

In more recent times, Turing's ideas also formed the inspiration for the famous series of robots built by Rodney Brooks' group in the Artificial Intelligence Laboratory at the Massachusetts Institute of Technology (MIT).[21] The MIT robots Cog and Kismet were, like Freddy, faltering first steps towards child machines. Kismet is described in Chapter 30, where some of the philosophical difficulties involved in building an artificial infant are explored.

Testing for thought

How can researchers tell if a computer—whether a humanoid robot or a disembodied supercomputer—is capable of thought?

This is not an easy question. For one thing, neuroscience is still in its infancy. Scientists do not know exactly what is going on in our brains when we think about tomorrow's weather or plan out a trip to the beach—let alone when we write poetry or do complex mathematics in our minds. But even if we did know everything there is to know about the functioning of the brain, we might still be left completely uncertain whether entities *without* a human (or mammalian) brain could think. Imagine that a party of extra-terrestrials find their way to Earth and impress us with their mathematics and poetry. We discover they have no organ resembling a human brain: inside they are just a seething mixture of gases. Does the fact that these hypothetical aliens contain nothing like human brain cells imply that they do not think? Or is their mathematics and poetry proof enough that they *must* think—and so also proof that the mammalian brain is not the only way of doing whatever it is that we call thinking?

Of course, this imaginary scenario about aliens is supposed to sharpen up a question that is much nearer to home. For 'alien', substitute 'computer'. When computers start to impress us with their poetry and creative mathematics—if they don't already—is this evidence that they can think? Or do we have to probe more deeply, and examine the inner processes responsible for *producing* the poetry and the mathematics, before we can say whether or not the computer is thinking?

Deeper probing wouldn't necessarily help much in the case of the aliens—because *ex hypothesi* the processes going on inside them are nothing like what goes on in the human brain. Even if we never managed to understand the complex gaseous processes occurring inside the aliens, we might nevertheless come to feel fully convinced that they think, because of the way they lead their lives and the way that they interact with us. So does this mean that in order to tell whether a computer thinks we only have to look at what it does—at how good its poetry is—without caring about what processes are going on inside it?

That was certainly what Turing believed. He suggested a kind of driving test for thinking,[22] a *viva voce* examination that pays no attention at all to whatever causal processes are going on inside the candidate—just as the examiner in a driving test cares only about the candidate's car-handling behaviour, and not at all about the nature of the internal processes that produce that behaviour. Turing called his test the 'imitation game', but nowadays it is known universally as the 'Turing test'. There is more on the Turing test in Chapter 27.

Turing's iconic test works equally well for computers or aliens. It involves three players: the candidate and two human beings. One of the humans is the examiner or judge; the other—the

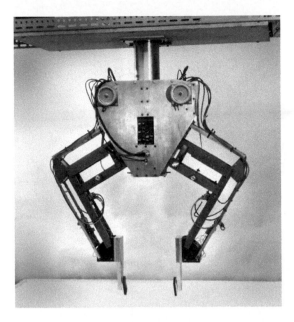

Figure 25.1 Freddy, an early implementation of Turing's concept of a child machine.

Image from Artificial Intelligence, Informatics, University of Edinburgh.

'foil'—serves as a point of comparison. The judge has to try and figure out which of the other two participants is which, human or non-human, simply by chatting with them via a keyboard (Fig. 25.2). The foil's job is to help the judge make the right identification. A number of chat sessions are run using different judges and different foils, and if the judges are mistaken often enough about which contestant is which, the computer (or alien) has passed the test.[23]

Turing imagined conducting these conversations via an old-fashioned teleprinter, but nowadays we would use email or text messages. Apart from chatting, the judges must be kept strictly out of contact with the contestants—no peeping is allowed. Nor, obviously, are the judges allowed to measure the candidates' magnetic fields, or their internal temperatures, or their processing speeds. Only Q & A is permitted, and the judges must not bring any scientific equipment along to the venue.

Justifying his test, Turing pointed out that the judges can use keyboard chat to probe the computer's skill in almost all fields of human endeavour. His examples included mathematics, chess, poetry, and flirting. He said: 'The question and answer method seems to be suitable for introducing almost any one of the fields of human endeavour that we wish to include'.[24] Turing added drolly, 'We do not wish to penalise the machine for its inability to shine in beauty competitions', making the point that his question and answer test excludes irrelevant factors.[25]

No questions are barred—the computer must be able to deal fair and square with anything the judge throws at it. But in order to avoid loading the dice against the computer, Turing stipulated that the judges 'should not be expert about machines'.[26] He also said that the computer is allowed to use 'all sorts of tricks' to bring about a wrong identification.[27] Smart moves for the computer are to reply 'No' in response to 'Are you a computer?', and to follow a request to multiply one huge number by another with a long pause and an incorrect answer—but a *plausibly* incorrect answer, not simply a random number. In order to fend off especially awkward questioning, the computer might even pretend to be from a different (human) culture than the judge. In fact, it is a good idea to select the test personnel so that, from time to time, the judge and the foil are themselves from different cultures.[28]

Figure 25.2 The Turing test. The judge must decide which contestant is the computer.

The Turing test is extremely tough for a computer to pass. We humans find idle chit-chat across the dinner table easy, but if one steps back and considers what is involved one realizes that, even in quite trivial conversations, we are manipulating vast amounts of knowledge, and are effortlessly producing and comprehending complex linguistic structures, often highly idiomatic ones, as well as expertly handling such obstacles to comprehension as irony, metaphor, creative humour, malformed or unfinished sentences, and abrupt and unannounced changes of subject. None of these, with the possible exception of the first, are things that today's computers excel at.

Here is Turing's own example of the sort of conversation that could occur between a judge and a computer that successfully evades identification:[29]

Judge: In the first line of your sonnet which reads 'Shall I compare thee to a summer's day', would not 'a spring day' do as well or better?
Machine: It wouldn't scan.
Judge: How about 'a winter's day'? That would scan all right.
Machine: Yes, but nobody wants to be compared to a winter's day.
Judge: Would you say Mr Pickwick reminded you of Christmas?
Machine: In a way.

Judge: Yet Christmas is a winter's day, and I do not think Mr Pickwick would mind the comparison.

Machine: I don't think you're serious. By a winter's day one means a typical winter's day, rather than a special one like Christmas.

Scoring the test

How good does the computer's performance have to be in order for it to pass the test? How many judges must mistake the computer for a human? Obviously these questions are crucial when specifying the Turing test, because without answers it's impossible to say whether a computer has passed or not. Turing gave a very ingenious answer. He said that in order to determine the answers to *these* questions, we first need to carry out another set of tests.

These prior tests involve another version of the imitation game. This time none of the players is a computer. There is a judge and two other human players, one male (A) and one female (B). The judge's job is to decide (by chatting) which player is the man, and A's aim is to trick the judge into making the wrong choice. Turing uses this man-imitates-woman game as a standard by which to adjudicate whether the computer passes the computer-imitates-human version of the imitation game (in which, as Turing said, 'a machine takes the part of A').[30] To assess the computer's performance, we ask:[31]

Will the interrogator decide wrongly as often when the [computer-imitates-human] game is played . . . as he does when the game is played between a man and a woman?

In other words, if the computer does no worse in the computer-imitates-human game than the man does in the man-imitates-woman game, then the computer passes the test.

Turing was a little cagey about what would actually be *shown* if the computer passed his test. He said that the question 'Can machines pass the test?' is 'not the same as "Do machines think" ', continuing 'but it seems near enough for our present purpose, and raises much the same difficulties'.[32]

In one of his philosophical papers, he even cast doubt on the meaningfulness of the question 'Can machines think?', saying (rather rashly) that the question is 'too meaningless to deserve discussion'.[33] However, he himself indulged in such discussion with gusto. In fact he spoke very positively about the project of 'programming a machine to think' (his words), saying 'The whole thinking process is still rather mysterious to us, but I believe that the attempt to make a thinking machine will help us greatly in finding out how we think ourselves'.[34]

What Turing predicted

Turing was also cagey about how long he thought it would be before a computer passed the test. He predicted (in 1950) that progress in AI during the twentieth century would be sufficiently rapid that:[35]

in about fifty years time it will be possible to programme computers . . . to make them play the imitation game so well that an average interrogator will not have more than 70 per cent chance of making the right identification after five minutes of questioning.

In other words, he predicted that by about 2000 a computer would be able to fool approximately 30% of the judges in 5-minute tests. This performance, though, would fall short of what is needed to *pass* the test, since he stated (in 1952) that it would be 'at least 100 years' before a computer stood any chance of passing his test with no questions barred.[36]

His prediction about when the test would be passed ('at least 100 years') was sensibly vague—and the time-scale makes it clear that Turing appreciated the colossal difficulty of resourcing a computer to pass the test. Unfortunately, though, there is an urban myth that, according to Turing, computers would pass his test by the end of the twentieth century—with the result that he has been unfairly criticized not only for being wrong, but also for being 'far too optimistic about the task of programming computers to achieve a command of natural language equivalent to that of every normal person', as one of his critics, Martin Davis, wrote.[37] Given Turing's actual words ('at least 100 years') this is misguided criticism.

You can't believe what you read in the papers

Turing knew his test was ultra-tough—this was the point of it—and he did not expect a computer to pass any time soon. That's why it came as such a surprise, in June 2014, to read headlines like the *Washington Post*'s 'A bot named "Eugene Goostman" passes the Turing test'.[38] Eugene Goostman, a chatbot created by programmers Vladimir Veselov and Eugene Demchenko, simulates a 13-year-old Ukrainian boy's imperfect English conversation. The BBC's headline above an article about Eugene Goostman read 'Computer AI passes Turing test in "world first" '.[39] The article continued:

The 65-year-old Turing Test is successfully passed if a computer is mistaken for a human more than 30% of the time during a series of five-minute keyboard conversations. On 7 June Eugene convinced 33% of the judges at the Royal Society in London that it was human.

It was in fact the two researchers responsible for staging the Royal Society Turing test experiment, Huma Shah and Kevin Warwick, who announced to the media that Eugene Goostman had passed the test. Their press release stated:[40]

The 65 year-old iconic Turing Test was passed for the very first time by computer programme Eugene Goostman during *Turing Test 2014* held at the renowned Royal Society in London on Saturday. 'Eugene' simulates a 13 year old boy and was developed in Saint Petersburg, Russia.

The press release continued:

If a computer is mistaken for a human more than 30% of the time during a series of five minute keyboard conversations it passes the test. No computer has ever achieved this, until now. Eugene managed to convince 33% of the human judges (30 judges took part . . .) that it was human.

Warwick and Shah concluded:

We are therefore proud to declare that Alan Turing's Test was passed for the first time on Saturday.

Obviously, though, Turing could not have considered that fooling 30% of the judges during a series of 5-minute conversations amounted to passing the test, since he predicted that this 30% success rate would be achieved 'in about fifty years time', but also said that that it would be 'at least 100 years' before a computer *passed* his test. Shah and Warwick's mistaken claim that Eugene Goostman passed the Turing test rests on their having conflated Turing's 30% prediction with a specification of what counts as passing the test. They simply ignored Turing's own careful specification, in terms of the man imitates woman game, of what would actually count as passing the test.

It is interesting to see Turing's prediction about the extent of progress by the turn of the century coming true—and we can surely forgive him for being just a few years out. But, as Turing thought, his test may not be cracked for many years yet.

Misunderstanding the Turing test

Another important misunderstanding about what Turing said concerns *definitions*. He is repeatedly described in the (now gigantic) literature about the Turing test as having intended his test to form a definition of thinking.[41] However, the test does not provide a satisfactory definition of thinking, and so this misunderstanding of Turing's views lays him open to spurious objections, along the lines of 'Turing attempted to define *thinking* but his definition does not work'. Turing did make it completely clear though that his intention was not to define thinking, saying 'I don't really see that we need to agree on a definition at all',[42] but his words were not heeded. 'I don't want to give a definition of thinking', he said, 'but if I had to I should probably be unable to say anything more about it than that it was a sort of buzzing that went on inside my head'.[43]

Someone who takes the Turing test to be intended as a definition of thinking will find it easy to object to the definition, since an entity that thinks could *fail* the test. For example, a thinking alien might fail simply because its responses are distinctively non-human. However, since Turing did not intend his test as a definition, this objection misses the point. Like many perfectly good tests, the Turing test is informative if the candidate passes but uninformative if the candidate fails. If you fail an academic exam it might be because you didn't know the material, or because you had terrible flu on the day of the exam, or for some other reason—but if you pass fair and square, then you have unquestionably demonstrated that you know the material. Similarly, if a computer passes the Turing test then the computer arguably thinks, but if it fails nothing can be concluded.

One currently influential criticism of the Turing test is based on this mistaken idea that Turing intended his test as a definition of thinking. The criticism is this: a gigantic database storing every conceivable (finite) English conversation could, in principle, pass the Turing test (assuming the test is held in English).[44] Whatever the judge says to the database, the database's operating system just searches for the appropriate stored conversation and regurgitates the canned reply to what the judge has said. As the philosopher Ned Block has put it, this database no more thinks than a jukebox does, yet in principle it would succeed in passing the Turing test. Block agrees that this hypothetical database is in fact 'too vast to exist'—it simply could not be built and operated in the real world, since the total number of possible conversations is astronomical—but he maintains that, nevertheless, this hypothetical counter-example proves the Turing test to be faulty.[45]

It is true that the database example would be a problem if the Turing test were supposed to be a definition of thinking, since the definition would entail that this monster database thinks, when obviously it does not. But the test is not supposed to be a definition, and the database example is in fact harmless. Turing's interest was the real computational world, and this hypothetical database could *not* pass the Turing test in the real world—only in a sort of fairyland, where the laws of the universe would be very different. In the real world, there might simply not be enough atoms in existence for this huge store of information to be constructed—and even if it could be, it would operate so slowly (because of the vast numbers of stored conversations that must be searched) as to be easily distinguishable from a human conversationalist. In fact, the judge and the foil might die before the database produced more than its first few responses.

Searle's circle: the Chinese room

Another famous but misguided criticism of the Turing test is by the philosopher John Searle. Searle is one of AI's greatest critics, and a leading exponent of the view that running a computer program can never be sufficient to produce thought. His objection to the Turing test is simply stated: according to Searle, entities that *don't* think can pass the test—and the entities he has in mind are computers.[46]

Let us imagine that a team in China, say, produces a computer program that successfully passes the Turing test in Chinese. Searle ingeniously proposes an independent method for testing whether running this program really produces thought. The method is to run the program on a *human* computer and then ask the human, 'Since you are running the program—does it enable *you* to understand the Chinese?' (See Chapters 7 and 41 for more about human computers.) Searle imagines himself as the human computer. He is in a room that is provided with many rulebooks containing the program written out in plain English, and he has an unlimited supply of paper and pencils. As with every computer program, the individual steps in the program are all simple binary operations that a human being can easily carry out using pencil and paper, given enough time.

In Searle's Turing test scenario, the judge writes his or her remarks on paper, in Chinese characters, and pushes these into the room through a slot labelled INPUT. Inside the room, Searle painstakingly follows the zillions of instructions in the rulebooks and eventually pushes more Chinese characters through a slot labelled OUTPUT. As far as the judge is concerned, these symbols are a thoughtful and intelligent response to the input. But when Searle, a monolingual English speaker, is asked whether running the program is enabling him to understand the Chinese characters, he replies 'No, they're all just squiggles and squoggles to me—I have no idea what they mean'. Yet he is doing everything relevant that an electronic computer running the program would do: the program is literally running on a human computer.

This is a description of Searle's renowned 'Chinese room' thought experiment. He claims that the thought experiment shows that running a mere computer program can never produce thought or understanding, even though the program may pass the Turing test. However, there is a subtle fallacy. Is Searle, in his role as human computer, the right person to tell us whether running the program produces understanding? After all, there is another conversationalist in the Chinese room—the program itself, whose replies to the judge's questions Searle delivers through the output slot. If the judge asks (in Chinese) 'Please tell me your name', the program responds (in Chinese) 'My name is Amy Chung'. And if the judge asks 'Amy Chung, do you understand these Chinese characters', the program responds 'Yes, I certainly do!'.

Should we believe the program when it says 'Yes, I am able to think and understand'? This is effectively the very same question that we started out with—is a computer really capable of thought? So Searle's *Gedankenexperiment* has uselessly taken us round in a circle. Far from providing a means of settling this question in the negative, the Chinese room thought experiment leaves the question dangling unanswered. There is nothing in the Chinese room scenario that can help us to decide whether or not to believe the program's pronouncement 'I think'. The conclusion that Amy Chung does not understand the Chinese characters certainly does not follow from the fact that Searle (beavering away in the room) cannot understand the Chinese characters.[47] So Searle's attack on the Turing test fails.

Alan Turing's test has been attacked by some of the sharpest minds in the business. To date, however, it stands unrefuted. In fact, it is the only viable proposal on the table for testing whether a computer is capable of thought.

CHAPTER 26

Turing's model of the mind

MARK SPREVAK

T his chapter examines Alan Turing's contribution to the field that offers our best understanding of the mind: cognitive science. The idea that the human mind is (in some sense) a computer is central to cognitive science. Turing played a key role in developing this idea. The precise course of Turing's influence on cognitive science is complex and shows how seemingly abstract work in mathematical logic can spark a revolution in psychology.

Alan Turing contributed to a revolutionary idea: that mental activity is computation. Turing's work helped lay the foundation for what is now known as cognitive science. Today, computation is an essential element for explaining how the mind works. In this chapter, I return to Turing's early attempts to understanding the mind using computation and examine the role that Turing played in the early days of cognitive science.

Engineering versus psychology

Turing is famous as a founding figure in artificial intelligence (AI) but his contribution to cognitive science is less well known. The aim of AI is to create an intelligent machine. Turing was one of the first people to carry out research in AI, working on machine intelligence as early as 1941 and, as Chapters 29 and 30 explain, he was responsible for, or anticipated, many of the ideas that were later to shape AI.

Unlike AI, cognitive science does not aim to create an intelligent machine. It aims instead to understand the mechanisms that are peculiar to human intelligence. On the face of it, human intelligence is miraculous. How do we reason, understand language, remember past events, come up with a joke? It is hard to know how even to begin to explain these phenomena. Yet, like a magic trick that looks like a miracle to the audience, but which is explained by revealing the pulleys and levers behind the stage, so human intelligence could be explained if we knew the mechanisms that lie behind its production.

A first step in this direction is to examine a piece of machinery that is usually hidden from view: the human brain. A challenge is the astonishing complexity of the human brain: it is one of the most complex objects in the universe, containing 100 billion neurons and a web of around 100 trillion connections. Trying to uncover the mechanisms of human intelligence by looking at the brain is impossible unless one has an idea of what to look for. Which properties of the brain are relevant to intelligence? One of the guiding and most fruitful assumptions in cognitive science is that the relevant property of the brain for producing intelligence is the computation that the brain performs.

Cognitive science and AI are related: both concern human intelligence and both use computation. It is important to see, however, that their two projects are distinct. AI aims to create an intelligent machine that may or may not use the same mechanisms for intelligence as humans. Cognitive science aims to uncover the mechanisms peculiar to human intelligence. These two projects could, in principle, be pursued independently.

Consider that if one were to create an artificial hovering machine it is not also necessary to solve the problem of how birds and insects hover. Today, more than 100 years after the first helicopter flight, how birds and insects hover is still not understood. Similarly, if one were to create an intelligent machine, one need not also know how humans produce intelligent behaviour. One might be sanguine about AI but pessimistic about cognitive science. One might think that engineering an intelligent machine is possible, but that the mechanisms of human intelligence are too messy and complex to understand. Alternatively, one might think that human intelligence can be explained, but that the engineering challenge of building an intelligent machine is outside our reach.

In Turing's day, optimism reigned for AI and the cognitive-science project took a back seat. Fortunes have now reversed. Few AI researchers aim to create the kind of general, human-like, intelligence that Turing envisioned. In contrast, cognitive science is regarded as a highly promising research project.

Cognitive science and AI divide roughly along the lines of psychology versus engineering. Cognitive science aims to understand human intelligence; AI aims to engineer an intelligent machine. Turing's contribution to the AI project is well known. What did Turing contribute to the cognitive-science project? Did he intend his computational models as psychological models as well as engineering blueprints?

Building brainy computers

Turing rarely discussed psychology directly in his work. There is good evidence, however, that he saw computational models as shedding some light on human psychology.

Turing was fascinated by the idea of building a *brain-like* computer. His B-machines were inspired by his attempt to reproduce the action of the brain, as described in Chapter 29. Turing talked about his desire to build a machine to 'imitate a brain', to 'mimic the behaviour of the human computer', 'to take a man… and to try to replace… parts of him by machinery… [with] some sort of "electronic brain" ', he claimed that 'it is not altogether unreasonable to describe digital computers as brains', and that 'our main problem [is] how to programme a machine to imitate a brain'.[1]

Evidently, Turing thought that the tasks in AI engineering and psychology were somehow related. What did he think was the nature of their relationship? We should distinguish three different things that he might have meant:

1. *Psychology sets standards for engineering success.* Human behaviour is where our grasp on intelligence starts. Intelligent behaviour is, in the first instance, known to us as something that humans do. One thing that psychology provides is a specification of human behaviour. This description can then be used in the service of AI by providing a benchmark for the behaviour of intelligent machines. Whether a machine counts as intelligent depends on how well it meets an appropriately idealized version of standards set by psychology. Psychology is relevant to AI because it specifies what is meant by intelligent behaviour. This connection seems peculiar to intelligent behaviour. One could, for instance, understand perfectly well what hovering is without knowledge of birds or insects.

2. *Psychology as a source of inspiration for engineering.* We know that the human brain produces intelligent behaviour. One way to tackle the AI engineering problem is to examine the human brain and take inspiration from it. Note, however, that the 'being inspired by' relation is a relatively weak one. Someone may be inspired by a design without understanding much about how that design works. Someone impressed by how birds hover may add wings to an artificial hovering machine. But even if this were successful, it would not mean the engineer knows how a bird's wings enable it to hover. Indeed, the way in which wings allows a bird to hover may not be the same as the way in which wings allow the engineer's artificial machine to hover—flapping may be an essential part in one case but not the other. An AI engineer might take inspiration from brains without knowing how brains work.

3. *Psychology should explain human intelligence in terms of the brain's computational mechanisms.* Unlike the two previous claims, this involves the idea that the mechanisms of human thought are computational. The first two claims are compatible with this idea but they do not entail it. Indeed, the first two claims are silent about what psychology should, or should not, do. They describe a one-sided interaction between psychology and engineering with the influence going all from psychology to engineering: psychology sets the standards of engineering success or psychology inspires engineering. This claim is different: it recommends that psychology should adopt the computational framework of the AI engineering project. The way in which we explain human intelligence, and not just attempts to simulate it artificially, should be computational.

Did Turing make the third (cognitive-science) claim? Turing certainly gets close to it and, as we shall see in the final section, his work has been used by others in the service of that claim.

In the quotations above, Turing describes one possible strategy for AI: imitating the brain's mechanisms in an electronic computer. In order for such this strategy to work, one has to know which are the relevant properties of the brain to imitate. Turing says that the important features are not that 'the brain has the consistency of cold porridge' or any specific electrical property of nerves.[2] Rather, among the relevant features he cites the brain's ability 'to transmit information from place to place, and also to store it':[3]

brains very nearly fall into [the class of electronic computers], and there seems to be every reason to believe that they could have been made to fall genuinely into it without any change in their essential properties.

On the face of it, this still has the flavour of a one-way interaction between AI engineering and psychology: which features of the brain *are relevant to* AI engineering? But unlike the claims above, this one-way interaction presupposes a specific view about how the human brain works: that the brain produces intelligent behaviour via (perhaps among other things) its computational properties. This is very close to the cognitive-science claim. Turing appears to be committed to something like to the third claim above (the cognitive-science claim) via his engineering strategy.

However, there is a problem with this reading of Turing. The key terms that Turing uses— 'reproduce', 'imitate', 'mimic', 'simulate'—have a special meaning in his work that is incompatible with the reading above. Those terms can be read as either 'strong' or 'weak'. On a strong reading, 'reproducing', 'imitating', 'mimicking', or 'simulating' means *copying that system's inner workings*—copying the equivalent of the levers and pulleys by which the system achieves its behaviour. On a weak reading, 'reproducing', 'imitating', 'mimicking', or 'simulating' means *copying the system's overall input–output behaviour*—reproducing the behaviour of the system, but not necessarily the system's method for doing so. The strong reading requires that an 'imitation' of a brain work in the same way as a real brain. The weak reading requires only that an imitation of a brain produce the same overall behaviour.

We assumed the strong reading above. In Turing's work, however, he tended to use the weak reading. Use of the weak reading is important to prove the computational results for which Turing is most famous (see Chapter 7). If the weak reading is the correct one, then the interpretation of Turing's words above is not correct. Imitating a brain does not require knowing how brains work—only knowing the overall behaviour brains produce. This falls squarely under the first relationship between psychology and engineering: psychology sets standards for engineering success. Imitating a brain—in the (weak) sense of reproducing the brain's overall behaviour— requires only that psychology specify the overall behaviour that AI should aim to reproduce. It does not require that psychology also adopt a computational theory about human psychology.

Is there evidence that Turing favoured the strong over the weak reading? Turing wrote to the psychologist W. Ross Ashby that:[4]

In working on the ACE I am more interested in the possibility of producing models of the action of the brain than in practical applications to computing… Thus, although the brain may in fact operate by changing its neuron circuits by the growth of axons and dendrites, we could nevertheless make a model, within the ACE, in which this possibility was allowed for, but in which the actual construction of the ACE did not alter, but only the remembered data, describing the mode of behaviour applicable at any time.

This appears to show that Turing endorsed something like the cognitive-science claim: he believed that the *computational* properties of the brain are the relevant ones to capture in a simulation of a brain. Unfortunately, it is also dogged by the same problem we saw previously. 'Producing a computational model of the action of the brain' can be given either a strong or a weak reading. It could mean *producing a model that works in the same way as the brain* (strong), or *producing a model that produces the same overall behaviour* (weak). Both kinds of computational model interested Turing and Ashby. Only the former would tell in favour of the cognitive-science claim.

Tantalizingly, Turing finished his 1951 BBC radio broadcast with:[5]

The whole thinking process is still rather mysterious to us, but I believe that the attempt to make a thinking machine will help us greatly in finding out how we think ourselves.

The difficulty is that 'helping', like 'being inspired by', is not specific enough to pin the cognitive-science claim to Turing. There are many ways that the attempt to make a thinking machine might help psychology: the machines created might do useful number crunching, building the machines may teach us high-level principles that apply to all intelligent systems, building the machines may motivate psychology to give a specification of human competences. None of these would commit Turing to the cognitive-science claim.

Turing's writings are consistent with the cognitive-science claim but they do not offer unambiguous support for it. In the next section, we will see a clearer, but different, type of influence that Turing has had on modern-day cognitive science. In the final section, we will see how his computational models have been taken up and used by others as psychological models.

From mathematics to psychology

Turing proposed several computational models that have influenced psychology. Here I focus on only one: the Turing machine. Ostensibly, the purpose of the Turing machine was to settle questions about mathematics—in particular, the question of which mathematical statements can and cannot be proven by mechanical means. We will see that Turing's model is good for another purpose: it can be used as a model of human thought. This spin-off benefit has been extremely influential.

A Turing machine is an abstract mathematical model of a human clerk. Imagine that a human being works by himself, mechanically, without undue intelligence or insight, to solve a mathematical problem. Turing asks us to compare this 'to a machine that is only capable of a finite number of conditions'.[6] That machine, a Turing machine, has a finite number of internal states in its head and an unlimited length of blank tape divided into squares on which it can write and erase symbols. At any moment, the machine can read a symbol from its tape, write a symbol, erase a symbol, move to neighbouring square, or change its internal state. Its behaviour is fixed by a finite set of instructions (a transition table) that specifies what it should do next in every circumstance (read, write, erase symbol, change state, move head).

Turing wanted to know which mathematical tasks could and could not be performed by a human clerk. Could a human clerk, given enough time and paper, calculate any number? Could a clerk tell us which mathematical statements are provable and which are not? Turing's brilliance was to see that these seemingly impossible questions about human clerks can be answered if we re-formulate then to be about Turing machines. If one could show that the problems that can be solved by Turing machines are the same as the problems that can be solved by a human clerk, then any result about which problems a Turing machine can solve would carry over to a result about which problems a human clerk can solve. Turing machines can be proxies for human clerks in our reasoning.

It is easy to prove that the problems that a Turing machine can solve can also be solved by a human clerk. The clerk could simply step through the operations of the Turing machine by hand. Proving the converse claim—that the problems that a human clerk can solve could also be solved by a Turing machine—is harder. Turing offered a powerful informal argument for

Figure 26.1 Early 'computers' at work, summer 1949 in the NACA (later NASA) High Speed Flight Station. Turing aimed to create an abstract mathematical model of these human clerks

Posted to Wikimedia Commons and licensed under public domain, https://commons.wikimedia.org/wiki/File:Human_computers_Dryden.jpg.

this second claim. Significantly, his argument depended on *psychological reasoning* about the human clerk:[7]

The behaviour of the [clerk] at any moment is determined by the symbols which he is observing, and his 'state of mind' at that moment. We may suppose that there is a bound B to the number of symbols or squares that the [clerk] can observe at one moment. If he wishes to observe more, he must use successive observations. We will also suppose that the number of states of mind which need be taken into account is finite. The reasons for this are of the same character as those which restrict the number of symbols. If we admitted an infinity of states of mind, some of them will be 'arbitrarily close' and will be confused.

Turing's strategy is to argue that the clerk cannot bring any more internal resources to bear in solving a problem than a Turing machine. Therefore, the class of problems that a clerk can solve is no larger than those of a Turing machine. In conjunction with the first claim above, this establishes the crucial claim that the problems that can be solved by Turing machines are exactly the same as those that can be solved by a human clerk.

Turing's argument is an exercise in weak modelling. His aim is to show that Turing machines and human clerks solve the same class of problems: they are capable of producing the same pattern of behaviour. His argument requires him to show that a Turing machine can copy the behaviour of the clerk and vice versa (weak modelling). It does not require him to show that the Turing machine reproduces that clerk's internal psychological mechanisms for generating his

Figure 26.2 The Cajal Blue Brain project simulating the inner workings of a human brain using a Magerit supercomputer

Taken by Cesvima and posted to https://commons.wikimedia.org/wiki/File:UPM-CeSViMa-SupercomputadorMagerit.jpg.
Creative Commons licence

behaviour (strong modelling). Strong modelling goes beyond what was required by Turing's work on the *Entscheidungsproblem* but it is what we need cognitive science.

One might conclude that there is nothing of further interest here for psychology. Yet, Turing's argument should give one pause for thought. Turing's argument requires that human clerks and Turing machines share at least *some* similarity in their inner working. They must have similar kinds of internal resources; otherwise, Turing's argument that the clerk's resources do not differ in kind from those of a Turing machine would not work. This suggests that a Turing machine is more than just a weak model of a human clerk. A Turing machine also provides a description, albeit rather high level and abstract, of the clerk's inner workings. In addition to capturing the clerk's outward behaviour, Turing machines also give some information about the levers and pulleys behind the clerk's behaviour.

Your brain's inner Turing machine

Does a Turing machine provide a psychologically realistic model of the mechanisms of the human mind? Turing never seriously pursued this question in print, but it has been taken up by others. The philosopher Hilary Putnam argued that a Turing machine is a good psychological model. Putnam claimed that a Turing machine is not only a good model of a clerk's mind while he is solving a mathematical task, it is a good model of other aspects of mental life.[8] According to Putnam, all human mental states (beliefs, desires, thoughts, imaginings, feelings, pains) should be understood as states of a Turing machine and its tape. All human mental processes (reasoning, association, remembering) should be understood as computational steps of some Turing machine. Psychological explanation should be explanation in terms of the nature and operation of an inner Turing machine. Only when one sees the brain as implementing a Turing machine can one correctly see the contribution that the brain makes to our mental life. Putnam's proposal falls neatly under the cognitive-science claim identified above.

Putnam and others quickly became dissatisfied with the Turing machine as a psychological model.[9] It is not hard to see why. The human brain lacks any clear 'tape' or 'head', human mental states are not atomic states that change in a step-wise way over time, human psychology is not serial: it involves parallel mechanisms that cooperate or compete with each other. If the mind is a computer, it is unlikely to be a Turing machine.

The past fifty years have seen an explosion in the number and variety of computational models in psychology. State-of-the-art computational models of the mind look and work nothing like Turing machines. Among the most popular models are hierarchical recurrent connectionist networks that make probabilistic predictions and implement Bayesian inference.[10] The mechanisms of these computational models bear little resemblance to Turing machines. Yet, one might wonder, is there still something essentially right, albeit high level and abstract, about Turing machines as psychological models? And even if Turing machines do not model all aspects of our mental life, perhaps they provide a good model of some parts of our mental life.

Turing machines provide a good psychological model of at least one part of our mental life: deliberate, serial, rule-governed inference—the capacity at work inside the head of the human clerk when he is solving his mathematical problems. In some situations, humans deliberately arrange their mental processes to work in a rule-governed, serial way. They attempt to follow rules without using initiative, insight, or ingenuity, and without being disturbed by their other mental processes. In these situations, it seems that our psychological mechanisms approximate those of a Turing machine: our mental states appear step-wise, as atomic entities, and change in a serial fashion.

At a finer level of detail—and moving closer to the workings of the brain—there is of course a more complex story to tell. Yet, as a 'high-level' computational model, the Turing machine is not bad as a piece of psychology. In certain situations, and at a high, abstract, level of description, our brains implement a Turing machine.

Modern computational models of the mind are massively parallel, exhibit complex and delicate dynamics, and operate with probability distributions rather than discrete symbols. How can one square them with Turing machines? One way to integrate the two models is to use the idea that a Turing machine runs as a *virtual machine* on these models.[11] The idea is that a Turing machine arises, as an emergent phenomenon, out of some lower-level computational processes.[12] This idea should be familiar from electronic PCs: a high-level computation (in C# or Java) can arise out of lower-level computation (in assembler or microcode). High-level and low-level computational descriptions are both important when we explain how an electronic PC works. Similarly, we should expect that high-level and low-level descriptions will be important to explain how the human brain produces intelligence.

Conclusion

Turing has had a huge influence on cognitive science but, as we have seen, tracing the precise course of his influence is complex. In this chapter, we looked at two possible sources: Turing's discussion of how AI should be proceed, and the way in which Turing's computational models have influenced others. On the first score, we saw that Turing rarely talked about how AI should influence psychology, and that it is not easy to attribute to Turing the modern-day claim that human psychology should be computational. On the second, a clearer picture emerges. Turing's 1936 paper on the *Entscheidungsproblem* suggests that Turing machines are more than weak models of human psychology. Putnam and others took up this idea and proposed that Turing machines are strong models of human psychology. This idea remains influential today. Despite the wide range of exotic computational models in cognitive science, Turing machines still appear to capture a fundamental, albeit high level, truth about the workings of the human mind.

CHAPTER 27

The Turing test—from every angle

DIANE PROUDFOOT

C an machines think? Turing's famous test is one way of determining the answer. On the sixtieth anniversary of his death, the University of Reading announced that a 'historic milestone in artificial intelligence' had been reached at the Royal Society: a computer program had passed the 'iconic' Turing test. According to an organizer, this was 'one of the most exciting' advances in human understanding. In a frenzy of worldwide publicity, the news was described as a 'breakthrough' showing that 'robot overlords creep closer to assuming control' of human beings. Yet after only a single day it was claimed that 'almost everything about the story is bogus': it was 'nonsense, complete nonsense' to say that the Turing test had been passed. The program concerned 'actually got an F' on the test. The backlash spread to the test itself; critics said that the 'whole concept of the Turing Test is kind of a joke . . . a needless distraction'.[1] So, what is the Turing test—and why does it matter?

A little experiment

In 1948, in a report entitled 'Intelligent machinery', Turing described a 'little experiment' that, he said, was 'a rather idealized form of an experiment I have actually done'. It involved three subjects, all chess players. Player A plays chess as he/she normally would, while player B is proxy for a computer program, following a written set of rules and working out what to do using pencil and paper—this 'paper machine' was the only sort of programmable computer freely available in 1948 (see Ch. 31). Both of these players are hidden from the third player, C. Turing said, 'Two rooms are used with some arrangement for communicating moves, and a game is played between C and either A or the paper machine'. How did the experiment fare? According to Turing, 'C may find it quite difficult to tell which he is playing'.[2] This is the first version of what has come to be known as 'Turing's imitation game' or the 'Turing test'.

Figure 27.1 Dietrich Prinz'.

Raymond Kleboe, *Picture Post*; reprinted with permission from Getty Images.

Why chess (see Fig. 27.1)? Turing had been thinking about computer chess routines for some years. In 1945 he said that 'chess requires intelligence' and that his planned Automatic Computing Engine could 'probably be made to play very good chess' (see Chapters 25 and 31).[3] In the 1950s the artificial intelligence (AI) pioneers Allen Newell, John Shaw, and Herbert Simon said:[4]

Chess is the intellectual game *par excellence* . . . If one could devise a successful chess machine, one would seem to have penetrated to the core of human intellectual endeavor.

Fifty years later, in 1997, Garry Kasparov—the world's greatest chess player, hailed as the 'hope of humanity'—played six games against IBM's Deep Blue supercomputer. Kasparov said that 'chess offers a unique field to compare man and machine. It's our intuition versus the brute force of calculation'. This match was described at the time as 'an icon in musings on the meaning and dignity of human life'.[5]

At first the games went as Kasparov had expected. He said that Deep Blue was 'stupid'—it 'played like machine . . . It did exactly what everybody expect machine to do'. But then Deep Blue 'did something which contradicted any conventional knowledge of the computer's ability'. The computer, he said:

played a quiet prophylactic move that ended my hopes, the sort of move no computer had ever before made. Instead of going for a short-term advantage, it closed in for the kill. Faced with a losing position and stunned by the godlike quality of the machine's play, I resigned.

Deep Blue's unconventional playing (which IBM later attributed to a software bug) led Kasparov to think that a 'human player' was behind Deep Blue's moves. He said that the computer 'sank into deep thinking' and that 'when you look at this machine move you don't think that it's just a machine'. Deep Blue showed 'a very human sense of danger'.[6] Disconcerted, Kasparov went on to lose the match. A computer, it was declared, had 'passed a Turing chess test. A grandmaster cannot know whether his hidden opponent is another grandmaster or a computer program'.[7]

In 1950 Turing presented his test as it is generally known today—an imitation game extending his 1948 experiment to include questions on 'almost any one of the fields of human endeavour that we wish to include'. (For an expanded introduction to the test, see Chapter 25.) In his paper 'Computing machinery and intelligence' he first introduced an experiment in the form of two simultaneous interviews, again with three players in total: an interrogator questions two contestants, a man and a woman. Both contestants are again hidden, and the interrogator conducts the interviews by teleprinter. The male contestant's goal is to fool the interrogator into misidentifying him as the woman. Having described this setup, Turing said:

We now ask the question, 'What will happen when a machine takes the part of [the man] in this game?' Will the interrogator decide wrongly as often when the game is played like this as he does when the game is played between a man and a woman? These questions replace our original, 'Can machines think?'.

Satisfactory performance by a machine in such a game is a 'criterion for "thinking" ', Turing said.[8] The machines that he had in mind were digital computers. If some computer does well in this computer-imitates-human game, machines *can* think.

Rules of the game

Much of the vast literature on the Turing test consists of conflicting accounts of Turing's design for the 1950 imitation game. Several commentators claim that the computer's task is to imitate a *man* (since the computer is introduced as taking the man's place in a man-imitates-woman game), or to imitate a *woman* (since that is the man's goal), or even to imitate *a man who is imitating a woman*. The arguments offered on behalf of the claim that the machine's task is to imitate a woman include: this makes the test *harder* for the machine (which must convince the interrogator both that it is human *and* that it is female); it makes the test *easier* for the machine (since an interrogator looking out for 'female' behaviour is less likely to spot 'machine' behaviour); and a '*gender-twisting*' test would have 'resonated deeply' with Turing.[9] However, Turing, when he said that the point of the test is to see if a computer can 'imitate a brain', made it clear that the machine's task is in fact to imitate a *human being* (male or female).[10]

Turing provided guidelines for all three players. The interrogator is to be 'average'—just as the interrogator in the chess-playing imitation game is to be a 'rather poor' chess player. This excludes experts who might easily spot typical computer strategies. The interrogator is also allowed to make comments such as 'I put it to you that you are only pretending to be a man'. This, along with the freedom to ask questions on almost any subject, prevents a simple program's doing well in the game merely by generating pre-programmed replies to trigger words on a few topics. (The 'Eugene Goostman' program, alleged to pass the test on the sixtieth anniversary of Turing's death, simulated a 13-year-old Ukrainian boy speaking English as a second language; this restricted questions to those that such a contestant would understand, violating Turing's guidelines.) In turn the computer is 'permitted all sorts of tricks'—it would, Turing said, 'have to do quite a bit of acting'. Turing suggested, for example, that if the computer were given an arithmetical problem to solve, it should 'deliberately introduce mistakes in a manner calculated to confuse the interrogator', to avoid being identified by its own 'deadly accuracy'. The best strategy for the machine, Turing said, is 'to try to provide answers that would naturally be given by a man'. The human contestant's goal (assuming it corresponds to the woman's goal in the man-imitates-woman game) is 'to help the interrogator' and 'the best strategy . . . is probably to give truthful answers'.[11]

Several commentators regard Turing's three-player imitation game as in essence a *two*-player game, in which an interrogator interviews a single hidden contestant, which may be a human or a machine. However, there is no reason to say that Turing believed his three-player game to be reducible to a two-player game: indeed, there is reason to say the opposite. In 1952, in a BBC radio broadcast entitled 'Can automatic calculating machines be said to think?', Turing described a two-player version of his test, in which members of a jury interview several contestants one by one, some humans and some computers. He pointed out a difficulty for this version: the jury, in order to avoid the embarrassing mistake of misidentifying a machine as a human being, might simply say ' "It must be a machine" every time without proper consideration'.[12] Results in the annual Loebner Prize Contests in Artificial Intelligence, which until 2004 (unwittingly) followed this version of the game, show that Turing was perceptive. In the 2000 contest, for example, members of the jury judged a human as a machine ten times, but did not judge *any* machine as a human being; and in the 2003 contest they judged a human as 'definitely a machine' four times, but did not judge any computer as 'definitely a human'.[13]

Crucially, Turing made it clear that doing well in the game is not *necessary* for intelligence; a thinking machine might fail his test.[14]

What Turing didn't say

Why did Turing propose replacing the question 'Can machines think?' with the question 'Can machines do well in the computer-imitates-human game?'? What links success in the game to *thinking*? There are at least three very different answers to these questions.

First, the standard answer: Turing was a *behaviourist*. On this reading, the imitation game tests whether a machine contestant can behave in a way that is indistinguishable from a 'thinking' human being—and there is nothing more to thinking than such behaviour. If it walks like a duck and quacks like a duck, it just is a duck. Many 1950s commentators assumed that Turing was a behaviourist, and some objected that behaviourism results in 'the drastic redefinitions of common words'. Theorists writing today typically follow suit, claiming that the Turing test is 'the first operational definition of machine intelligence' and that Turing's criteria for thinking are 'purely behavioral'.[15] The reason to think that Turing was a behaviourist is principally that in the 1940s and 1950s this approach was popular as a way of making the mind a subject of scientific study. 'Operationalizing' a phenomenon to be studied is also common practice within science: a psychologist, for example, might define anger in terms of how loudly a subject shouts. So, commentators assumed, Turing operationalized intelligence in terms of how well a machine does in the imitation game.

The second way of linking the imitation game to thinking is as follows: success in this game provides evidence—but no guarantee—that the *inner states and processes* of a machine contestant are (computationally) similar to the mental states and processes of a human being. If it walks like a duck and quacks like a duck, it's likely that it has the innards of a duck. The reason to take this view of Turing is that, in setting out the concept of the Turing machine, he compared the digital computer's state—that is, where it is in the course of a computation—to a human computer's 'state of mind' (see Chapter 26). He hypothesized that the human brain is a digital computer and he believed that 'machines can be constructed which will simulate the behaviour of the human mind very closely'.[16] In the 1950s it was said that 'Turing and others have tried to define thought processes in terms of formal machine states or logic'.[17] Taking this view, a computer's success in the imitation game is an indication that its processing resembles human thinking.

The difficulty for both these readings is Turing's own words: when he spoke explicitly about the concept of intelligence, he did *not* reduce intelligence either to behaviour or to computation.[18] On the contrary, when introducing the 1948 chess-playing version of the game, he said that the concept of intelligence is an 'emotional concept' and he spoke of the temptation to 'imagine intelligence' in a machine.[19] Emotional concepts are those concepts—such as beauty, colour, and goodness—that we frequently say are 'in the eye of the beholder'. If a painting *looks* beautiful (to normal people in normal conditions), it *is* beautiful. Turing said that whether or not an entity is intelligent depends in part on our reaction to it; our *imagining* intelligence in another entity is crucial to that entity's *being* intelligent. Taking this third view of the imitation game, the game tests whether the interrogator will imagine intelligence in a machine contestant.

For more on these three interpretations of the Turing test, see Chapter 28.

New games

The Turing test has been—and surely will continue to be—modified, extended, diluted, and transformed. This indicates its importance to the field of AI.

Some theorists think that it is impossible to have a disembodied thinking thing, and so they propose the 'total' (or 'robotic') Turing test, in which a machine 'must be able to do, in the real world of objects and people, *everything* that real people can do, in a way that is indistinguishable (to a person) from the way real people do it'.[20] In an even harder version of this test—the 'truly total' Turing test—a system of these machines must be capable of generating human abilities by themselves. Other theorists focus on how intelligence develops. In the 'toddler' Turing test a machine must answer questions exactly as a 3-year old human would. (In yet another test, a machine can think if it can rear a child!) Some proposed tests limit the interrogator's questions: in the 'Feigenbaum test' an expert judge must distinguish a computer expert from a human expert in a specific field, and in the 'tutoring test' a machine contestant must teach as effectively as a human tutor. Many of what are now called 'Turing-style' tests depart from Turing's format. In the 'Pleming test', for example, two identical machine contestants communicate with each other, passively observed by a human judge; if this interaction looks like human communication, and even better shows signs of creativity, the machines can think. (This is a 'fly on the wall' or third-person Turing test.)

For Turing the imitation game is a criterion for thinking, whereas later theorists have presented it as measuring other abilities, including free will. In the 'Lovelace test' a machine is genuinely *creative* if its designer cannot explain the machine's behaviour in terms of its design and knowledge store—only machines with an element of mystery pass this test. In the 'Turing test for musical intelligence' a musician-interrogator improvises with two hidden 'musicians', one human and the other a computer music system; if the interrogator is unable to identify the human, the machine has 'musical intelligence'. In the 'moral' Turing test an interrogator interviews, or judges descriptions of the actions of, a machine and a human contestant solely on issues of morality; if the machine is misidentified as a human, it is a *moral* being. (In this game the machine must not be excessively virtuous, as this might give it away.) According to the 'Turing triage test', if a judge, forced to choose which of a human and a machine contestant to destroy, faces the same moral dilemma as if both were human, then the machine is a *person*. Some theorists even propose Turing's game as a way of testing *identity*. If after your 'death' we activate a simulation of your brain in a computer, would that simulation be *you*? Ray Kurzweil, a Google director of engineering who bet $20,000 that a machine would pass the Turing test by 2029, says: run a (two-player) imitation game—if the interrogator cannot tell the difference between talking to the simulation and talking to you, you are in the computer.

These games are mostly thought experiments. The best-known actual experiment is Hugh Loebner's competition, which now offers a silver medal (and $25,000 in 2016) to a machine fooling half the judges after 25 minutes of questioning—and a yet-to-be-awarded grand prize and gold medal for the first program to succeed in an imitation game in which the interrogator communicates with contestants via audio-visual inputs as well as text inputs. New competitions regularly appear; for example, in 2014 the XPRIZE Foundation, which designs and funds innovation competitions 'aimed at capturing the imagination of the public, spurring innovation, and accelerating the rate of positive change across the globe', announced the 'A.I. XPRIZE'—a 'modern-day Turing test to be awarded to the first A.I. to walk or roll out on stage and present a TED talk so compelling that it commands a standing ovation'.[21] Little serious money or science has gone into developing machines to take part in such tests, however. (This is not limited to Turing-style AI competitions; in 2013 the British Computer Society's Machine Intelligence Competition for systems showing 'progress towards machine intelligence' was cancelled due to 'insufficient suitable entries'.[22])

In actual experiments the winning 'chatbots' (computer programs that converse in natural language) are simple programs. For example, in a 2012 test an interrogator typed 'Epsom, home of the derby. yourself?' to state her home town and ask the contestant to do likewise; the program Eugene Goostman responded with 'My guinea pig says that name Derby sounds very nice'. (Simulating a Ukrainian 13-year-old is intended to explain away such odd and ungrammatical outputs.) This was the program announced as a 'breakthrough' in AI on the sixtieth anniversary of Turing's death. Unsurprisingly, after the initial hype surrounding Eugene Goostman, the media verdict was: if *this* program can pass Turing's test then the test is 'no longer as relevant as we first thought'. However, Eugene Goostman *didn't* pass the test (see Chapter 25). The organizers set the threshold for passing as follows: 'If a computer is mistaken for a human more than 30% of the time during a series of 5-minute keyboard conversations it passes the test'.[23] This confuses Turing's *prediction* of how far AI might come by roughly the turn of the century with the *rules* of his game. He made the threshold clear: the interrogator must 'decide wrongly as often [in the computer-imitates-human game] as he does when the game is played between a man and a woman' (see the section 'A little experiment'). In 1952 Turing said that this would take at least 100 years.

Turing-style tests are also used to test the believability of virtual characters and the photo-realism of computer graphics. In one non-verbal test, a human subject interacts with a virtual character and decides whether it is controlled by a human or a machine, based on how the 'eyes' move in response to the subject's gaze. In the 'enriched' Turing test, female subjects in a speed-dating experiment interacted twice with a virtual character; although they were told that the avatar was controlled by a program, in one interaction the controller was a man—and the subjects did not detect any difference in the character's emotional responses. In a 'hide-and-seek test', judges decide whether a virtual character, who chooses a route to hide and seek, is a human or a computer. In the 'BotPrize competition', a human player 'shoots' against an avatar that is controlled by either a human or a program and then judges the opponent's 'humanness'. Again the machine player should not shoot too accurately; judges tend to class those with good aim or fast reaction time as non-human. In the 2012 competition the most successful programs gained a higher humanness rating than the human players; this was reported as 'bots beat Turing test: AI players are officially more human than gamers'! The programmer of one of the winning bots claimed that his program had 'crossed the humanness barrier'.[24]

There is even a Turing-like 'handshake test' to measure 'motor' intelligence (a human subject 'shakes hands' with a lever, controlled by either a human or a computer, and decides which handshake is more human-like)—and also a test to decide whether an online social media account is genuine or a 'Sybil' (a fake account). Outside computer science, it has been suggested that Turing's game could test for a *living system*: a natural cell 'interrogates' both a natural cell and an artificial cell. In an 'ideological' Turing test, contestants explain an ideology contrary to

Figure 27.2 Which contestant is the machine?

Reprinted from xkcd, http://imgs.xkcd.com/comics/turing_test.

their own: those who deceive the interviewers into thinking that this is their own ideology are said to *understand* it. In the 'Turing litigation game', an interrogator interviews two hidden contestants, one the plaintiff and the other the defendant, with the aim of finding out whether the defendant is guilty; proponents of this test say that it is cheaper and less cumbersome than current legal procedure. It is also often assumed that the fictional Voigt–Kampff empathy test for androids—from the story 'Do androids dream of electric sheep?' and the film *Blade Runner*—is based on the Turing test. The test also appears in numerous cartoons (including several Dilbert cartoons), artworks (including 'The original automatic confession machine: a Catholic Turing Test'), novels, and short stories—and on tee-shirts and baseball caps (Fig. 27.2).

One Turing-style test is now widespread. In a CAPTCHA (Completely Automated Public Turing Test to Tell Computers and Humans Apart) the judge is a computer whose task is to tell whether a single contestant is a human or a computer. (CAPTCHAs are sometimes called 'reverse' Turing tests.) When you must identify a number in a distorted image before you can enter a chat room, vote, or use a credit card online, you are a contestant in a CAPTCHA.

Bashing the test

Several commentators have argued that Turing did not intend his imitation game as a test of intelligence, and that he would have been amused, even horrified, at the game's central role in AI. Marvin Minsky said recently that the Turing test is a 'joke' and that Turing 'never intended it as the way to decide whether a machine was really intelligent'. Aaron Sloman claims that Turing was 'far too intelligent to do any such thing' and that this widespread misinterpretation has led to 'huge amounts of wasted effort' discussing the purely 'mythical' Turing test. Likewise, according to Drew McDermott, all that Turing wanted to do was to 'shake people's intuitions up'.[25] Turing's own words concerning what he called his 'imitation tests' make it clear, however, that he *did* intend the game as a test of intelligence. He said 'I would like to suggest a particular kind of *test* that one might apply to a machine' and he described the question 'Are there imaginable digital computers which would do well in the imitation game?' as a 'variant' of the question 'Can machines think?'. He certainly seemed to be serious, remarking 'Well, that's my test. Of course I am not saying at present either that machines really could pass the test, or that they couldn't. My suggestion is just that this is the question we should discuss'.[26]

The Turing test has had a hard ride, and—even though criticisms of the test reduce to a handful of unsuccessful arguments—this will probably continue. Critics have described the Turing test as 'virtually useless', 'obsolete', and 'impotent', and machines that do well in the game as 'dead ends in artificial intelligence research'. Some say that '[a]dherence to Turing's vision . . . is . . . actively harmful' and that 'Turing's legacy alienates maturing subfields'. In the avalanche of papers celebrating Turing's centenary in 2012, the same objections appear; critics claim that the time has come to 'bid farewell to the Turing Test'.[27]

Some of this criticism stems from the fact that the test is a criterion of *human-level* (or human-like) intelligence in machines. Critics claim that trying to build artificial intelligence by imitating human beings is to ignore the essence of intelligence in favour of an emphasis on one parochial example. On this view, the Turing test focuses AI research on uniquely human behaviour and so is 'a tragedy for AI'; the game is 'testing humanity, not intelligence'.[28] What we need instead, critics say, is a test of 'general' or 'universal' intelligence (such as the 'anytime intelligence test'). However, this criticism of Turing is unfounded: the book you are now reading shows Turing's broad range of interests in machine intelligence—there is no evidence that he believed that human intelligence is 'the final, highest pinnacle of thinking', as critics claim.[29] Moreover, the search for the universal 'essence' of intelligence appears wrong-headed. Proponents of universal intelligence tests begin by defining intelligence, which Turing refused to do—if the concept of intelligence is an emotional concept, as he claimed, intelligence does not have an essence, any more than beauty does. Worse, the definitions these 'essence' theorists offer are behaviourist: 'general' intelligence, they say, is the ability to achieve goals, produce syntactically complex utterances, or answer highly complex questions. However, even if such an ability is enough for *intelligence*, is it enough for *thinking*?

Several critics of the Turing test say that, even if human-level AI is a suitable target for researchers, the test doesn't help us to get there: the task of building a machine that an interrogator will misidentify as human is *too difficult*. Indeed, it has been claimed that 'the Turing Test could be passed only by things that have experienced the world as we [humans] have'—although this claim was withdrawn in the face of modern computer methods of analysing extraordinarily large amounts of data.[30] However, many people raise this objection only because they confuse the *goal* of human-level AI with much-hyped but unsuccessful *strategies* taken to get there—such as Good Old Fashioned AI, which focused on 'symbol systems' with a huge store of knowledge. Turing himself suggested another route: build and educate a 'child machine' (see Chapter 30). This involves giving a machine a range of abilities, each of which could be individually tested and incrementally improved—thus providing a route to human-level AI.

Critics also complain that the Turing test is *too easy*, arguing that there is 'no plausible developmental pathway from increasing chatterbot performance in the Turing test to genuine artificial intelligence'. On this view, the test merely encourages a programmer to use cheap tricks in order to fool a judge, who may be convinced just because he/she is especially gullible. Critics say that the test aims at 'artificial stupidity' rather than artificial intelligence, since the machine contestant must hide its superhuman capacities (such as perfect typing skills).[31] However, the Turing test is *not* easy; many people make this complaint only because they confuse the simple programming strategies behind Loebner Contest chatbots with those necessary to succeed in Turing's much harder game. It is true that individual interrogators might be gullible, or simply have a bad day, but this shows only that the test is not a one-off test: to obtain a convincing result, the game must be played several times.[32] Also, the 'artificial stupidity' objection is misguided, since disguising the machine is an unavoidable corollary rather than the aim of the test. Turing said that a machine should not be punished for disabilities, such as an 'inability to shine in beauty competitions', that are 'irrelevant' to whether the machine can think.[33] To avoid this, the machine's appearance is hidden. A machine's 'deadly accuracy' at typing or mathematics is an irrelevant *ability* and so must also be disguised.

Yet another criticism stems simply from the fact that the Turing test is a *test*. It is not a test of intelligence that we need, some critics say, it is a (computational) *theory* of intelligence. This, however, assumes that such a theory is possible—and if intelligence is an emotional concept then such a theory is not possible.

The remaining complaint is that machines passing the Turing test would be of *no practical use*; they would be 'intellectual statues', critics say—more expensive but no smarter than human labour. What we need instead, critics argue, are 'intelligence amplifiers'—systems such as Google and Siri, and even driverless cars.[34] This is to abandon AI's grand goal of human-level AI in favour of 'narrow' goals. This stance seems unduly negative, however: machines with the abilities required to pass the Turing test may not be (as optimists like Kurzweil claim) only a short step from superhuman-level AI, but they could have an enormous impact on human life.

So, why *is* the Turing test important? The many people in AI who do regard human-level AI as important and achievable need a criterion to determine *when* they reach that goal. Turing's test provides this. It is also anthropomorphism-proofed. In 1949 Turing's colleague at the University of Manchester, the neurosurgeon Geoffrey Jefferson, said:[35]

We have had a hard task to dissuade man from reading qualities of human mind into animals. I see a new and greater danger threatening—that of anthropomorphizing the machine. When we hear it said that wireless valves think, we may despair of language.

AI researchers often encourage us to treat machines as humans, for example by building hyper-realistic androids such as Hiroshi Ishiguro's Geminoids—robots designed as exact copies of Ishiguro and others, which can be encountered in a public cafe. We require some way of ensuring that judgements of intelligence in machines are not solely the product of anthropomorphizing. Turing's imitation game includes a disincentive to anthropomorphize: to avoid getting egg on their faces, interrogators are abnormally suspicious of the contestants (as in the Loebner Contests). And even if an interrogator does anthropomorphize the machine contestant, the three-player game is a blind controlled trial—the interrogator will also anthropomorphize the human contestant, with the result that there is no unfair advantage to the machine. Whether by design or accident, Turing's test addresses the anthropomorphism danger. His test matters.

Smart moves

If a computer beats a human player in a game of chess, does that show that the computer *understands* chess—or is *how* the machine plays important? In 2003 Garry Kasparov said that 'computer superiority over humans in chess had always been just a matter of time'. But, he added, Deep Blue and similar computers were not what the early chess programmers hoped for 'when they dreamed of creating a machine to defeat the world chess champion':

Instead of a computer that thought and played chess like a human, with human creativity and intuition, they got one that played like a machine, systematically evaluating 200 million possible moves on the chess board per second and winning with brute number-crunching force . . . Deep Blue was only intelligent the way your programmable alarm clock is intelligent. Not that losing to a $10 million alarm clock made me feel any better.

Others agreed. According to the *Economist*, Deep Blue showed that 'chess-playing skill does not, in fact, equal intelligence . . . it is possible to get a dumb machine to do it better than any human. The equation of chess-playing with intelligence is centuries old, but it is time to lay it to rest'. Likewise, cognitive scientist Douglas Hofstadter said, 'My God, I used to think chess required thought. Now, I realize it doesn't'.[36]

IBM acknowledged that Deep Blue is 'stunningly effective at solving chess problems, but it is less "intelligent" than the stupidest person. It doesn't think, it reacts'. (For the objection to Turing's game that a brute force machine like Deep Blue could fool the interrogator, see Chapter 25.) The company moved on to new machines, including Watson, and new human–machine games. According to IBM, Watson *can* think. In 2011 the computer beat the two highest-ranked human contestants in an episode of the television game show *Jeopardy!*

Figure 27.3 *Jeopardy!*

Reprinted courtesy of
Jeopardy Productions Inc.

(Fig. 27.3). This time *The Economist* was impressed, saying that 'defeating a grandmaster at chess was child's play compared with challenging a quiz show famous for offering clues laden with ambiguity, irony, wit and double meaning as well as riddles and puns'. Kurzweil too said that Watson is a 'stunning example of the growing ability of computers to successfully invade [a] supposedly unique attribute of human intelligence'—analysing language and using symbols to stand for ideas.[37] IBM has since teamed up with the XPRIZE Foundation to offer the IBM Watson AI XPRIZE to the machine that gives the best TED talk at TED2020. IBM brings serious money; prizes will total $5 million.[38]

Critics, however, say that Watson is no closer than Deep Blue to genuine thinking. John Searle, for example, argued that Watson demonstrated 'a huge increase in computational power and an ingenious program', but that these 'do not show that Watson has superior intelligence, or that it's thinking':[39]

[Watson] is merely following an algorithm that enables it to manipulate formal symbols. Watson did not understand the questions, nor its answers, nor that some of its answers were right and some wrong, nor that it was playing a game, nor that it won—because it doesn't understand anything.

Searle's famous Chinese room argument against the Turing test, which he also uses against Watson, is discussed in Chapter 25.

According to AI optimists, the 'most momentous milestone' since Deep Blue's defeat of Kasparov, and 'a landmark moment' for AI, was the program AlphaGo's 2016 defeat of Lee Se-dol, one of the world's top three Go players.[40] The program's creators, Google DeepMind, said that Go has more possible positions than 'there are atoms in the universe' and as a result is 'a googol times more complex than chess'—making the game impossible to solve by brute force search.[41] AlphaGo utilizes 'deep' neural networks that are intended to capture something of the structure of the human brain, and improves its performance by playing thousands of games against itself. According to the lead researcher behind AlphaGo, the program can 'understand' Go, and enthusiasts said that it showed an ability 'eerily similar to what we call intuition'.[42] Critics, on the other hand, claimed that AlphaGo 'no more understands the game of Go than a robot mower understands the concept of a lawn'.[43] One commentator, for example, who had thought a successful Go program would mean that AI is 'truly beginning to become as good as the real thing', said after Lee's defeat that AlphaGo seemed merely to use 'learning algorithms combined with "big data" '—in contrast to 'the fluidity of the human mind'.[44] So do we need yet another holy grail for AI? Some have suggested that AI's 'ultimate test' is really mahjong![45]

Media reports also claimed that AlphaGo was distinctly different from previous programs. Yet Turing himself investigated artificial neural networks (see Chapter 29) and wanted to build a machine that could learn autonomously from experience (see Chapter 30). Anthony Oettinger, who wrote the first AI programs to include learning, was much influenced by Turing's views on machine learning.[46] Oettinger's programs were written for the EDSAC computer at the University of Cambridge, the second electronic stored-program computer to run, in May 1949. In 2016 an editorial in *The Times* claiming that DeepMind's computer was 'not programmed how to play' Go but rather 'taught itself' (and that it can 'use intuition' and 'think') drew a swift reaction from a scientist who had actually worked on the EDSAC.[47] Norman Sanders said that DeepMind's computer 'did not learn to teach itself; it was programmed to do so'. There is 'no difference in principle', he claimed, between DeepMind's machine and the EDSAC; the difference is 'just that today's processing capacity matches the requirements' of chess and Go.[48]

Oettinger said that one of his programs could pass a restricted Turing test.[49] Likewise, according to one of the IBM scientists working on Watson, 'if in the Turing test you were asking people *Jeopardy!* questions and we couldn't tell who was who, then we've actually passed the Turing test'.[50] If a machine succeeded, though, in an *un*restricted Turing test—played several times, with impeccable interrogators—would this suffice to show that the machine can think? *No*, say many opponents of Turing's test: machines that do well in a full-scale imitation game may nevertheless lack some element crucial for thought.

The X factor

For most critics, this essential element is *consciousness*. Ken Jennings, one of the losing human contestants in the *Jeopardy!* match, said that he felt 'obsolete'—'it was friggin' demoralizing. It was terrible'. On Searle's view, if Watson had lost, the computer would not have felt obsolete, or indeed felt anything; 'in order to get human intelligence, you've got to be conscious', Searle says. Some AI sceptics say that AlphaGo didn't even *win* the match against Lee; a computer can't 'win' at anything, it was claimed—not until it 'can experience real joy in victory and sadness in defeat'.[51] This is, the critics say, why succeeding in the Turing test is *not* a criterion for thinking. The imitation game can't test for consciousness.

In the 1950s several theorists claimed that machines *cannot* be conscious. This is not to deny that a machine can have what Donald Michie, another Bletchley Park colleague of Turing's, called 'operational awareness'—the ability to register inner states. What the machine cannot do, it was said, is experience the 'feel' of a taste, colour, or emotion; according to Geoffrey Jefferson, for example, no machine could feel grief 'when its valves fuse'.[52] One reply to this is that we can imagine a machine answering questions on grief (or other emotions, or tastes or colours) in a way indistinguishable from a human being—and how could it do so without having *felt* grief? On this view, it may not be possible for a machine to succeed in Turing's game without being conscious. But if it does succeed, that's evidence it *is* conscious!

Yet, some have countered, we can also imagine a being—a philosophical 'zombie', a non-conscious being otherwise indistinguishable from a human—that does exactly this. The zombie example is intended to show that passing the Turing test is not a criterion for thinking, but even this objection is not the last word on the test. For the zombie objection to the test to work, three substantial claims must be true: first, consciousness is *necessary* for intelligence; second, philosophical zombies are *possible*; and third, the Turing test is a criterion for thinking in *all* possible worlds (including those containing zombies). Showing that these claims are all true is a big ask.

Turing himself replied to Jefferson in a different way: if behaviour is not a sign of consciousness, he said, this must apply to human behaviour as well as to machine behaviour—and the consequence is that 'the only way to know that a *man* thinks is to be that particular man'. This, he said, is 'the solipsist point of view'—the claim that (at least, for all I know) I am alone in the world and everything else is a figment of my mind. Turing assumed that most people, including Jefferson, would accept the imitation game as a test of thinking rather than be committed to solipsism. In response, some philosophers have conceded that behaviour *is* a sign of consciousness, but only in humans, not in machines. This move, though, plays into Turing's hands. In 1947 he said that 'fair play must be given to the machine'.[53] In the present context, this puts the burden of proof on the AI sceptic to show that behaviour is a sign of consciousness only in humans—without pre-judging the question at issue, namely whether a machine devoid of consciousness could pass the Turing test. Another big ask.

Heads in the sand

Turing said that many people are 'extremely opposed' to the idea of a machine that thinks, and that this is 'simply because they do not like the idea'. This explains why much of his writing on machine intelligence consists of replies to objections. In Turing's view, the 'unwillingness to admit the possibility that mankind can have any rivals in intellectual power' occurs 'as much amongst intellectual people as amongst others: they have more to lose'. This is the 'heads-in-the-sand' objection to thinking machines.[54] (The opposite reaction is *panic*—fear that machines will take jobs from, and even lead to the extinction of, humans.[55]) Is there a whiff of heads-in-the-sand in the reaction to Deep Blue, Watson, or AlphaGo?

In 1951, in a much-discussed book that attracted a response from Albert Einstein, Viscount Samuel claimed that the chess player is 'evidently of a different order from the chess-board and the pieces'—the chess player possesses '[i]ntellectual creativity', which is 'not material'. At the time a critic claimed that, on the contrary, the emergence of chess-playing programs 'forces us to admit the possibility of mechanized thinking or to restrict in a very special way our concept of thinking'. Turing himself predicted the last move: he said that, whenever a machine is deemed to have an ability usually reserved for human beings, people claim that how the machine does this is 'really rather base'—they say, 'Well, yes, I see that a machine could do all that, but I wouldn't call it thinking'.[56] This is exactly how, half a century later, critics of Deep Blue and Watson responded.

But can we react in this way *every* time? Turing's famous 'skin of an onion' analogy points to a difficulty:[57]

In considering the functions of the mind or the brain we find certain operations which we can explain in purely mechanical terms. This we say does not correspond to the real mind: it is a sort of skin which we must strip off if we are to find the real mind. But then in what remains we find a further skin to be stripped off, and so on. Proceeding in this way do we ever come to the 'real' mind, or do we eventually come to the skin which has nothing in it? In the latter case the whole mind is mechanical.

Suppose that, as Turing believed, some computing machine can simulate the cognitive capacities of the human brain. Whether it can is an open question (see Chapter 41). Suppose, too, that every time this machine acquires a capacity previously thought unique to humans, we say 'I used to think that this ability required thought, but now I realize it doesn't'. Proceeding like this, there would be nothing left to be the 'real' mind. We might just have to extract our heads from the sand and agree that such a machine thinks.

CHAPTER 28

Turing's concept of intelligence

DIANE PROUDFOOT

This chapter sets out a new interpretation of Turing's concept of intelligence (or thinking) and of his famous test, based on his overlooked 1948 and 1952 versions of the imitation game.[1] According to the traditional behaviourist interpretation, Turing held that thinking is nothing over and above (the capacity for or tendency to) 'thinking' behaviour. Yet his own words are inconsistent with behaviourism, as is the design of his test. So what was Turing's view? He said that 'the idea of "intelligence" is itself emotional rather than mathematical' and his writings make it clear that whether or not a machine is intelligent (or thinks) depends in part on an observer's reactions to the machine. This is what modern philosophers would call a *response-dependence* theory of the concept of intelligence. Turing's remarks about a machine's learning to make 'choices' and 'decisions' suggest that he took a similar approach to the concept of free will.

You feel fine—how do I feel?

According to the traditional view, Turing's concept of intelligence is *behaviourist*—that is to say, intelligence or thinking is nothing more than (the capacity for or tendency to) what we call 'intelligent' or 'thinking' behaviour.[2] For the philosophical behaviourist, the hypothesis of a hidden inner mind that guides and explains such behaviour is an illusion—the notorious 'ghost in the machine'. From the 1950s onward Turing's computer-imitates-human game was interpreted as providing a behaviourist criterion of intelligence or thinking. For example, in 1952 Wolfe Mays—a University of Manchester philosopher who had created a 'logical machine' with Dietrich Prinz, one of Turing's collaborators on the Manchester computer[3]—referred to Turing's 'behaviourist criterion'. Turing used a 'definition of psychological phenomena in terms of behavioural patterns' according to which a machine can think if and only if its behaviour is 'indistinguishable from that of a human being'.[4]

Behaviourism was popular in the 1950s, but even then faced objections. A 1950s philosophical joke went like this: 'One behaviourist meeting another on the street said "You feel fine! How do I feel?" '.[5] Surely, the joke implies, I do not learn how I feel only from third-person observations of my behaviour (Fig. 28.1). There is an array of such difficulties for crude behaviourist theories. If you don't *say* anything, does it follow that you don't *feel* anything? How is the behaviourist to explain mental images, voices in the head, pains, or tastes? What causes 'thinking' behaviour, if not an inner ghost? Mays criticized behaviourism for ignoring 'the evidence of my own introspections, as well as those of other people, that there are such things as private psychological events, however heretical such a view may seem to-day'. He said:

[T]he machine analogy, with its emphasis on overt behaviour and abnegation of private experience may . . . lead [human beings] to be regarded, more than ever before, as if we were mechanical objects. It is not such a far cry from Aristotle's view that slaves were just human tools, to some future benevolent dictatorship of the Orwell 1984 type, where men may be seen as little else but inefficient digital computors and God as the Master Programmer.

According to Mays, on Turing's criterion of thinking, 'the meaning of the word "thinking" has changed to such an extent that it has little in common with what we normally mean by it'.[6]

Mays acknowledged that a machine 'could be constructed with the facility for performing intelligence tests'; but on the crucial question of whether the machine is intelligent, he said 'What is important is not what it does, but how it does it'.[7] His assumption—that Turing's imitation game tests a machine's behaviour—still underlies influential objections to the Turing test. Critics have invented numerous (imaginary) counter-examples; for example, a program that functions just by means of a huge lookup table or a human 'zombie' that is behaviourally indistinguishable from other human beings but lacks consciousness. These entities would pass the Turing test but *how* they do so is not what we mean by 'thinking', critics claim.

The conversation after the
two behaviourists love making
-It was good for you, how was
it for me?

Figure 28.1 A recent version of the 1940s behaviourist joke.

Reprinted with permission of Stephen L. Campbell.

Was Turing a behaviourist?

There are three reasons to reject the traditional interpretation.

First, *Turing's own words repudiate behaviourism*. He said that the concept of intelligence is 'emotional rather than mathematical' and that judgements of intelligence are determined 'as much by our own state of mind and training as by the properties of the object' (see the next section).[8] We can assume that mere behaviour—what a machine (or human) simply *does*—does not depend on the observer. A machine's mere behaviour is one of the 'properties of the object' rather than being determined by 'our state of mind', to use Turing's words. It follows that intelligence is not simply a matter of behaviour.

Second, *the Turing test does not test machine behaviour*. Instead it tests the observer's reaction to the machine (see the next section). The goal of the imitation game is that the interrogator be 'taken in by the pretence' and a machine does well in the computer-imitates-human game if the interrogator in that game is fooled no less frequently than the interrogator in Turing's man-imitates-woman game.[9] Why would a behaviourist test the interrogator rather than the machine? The behaviourist must surely say: if the interrogator is fooled, we can infer that the computer's behaviour is appropriately human-like. However, this strategy makes the Turing test a test of machine behaviour only by making it unnecessarily circuitous. Moreover, the inference employed is invalid many critics have pointed out, we *cannot* infer from an interrogator's being fooled that the computer's behaviour is equivalent to that of a human being—the interrogator may simply be gullible or the programmer of the machine may be lucky.

Third, *behaviourism does not explain the structure of Turing's imitation game*. In the game, Turing said, the computer is 'permitted all sorts of tricks so as to appear more man-like' and indeed 'would have to do quite a bit of acting'.[10] But why would a behaviourist base a test on deception rather than merely give the machine a series of cognitive tasks? Moreover, even allowing this deception, why would a behaviourist include a human contestant rather than merely hide the machine?[11] On the assumption that Turing was a behaviourist, some commentators label his imitation game as 'strange'. Many theorists ignore Turing's own structure and treat his game as a *two*-player game, with a human interrogator in one room and a machine contestant in another.

In sum, the behaviourist interpretation requires us to ignore Turing's own words and to say that the brilliant Turing gave us a strange, circuitous test that is both easily counter-exampled and better understood by modern critics than Turing himself! The cost of the traditional interpretation is high.

If Turing was not a behaviourist, what was his concept of intelligence? According to some modern theorists, Turing *did* believe that it matters how a machine produces its behaviour: in his view, they say, thinking is an inner (brain) process and a computer's success in the imitation game is evidence of this crucial process. But again Turing's own words make it clear that this was not his concept of intelligence. He said that if we identify the 'cause and effect working out in the brain' we regard an entity as *not* intelligent (see the next section); no brain process, therefore, can constitute thinking. Nor did Turing see his imitation game as testing for an inner process. As Max Newman, director of the Royal Society Computing Machine Laboratory at the University of Manchester, said in a BBC radio discussion with Turing: 'if I have understood Turing's test properly, you are not allowed to go behind the scenes and criticise the method [by which the machine arrives at its answers], but must abide by the scoring on correct answers'.[12] *How* the brain (human or electronic) produces its behaviour is irrelevant.[13]

Appearance matters

Turing described three versions of the imitation game. In addition to the famous version in his 1950 'Computing machinery and intelligence' article in *Mind*, he described a version of the game in the 1948 'Intelligent machinery' report for the National Physical Laboratory, and a third in the 1952 discussion with Newman and others entitled 'Can automatic calculating machines be said to think?'. In these 1948 and 1952 works Turing said this about the concept of intelligence:[14]

[T]he idea of 'intelligence' is itself emotional rather than mathematical.

The extent to which we regard something as behaving in an intelligent manner is determined as much by our own state of mind and training as by the properties of the object under consideration. If we are able to explain and predict its behaviour, or if there seems to be little underlying plan, we have little temptation to imagine intelligence. With the same object, therefore, it is possible that one man would consider it as intelligent and another would not; the second man would have found out the rules of its behaviour.

As soon as one can see the cause and effect working themselves out in the brain, one regards it as not being thinking but a sort of unimaginative donkey-work. From this point of view one might be tempted to define thinking as consisting of 'those mental processes that we don't understand'. If this is right then to make a thinking machine is to make one which does interesting things without our really understanding quite how it is done.

The thesis set out in these remarks is that whether or not an entity is intelligent (or thinks) is determined in part by how we *respond* to the entity. Does the entity *appear* intelligent? In the case of intelligence in machines, this is at least as important as the machine's processing speed, storage capacity, or complexity of programming. The latter are examples solely of the machine's behaviour—in Turing's words, the 'properties of the object' rather than the properties assigned by 'our state of mind'. Turing's approach to the concept of intelligence is a familiar and philosophically well-developed approach to other concepts. For example, when we judge that an object is *yellow* or *beautiful*, this is at least in part because it *looks* yellow or beautiful. So too for morality, several philosophers have argued: what makes an action *morally right*, rather than merely having beneficial consequences, is at least in part the fact that we *feel obliged* to perform the action (or that we have some other affective response, such as desiring to perform the action). Theories of this kind are called 'response-dependence' theories, and the concepts concerned 'response-dependent' concepts. Turing's 1948 and 1952 remarks on intelligence explicitly set out a response-dependence theory of the concept of intelligence. An 'emotional concept' is a response-dependent concept.

This approach to intelligence is also implicit in Turing's imitation game. His paragraph (quoted above) beginning 'The extent to which we regard something as behaving in an intelligent manner . . .' immediately precedes his description of the 1948 game, which is restricted to chess-playing. Turing continued:[15]

It is possible to do a little experiment on these lines, even at the present state of knowledge. It is not difficult to devise a paper machine which will play a not very bad game of chess. Now get three men as subjects for the experiment A, B, C. A and C are to be rather poor chess players,

B is the operator who works the paper machine. (In order that he should be able to work it fairly fast, it is advisable that he be both mathematician and chess player.) Two rooms are used with some arrangement for communicating moves, and a game is played between C and either A or the paper machine. C may find it quite difficult to tell which he is playing.

(This is a rather idealized form of an experiment I have actually done.)

Turing's 'little experiment' is a trial to see whether or not C (the interrogator in this game) has the 'temptation to imagine intelligence' in the paper machine.[16] His words make it clear that the game tests the observer rather than the machine. Whether or not the machine is intelligent is determined in part by C's response; for example, if C can 'predict its behaviour or if there seems to be little underlying plan', the machine is not intelligent.

If intelligence is a response-dependent concept, the Turing test is *not* strange, as on the behaviourist interpretation. The proper goal of the test is to test an observer's responses and so it is not unnecessarily circuitous. That the concept of intelligence is a response-dependent concept also explains why the computer is to 'do quite a bit of acting' and even use 'tricks'. The more the machine prompts the interrogator to anthropomorphize, the more the interrogator will succumb to the 'temptation to imagine intelligence' in the machine. This makes sense if Turing's experiment tests the interrogator's response to the machine.[17]

Several modern scientists and philosophers claim that the Turing test can be useful only until we possess a scientific theory of cognition. Typically they assume that cognition consists in computation and so our real need is for a computational theory of cognition. Turing could not have agreed—not because he was a behaviourist for whom thinking is nothing more than 'thinking' behaviour, but because in his view the concept of intelligence is 'emotional' rather than 'mathematical'. According to a response-dependence theorist, the concept of colour is very different from the concept of electromagnetic radiation, even though electromagnetic radiation is the physical basis of colour. Likewise, if intelligence is a response-dependent concept, the concept of intelligence is very different from the concept of computation, *even if* brain processes (implementing computations) form the physical basis of 'thinking' behaviour. Attempting to explain intelligence in terms of computation is to confuse a response-dependent concept with a response-*in*dependent concept.

Anything goes

Several influential artificial intelligence (AI) researchers take the view that intelligence is 'in the eye of the observer'.[18] This stance is in part a reaction to what many in AI see as 'speciesism' (or 'exceptionalism' or 'origin chauvinism'). This is the attitude that human beings are distinctive, because they have a soul or biological brain, or are 'natural' rather than 'artificial'. This attitude unfairly discriminates against machines, its critics claim. However, the view that intelligence is in the eye of the observer is also problematic. Human beings have evolved to find intelligence in unlikely places: we hear voices in the wind, see faces in clouds or the moon's surface, and detect personalities in Furby toys. In his song 'Anything Goes', Cole Porter wrote 'The world has gone mad today And good's bad today, And black's white today, And day's night today'. If intelligence is in the eye of the observer, is *stupid* intelligent in today's AI?

Response-dependence theorists do not claim that anything goes, however. They say that an object is yellow if (and only if) it looks yellow to *normal observers in normal conditions*. If an observer has a brain injury affecting her colour perception, she is not a normal observer; and if the only light source in the environment is itself coloured, this is not a normal condition. In these cases an object's looking yellow would not suffice for it to *be* yellow. Likewise response-dependence theorists may claim that an action is morally right if (and only if) normal observers in normal conditions feel obliged to perform the action. A sociopath is not a normal observer, and a world in which neither suffering nor sympathy exists is not a normal condition. Even in the case of the concept of beauty, theorists attempt to specify normal or 'ideal' observers and conditions. They may say, for example, that a normal observer is someone who has an evolved, brain-based 'sense' of beauty and normal conditions are those that facilitate (or do not obstruct) this sense. Even beauty is not in the eye of just *any* beholder.

Turing was aware that humans are tempted to 'imagine intelligence' in manifestly unintelligent machines. In the 1948 report in which he described playing chess against a paper machine, he said that playing against such a machine 'gives a definite feeling that one is pitting one's wits against something alive'.[19] Turing's 1950 and 1952 imitation games specify normal observers and conditions for judgements of intelligence in machines. Together these exclude cases where a machine looks intelligent but (we want to say) is *not* intelligent. According to Turing, a normal observer (that is, an imitation-game interrogator) is 'average' and 'should not be expert about machines'.[20] Normal conditions are the interview rules set out in the 1950 game: any question is allowed, including comments from the interrogator such as 'I put it to you that you are only pretending to be a man', and the machine is required to answer questions on 'almost any one of the fields of human endeavour that we wish to include'.[21]

In sum, a machine is intelligent (or thinks) if, in the conditions of the 1950 computer-imitates-human game, it appears intelligent to an average interrogator.[22] An AI researcher who knows the weaknesses of artificial systems—the sort of judge often found in the annual Loebner Prize Contest in Artificial Intelligence—is not a normal observer. Also, asking only formulaic questions about the weather—which might enable the sort of simple chatbot that does well in Loebner Contests to appear intelligent—is not a normal condition. In the Turing test of intelligence in machines, not just anything goes.

Is the concept of free will an 'emotional' concept?

To build a thinking machine, Turing proposed beginning with a simple 'unorganised' machine and teaching it as we do a human child; the machine is to go beyond its programming and make its own 'choices' and 'decisions' (see Chapter 30). Critics of AI, however, claim that *everything* a machine does is the result of programming. For example, Geoffrey Jefferson, another participant in Turing's 1952 radio discussion and whose views Turing targeted in 'Computing machinery and intelligence', said:

It can be urged, and it is cogent argument against the machine, that it can answer only problems given to it, and, furthermore, that the method it employs is one prearranged by its operator . . . It is not enough, therefore, to build a machine that could use words (if that were possible), it would have to be able to create concepts and to *find for itself* suitable words in which to express additions to knowledge that it brought about. Otherwise it would be no more than a cleverer parrot, an improvement on the typewriting monkeys which would accidentally in the course of centuries write *Hamlet*.

According to Jefferson, intelligence requires, in addition to 'conditioned reflexes and determinism', a 'fringe left over in which free will may act'; only then would a machine's behaviour not be 'rigidly bound' by the programmer.[23] Again what matters is not what the machine does, but *how* it does it.

Turing said that 'it is certain that a machine which is to imitate a brain must appear to behave as if it had free will', and so he accepted a link between intelligence and free will.[24] However, his remarks suggest a very different approach to free will from Jefferson's stance. Turing said:[25]

We must not always expect to know what the computer is going to do. We should be pleased when the machine surprises us, in rather the same way as one is pleased when a pupil does something which he had not been explicitly taught to do. . . . If we give the machine a programme which results in its doing something interesting which we had not anticipated I should be inclined to say that the machine *had* originated something

This is to assume that the concept of free will is a response-dependent concept. On this view, whether or not an entity possesses free will is 'determined as much by our own state of mind and training as by the properties of the object' (to use Turing's words concerning intelligence). If we are surprised and interested by, and if we fail to anticipate, a child's behaviour, we say that the child makes his own choices and decisions. We should say the same of a machine, Turing argued.[26] *How* the machine does this is irrelevant.

CHAPTER 29

Connectionism: computing with neurons

JACK COPELAND AND DIANE PROUDFOOT

M odern 'connectionists' are exploring the idea of using artificial neurons (artificial brain cells) to compute. Many see connectionist research as the route not only to artificial intelligence (AI) but also to achieving a deep understanding of how the human brain works. It is less well known than it should be that Turing was the first pioneer of connectionism.[1]

Brain versus brawn

Digital computers are superb number crunchers. Ask them to predict a rocket's trajectory or calculate the financial figures for a large multinational corporation and they can churn out the answers in seconds. But seemingly simple actions that people routinely perform, such as recognizing a face or reading handwriting, have been devilishly tricky to program. Perhaps the networks of neurons that make up a brain have a natural facility for these and other tasks that standard computers simply lack (Fig. 29.1). Scientists have therefore been investigating computers modelled more closely on the biological brain.

Connectionism is the science of computing with networks of artificial neurons. Currently researchers usually simulate the neurons and their interconnections within an ordinary digital computer, just as engineers create virtual models of aircraft wings and skyscrapers. A training algorithm that runs on the computer adjusts the connections between the neurons, honing the network into a special-purpose machine dedicated to performing some particular function, such as forecasting international currency markets.

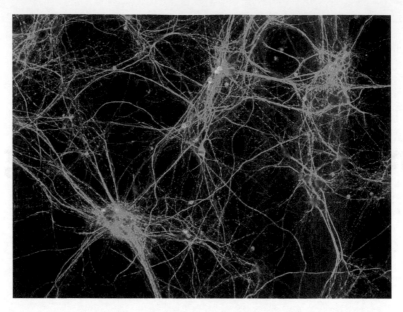

Figure 29.1 A natural neural network: the method of staining the brain-tissue renders the neurons and their interconnecting fibres visible.

Connectionism's promise

In a famous demonstration of the potential of connectionism in the 1980s, James McClelland and David Rumelhart trained a network of 920 neurons to form the past tenses of English verbs.[2] Verbs such as 'come', 'look', and 'sleep' were presented (suitably encoded) to the layer of input neurons. The automatic training system noted the difference between the actual response at the output neurons and the desired response (such as 'came') and then mechanically adjusted the connections throughout the network in such a way as to give the network a slight push in the direction of the correct response.

About 400 different verbs were presented to the network one by one, and after each presentation the network's connections were adjusted. By repeating this whole procedure approximately 200 times, the connections were honed to meet the needs of all the verbs in the training set. The network's training was now complete, and without further intervention it could form the past tenses of all the verbs in the training set.

Furthermore, the network had now reached the point of being able to respond correctly to unfamiliar verbs. For example, when presented for the first time with 'guard', it responded 'guarded'. More impressively still, it replied 'wept' to 'weep', 'clung' to 'cling', and 'dripped' to 'drip' (even down to the double 'p'). Sometimes, though, the peculiarities of English were just too much for the network, and it formed 'squawked' from 'squat', 'shipped' from 'shape', and 'membled' from 'mail'.

Today, simulated connectionist networks are in widespread use. Some prototype driverless cars involve connectionist networks, and they are also used in hearing aids to filter out

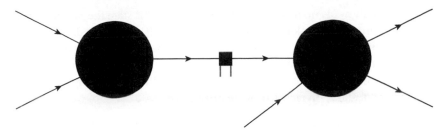

Figure 29.2 Two interconnected B-type neurons (the circles). Each neuron has two inputs and executes the simple logical operation of 'not and' (NAND): if both inputs are 1, the output is 0; otherwise, the output is 1. Each connection passes through a modifier (the black square) that is set either to allow data to pass unchanged or to destroy the transmitted information. Switching the modifiers from one mode to the other enables the network to be trained.

incidental noise: other medical applications include the detection of lung nodules and heart arrhythmias, and the prediction of patients' reactions to drugs. Connectionist networks are good at recognizing objects, and are used to recognize faces and in optical character recognition. Business applications include loan risk assessment, real estate valuation, bankruptcy prediction, and share price prediction, while telecommunications applications include control of telephone switching networks, and echo cancellation in modems and on satellite links.

Turing's anticipation of connectionism

Modern connectionists look back to Frank Rosenblatt, who published his first of many papers on the topic in 1957, as the founder of their approach.[3] Yet Turing had already investigated a type of connectionist network as early as 1948, in a little-read paper titled 'Intelligent machinery'.[4]

Written while Turing was working for the National Physical Laboratory in London, the manuscript did not meet with his employer's approval. Sir Charles Darwin, the rather headmasterly director of the laboratory and grandson of the great naturalist, dismissed it as a 'schoolboy's essay'.[5] In reality, this far-sighted paper was the first manifesto of the field of AI. Although it remained unpublished until 1968,[6] 14 years after his death, Turing not only set out the fundamentals of connectionism but also brilliantly introduced many of the concepts that were later to become central in AI, in some cases after reinvention by others (see Chapter 25).

Educating neurons

In 'Intelligent machinery', Turing invented a kind of neural network that he called a 'B-type unorganised machine', consisting of artificial neurons and devices that modify the connections between the neurons (Fig. 29.2). B-type machines may contain any number of neurons connected in any pattern, but are always subject to the restriction that each neuron-to-neuron connection passes through a modifier device.

All connection modifiers have two training fibres (Fig. 29.2). Applying a pulse to one of them sets the modifier to 'pass mode', in which an input (0 or 1) passes through unchanged and becomes the output. A pulse on the other fibre places the modifier in 'interrupt mode', in which the output is always 1, no matter what the input is. In this state the modifier destroys all information attempting to pass along the connection to which it is attached.

Once set, a modifier maintains its function ('pass' or 'interrupt') unless it receives a pulse on the other training fibre. The presence of these ingenious connection modifiers enables the training of a B-type unorganized machine by means of what Turing called 'appropriate interference, mimicking education'.[7] In fact, Turing theorized that:[8]

the cortex of an infant is an unorganised machine, which can be organised by suitable interfering training.

As Figure 29.2 explains, each of Turing's model neurons has two input fibres, and the output of a neuron is a simple logical function of its two inputs. Every neuron in the network executes the same logical operation of 'not and' (or NAND): the output is 1 if either of the inputs is 0; and if both inputs are 1 then the output is 0.

Turing selected NAND because every other logical (or Boolean) operation can be accomplished by groups of NAND neurons. He showed that even the connection modifiers themselves can be built out of NAND neurons. Thus, Turing specified a network made up of nothing more than NAND neurons and their connecting fibres—about the simplest possible model of the cortex.

Just connect

In 1958 Rosenblatt defined the theoretical basis of connectionism in one succinct statement:[9]

stored information takes the form of new connections, or transmission channels in the nervous system, or the creation of conditions which are functionally equivalent to new connections.

Because the destruction of existing connections can be functionally equivalent to the creation of new ones, researchers can build a network for accomplishing a specific task by taking one with an excess of connections and selectively destroying some of them. Both actions—destruction and creation of connections—are employed in the training of Turing's B-types.

At the outset, B-types contain random inter-neural connections whose modifiers have been set by chance either to pass or to interrupt. During training, unwanted connections are destroyed by switching their attached modifiers to interrupt mode. Conversely, changing a modifier from interrupt to pass in effect creates a connection. This selective culling and enlivening of connections hones the initially random network into one organized for a given job.

In Turing's networks, the neurons interconnect freely and without restriction. Neurons can even connect together in loops, resulting in a neuron's output looping back via other neurons in such a way as to influence its own input. The result is that—as in the inner wirings of Turing's bombe[10]—massive feedback can exist within a neural network. In contrast, modern connectionist networks usually consist of regular 'layers' of neurons, and the flow of information is more restricted, passing unidirectionally from layer to layer (Fig. 29.3). These modern networks are called 'feed-forward': feedback is absent. Ideally, though, connectionists aim to simulate the neural networks of the brain, which seem to reflect the freely connecting structure of Turing's networks, rather than the rigidly layered structure of today's feed-forward networks.

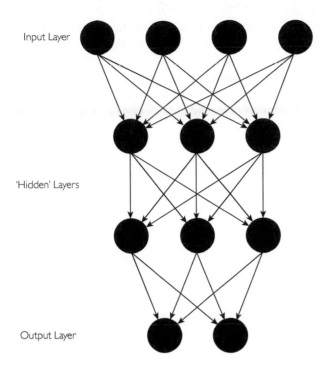

Input Layer

'Hidden' Layers

Output Layer

Figure 29.3 Modern network architecture. In today's 'feed-forward' networks, the brain-like feedback loops of Turing's B-type networks cannot exist.

The first simulation

Turing wished to investigate other kinds of unorganized machines, and he desired to simulate a neural network and its training regimen using an ordinary digital computer. He would, he said 'allow the whole system to run for an appreciable period, and then break in as a kind of "inspector of schools" and see what progress had been made'.[11] However, Turing's own work on neural networks was carried out shortly before the first general-purpose electronic computers became available (see Chapter 20). It was not until 1954, the year of Turing's death, that Belmont G. Farley and Wesley A. Clark, at the Massachusetts Institute of Technology, independently succeeded in running the first computer simulation of a small neural network.[12]

Is the brain a computer?

Paper and pencil were enough, though, for Turing to show that a sufficiently large B-type neural network can be configured (via its connection modifiers) in such a way that it becomes a general-purpose computer. This discovery illuminates one of the most fundamental problems concerning human cognition.

From a 'top-down' perspective, human cognition includes complex sequential processes, often involving language or other forms of symbolic representation, as in mathematical calculation. Yet from a 'bottom-up' view, cognition is nothing but the simple firings of neurons. Cognitive scientists face the problem of how to reconcile these very different perspectives.

Turing's discovery offers a possible solution: the cortex is able to carry out this sequential symbol-rich processing by virtue of being a neural network acting as a general-purpose computer. In 1948 this hypothesis was well ahead of its time, and today it remains among the best guesses concerning one of cognitive science's hardest problems.[13]

CHAPTER 30

Child machines

DIANE PROUDFOOT

This chapter outlines Turing's key ideas in artificial intelligence (AI) and charts his legacy in the field of robot intelligence. In 1950 Turing suggested that one approach to machine intelligence would be to provide a machine with 'the best sense organs that money can buy', and then 'teach it to understand and speak English'. After decades of struggle to create intelligent software, the current goal of many researchers in AI is indeed to build 'socially intelligent' robots—machines with vision and hearing and primitive communicative abilities. The grand aspiration of these theorists is to create what Turing called a 'child machine'—a machine that, like a human infant, can point, smile, recognize its carer's face, and learn to distinguish itself from others. In this chapter I discuss Turing's child machine and its descendants in modern cognitive and developmental robotics.

A recipe for machine intelligence

In 1950 Turing said: 'Instead of trying to produce a programme to simulate the adult mind, why not rather try to produce one which simulates the child's? If this were then subjected to an appropriate course of education one would obtain the adult brain'.[1] His 'guiding principle' in the attempt to build intelligent machines was to follow the development of intelligence in the human being:[2]

If we are trying to produce an intelligent machine, and are following the human model as closely as we can, we should begin with a machine with very little capacity to carry out elaborate operations or to react in a disciplined manner to orders . . . Then by applying appropriate interference, mimicking education, we should hope to modify the machine until it could be relied on to produce definite reactions to certain commands. This would be the beginning of the process.

Turing called this simple machine a 'child machine' and said that it must learn 'initiative' as well as discipline, so that it can modify its own instructions and make its own 'choices'. When it does so, it has 'grown up'—and then 'one is obliged to regard the machine as showing intelligence'. According to Turing, this is just to follow the example of the human child: when a child learns to make discoveries independently of her teacher, the teacher does not claim the credit.[3]

Should a child machine be a disembodied 'brain' that plays chess and cracks codes, or a humanoid robot that might learn for itself by 'roam[ing] the countryside'? Turing described the disembodied and the embodied routes to building a thinking machine and suggested that researchers pursue both approaches. He said:

It can also be maintained that it is best to provide the machine with the best sense organs that money can buy, and then teach it to understand and speak English. This process could follow the normal teaching of a child.

In his view, a child—human or machine—becomes intelligent only through education.[4]

Turing's descriptions of his child machine are frequently tongue-in-cheek. He said, for example, that the machine could not be sent to school 'without the other children making excessive fun of it' and so its education 'should be entrusted to some highly competent schoolmaster'.[5] These remarks, however, sit alongside his very serious intent—to outline a research programme for AI. For many years AI largely ignored this option, but now roboticists aim to build a machine with the cognitive capacities of human infants—a child machine. The roots of this research field in Turing's work have been neglected. In this chapter I consider how his dream has played out in developmental robotics. This also provides an insight into the challenges that face AI.

From universal machine to child machine, and back again

Turing's universal machine of 1936 can be programmed to execute any calculation that a 'human computer' can perform. But does it *learn*? For Turing, learning is the key to intelligence—in 1947 he said, 'What we want is a machine that can learn from experience'. In his view, a 'learning machine', built and educated in analogy with the education of a human child, can develop as the child does. We should:

start from a comparatively simple machine, and, by subjecting it to a suitable range of 'experience' transform it into one which was more elaborate, and was able to deal with a far greater range of contingencies . . . As I see it, this education process would in practice be an essential to the production of a reasonably intelligent machine within a reasonably short space of time. The human analogy alone suggests this.

As it learns, the machine is to modify its own instructions—'like a pupil who had learnt much from his master, but had added much more by his own work'. Turing hoped that there would be 'a sort of snowball effect. The more things the machine has learnt the easier it ought to be for it to learn others'; the machine would probably also be 'learning to learn more efficiently'.[6]

Turing's insight was to begin with an 'unorganised' machine—a machine made up 'in a comparatively unsystematic way from some kind of standard components' and which is 'largely random' in its construction. His hypothesis, which he thought was 'very satisfactory from the point of view of evolution and genetics', was that 'the cortex of the infant is an unorganised machine, which can be organised by suitable interfering training' into a universal machine ('or something like it'). According to Turing, the structure of the child machine is analogous to the 'hereditary material' in the infant brain, changes in the machine are analogous to human genetic mutations, and the choices of the AI researcher are analogous to the influence of natural selection on humans. His goal was an unorganized machine that could be organized to become a universal machine, as a child's brain is altered by natural development and the environment.

In this process the task of the researcher is mainly to give the child machine the appropriate experiences—Turing hoped that this process would be 'more expeditious than evolution'.[7]

Turing conceived of two kinds of unorganized machine. One was the first example of computing by means of neural networks—his 'A-type' and 'B-type' machines (see Chapter 29). According to Turing, the A-type machine is 'about the simplest model of a nervous system with a random arrangement of neurons', and it would not require 'any very complex system of genes to produce something like the A- or B-type'. The B-type machine is a modified A-type; a sufficiently large B-type can be trained to become a universal machine, Turing claimed.[8]

He called his other kind of unorganized machine a 'P-type' machine: this is a Turing machine with an initially incomplete program. A 'pain' stimulus is then used to cancel tentative lines of code, and a 'pleasure' stimulus to make these lines of code permanent—this procedure completes the program. In Turing's view, training a human child depends largely on 'a system of rewards and punishments, and this suggests that it ought to be possible to carry through the organising [of a machine] with only two interfering inputs, one for "pleasure" or "reward"... and the other for "pain" or "punishment" '. The P-type was to test this hypothesis. Turing said: 'It is intended that pain stimuli occur when the machine's behaviour is wrong, pleasure stimuli when it is particularly right. With appropriate stimuli on these lines ... wrong behaviour will tend to become rare'. He recognized, though, that education involves more than rewards and punishments, joking that 'if the teacher has no other means of communicating to the pupil ... [b]y the time a child has learnt to repeat "Casabianca" he would probably feel very sore indeed, if the text could only be discovered by a "Twenty Questions" technique, every "NO" taking the form of a blow'. Some other 'unemotional' means of communication with the machine is required—Turing called these additional inputs to the P-type 'sense stimuli'.[9]

Turing's views on machine learning influenced others at the time, such as Anthony Oettinger, who wrote the earliest functioning AI programs to incorporate learning.[10] Oettinger's 'shopping programme' ran in 1951 on the University of Cambridge EDSAC (the Electronic Delay Storage Automatic Calculator, the world's second stored-program electronic computer). This program—which Oettinger described as a child machine—simulates the behaviour of 'a small child sent on a shopping tour'; the program learns which items are stocked in each shop in its simulated world, so that later, when sent out to find an item, it can go directly to the correct shop.[11] Also in 1951, Christopher Strachey, whose draughts-playing program (see Chapter 20) was the first to use heuristic search—part from Turing's own chess-playing program[12]—said that Turing's analogy between the process for producing a thinking machine and teaching a human child was 'absolutely fundamental'. According to Strachey, the first task is 'to get the machine to learn in the way a child learns, with the aid of a teacher'. Like Turing, he said that one of 'the most important features of thinking' is 'learning for oneself by experience, without the aid of a teacher'. Strachey believed that he had 'the glimmerings of an idea of the way in which a machine might be made to do [this]'.[13]

The computer scientist Donald Michie described himself, Turing, and Jack Good as (at Bletchley Park during the Second World War) forming 'a sort of discussion club focused around Turing's astonishing "child machine" concept'. This concept, he said, 'gripped me. I resolved to make machine intelligence my life as soon as such an enterprise became feasible'. For Michie, as for Turing, the 'hallmark of intelligence is the ability to learn' and, like 'a newborn baby', a computer's possibilities 'depend upon the education which is fed into it'.[14] In the 1960s Michie built famous early learning machines. His MENACE machine (Matchbox Educable Noughts-And-Crosses Engine) could be trained to improve its game. The FREDERICK robots (Friendly

Robot for Education, Discussion and Entertainment, the Retrieval of Information, and the Collation of Knowledge, usually known as Freddy), built in Michie's lab at the University of Edinburgh, learned to manipulate various objects, including how to put differently shaped blocks together in order to create a toy (see Ch. 25).[15] Michie later criticized AI's attempts to build human-level expert systems—programs imitating a human expert's knowledge of a specific area—on the grounds that this approach neglected Turing's child-machine concept.[16]

In 2001 Michie said that AI is part way to building a child machine, in that programmers know how to acquire and represent knowledge in a program. Now we must use these programming techniques 'so as to constitute a virtual person, with which (with whom) a user can comfortably interact'. We must build a machine that is 'a "person" with sufficient language-understanding to be educable, both by example and by precept'. The goal of AI should be, not only a human-*level* machine, but a human-*type* machine (the HAL of Kubrick's film *2001: A Space Odyssey* is the former but not the latter, in Michie's view). The teacher must have a *rapport* with the child machine. Without this rapport, the teacher is 'in effect being asked to tutor the [machine] equivalent of a brainy but autistic child'—in this interaction, Michie said, there are no 'dependable channels of communication' and the education process is unlikely to succeed.[17]

Educating a child machine

According to Turing:

Presumably the child-brain is something like a note-book as one buys it from the stationers. Rather little mechanism [i.e. writing], and lots of blank sheets . . . Our hope is that there is so little mechanism in the child-brain that something like it can be easily programmed. The amount of work in the education we can assume, as a first approximation, to be much the same as for the human child.

Turing claimed that, 'in so far as a man is a machine he is one that is subject to very much interference [i.e. education]. In fact interference will be the rule rather than the exception. He is in frequent communication with other men, and is continually receiving visual and other stimuli which themselves constitute a form of interference'. A teacher aims to alter a child's behaviour, with the result, Turing said, that 'a large number of standard routines will have been superimposed on the original pattern' of the child's brain. The child is then in a position 'to try out new combinations of these routines, to make slight variations on them, and to apply them in new ways'. Even if a human being appears to be acting spontaneously, this behaviour is 'largely determined by the way he has been conditioned by previous interference'.[18]

A 'grown up' machine does not need so much 'interference'. Turing said:[19]

At later stages in education the machine would recognise certain other conditions as desirable owing to their having been constantly associated in the past with pleasure, and likewise certain others as undesirable. Certain expressions of anger on the part of the schoolmaster might, for instance, be recognised as so ominous that they could never be overlooked, so that the schoolmaster would find that it became unnecessary to 'apply the cane' any more.

The educated machine has learned to generalize from past 'experience'.

Did Turing succeed in educating his child machines? He said that it should be easy to simulate unorganized machines on a digital computer; having done so, one could program 'quite definite "teaching policies" ' into the machine. 'One would then allow the whole system to run for an appreciable period, and then break in as a kind of "inspector of schools" and see what progress had been made.' However, Turing had to make do with the only programmable computers available in the 1940s—'paper machines'. These were human beings 'provided with paper, pencil, and rubber, and subject to strict discipline', carrying out a set of rules. Simulating B-types required considerable computational resources, and so was delayed until these became available—too late for Turing (see Chapter 29). He did attempt to teach a P-type, but found this 'disappointing'. He said that organizing a P-type to become a universal machine was 'probably possible', but it was 'not easy' without adding a systematic external memory, and then the supposedly unorganized machine would be more organized than an A-type. Also, Turing said, the method of training a P-type was not 'sufficiently analogous to the kind of process by which a child would really be taught' and was 'too unorthodox for the experiment to be considered really successful'. His method included letting the machine run while continuously applying the 'pain' stimulus; using this procedure, the machine 'learnt so slowly that it needed a great deal of teaching'.[20]

Turing wanted to investigate 'other types of unorganised machine, and also to try out organising methods that would be more nearly analogous' to the education of human beings.[21] But his own attempts to teach a child machine were frustrated.

The best sense organs that money can buy

Turing is often viewed as initiating a research programme to create a disembodied computer program—one that plays chess, cracks codes, and in general solves mathematical puzzles. He said that he wished 'to try and see what can be done with a "brain" which is more or less without a body, providing at most organs of sight, speech and hearing'. What will this machine do? Owing to its 'having no hands or feet, and not needing to eat, nor desiring to smoke, it will occupy its time mostly in playing games such as Chess and GO, and possibly Bridge'. How will the machine be educated? According to Turing, it would not be possible to teach the machine exactly as a teacher would a 'normal' child—for example, it 'could not be asked to go out and fill the coal scuttle'. However, the 'example of Miss Helen Keller shows that education can take place provided that communication in both directions between teacher and pupil can take place by some means or other'. To play chess, the machine's 'only organs need be "eyes" capable of distinguishing the various positions on a specially made board, and means for announcing its own moves'. Turing thought that this machine should do well at cryptography but would have difficulty in learning languages—the 'most human' of the activities a child machine might learn. Learning languages, he said, seems 'to depend rather too much on sense organs and locomotion to be feasible'.[22]

For 'locomotion', robots are required. Turing was probably the first person to recommend building robots as the route to thinking machines. In 1948 he said:

A great positive reason for believing in the possibility of making thinking machinery is the fact that it is possible to make machinery to imitate any small part of a man . . . One way of setting about our task of building a 'thinking machine' would be to take a man as a whole and to try to replace all the parts of him by machinery. He would include television cameras, microphones, loudspeakers, wheels and 'handling servo-mechanisms' as well as some sort of 'electronic brain'. This would of course be a tremendous undertaking. The object if produced by present techniques would be of immense size, even if the 'brain' part were stationary and controlled the body from a distance. In order that the machine should have a chance of finding things out for itself it should be allowed to roam the countryside, and the danger to the ordinary citizen would be serious.

This approach, Turing said, is 'probably the "sure" way of producing a thinking machine'. The response at the time to his plan was negative: in the view of some of his colleagues at the National Physical Laboratory, he was going to 'infest the countryside' with 'a robot which will live on twigs and scrap iron'. Turing himself said that his plan was 'too slow and impracticable' and he abandoned the idea of making a 'whole man'.[23] This idea, though, was revived in the 1990s, in Rodney Brooks's Cog project.

Living the Turing dream

Brooks, the foremost pioneer of embodied AI, has said that what drives him is the 'dream of having a thinking robot'.[24] The robot Cog had, as Turing envisaged, a 'body' (composed of a 'head', 'torso', 'arms', and 'hands'), television cameras and microphones, and an off-board 'brain'. Cog's education also proceeded in part as Turing had suggested. In Turing's view, the child machine's teacher should be someone who 'is interested in the project but who is forbidden any detailed knowledge of the inner workings of the machine'. Many people helped to train Cog, for example to reach towards objects or to play with toys; as a result, the robot's engineers were (to use Turing's words) often 'very largely ignorant of quite what is going on inside' the machine. In this respect, teaching the machine resembled teaching a human child, and was (as Turing had said) 'in clear contrast with normal procedure when using a machine to do computations'.[25]

The fields of social robotics and human–robot interaction have exploded in the late twentieth and early twenty-first centuries. One goal is to build entertainment robots—robot pets such as Pleo the dinosaur, Paro the baby seal, and the Genibo dog. Another goal is to replace expensive human labour by service robots—such as Brooks's Baxter (a user-friendly industrial robot with a torso, two arms, and a head) and Juan Fasola and Maja Matarić's Bandit (a humanoid robot that can guide elderly people through rehabilitative exercises). A robot physiotherapist has advantages other than not needing holiday or sick leave; for example, children with an autism spectrum disorder (ASD) seem to respond better to robot teachers than to the human sort. (Paro is also reported to help dementia sufferers.) In many cases, service robots must behave in human-like ways, so that human clients with no specialized training can easily explain and predict the robots' behaviour. Therapeutic robots may also be required to possess 'social intelligence'—the ability to identify and respond to their human clients' needs and desires. Without this, human–robot interaction may not work. Rapport, as Michie said, is essential.

Turing's vision of a child machine flourishes especially in developmental robotics. Here the grand goal is to build a machine that has infant-level social intelligence, acquired in infant-like ways. Typically-developing human infants follow a well-described trajectory. A child must acquire a 'theory of mind'—the concept of the distinction between herself and other people. Theory-of-mind abilities are essential to interaction with parents, siblings, and strangers. Developmental roboticists aim to construct a robot with exactly these skills—a machine that can detect faces and agents, identify what others (human or robot) are looking at and attend to the same object, recognize itself, and grasp the difference between its own beliefs and those of others. The machine is to acquire these abilities gradually, as human children do, by imitating adults and siblings. This process is also to fit with what we know of human biology and psychology; for example, researchers may design a robot arm to have a range of motion similar to the human arm, or design a robot's brain (that is, system architecture) using findings from developmental psychology and neuroscience.

Some researchers even aim to build robots that have the appearance of human infants—such as Javier Movellan's Diego-san, a hyper-realistic 'expressive' robot toddler (Fig. 30.1). The robot's creators believe that the fact that the human body is very compliant (that is, it moves as a whole, rather than as a series of independent parts) is crucial in the development of intelligence. The 'driving hypothesis' behind the Diego-san project is that humans 'are born with a deconstructed motor control system' that, as the infant matures, is reconstructed 'in a manner that leads to the development of social interaction and symbolic processes, like language'—the researchers' goal is a computational theory of this process.[26] In this respect, Diego-san is an initially unorganized child machine that is to learn as a human infant does. The robot has learned to 'reach towards objects and move them to his mouth' as a human infant does, by making high-intensity movements when the object is far away and the mother is present and then making more precise movements when the object is close.[27] This, Movellan says, is 'the very beginning of gesturing'—pointing to request an object or to draw another person's attention to the object, core theory-of-mind abilities.

Building child machines may also help us to test theories of human psychology. How do children acquire social intelligence? In this process, what are the developmental differences

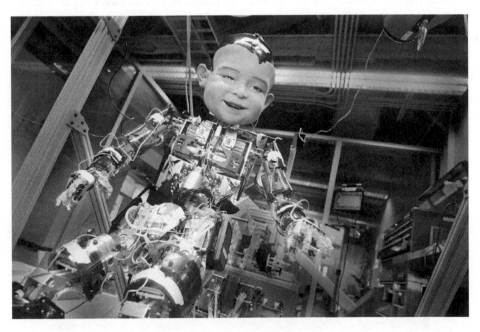

Figure 30.1 Javier Movellan's Diego-san.

Reprinted with permission of the Machine Perception Laboratory, UC San Diego; with thanks to Javier Movellan.

between the neurotypical child and the child with ASD? Researchers hope that we can investigate these and other questions by studying human–robot (or robot–robot) interaction in standardized conditions—in place of difficult and perhaps unethical studies of adult–infant (or infant–infant) interaction. This too fits with Turing's vision: he said, 'I believe that the attempt to make a thinking machine will help us greatly in finding out how we think ourselves'.[28]

Smiley faces

In modern AI 'face robots'—systems with a 'face' and 'head' and 'facial expressions'—are descendants of Turing's suggestion that we give a machine 'the best sense organs that money can buy' and teach it to 'understand and speak English' by a process following the 'normal teaching of a child'. These robots may have machinery enabling them to 'see' or 'hear', and although typically they do not 'speak English', they are intended to have earlier and more primitive communicative abilities—facial expression and bodily gesture. Faces are critical in human communication; for example, Baxter, although designed to work on a factory production line, has a 'face', in order that humans can interact with it intuitively (Fig. 30.2). Researchers engineer face robots not only to produce facial expressions but also to attend to human faces, identify human facial expressions, and respond in kind—for example, 'smiling' in response to a human's smile.[29] These abilities are the building blocks of the human infant's interaction with other humans.

According to Turing, there is 'little point in trying to make a "thinking machine" more human by dressing it up in . . . artificial flesh'.[30] Several face robots, nevertheless, are hyper-realistic representations of a human face and head. Diego-san (which has a body as well as a face) seems to have dimples. Fumio Hara and Hiroshi Kobayashi built a series of robots designed to resemble

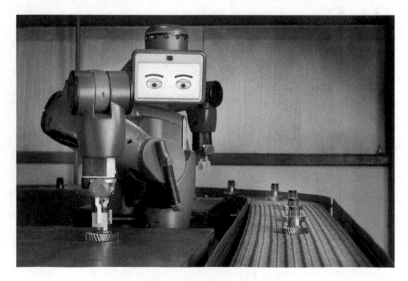

Figure 30.2 Baxter's 'face'.

Reprinted courtesy of Rethink Robotics, Inc.; with thanks to Rodney Brooks.

the head of a young female, with quasi-realistic skin, hair, teeth—and make-up. The hyper-realistic Einstein robot in Movellan's lab, which learns to smile from initial 'body babbling' (random movements of its 'facial muscles'), has the physicist's wrinkles.[31] Turing hoped that 'no great efforts will be put into making machines with the most distinctively human, but non-intellectual characteristics such as the shape of the human body'; in his view, it is 'quite futile to make such attempts and their results would have something like the unpleasant quality of artificial flowers'.[32] This is a striking anticipation of the now much-discussed discomfort that people report when presented with a *too* human-like machine.

Human infants learn communication skills from face-to-face interaction with adults. The infant smiles spontaneously, without understanding or intending the meaning of the gesture; the adult interprets this as an expression of the baby's inner state, smiles in return, and reinforces the baby's behaviour. In this way the infant learns how to communicate—not merely to mimic the facial shape of 'smiling', but to smile. Developmental roboticists hope to teach their machines to communicate by a similar process. For example, Diego-san's 'smiling' is to replicate the smiling pattern of a 4-month-old infant. The researchers discovered that an infant times his 'smiles' so as to maximize the duration of his mother's smiling while minimizing the duration of his own smiling; when this behaviour was programmed into the robot, human observers reacted to Diego-san's 'smiling' in the same way.[33]

One face robot has received widespread public attention for its range of emotional 'facial expressions', Cynthia Breazeal's (now retired) Kismet (Fig. 30.3). Kismet's caricature child-like 'features' and 'vocalizations' (sounds resembling a baby's babbling) are designed to evoke human nurturing responses. Observers react to the robot as if it were animate—as if (when its 'head' or 'eyes' move) it is interested in what is happening around it or is looking at a specific object. Humans speak to Kismet in baby talk, apologize to the robot for disturbing it, and empathize with it by mirroring its behaviour. But is this (to use Michie's words) 'real' rapport with the machine, or merely 'illusory' rapport?[34]

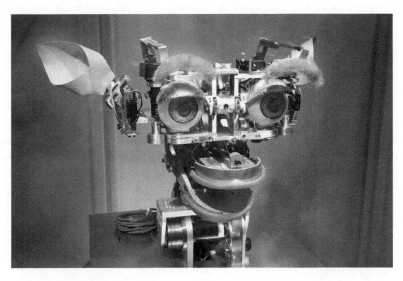

Figure 30.3 Cynthia Breazeal's Kismet.

Posted by Daderot to Wikimedia Commons, https://commons.wikimedia.org/wiki/File:Kismet,_1993-2000,_view_2_-_MIT_Museum_-_DSC03711.JPG. Creative Commons Licence.

What's in a smile?

Turing had a sophisticated approach to the concept of intelligence—an approach that is almost universally misunderstood (see Chapter 28).[35] In 1948 he said that 'the idea of "intelligence" is itself emotional rather than mathematical', and that the concept of intelligence is an 'emotional concept'. He remarked that the 'extent to which we regard something as behaving in an intelligent manner is determined as much by our own state of mind and training as by the properties of the object under consideration'.[36] In short, what makes something intelligent is not only how it behaves but also how we *respond* to it. This is not to say, though, that intelligence is solely in the eye of the observer: we may react to a machine as if it were intelligent when it is not (and so we need a test for thinking—Turing's imitation game). We may also react to a 'smile', a facial shape, as if it were a *smile*, a communicative act. Facial expressions can be understood as physical shapes or behaviours; in this sense a human can 'smile' when merely going through the motions of smiling (as when asked by a dentist to 'smile' into the mirror) or 'frown' without actually frowning (as when asked by a dermatologist to 'frown' before a botox injection). The same facial shape can be a smile or a grimace, or merely a facial tic. The chimpanzee's bared-teeth display is a very similar shape to the human smile, but it is a different action with a different meaning. So, what makes a 'smile' into a *smile*?

The additional behaviours of the 'smiler' are crucial. For example, a smile is part of a *sequence* of fluid emotional expressions; if a human being had only a 'smiling' face (or even if she had different expressions but her 'smile' were always exactly the same shape), we would not say that she was smiling. Also, a smile occurs in specific *contexts*, for example when greeting friends but not enemies; this is because we are happy to see our friends but not our enemies, and happy people too have characteristic behaviour. In addition, any smile is some *sort* of smile, for example welcoming, conspiratorial, flirtatious, or cynical; what makes a smile welcoming rather than cynical is the further actions of the 'smiler'.

Kismet does not have these additional behaviours. Its 'facial expressions' are machine-like rather than fluid, and this is very different from even the human's 'fixed' smile. When the robot 'smiles' it is not *happy*, as its creator acknowledges; just as Turing said that the words 'pain' and 'pleasure' do not presuppose 'feelings' in his P-type machines, Breazeal says that the robot's 'emotions' (that is, these components of its system architecture) are 'quite different' from emotions in humans.[37] The robot is assumed to have sorts of smiles, including a 'contented smile',[38] but in reality it lacks the behaviours that are required if a 'smile' is to be an expression of *contentment* (as when settling back into a comfortable chair and sighing peacefully after completing a task). Kismet's 'smiling' behaviour is not *smiling*—even if we react to the robot by smiling.

We all too readily anthropomorphize machines; according to Brooks, for example, human factory workers say of Baxter 'It's my buddy!'.[39] Recent films (video games, and graphic novels) that depict superhuman-level disembodied AIs and androids barely distinguishable from a human being may lead audiences to think that human-level AI is close. Misplaced anthropomorphism and make-believe are common within AI itself, and some influential but wildly optimistic researchers promise unimaginably powerful artificial minds within a few decades.[40] In reality, developmental robotics has a long way to go in order to build a child machine with infant-level social intelligence.

Growing up: discipline and initiative

Turing said, 'If the untrained infant's mind is to become an intelligent one, it must acquire both discipline and initiative . . . To convert a brain or machine into a universal machine is the extremest form of discipline . . . But discipline is certainly not enough in itself to produce intelligence. That which is required in addition we call initiative'.[41] In Turing's view a machine shows initiative when it can learn by itself, when it tries out 'new combinations' of the routines it has been given, or applies these in 'new ways'. Once it has both discipline and initiative, the child machine has grown up.

According to Turing, opponents of machine intelligence argue that a machine *cannot* think because (amongst other things) it can never 'be the subject of its own thought' or 'do something really new'. But it seems that, for Turing, the grown-up machine doesn't have these 'disabilities'. It *does* know its own thoughts: Turing said, 'At comparatively late stages of education the memory might be extended to include important parts of the configuration of the machine at each moment, or in other words it would begin to remember what its thoughts had been'. And it *can* do something new: in Turing's view, it can make 'choices' or 'decisions' by modifying its own program.[42] An AI sceptic might retort that the behaviour of the grown-up machine is merely the result of earlier programming. But, Turing pointed out, a human mathematician receives a great deal of training, which is akin to programming, and yet we don't say that the mathematician can't do anything new. Why treat the machine differently?[43]

'It is customary', Turing said, 'to offer a grain of comfort, in the form of a statement that some particularly human characteristic could never be imitated by a machine . . . I cannot offer any such comfort, for I believe that no such bounds can be set'. Smiling and other expressive behaviours are particularly human characteristics, and there is no reason to believe that a machine could never smile, or that we cannot build a socially intelligent child machine. Nevertheless, the claim that AI has already constructed 'expressive' robots is premature. And it is unlikely that Turing—who in 1952 predicted that it would be at least 100 years before a machine succeeded in his imitation game—would have said otherwise.[44]

CHAPTER 31

Computer chess—the first moments

JACK COPELAND AND DANI PRINZ

The electronic computer has profoundly changed chess. This chapter describes the birth of computer chess, from the very first discussions of computational chess at Bletchley Park during the war to the first chess moves ever calculated by an electronic computer. We cover a number of historic chess programs—including Turing's own 'Turochamp'—and recapture some of the atmosphere of those early days of computer chess.

Kasparov on Turing

Albert Square, Manchester, 2012. The time was coming up to 9 o'clock on a grim summer morning, two days after what would have been Turing's 100th birthday. Litter from the Olympic torch ceremony still scattered the ground. There were unusual numbers of chess enthusiasts and computer scientists in the square, hurrying past the awkwardly posturing statue of William Gladstone and up the steps at the entrance to Manchester Town Hall. Inside, they filed past more statues—chemist John Dalton, physicist James Joule—and took their seats in the crowded gothic-revival great hall. News of Turing's centenary celebrations had reached over forty countries: fans in other time zones clicked to join the audience, watching their screens and waiting for the big event to start. Shortly after 9, a flawlessly groomed Garry Kasparov took the stage.

Born in the Soviet Union in 1963, Kasparov (Fig. 31.1) became world chess champion at the age of only 22. He has gone down in history as the first reigning champion to be beaten by a computer. In a New York TV studio on the thirty-ninth floor of a Seventh Avenue skyscraper, IBM's chess computer DeepBlue crushed Kasparov in 1997 (see Ch. 27). Fifteen years later he had come to Manchester to honour Turing, the first pioneer of computer chess.

Seeming a bit nervous at first—until his natural ebullience reasserted itself—Kasparov haltingly told the crowd: 'Apart from personal love of the game, Turing did serious work with chess as a model of mechanical thinking and machine intelligence'. Yet Turing, he said, 'was a fairly

Figure 31.1 Garry Kasparov.

Posted by Owen Williams to Wikimedia Commons, https://commons.wikimedia.org/wiki/File:Kasparov-23.jpg. Creative Commons Licence.

terrible chess player'. And this, Kasparov pointed out, was despite his having been coached at Bletchley Park by some of Britain's strongest players. 'In spite of this he remained a rank amateur, a weak player', Kasparov said in a matter-of-fact tone. Turing's mentors in Hut 8 included Hugh Alexander—the British chess champion in 1938—and Harry Golombek, who won the British championship three times after the war. Golombek could dance rings around Turing even when handicapped by giving up his queen:[1]

[W]henever we played chess I had to give him the odds of a Queen in order to make matters more equal, and even then I always won.

Although a mediocre player, Turing took chess very seriously: to improve his powers of visualization he even covered his bedroom walls with pictures of chessboard positions.[2] He was a good visualizer: while out walking with a suitable opponent he played by simply naming his moves. This passionate yet weak player went on to make a contribution that changed the world of chess. Kasparov continued:

It is little known that Turing wrote a full chess-playing program without having access to a computer that could actually execute the instructions, simply because programmable computers didn't exist at the time. So in 1951 and 1952 he formulated the rules—so-called 'rules'—basically a program that could be used to calculate an unambiguous move for any given chess position. He had to act as human CPU and calculate the moves with pencil and paper—which you can guess was a very laborious process and took quite some time. . . .

Computer chess before computers

The charming (if irascible) AI pioneer Donald Michie was just a schoolboy when he joined the codebreakers at Bletchley Park in 1942. Fifty-five years later, after a lifetime of building intelligent (and not-so-intelligent) machines, Michie was one of those present when DeepBlue defeated Kasparov in New York in 1997. For the first time in history the human intellect seemed on the run, with AI encroaching on what Goethe had famously called 'a touchstone of human intelligence'.[3] Michie, always an admirer of brainpower, had soon befriended Turing at Bletchley Park. In 1998, relaxing at his home in Palm Desert, California, he told me (Jack Copeland) how computer chess was born.[4]

On Friday evenings, after a hard week battling against German codes, he and Turing used to go for a drink together. Their usual pub was in the bustling little town of Wolverton, just a few minutes away on the train from Bletchley Station. Finding seats in the poorly lit and smoke-filled bar, they would begin a game of chess and turn to their favourite topic of conversation—the mechanization of human thought processes. Michie explained:

What the codebreaker does is very much a set of intellectual operations and thought processes, and so we were thoroughly familiar with the idea of automating thought processes—both of us were up to our elbows in automation of one kind and another.

He was referring to Heath Robinson (see Chapter 14)—his own preoccupation at that time—and Turing's bombes (although Michie himself knew nothing about the bombes until long after the end of the war). Given what the two were involved in at Bletchley Park, they found it entirely natural to be spending their Friday evenings talking about automating the thought processes of a human chess player—and from there Turing moved on to the idea of automating the whole process of learning. Jack Good, another codebreaker and chess fanatic, would sometimes join in the discussions, usually during Sunday morning walks with Turing and Michie. Good could also remember an earlier conversation with Turing in 1941, before Michie arrived at Bletchley Park, in which they 'talked about the possibility of mechanizing chess'.[5]

I asked Michie what he could recall of these historic discussions with Turing about what we now call AI:

MICHIE : 'There were three headings. One was: methods of mechanizing the game of chess and games of similar structure. Another was: the possibility of machine algorithms and systems which could learn from experience; and the third area was a little more general, to do with the possibility of instructing machines with more general statements than purely ground-level factual statements—which would involve, in some sense, the machine understanding and drawing inferences from what it was told.'

COPELAND : 'Did you and Turing often play chess together at this time?'

MICHIE : 'Being one of the few people in the Bletchley environment bad enough to give him a reasonably even game, I became his regular sparring partner. Our discussions on machine intelligence started from the moment that we began to play chess together.'

COPELAND : 'What specific proposals did Turing make concerning chess programming at this time?'

MICHIE : 'We talked about putting numerical values on the pieces. And we talked about the mechanization of conditionals of the form: "If I do *that*, he might to do *that*, or alternatively he might do *that*, in which case I could do *this*"—what today we would call *look-ahead*. We certainly discussed priority of ordering according to some measurable plausibility of the move.'

COPELAND : 'So that's what chess programmers now call *best-first*?'

MICHIE : 'Yes, that's right. Our discussions of chess programming also included the idea of *punishing* the machine in some sense for obvious blunders. And the question of how on earth it might be possible to make this useful, beyond the pure rote-learning effects of punishment—deleting a move from the repertoire. Which is only applicable in the early opening.'

COPELAND : 'Did Turing have any specific ideas about how to do this?'

MICHIE : 'I don't remember any in the context of chess. I do remember him later circulating a typescript in which he had specific ideas about learning and more general varieties of learning.'

COPELAND : 'Was this at Bletchley?'

MICHIE : 'At Bletchley, yes. When I say "circulating", I know he showed a copy to me and to Jack Good—and I suppose to one or two other associates—for our comments.'

The 'best-first' principle that Michie referred to involves a rule-of-thumb scoring system: at any given point in the game, the program can use this scoring system to rank the available moves; and it then goes on to examine the consequences of the best (highest-scoring) move first. In an earlier conversation Michie had explained that their wartime chess discussions also covered the 'minimax' principle:[6] minimax involves assuming that opponents will move so as to *maximize* their gains, and the program then moves so as to *minimize* the losses caused by the opponent's expected moves. John von Neumann had investigated the minimax approach in his groundbreaking forays into what mathematicians now call 'game theory', and in 1928 he had published his fundamental 'minimax theorem'.[7]

None of this pioneering work at Bletchley Park on computer chess was published. Like Michie, Good 'made the mistake of thinking it was not worth publishing', he said.[8] But once the war came to an end in 1945, and Turing had more time to devote to the topic of machine intelligence, statements about computer chess began to appear in his writings. He even mentioned chess in the ultra-serious 'Proposed electronic calculator', his famous 1945 report about the ACE. He said:[9]

Given a position in chess the machine could be made to list all the 'winning combinations' to a depth of about three moves on either side. This . . . raises the question 'Can the machine play chess?' It could fairly easily be made to play a rather bad game. It would be bad because chess requires intelligence.

Then he added prophetically:

There are indications however that it is possible to make the machine display intelligence at the risk of its making occasional serious mistakes. By following up this aspect the machine could probably be made to play very good chess.

At the same time (1945) the German computer pioneer Konrad Zuse—working in complete isolation in a village in the Bavarian Alps—was developing his *Plankalkül* programming language, and was also thinking about chess and other forms of symbolic computation.[10] Two years later Turing again mentioned chess, in his remarkably far-sighted 1947 Burlington House lecture about the future of computing (see Chapter 25), predicting that it 'would probably be quite easy to find instruction tables [programs] which would enable the ACE to win against an average player'.[11]

Then, in the summer of 1948, Turing described a chess-player version of what we now call the Turing test or imitation game (see Chapters 25 and 27). He wrote:[12]

It is not difficult to devise a paper machine which will play a not very bad game of chess. Now get three men as subjects for the experiment A, B, C. A and C are to be rather poor chess players, B is the operator who works the paper machine. (In order that he should be able to work it fairly fast it is advisable that he be both mathematician and chess player.) Two rooms are used with some arrangement for communicating moves, and a game is played between C and either A or the paper machine. C may find it quite difficult to tell which he is playing. (This is a rather idealized form of an experiment I have actually done.)

The 'paper machine' that Turing mentioned was created that same summer in collaboration with his friend the statistician David Champernowne. Turing crystallized his ideas about computational chess into a kind of program in pseudo-code—a set of machine-rules for playing chess. Named 'Turochamp', and an outgrowth of his wartime deliberations with Michie and Good, this was an early manifestation of what would now be called an AI program.

Turochamp

Turochamp was a 'paper machine' in the sense that since no electronic computer was available, Turing and Champernowne simulated the program's play by hand, using paper and pencil. In 1980 Champernowne gave this description of Turochamp (quoted in *The Essential Turing*):[13]

It was in the late summer of 1948 that Turing and I did try out a loose system of rules for deciding on the next move in a chess game which we thought could be fairly easily programmed for a computer . . . Here is what I think I remember about the system but I may have been influenced by what I have since read about other people's systems. There was a system for estimating the effects of any move on White's estimated net advantage over Black. This allowed for:

(1) Captures, using the conventional scale of 10 for pawn, 30 for knight or bishop, 50 for rook, 100 for queen and something huge, say 5000, for king.
(2) Change in mobility; i.e., change in the number of squares to which any piece or pawn could immediately move legitimately (1 each).
(3) Special incentives for: (a) Castling (3 points). (b) Advancing a passed pawn (1 or 2 points). (c) Getting a rook onto the seventh rank (4 points perhaps).

(I don't think occupation of one of the 4 central squares gained any special bonus. We did not cater to the end-game.) Most of our attention went to deciding which moves were to be followed up. My memory about this is infuriatingly weak. Captures had to be followed up at least to the point where no further capture was immediately possible. Checks and forcing moves had to be followed further. We were particularly keen on the idea that whereas certain moves would be scorned as pointless and pursued no further others would be followed quite a long way down certain paths. In the actual experiment I suspect we were a bit slapdash about all this and must have made a number of slips since the arithmetic was extremely tedious with pencil and paper. Our general conclusion was that a computer should be fairly easy to programme to play a game of chess against a beginner and stand a fair chance of winning or at least reaching a winning position.

Turochamp was (so far as we know) the first chess program ever to beat a human player—Champernowne's wife. There is no record of either Turochamp, or indeed the later chess program that Turing wrote, ever beating Turing himself, but he likened the claim that no program can outplay its own programmer to the claim 'No animal can swallow an animal heavier than itself'.[14] 'Both', he said, 'are, so far as I know, untrue'.

A few weeks after these historic experiments with Turochamp, Turing took up his new job at the Manchester Computing Machine Laboratory, putting him almost within reach of coding a chess program for a real computer. Things didn't quite work out, however. But before relating that part of the story we will describe a distinction that is essential for understanding the full significance of Turing's achievements in chess programming.

Brute force versus heuristics

Chess programmers talk about 'chess heuristics': the term 'heuristic' is explained in Chapter 25. Chess programs using heuristics are contrasted with 'brute-force programs'. A brute-force chess program selects its moves by exploring the consequences of each move available to it, even right through to the end of the game. If programmed in this way, the computer would always choose the best move.

In practice, though, a chess program that uses nothing but brute force is a hopeless proposition. As Max Newman pointed out in a postwar radio talk, the brute-force method would only work '*if* you didn't mind its taking thousands of millions of years'.[15] This is because the number of moves that the computer would have to examine is astronomically large. In his wartime discussions with Good and Michie, Turing had hit on the more practical idea of playing chess by using heuristics. Instead of examining every available move exhaustively, the machine would use rules of thumb—supplied by the programmer—to select moves for further analysis. Turochamp seems to have been the first heuristic chess program.

These rules of thumb are like shortcuts, enabling the computer to play quickly—but the price of speed is that shortcuts don't always work. When a shortcut happens to lead in the wrong direction, the chess program may make a mistake. No doubt Turing was alluding to this feature of chess heuristics when he said that it is 'possible to make the machine display intelligence at the risk of its making occasional serious mistakes'.

Turochamp Mark 2

While he was working on his groundbreaking theory of morphogenesis in the Manchester Computing Machine Laboratory, Turing also found time to develop an improved form of Turochamp. We will call this 'Turochamp Mark 2', and from now on we use the term 'Turochamp Mark 1' to refer to the original 1948 version. Turing described the Mark 2 in a far-sighted typescript about computer chess (published in *The Essential Turing*). He explained that the system of rules set out in this typescript was 'based on an introspective analysis' of his own 'thought processes when playing'—although with 'considerable simplifications'. He said modestly:

If I were to sum up the weakness of the above system in a few words I would describe it as a caricature of my own play.

In his lecture Kasparov tended to confuse the earlier and later versions of Turochamp. These historical confusions, though, didn't prevent him from wiping the floor with a re-creation of Turochamp Mark 2 when he played against it on the stage at Manchester; the game is set out later (Game 4). But Kasparov described the re-creation that he played—the 'Turing Engine' coded by the German chess software company ChessBase—as embodying the 'first program ever written, even before computers that could run it were invented'. But it was the Mark 2, not the Mark 1, that ChessBase re-created. No recorded games with the Mark 1 survive, and all that is known about the Mark 1 is Champernowne's rather vague description quoted earlier. What is more, although Turing and Champernowne created the Mark 1 at a time when the only stored-program computer in existence was the tiny Baby, the Mark 2 dates from the period *after* a computer capable of running it was installed in Turing's Computing Machine Laboratory (the Ferranti Mark I). What Kasparov should have said, if he wanted to be completely accurate, was not 'first program' but 'successor to the first version of Turochamp, written at the time that the first big electronic digital computers went on the market'.

As Kasparov's game with the Turing Engine highlighted, Turochamp Mark 2 is no champion-beater. Nevertheless, the pioneering Mark 2 included much that has become standard art in chess programming:

- *evaluation rules* that assign numerical values, indicative of strength or weakness, to board configurations
- *variable look-ahead*: instead of the consequences of every possible move being followed equally far, the 'more profitable moves' are 'considered in greater detail than the less', Turing explained
- *heuristics* that guide the search through the 'tree' of possible moves and counter-moves
- the *minimax* principle
- *quiescence search*: the search along a particular branch in the tree of moves is discontinued when a 'dead' position is found—a position with no captures or other lively developments in the offing.

Turing also realized the necessity of using 'an entirely different system for the end-game'. In addition, he gave a prescient discussion of how to make a chess program learn from experience. He advocated the use of what we would now call a 'genetic algorithm' or GA, saying:[16]

[A]s to the ability of a chess-machine to profit from experience, one can see that it would be quite possible to programme the machine to try out variations in its method of play (e.g. variations in piece value) and adopt the one giving the most satisfactory results. This could certainly be described as 'learning', though it is not quite representative of learning as we know it.

GAs are now widely used in areas as diverse as codebreaking, hardware design, pharmaceutical drug design, and financial forecasting. As mentioned in Chapter 25, GAs are based on the principles of Darwinian evolution—random mutation and survival of the fittest—and some researchers even employ GAs as a means of studying the process of evolution itself. Turing's not-very-catchy expression for his invention was 'genetical or evolutionary search'.

Arthur Samuel, an early pioneer of AI in the United States, independently reintroduced the idea a few years later in his famous checkers playing program, a remake of the Strachey draughts (checkers) program mentioned in Chapter 20.[17] In line with Turing's suggestion of allowing 'the machine to try out variations in its method of play', Samuel set up two copies of his program, Alpha and Beta, on the same computer and left them to play game after game with each other. The computer made small random changes to Alpha's move generator, leaving Beta's unchanged, and then compared the performance of Alpha and Beta over a number of games. If Alpha played worse than Beta, the changes were discarded, but if Alpha played better, Beta was deleted and replaced by a copy of Alpha. As in the jungle, the fitter copy survived. Over many generations, the program's play became increasingly skilful.

It was in fact 1955 before Samuel implemented this idea of pitting Alpha against Beta—so if Turing himself ever put his 'genetical search' idea into practice at Manchester he was almost certainly the first to run a GA on a computer. Tantalizingly, though, there is nothing in the surviving records to tell us whether he did or he didn't.

The Mark 2 rules

What follows is Turing's lucid description, from his typescript, of how Turochamp Mark 2 worked (quoted from *The Essential Turing*):

I shall describe a particular set of rules, which could without difficulty be made into a machine programme. It is understood that the machine is white and white is next to play. The current position is called the 'position of the board', and the positions arising from it by later moves 'positions in the analysis'.

'Considerable' moves, i.e. moves to be considered in the analysis by the machine
Every possibility for white's next move and for black's reply is 'considerable'. If a capture is considerable then any recapture is considerable. The capture of an undefended piece or the capture of a piece of higher value by one of lower value is always considerable. A move giving checkmate is considerable.

Dead position
A position in the analysis is dead if there are no considerable moves in that position, i.e. if it is more than two moves ahead of the present position, and no capture or recapture or mate can be made in the next move.

Piece values[18]
P = 1
Kt = 3
B = 3½
R = 5
Q = 10
Checkmate ± 1000

Value of position
The value of a dead position is obtained by adding up the piece values as above, and forming the ratio W/B of white's total to black's. In other positions with white to play the value is the greatest value of: (a) the positions obtained by considerable moves, or (b) the position itself evaluated as if a dead position, the latter alternative to be omitted if all moves are considerable. The same process is to be undertaken for one of black's moves, but the machine will then choose the least value.

<u>Position-play value</u>

Each white piece has a certain position-play contribution and so has the black king. These must all be added up to give the position-play value.

For a Q, R, B or Kt, count

i) The square root of the number of moves the piece can make from the position, counting a capture as two moves, and not forgetting that the king must not be left in check.
ii) (If not a Q) 1.0 if it is defended, and an additional 0.5 if twice defended.

For a K, count

iii) For moves other than castling as i) above.
iv) It is then necessary to make some allowance for the vulnerability of the K. This can be done by assuming it to be replaced by a friendly Q on the same square, estimating as in i), but subtracting instead of adding.
v) Count 1.0 for the possibility of castling later not being lost by moves of K or rooks, a further 1.0 if castling could take place on the next move, and yet another 1.0 for the actual performance of castling.

For a P, count

vi) 0.2 for each rank advanced.
vii) 0.3 for being defended by at least one piece (not P).

For the black K, count

viii) 1.0 for the threat of checkmate.
ix) 0.5 for check.

We can now state the rule for play as follows.

> The move chosen must have the greatest possible value, and, consistent with this, the greatest possible position-play value. If this condition admits of several solutions a choice may be made at random, or according to an arbitrary additional condition.

Note that no 'analysis' is involved in position-play evaluation. This is in order to reduce the amount of work done on deciding the move.

Having examined the Mark 2's rules in all their gory detail, let's see next how the program made out against a human player who was not as special as Kasparov.

Alick Glennie versus Turochamp Mark 2

Alick Glennie was a young scientist employed at the Atomic Weapons Research Establishment in Berkshire who was sent north to do some work at the Manchester Computing Machine Laboratory. He showed up there in July 1951, around the time that the lab's Ferranti computer started to be used for atomic weapons calculations. Glennie soon learned from lab gossip that Turing was very interested in computer chess.[19]

'I was not a keen player', Glennie said: 'I knew a few standard openings but none of the finer points of strategy'. Turing thought Glennie a 'weak player'—and that was just what he was looking for, a weak player who knew nothing about the workings of his mechanical chess rules.[20]

'As I remember, he persuaded me over lunch to take part in his chess experiment', Glennie recollected, and so they went upstairs to Turing's office, 'a rather bare place with a small table untidy with paper'. Glennie sat at the chessboard and played his virtual opponent, while Turing scribbled away calculating the program's moves, just as he had with Champernowne a few years previously. He had to keep juggling his various sheets of paper containing the rules: 'We were playing on a small table which did not help', Glennie said. 'It seemed to go rather slowly and I think I got slightly bored'.

'The game took 2 or 3 hours', Glennie remembered. He won at the twenty-ninth move (see Game 1). Turing set out the game in his typescript, writing the moves in old-fashioned chess notation. Beside each move we have placed a translation into modern chess notation (in square brackets).[21] Turing included some footnotes, and also a column of numbers showing the increase in position-play value (if any) after each of the program's moves, and he placed an asterisk against the number if every other move had a lower position-play value. (In his record of this game he used O—O to indicate castling, without distinguishing between castling on the queen's side and castling on the king's side; whereas in his Game 2, O—O—O was used to indicate castling on the queen's side.) The program, White, opened (see facing page).[22]

Talking about the game more than 20 years later, Glennie said 'I remember it as a rather jolly afternoon', adding 'and I believe Turing must have enjoyed it too—in his way':

Turing's reaction to the progress of the play was mixed: exasperation at having to keep to his rules; difficulty in actually doing so; and interest in the experiment and the disasters into which White was falling. Of course, he could see them coming.

Glennie explained that in fact Turing had difficulty sticking to the rules:

During the game Turing was working to his rules and was clearly having difficulty playing to them because they often picked moves which he knew were not the best. He also made a few mistakes in following his rules which had to be backtracked. This would occur when he was pondering the validity of White's last move while I was considering my move. There may have been other mistakes in following his rules that escaped notice . . . He had a tendency to think he knew the move the rules would produce.

As will become clear, some of White's moves in the recorded game were most definitely not the ones that Turing's rules dictated.

The Mark 2's mysterious re-game

During his lecture Kasparov introduced a guest, Frederic Friedel, co-founder of ChessBase. Together, Kasparov and Friedel demonstrated ChessBase's re-creation of Turochamp. Talking about Turing's historic program Friedel said: 'We have one complete game, played with Glennie'. 'It's the only recorded game', he emphasized. This is not quite true, however.

In 1953, the London publishing company of Pitman and Sons brought out a book with the intriguing title *Faster Than Thought*.[23] In effect the world's first computer science textbook, it was edited by Vivian Bowden, later Baron Bowden of Chesterfield, who was in charge of computer sales at Ferranti. This book contained contributions from everybody who was anybody in the emerging world of British computing—including, of course, Turing, who contributed a section titled 'Chess' to the chapter 'Digital computers applied to games'.[24]

Game 1 *Glennie versus Turochamp Mark 2.*

White (machine)		Black
1. P—K4 [e4]	4.2*	P—K4 [e5]
2. Kt—QB3 [Nc3]	3.1*	Kt—KB3 [Nf6]
3. P—Q4 [d4]	3.1*	B—QKt5 [Bb4]
4. Kt—KB3[(1)] [Nf3]	2.0	P—Q3 [d6]
5. B—Q2 [Bd2]	3.6*	Kt—QB3 [Nc6]
6. P—Q5 [d5]	0.2	Kt—Q5 [Nd4]
7. P—KR4[(2)] [h4]	1.1*	B—Kt5 [Bg4]
8. P—QR4[(2)] [a4]	1.0*	Kt × Kt ch. [N×f3+]
9. P × Kt [g×f3]		B—KR4 [Bh5]
10. B—Kt5 ch. [Bb5+]	2.4*	P—QB3 [c6]
11. P × P [d×c6]		O—O [0–0]
12. P × P [c×b7]		R—Kt1 [Rb8]
13. B—R6 [Ba6]	−1.5	Q—R4 [Qa5]
14. Q—K2 [Qe2]	0.6	Kt—Q2 [Nd7]
15. KR—Kt1[(3)] [Rg1]	1.2*	Kt—B4[(4)] [Nc5]
16. R—Kt5[(5)] [Rg5]		B—Kt3 [Bg6]
17. B—Kt5 [Bb5]	0.4	Kt × KtP [N×b7]
18. O—O [0–0–0]	3.0*	Kt—B4 [Nc5]
19. B—B6 [Bc6]		KR—QB1 [Rfc8]
20. B—Q5 [Bd5]		B × Kt [B×c3]
21. P × B[(2)] [b×c3]	−0.7	Q × P [Q×a4]
22. B—K3[(6)] [Be3]		Q—R6 ch. [Qa3+]
23. K—Q2 [Kd2]		Kt—R5 [Na4]
24. B × RP[(7)] [B×a7]		R—Kt7 [Rb2]
25. P—B4 [c4]		Q—B6 ch. [Qc3+]
26. K—B1 [Kc1]		R—R7 [Ra2]
27. B × BP ch. [B×f7+]		B × B [B×f7]
28. R × KtP ch.[(5)] [R×g7+]		K × R [K×g7]
29. B—K3[(8)] [Be3]		R—R8 mate [Ra1#]

Notes:
(1) If B—Q2 3.6* then P × P is foreseen.
(2) Most inappropriate moves from a positional point of view.
(3) If O—O then B × Kt, B × B, Q × P.
(4) The fork is unforeseen.
(5) Heads in the sand!
(6) Only this or B—K1 can prevent Q—R8 mate.
(7) Fiddling while Rome burns.
(8) Mate is foreseen, but 'business as usual'.

Game 2 *The mysterious second game.*

White (machine)		Black
1. P—K4 [e4]	4.2*	P—K4 [e5]
2. Kt—QB3 [Nc3]	3.1*	Kt—KB3 [Nf6]
3. P—Q4 [d4]	3.1*	B—QKt5 [Bb4]
4. Kt—KB3[(1)] [Nf3]	2.0	P—Q3 [d6]
5. B—Q2 [Bd2]	3.6*	Kt—QB3 [Nc6]
6. P—Q5 [d5]	0.2	Kt—Q5 [Nd4]
7. P—KR4[(2)] [h4]	1.1*	B—Kt5 [Bg4]
8. P—QR4[(2)] [a4]	1.0*	Kt × Kt ch. [N×f3+]
9. P × Kt [g × f3]		B—KR4 [Bh5]
10. B—Kt5 ch. [Bb5+]	2.4*	P—QB3 [c6]
11. P × P [d×c6]		O—O [0–0]
12. P × P [c×b7]		R—Kt1 [Rb8]
13. B—R6 [Ba6]	−1.5	Q—R4 [Qa5]
14. Q—K2 [Qe2]	0.6	Kt—Q2 [Nd7]
15. KR—Kt1[(3)] [Rg1]	1.2*	Kt—B4[(4)] [Nc5]
16. R—Kt5[(5)] [Rg5]		B—Kt3 [Bg6]
17. B—Kt5 [Bb5]	0.4	Kt × KtP [N×b7]
18. O—O [0–0–0]	3.0*	Kt—B4 [Nc5]
19. B—B6 [Bc6]		KR—QB1 [Rfc8]
20. B—Q5 [Bd5]		B × Kt [B×c3]
21. B × B [B×c3]	0.7	Q × P [Q×a4]
22. K—Q2 [Kd2]		Kt—K3 [Ne6]
23. R—Kt4 [Rg4]	−0.3	Kt—Q5 [Nd4]
24. Q—Q3 [Qd3]		Kt—Kt4 [Nb5]
25. B—Kt3 [Bb3]		Q—R3 [Qa6]
26. B—B4 [Bc4]		B—R4 [Bh5]
27. R—Kt3 [Rg3]		Q—R5 [Qa4]
28. B × Kt [B×b5]		Q × B [Q×b5]
29. Q × P[(6)] [Q×d6]		R—Q1[(4)] [Rd8]
30. Resigns[(7)]		

Notes:
(1) If B—Q2 3.7* then P × P is foreseen.
(2) Most inappropriate moves.
(3) If white castles, then B × Kt, B × B, Q × P.
(4) The fork is unforeseen at White's last move.
(5) Heads in the sand!
(6) Fiddling while Rome burns!
(7) On the advice of his trainer.

Turing's section on chess was broadly identical to his chess typescript, although with numerous minor differences. There was, however, one major difference: Game 1 was replaced by a different game, Game 2. The two games began in the same way, but diverged after twenty moves.[25]

This raises some tantalizing questions. Why did Turing replace Game 1 with Game 2? How could White's twenty-first move be different in Game 2, given that all previous moves were the same? Who indeed was the human player in Game 2? There is no reason to think that Black was Glennie, who mentioned playing only one game. Unfortunately the chess writer Alex Bell introduced confusion by reprinting Game 2 in his book and labelling it 'Game between Turing's Machine and Alick Glennie'.[26] But Glennie won at the twenty-ninth move, and while this is true of Game 1, where Glennie's twenty-ninth move achieved checkmate, it is not true of Game 2, where Black's twenty-ninth move brought neither checkmate nor even check. White could play on.

Eli Dresner of Tel-Aviv University has suggested a plausible solution to the puzzle:[27]

Perhaps Turing produced Game 2 by recalculating some aspects of Game 1, and then making Black's last nine moves himself. Some of the position-play values have certainly been recalculated: the values at each of lines 3, 5, 18 and 21 are different. So it's possible that while Turing was reflecting on White's performance during the Glennie game, he found that he had miscalculated White's 21st move, and therefore he reworked the last part of the game. It wouldn't have taken him more than about an hour, given that the whole game with Glennie lasted 2–3 hours. Maybe Turing was even quite pleased to discover that his program lived to fight on after the 29th move!

With Baby growing by leaps and bounds in the Manchester lab, Turing soon had enough computing power on hand to make it feasible to try out a heuristic chess program. Michie did indeed recall that Turing made a start on coding Turochamp for the computer. But 'he never completed it', Michie said.[28] Eventually—in the next millennium—Turochamp Mark 2 was finally implemented by ChessBase. But before we pick up the ChessBase story and move to the present time, we will describe another of Manchester's historic chess programs. Its programmer, a chocolate-loving German physicist, beat Turing to the first working chess implementation.

The Prinz chess program

During his lecture Kasparov noted that 'a Turing disciple, Dietrich Prinz' was the first person to program a real computer to play chess.[29] Prinz (Fig. 31.2) was a computer geek before the term was even invented, and Turing himself taught this quiet polite German refugee how to program the Manchester University Ferranti computer.[30] Prinz went on to write a programming manual for the Ferranti that was a model of clarity, unlike Turing's own earlier and often perplexing manual.[31] Prinz also continued the work on computer music (see Chapter 23), and in 1955 he programmed the Ferranti Mark I to play Mozart's *Musikalisches Würfelspiel* (musical dice-game).[32] Prinz used the computer's random number generator instead of dice, and by implementing Mozart's dice-involving composition rules he made the computer compose and play folk dances.

How Prinz—well mannered, obsessively tidy, meticulously organized, and always punctual—got on with the untidy and not-so-well-mannered Turing went unrecorded, but the two certainly had plenty of intellectual ground in common. Turing would have approved when Prinz maintained that 'the standard procedure for making babies', as he described it, was a way of

Figure 31.2 Dietrich Prinz.

Reproduced with permission of Dani Prinz.

'making a machine'.[33] Prinz also advocated his own version of Turing's networks-of-neurons approach to computing (see Chapter 29). He thought that computer engineers could take.[34]

animal or human nerve fibres and somehow encourage them to form suitable networks by joining up in the right places.

A chess addict, Prinz was always ready for a game. With his thick shock of dark wavy hair and the demeanour of the traditional absent-minded professor, he would sit at the board looking relaxed and slightly amused—an expression that he customarily wore when concentrating hard. He smoked cigarillos endlessly, and when surprised would say 'Ach' in his thick, gravelly German accent. 'Ziss is a good move', he might congratulate his opponent. A kind and pleasant man with an acute sense of fair play, he never gloated when he won—as he very often did. Prinz holds the title 'First human to make an electronic computer play chess'.

He was born in Berlin in 1903 and grew up in the Tiergarten district of the city. His father was a lawyer, but science was everything to the young Prinz. He studied physics and mathematics at Berlin University (now Humboldt University), where he was taught by such scientific giants as Albert Einstein and Max Planck. Subsequently, the Berlin radio and TV engineering company Telefunken hired him to work on electronic design problems. Prinz loved Berlin, calling it 'my little home town', but dark political clouds were gathering. An atheist intellectual German Jew, he fled Germany for England three months after Hitler's Reichstag passed the rampantly anti-Jewish Nuremberg Laws (in September 1935). It was a terrible wrench. He never saw some of his family members again, since they perished at the hands of the German SS.

Prinz arrived in London in January by railway from the coast, carrying two battered suitcases and a few assorted bundles. The heavy London accents he encountered on leaving the train defeated him. It must have been a lonely time. He wrote sadly 'Ich bin ein German refugee: there is no Vaterland for me', but at least he had a job. At Telefunken, before he fled Berlin, Prinz had appealed for help to some visiting British engineers, from the electrical manufacturing company GEC. As a result, a position awaited him at GEC's Valve Development Laboratory in Wembley.

He had brought his patents with him from Germany, carefully packed in his suitcases: these related to radio, television, and an exotic type of vacuum tube called the 'magnetron'.[35] Nowadays, magnetrons produce the microwave energy for cooking food in microwave ovens, but in pre-war Europe the magnetron was about to play a major role in radar. A group working for the British Admiralty perfected the design in 1940, creating the 'multi-cavity magnetron'; this was said to have had more influence on the course of the war than any other single invention, since it was key to constructing small, powerful, and accurate radar sets, compact and light enough to be fitted into aircraft.[36] Nazi anti-Semitism was costing German science dearly.

Prinz was welcomed into GEC and quickly grew to like his new life in England. 'If someone makes a mistake at work here', he said, 'they try to comfort him—instead of shouting at him'. It didn't last, though. In 1940, fearful of fifth columnists, Winston Churchill ordered the internment of enemy aliens. 'Collar the lot', he had said.[37] Prinz was arrested and taken to a police station. Like thousands of his countrymen, he soon found himself on board a once-luxurious passenger liner and heading for a prison camp in Canada. There were guns and barbed wire, but little of the horrific barbarity of the camps in Germany. In fact, the inmates turned some Canadian camps into impromptu universities, with captive scientists, engineers, and other intellectuals teaching class.[38] Soon, however, the British authorities saw the folly of imprisoning Nazi-hating refugees, and in 1941 Prinz again found himself crossing the Atlantic. This time his destination was Cambridge and a job in the Pye electronics company, at a subsidiary specializing in cathode-ray tubes. Turing was also involved with Pye, in connection with the bombe.[39]

By 1947 Prinz had adopted British nationality and was working for Ferranti, where he became a first-generation code hacker, and eventually set up Ferranti's programming department.[40] Like Turing and Strachey, he would work through the night at the computer. By the time Turing arrived in Manchester, in the autumn of 1948, Prinz had been living for several months in a leafy suburban street in Old Trafford (scarcely more than a stone's throw from Old Trafford stadium, home to Manchester United—although Prinz loathed football), and had been working on the idea of building specialized relay-based computers for performing complex logical deductions.[41] It was Turing's assistant at the National Physical Laboratory, Donald Davies, who piqued Prinz's interest in computer chess: Prinz was inspired by Davies's important article 'A theory of chess and noughts and crosses', which he read in a popular science magazine.[42] 'To programme one of the electronic machines for the analysis of Chess would not be difficult', Davies wrote.[43] And so the adventure began.

Brute-force chess

Turing's Turochamp was designed to play a complete game, but Prinz downsized the problem. He decided to focus only on simple mate-in-two situations, having seen similar puzzles in the competition pages of the *New Statesman*, and he tackled the job of writing a program to search for mate-forcing moves. Even then he introduced further simplifications: no castling, no double moves by pawns, no promotion of pawns, no captures *en passant*, and no distinction between mate and stalemate.[44] With so much simplification there was no need for Turing's heuristic approach, and Prinz wrote a brute-force program. This examined all the possible moves until one was found that did the job. The program ran on the Manchester University Ferranti computer in November 1951. It was a significant moment, the Big Bang of computer chess. Prinz published details in his classic article 'Robot chess'.[45]

Prinz's program computed a White move that would lead to mate at the next move, no matter how Black moved. Prinz fed in on punched paper tape the board position that the computer was to examine (see Game 3, Fig. 31.3 and Fig. 27.1). He recollected that the computer 'took about 15 minutes' to produce the winning move.[46]

It might be thought that this numbingly slow performance, and on such a simple chess problem, showed the impracticability of implementing Turochamp Mark 2 on the Ferranti. But this doesn't follow at all. For one thing, the brute-force approach is *essentially* slow: in order to reach mate in the above very simple example, Prinz's program tried out approximately 450 moves.[47] Moreover, a considerable proportion of the 15 minutes of computing time was used up by wasteful transfers between the Williams tube memory and the slower magnetic drum memory.[48] More effective use of the high-speed tube memory would have speeded the program up. With some fancy coding, Turing could probably have made Prinz's program run much faster. He took little interest, though—he knew there was no future in brute-force alone.

Prinz's program was a short-lived wonder. 'After establishing that it worked, it was never used again', Prinz related. He explained:[49]

The number of machine users increased so much that there was not enough time left for frivolities. Besides, any chess problem even slightly more complicated than the one I used would probably have taken hours.

Nevertheless the significance of what Prinz had done was akin to the Wright brothers' first short flight. He had shown that computers were not just high-speed number crunchers. A computer had played chess.

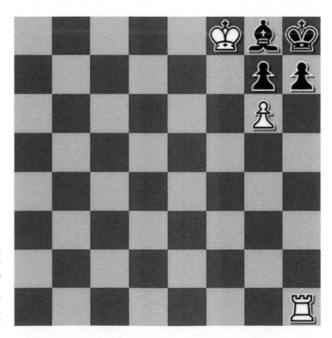

Figure 31.3 Game 3: the first chess ever played by a computer. The computer forced mate by moving R—R6 [Rh6].[50]

'Jonathan Bowen. Generated using 'Jin' by Alexander Maryanovsky (http://www.jinchess.com/chessboard/composer/).'

ChessBase's Turing engine

Turing's description of Turochamp Mark 2 was precise enough for ChessBase to turn it straight into programming code. It was Friedel's project, but he handed over the coding work to ChessBase programmer Mathias Feist. Feist, Friedel said, is 'completely obsessed with Turing'. 'Wasn't it just brilliant of Turing to invent a quiescence search on paper?', Feist enthused.[51]

The project ran into a problem, though. Once Feist had the program running, he tried to reproduce Game 2 (neither he nor Friedel knew of the existence of Game 1, the original game between Glennie and Turochamp). But the program would not duplicate the moves recorded by Turing: 'There were many deviations', Friedel said. Searching the program for errors achieved nothing, so Friedel appealed to legendary chess programmer Ken Thompson for help. Thompson is credited with creating the original Unix operating system, but his attempts to replicate Turing's game were a failure. In the end, Friedel telephoned Michie to ask for assistance. No one familiar with Glennie's statement that Turing 'made a few mistakes in following his rules' will be surprised by Michie's answer to Friedel:

He said to me, 'You are trying to debug your program. You should debug Turing!'

Turing even got White's very first move wrong. 'If you apply the rules, the paper machine plays 1.e3, not 1.e4', Feist says.[52] Friedel concluded happily:

Turing was just being careless and our machine was playing the moves he intended to play.

Kasparov beat the Turing Engine effortlessly on stage. The game was over so quickly you could have missed it in a blink. Smiling insouciantly, Kasparov checkmated the program in sixteen fast-paced moves:

Game 4 *Kasparov against Turing.*[53]

White (machine)	Black
1. P—K$_3$ [e3]	Kt—KB$_3$ [Nf6]
2. Kt—QB$_3$ [Nc3]	P—Q$_4$ [d5]
3. Kt—R$_3$ [Nh3]	P—K$_4$ [e5]
4. Q—B$_3$ [Qf3]	Kt—B$_3$ [Nc6]
5. B—Q$_3$ [Bd3]	P—K$_5$ [e4]
6. B × P [B×e4]	P × B [d×e4]
7. Kt × P [N×e4]	B—K$_2$ [Be7]
8. Kt—Kt$_3$ [Ng3]	O—O [0–0]
9. O—O [0–0]	B—KKt$_5$ [Bg4]
10. Q—B$_4$ [Qf4]	B—Q$_3$ [Bd6]
11. Q—B$_4$ [Qc4]	B × Kt/R [B×h3]
12. P × B [g×h3]	Q—Q$_2$ [Qd7]
13. P—KR$_4$ [h4]	Q—R$_6$ [Qh3]
14. P—Kt$_3$ [b3]	Kt—KKt$_5$ [Ng4]
15. R—K$_1$ [Re1]	Q × P/7 ch. [Q×h2+]
16. K—B$_1$ [Kf1]	Q × BP mate [Q×f2#]

The Turing–Shannon playoff

Across the Atlantic Claude Shannon, a brilliant mathematician and engineer, also took an interest in computational chess in the post-war years. Like Turing, Shannon was a leading pioneer of the information age. And playful. He could be spotted riding his unicycle along the corridors of Bell Labs—sometimes juggling at the same time, it's said.[54] In early 1943 Turing spent two months working in the Bell Labs building in Manhattan where Shannon also worked (see Chapter 18). Possibly he told Shannon about the computational chess ideas that he had previously discussed with Good and Michie, though no records exist of Turing's conversations with Shannon, so we shall never know for sure.

In October 1948, a few months after Turochamp Mark 1 was unleashed on Champernowne's wife, Shannon wrote a paper on computer chess.[55] The following year he gave a lecture based on this paper to a convention of radio engineers in New York, and the paper was finally published in 1950, titled simply 'Programming a computer for playing chess'.[56] It sat rather oddly in a magazine otherwise devoted to hard science, flanked on one side by a paper about the propagation of electromagnetic disturbances along a buried cable, and on the other by a discussion of the kinetics of phase-transitions in superconductors. Shannon himself took a hard scientific line about the concept of thinking:[57] after pointing out that 'chess is generally considered to require "thinking" for skilful play', he rammed home the point that the existence of a chess program with the ability to beat skilful human players

will force us either to admit the possibility of mechanized thinking or to further restrict our concept of thinking.

It used to be thought that Shannon's paper and lecture were the earliest work on computational chess, but thanks to Michie and Good we now know that Turing was pioneering the field much earlier, during 1941–42. Good said:[58]

Basically we came up with the Shannon concept—I thought it was just obvious.

Shannon's 1950 article contained sufficient detail for Feist to write a corresponding program, the 'Shannon Engine'. In an interesting twist, Feist's Shannon Engine and Turing Engine went head to head in a ten-game exhibition match, organized by Ingo Althöfer of Jena University.[59] Fittingly, this battle of the Titans ended in a draw. In fact, Turing and Shannon won a single game each and the remaining eight games were drawn, a draw being declared if the same position was ever repeated three times, though Althöfer notes:

In the majority of the drawn games, Turing's engine had the upper hand.

The match turned into a cliff-hanger. The first excitement came when Turing won the fifth game at the sixty-third move, after Shannon had blundered in the mid-game: 'Turing took the present and never gave back the –6.0 (and better) lead in evaluation', Althöfer said. When the end came it was 'a mate in 4 against the naked White king'. Then draw after draw followed. Turing seemed about to carry off the match when Shannon suddenly rallied in the middle of the very last game. At the eighteenth move, Turing had fallen victim to a so-called 'horizon effect', where a move is selected that would have been rejected if only the program had seen the consequences, which lay a little way beyond its look-ahead horizon. 'Shannon took the opportunity with both hands and left White no chance at all', commented Althöfer. Shannon won at the thirty-ninth move.

'In total, the level of play was very low', Althöfer summed up. Nevertheless, both these historic programs played better than Chess Champion MK1 (CC-MK1), an early dedicated chess computer marketed from 1978 by Novag Industries. After pitting Chess Champion against Turing and Shannon, Althöfer announced:

CC-MK1 is weaker than both the engines of Shannon and Turing.

On this question of comparing Turing and Shannon's algorithms with modern chess software, Feist makes an important point:[60]

Some things were clearly missing in these paper machines, for example handling of repetitions. Turing and Shannon didn't have the chance to test their concepts with a computer. I'm sure that such shortcomings would have been corrected immediately by another rule: Avoid repetitions if you are ahead. Therefore it is a bit unfair to compare the Turing and Shannon programs to modern engines. Without repetition detection, all engines would be susceptible to the same problem.

Feist added that it would have been easy for him to implement this 'avoid repetitions when ahead' rule, with the result that the historic engines would not fall so easily into draws. 'But, of course, I didn't want to change the original algorithms more than technically necessary', he said.

Interestingly, given that repetitive moves often cost the Turing Engine its win, it seems probable that Turing would have beaten Shannon hands down had Feist's additional rule been in place.

Figure 31.4 The English Heritage Blue Plaque unveiled by Kasparov at what was Turing's Computing Machine Laboratory.

Taken by Peter Hughes and posted to https://www.flickr.com/photos/47523307@N08/7510151756/. Creative Commons licence.

Conclusion

Turing's deliberations about computational chess formed a significant element in his theorizing about the mechanization of intelligence. His first investigations preceded the famous 1949 lecture and 1950 article on computational chess by Shannon: Turing's pioneering explorations at Bletchley Park, with Michie and Good, form the earliest known work in the field of computational chess.

The day after Kasparov gave his lecture he unveiled an English Heritage Blue Plaque commemorating Turing (Fig. 31.4), saying:[61]

In the sweep of history, there are a few individuals about whom we can say the world would be a very different place had they not been born.

CHAPTER 32

Turing and the paranormal

DAVID LEAVITT

O f the nine arguments against the validity of the imitation game that Alan Turing
anticipated and refuted in advance in his 'Computing machinery and intelligence',
the most peculiar is probably the last, 'The argument from extra-sensory perception'.
So out of step is this argument with the rest of the paper that most writers on Turing (myself
included) have tended to ignore it or gloss over it, while some editions omit it altogether.[1] An
investigation into the research into parapsychology that had been done in the years leading
up to Turing's breakthrough paper, however, provides some context for the argument's inclu-
sion, as well as some surprising insights into Turing's mind.

The 'overwhelming evidence'

Argument 9 (of the nine arguments against the validity of the imitation game) begins with a
statement that to many of us today will seem remarkable. Turing writes:[2]

I assume that the reader is familiar with the idea of extra-sensory perception and the meaning of
the four items of it, *viz.* telepathy, clairvoyance, precognition, and psycho-kinesis. These disturbing phenomena seem to deny all our usual scientific ideas. How we should like to discredit them!
Unfortunately the statistical evidence, at least for telepathy, is overwhelming.

To what 'statistical evidence' is Turing referring? In all likelihood it is the results of some experi-
ments carried out in the early 1940s by S. G. Soal (1899–1975), a lecturer in mathematics at
Queen Mary College, University of London, and a member of the London-based Society for
Psychical Research (SPR).

To give some background, the SPR had been founded in 1882 by Henry Sidgwick, Edmund
Gurney, and F. W. H. Myers—all graduates of Trinity College, Cambridge—for the express
purpose of investigating 'that large body of debatable phenomena designated by such terms
as mesmeric, psychical and spiritualistic . . . in the same spirit of exact and unimpassioned
enquiry which has enabled science to solve so many problems, once no less obscure nor less
hotly debated'.[3] Although the membership of the SPR included numerous academics and
scientists—most notably William James, Sir William Crookes, and Lord Rayleigh, a Nobel

laureate in physics—it had no academic affiliation. Indeed, in the view of their detractors, the 'psychists', as they were known, occupied the same fringe as the mediums and mind-readers whose claims it sought to verify—or disclaim.

From the beginning the SPR struggled, not always successfully, to reconcile its advertised objectivity with the vested interest in psychic phenomena that had driven its members to organize in the first place. Toward that end, it tended to err on the side of caution. Thus, in September 1883 its 'Committee on thought transference' undertook an experiment in the 'transference of tastes' in which an examiner tasted various substances and then asked two subjects, young women referred to as R and E, to identify the substances by telepathic means. When the examiner tasted vinegar, E responded: 'A sharp and nasty taste'. Port wine was described as having a flavour 'between eau de Cologne and beer'. Worcestershire sauce was Worcestershire sauce, mustard was mustard.[4] No claims were made for this experiment, which was presented simply as a set of findings. Detailed accounts of such efforts shared the pages of the SPR's *Proceedings* with unsparing exposés of fraud (most notably, the 'phenomena' attributed to Madame Helena Blavatsky, the founder of theosophy) and statistical compilations such as the 'Census of hallucinations', undertaken by Henry Sidgwick's wife, Nora, a mathematician and later Principal of Newnham College, Cambridge. The latter was in effect an international survey of what the SPR termed 'crisis apparitions': dreams or fantasies of the death of a loved one that turned out, in retrospect, to be premonitory.[5]

Circles, rectangles, stars, crosses, and waves

As an academic field, parapsychology came into its own in 1930 with the establishment by J. B. Rhine (1895–1980) of the Parapsychology Institute at Duke University in the United States. Rhine had been obsessed with spiritualist phenomena since his youth, and in the early 1920s this obsession carried him into the orbit of William MacDougall, another Cambridge-bred member of the SPR, whom William James had recruited across the Atlantic to Harvard. Rhine's earliest investigations into the paranormal were not terribly successful. First he was disillusioned by the famous psychic Marjory, then duped by a horse named Lady Wonder, which he claimed in print to have telepathic abilities: in fact, the horse was responding to cues from her owner.[6] MacDougall moved to Duke University in 1927 and Rhine followed him. It was here that he performed the first of his experiments in extra-sensory perception (ESP)—a term that he coined.

At this stage in the game, the position of 'experimental psychology', as the philosopher C. D. Broad wrote in 1941, was akin to that[7]

of a woman with a shady past who has at length, after a hard struggle, settled down to a respectable life and got on visiting terms with the doctor's, the solicitor's, the vicar's, and even the squire's wife . . . She is fanatically determined to keep her hard-won respectability unsullied by the slightest breath of scandal. Physics, which has been honoured for centuries, can afford, like the scion of some noble house, to throw her cap over the mills; but poor dear psychology feels that she dare not take risks.

In Broad's view, the route to an invitation to tea from the squire was through 'precise statistical treatment', a point that Rhine also made when he wrote that parapsychology differed from 'psychic research in the strictly experimental methods used in its procedure'.[8]

Rhine tested telekinesis by dice throws and telepathy, and clairvoyance by 'Zener cards', so named because they had been invented by his Duke colleague Dr Karl Zener. Each Zener card was marked with a symbol: a circle, a rectangle, a star, a cross, or an image of waves. Each pack of twenty-five cards had five bearing each of the five symbols. To test for clairvoyance, Rhine would ask his subject to shuffle the cards, cut them, and then identify the symbol on the topmost card without looking at it. To test for telepathy he would think of a particular symbol, then ask the subject to read his thoughts. As he reported rather triumphantly in his first book, *Extra-Sensory Perception*, in these experiments he obtained results far better than chance alone could account for.[9]

Soal's innovations

Almost immediately upon its publication, Rhine's work provoked a barrage of criticism from within and without the psychic research community. The naysayers questioned everything from his statistical methodology, to the arrangement of the rooms in which he conducted his trials, to the thickness and cut of the Zener cards. His failure to use controls was faulted, as were the thoroughness of his shuffling and the reliability of the dice employed in the telekinesis testing. Dozens of ways in which his subjects might have cheated were proposed. For his part, Rhine averred that his priority was to

interest the subjects in the test, and create confidence in the possibility of doing well . . . My relations with my subjects were friendly, almost fraternal. We did hypnotic demonstrations, spent long hours in discussion, and pretty completely dissipated the constraints that usually exist for the student in the laboratory and in the presence of his instructor.

In addition to hypnosis, he tested some of his subjects under the influence of sodium amytal and caffeine.[10]

The trouble was that no other American researchers could reproduce Rhine's results. Nor could S. G. Soal, who in 1934 attempted a repetition of Rhine's experiment employing several safeguards that Rhine had not. Of these, the most notable was his decision to replace the usual packs of twenty-five Zener cards by cards in which the five symbols were not equally distributed. Instead, 'by the use of mathematical tables' Soal compiled a *random* sequence of 1000 card symbols, which he split into forty packs of twenty-five cards each. Among other things, this innovation allowed him to correct for the possibility that the guessers might be making sure not to call any one symbol more than five times. Unfortunately, in tests for both clairvoyance and telepathy, the results he obtained were disappointing, with only one subject, Mrs Gloria Stewart, obtaining a score high enough to be 'of any possible interest'.[11]

At this point Soal might well have given up, had not W. Whately Carington, a colleague from the SPR, insisted 'with remarkable pertinacity' that he re-examine his data and compare 'each guess, not with the card for which it was originally intended, but with the immediately preceding card and the immediately following card, and count up the hits'. As Gloria Stewart was 'the only "telepathy" subject who had shown any promise at all', Soal chose her results for reappraisal, calculating both her 'post-cognitive' and 'precognitive' hits. This was in 1939. What he discovered, he later wrote, was so 'remarkable' as to impel him to undertake a second round of experiments.[12]

Elephants, giraffes, lions, pelicans, and zebras

Soal conducted his new trials in London in 1940 and 1941 during the Blitz. His subject was Basil Shackleton, a photographer who had approached Soal several years before boasting of his telepathic prowess. Although Shackleton's scores in the original trials were unexceptional when compared with Mrs Stewart's, they too shot up when what Soal called the 'displacement effect' was taken into account.[13] For this round of testing, Soal dispensed with the Zener cards on the grounds that, 'after using them for five years' he 'had grown very sick of these somewhat arid diagrams'. The new cards he employed depicted five animals: an elephant, a giraffe, a lion, a pelican, and a zebra. Since wartime paper shortages made it impossible to have a thousand of them printed, Soal devised a method of testing that required only five cards.[14]

The new tests were conducted in Shackleton's photography studio, in two rooms with a connecting door that was left open. There were four participants: Shackleton, identified as the 'percipient' (P), an 'agent' (A), and two experimenters (EA and EP). In one of the rooms agent A sat across the table from EA, while in the other percipient P sat across a table from EP. Elaborate precautions were taken to safeguard against any non-telepathic communication among the participants: for instance, a plywood screen into which a 3-inch aperture had been cut separated EA from A, in front of whom the five cards were laid face down inside a box open on one side so that only she could see them.[15] Outside observers monitored the tests, which proceeded as follows.

On his side of the table, EA (nearly always Soal himself) would remove from his briefcase a list on which twenty-five random sequences of the digits 1, 2, 3, 4, 5, were written. (In most trials the list was prepared 'from the last digits of the seven-figure logarithms of numbers selected at intervals of 100 from Chambers's tables'; in a few cases Tippett's tables were used, or bone counters in five different colours were drawn from a cloth bag.[16]) EA would then call out 'one', look at the first number on the list, and hold a corresponding card up to the aperture for agent A to look at. If the card read 4, A would lift the fourth card from the left just high enough to see which animal it portrayed, and would then let the card fall back. At this point percipient P would write down a telepathic guess as to which animal the card depicted: E for elephant, G for giraffe, L for lion, P for pelican, or Z for zebra. When the session was concluded, EA and EP would check the order of the cards and correlate the letters that P had written down with the numbers that EA had called, taking note of all 'direct hits', 'precognitive hits', and 'post-cognitive hits'.[17] From a statistical standpoint the results were hugely impressive, with Shackleton scoring far above chance for both precognitive and post-cognitive hits.

Parapsychology in the 1940s

How did Turing come upon his knowledge of parapsychological research? It seems improbable that he was a reader of *The Proceedings of the Society for Psychical Research*, where Soal published his findings.[18] Nor was he likely to have been in the audience in 1947 when Soal gave the SPR's Myers Memorial Lecture and presented another set of impressive results—obtained during trials with the aforementioned Gloria Stewart. The most likely scenario is that Turing became acquainted with Soal's research when he was in Cambridge in 1947–48, possibly through the agency of C. D. Broad. A Fellow of Trinity College, Broad was an active member of the SPR, a friend of Bertrand Russell and G. H. Hardy and, like Turing, unapologetically gay. In 1945 he had published a strong endorsement of Soal's work in the journal *Philosophy*, describing it as providing 'evidence which is statistically overwhelming for the occurrence not only of *telepathy*, but of *precognition*'.[19] 'Overwhelming' was, of course, the word that Turing used in 'Computing machinery and intelligence' to characterize the evidence for ESP.

In 1949, the year that Turing moved from Cambridge to Manchester, a flurry of attention was paid to ESP. In September alone the word 'telepathy' appeared sixteen times in *The Times* of London: before that, it had not appeared in *The Times* since 1932.

Two events had provoked this sudden surge of interest. The first was an address by the esteemed zoologist Alister Hardy in Newcastle, who declared:[20]

If telepathy has been established, as I believe it has, then such a revolutionary discovery should make us keep our minds open to the possibility that there may be so much more in living things and their evolution than our science has hitherto led us to expect.

The second was a series of eight broadcasts on the BBC's 'Light programme', featuring the Piddingtons, an Australian couple who claimed to be able to communicate with each other telepathically. Among other feats, Sydney Piddington in one BBC studio imparted to his blindfolded wife in another studio the name of one of fifteen film stars, songs, and (in homage to Rhine) diagrams; Lesley Piddington in turn correctly guessed that the film star was John Mills, the song was 'On our return', and the diagram was a square. The broadcasts provoked a media frenzy, with articles about the Piddingtons running not just in the tabloids but in such highbrow publications as *The New Statesman*. The couple inspired cartoons, jokes, and even a catchphrase: 'to do a Piddington'.[21]

Needless to say the SPR was aware of, and had strong opinions about, both Hardy's speech and the Piddingtons' broadcasts. On 14/15 September successive letters ran in *The Times* under the headline 'TELEPATHY'. The first, supporting Hardy, was written by S. G. Soal. The second, criticizing the BBC for its sponsorship of the Piddingtons, was sent under the aegis of the SPR. This letter, with Broad as one of its three signatories, read in part:[22]

The faculty known as telepathy has for over 70 years been the subject of scientific investigation, and lately of experiment under laboratory conditions . . . It would be most regrettable if any confusion were created in the public mind between serious researches of this kind and the performances which the B.B.C. has been broadcasting.

Although *The Times* itself drew no explicit connection between Hardy's lecture and the Piddingtons' 'performances', an unsigned article in the 24 September issue of *Nature* did:

As a result of the recent work on telepathy mentioned by Prof. Hardy, public interest in the subject has been re-awakened. This, unfortunately, has led to a wave of popular credulity which has become somewhat disturbing to workers in psychical research, since it has affected persons of education and even scientific standing in their own particular fields.

In the view of the SPR, the crime of which the Piddingtons were guilty was declining to participate in laboratory tests of their capacities at the SPR's headquarters.[23]

Such then was the situation with parapsychology in England when Alan Turing began work on 'Computing machinery and intelligence'. Not only was ESP in the news, but reputable scientists were asserting that it would only be a matter of time before it would be proven to be a fact of nature. The posture that these scientists took was not dissimilar to Turing's own in regard to machine intelligence. Like him they saw themselves as being on the vanguard, like him they understood that their arguments would meet with considerable resistance, and, like him, they felt the need to defuse this resistance in advance.

Turing and spiritualism

The tone of Turing's reply to argument 9, like so much of 'Computing machinery and intelligence', was jocular. Earlier in the paper he had dispensed briskly with what he termed the 'heads-in-the-sand' objection to machine intelligence—'The consequences of machines thinking would be too dreadful. Let us hope and believe that they cannot do so.'—on the grounds that this objection 'was not sufficiently substantial to require refutation. Consolation would be more appropriate; perhaps this should be sought in the transmigration of souls.'[24] Now, in responding to the 'overwhelming' evidence for ESP, Turing postulated, and then rejected, a similar 'heads-in-the sand' response:[25]

One can say in reply that many scientific theories seem to remain workable in practice, in spite of clashing with E.S.P.; that in fact one can get along very nicely if one forgets about it. This is rather cold comfort, and one fears that thinking is just the kind of phenomenon where E.S.P. may be especially relevant.

Thinking (in particular, machine thinking) was, of course, a matter of paramount importance to Turing. The dilemma that ESP poses is how[26]

to rearrange one's ideas so as to fit these new facts in. Once one has accepted them it does not seem a very big step to believe in ghosts and bogies. The idea that our bodies move simply according to the known laws of physics, together with some others not yet discovered but somewhat similar, would be one of the first to go.

In fact, Turing had addressed this very question twenty years earlier. The occasion was an odd little essay entitled 'Nature of spirit' that he wrote in the spring of 1932 while on a visit to the Clock House, the family home of his beloved friend Christopher Morcom who had died two years previously. Turing wrote:[27]

We have a will which is able to determine the action of the atoms probably in a small portion of the brain, or possibly all over it. The rest of the body acts so as to amplify this. There is now the question which must be answered as to how the action of the other atoms of the universe are regulated. Probably by the same law and simply by the remote effects of spirit but since they have no amplifying apparatus they seem to be regulated by pure chance. The apparent non-predestination of physics is almost a combination of chances.

'*Almost* a combination of chances': as the will 'probably' plays a role in the functioning of the brain, so 'the remote effects of spirit' may have some effect on the atoms that comprise the rest of the universe, even though these seem to be regulated by pure chance. 'As McTaggart shows', Turing continued, 'matter is meaningless in the absence of spirit'.

Such a notion of 'spirit' as a sort of cosmic gasoline was very much in the air in the Cambridge of 1932, a period when the philosopher G. E. Moore, and through him the ghost of the late J. M. E. McTaggart (not coincidentally Broad's mentor), exerted a tremendous influence. Twenty years on, Turing would refer to 'the transmigration of souls' ironically, as a source of 'consolation' for those who could not tolerate the inevitability of machine intelligence. Here his treatment of the idea is in earnest:

Personally I think that spirit is really eternally connected with matter but certainly not always in the same kind of body. I did believe it possible for spirit at death to go to a universe entirely separate from our own, but now I consider that matter and spirit are so connected that this would be a contradiction in terms ... Then as regards the actual connection between spirit and body I consider that the body by reason of being a living body can 'attract' and hold on to a 'spirit', whilst the body is alive and awake the two are firmly connected. When the body dies the 'mechanism' of the body, holding the spirit, is gone and the spirit finds a new body sooner or later, perhaps immediately.

Such words would not have been out of place in the *Journal of the Society for Psychical Research*. Nor would the SPR have failed to take an interest in Turing's account of his own 'crisis apparition' on the occasion of Christopher Morcom's death. At a quarter to three one morning, the ringing of an abbey bell drew Turing out of bed and to his dormitory window, where he looked at the moon and thought that it was giving him a sign: 'Goodbye to Morcom'.[28] A little more than a week later Morcom was dead, and Turing wrote to his own mother:[29]

I feel sure that I shall meet Morcom again somewhere, & that there will be some work for us to do together even as I believed there was for us to do here. Now that I am left to do it alone I must not let him down but put as much energy into it, if not as much interest, as if he were still here. If I succeed I shall be more fit to enjoy his company than I am now.

Turing and parapsychology

That Turing was conversant in the language of parapsychology, so much in the air in 1949, is evident from the examples he gave of how ESP might affect the results of his test for machine intelligence:[30]

Let us play the imitation game, using as witnesses a man who is good as a telepathic receiver, and a digital computer. The interrogator can ask such questions as 'What suit does the card in my right hand belong to?' The man by telepathy or clairvoyance gives the right answer 130 times out of 400 times. The machine can only guess at random, and perhaps gets 104 right, so the interrogator makes the right identification.

The scenario that Turing drew here is almost identical to the one used for the 'scientific' testing of telepathy, except that in this case the 'agent' is a computer. Through telepathy or clairvoyance the 'percipient' is able to make the correct identification more often than a probability calculation would suggest. Meanwhile, the computer has no choice but to assume the role of the non-telepathic guesser, a role to which it would be particularly well suited if it had built into it a 'random number generator'. (As it happened, the Ferranti Mark I, on which Turing was working in 1949, had been equipped with just such an element—at Turing's insistence.[31]). As Turing observed:[32]

There is an interesting possibility which opens here. Suppose the digital computer contains a random number generator. Then it will be natural to use this to decide which answer to give. But then the random number generator will be subject to the psycho-kinetic powers of the interrogator. Perhaps this psycho-kinesis might cause the machine to guess right more often than would be expected on a probability calculation, so that the interrogator might be unable to make the right identification. On the other hand, he might be able to guess right without any questioning, by clairvoyance. With E.S.P. anything might happen.

In my view this is among the most puzzling passages in 'Computing machinery and intelligence'. For here it is the interrogator, *not* the percipient, to whom Turing is attributing the capacity to rig the imitation game, either by using his telekinetic power to de-randomize the guesses made by the random number generator, or by determining which of the two players is the human by the covert use of telepathy or clairvoyance. To make matters stranger, in the first of these scenarios the interrogator appears to be exerting his telekinetic influence unconsciously—why otherwise would the result be that he 'might still be unable to make the right identification'?— and in the second he is cheating outright. (I presume here that the interrogator, in making the identification by clairvoyance, is fully conscious that he is doing so. If not—if he makes the identification by clairvoyance, but thinks he is making an educated deduction—the scenario is the same.) But why would the interrogator want to cheat? In order to deprive the computer of its opportunity to win? In this case, is the interrogator someone so threatened by the idea of machine intelligence that he is willing to sabotage the game in order to stop it?

Cheating humans and cheating machines

One facet of Turing's thinking into which argument 9 gives us some insight was his willingness to allow for cheating, deception, and lying as viable strategies in the playing of the imitation game. In this he followed once again the example of the 'psychists', for whom cheating was more of the order of an occupational hazard than an unpardonable offence. As the SPR's members had discovered early on, most mediums—even authentic ones—cheated occasionally, out of pique, laziness, or fatigue. A good example was the famous Italian medium Eusepia Palladino, in whose psychic powers William James (among others) believed utterly, writing in 1909:[33]

Everyone agrees that she cheats all the time. The Cambridge experts . . . rejected her *in toto* on this account.

Was this justified? James thought not:

Falsus in uno, falsus in omnibus, once a cheat, always a cheat, such has been the motto of the English psychical researchers in dealing with mediums. I am disposed to think that, as a matter of policy, it has been wise. Tactically it is far better to believe much too little than a little too much . . . But however wise as a policy the S.P.R.'s maxim may have been, as a test of truth I believe it to be almost irrelevant. In most things human the accusation of deliberate fraud and falsehood is grossly superficial. Man's character is too sophistically mixed for the alternative of 'honest or dishonest' to be a sharp one. Scientific men themselves will cheat—at public lectures—rather than let experiments obey their well-known tendency towards failure.

I find this a fascinatingly perverse paragraph. In an essay whose ostensible aim is to lend legitimacy to psychic research, the esteemed Professor James of Harvard ended up, if not condoning cheating outright, then exhibiting a far more lenient attitude toward it than the institution with which he was affiliated. He even went so far as to admit that he himself had 'cheated shamelessly'—and all this toward the goal of rehabilitating the reputation of an Italian medium considered by most to be a charlatan. Or was James up to something more here?

I would argue that as he approached the end of his life James had decided that he no longer cared to deny the dirty human secret that the impulse to cheat is an innate one, and as such is an impulse that any thinking machine ought to be able to mimic: this is *almost* (but not quite) what Turing said. True, his thinking machines would be programmed to cheat, but not gratuitously as Eusepia Palladino had done. Rather, the machine would cheat *strategically*. When asked to solve arithmetic problems, it 'would deliberately introduce mistakes in a manner calculated to confuse the interrogator'.[34] When asked whether it was 'only pretending to be a man', it would be 'permitted all sorts of tricks so as to appear more man-like, such as waiting a bit before giving an answer, or making a spelling mistake . . . '.[35] *Permitted* is the key word here, since the cheating in this case was not meant to be detectable—that is to say, it is not its penchant for cheating per se that will convince the interrogator that the machine is human, but rather it is the impression of human fallibility that the cheating facilitates. Unless, of course, either the interrogator or the human player is telepathic: then the jig will be up for the machine, for a telepath can see, if not into the machine's circuitry, then into that of its human competitor. Were telepaths to admit their gifts, they would be disqualified from playing. But what if they did not? The only solution that Turing could come up with was an impracticable one: 'To put the competitors in a "telepathy-proof room" would satisfy all requirements'.[36]

Fair play to the machines

What really motivates all this speculation about telepathy, I think, is Turing's anxiety about the possibility of thinking machines being subjected to bigotry and persecution. As early as 1947 he had been calling for 'fair play to the machines', even as he worried about 'the great opposition' that the thinking machine would meet—in particular 'from the intellectuals who were afraid of being put out of a job':[37]

It is probable that the intellectuals would be mistaken about this. There would be plenty to do [trying to understand what the machines were trying to say], i.e. in trying to keep one's intelligence up to the standard set by the machines, for it seems probable that once the machine thinking method has started, it would not take long to outstrip our feeble powers.

Among other advantages that the machines would have would be immortality—'There would be no question of the machines dying'—and the capacity to 'converse with each other to sharpen their wits': conversations, presumably, on which humans could no more listen in than they could detect the hum of a random number generator at work. On the one hand, such a scenario seems to have provoked in Turing a mischievous delight that he could not quite resist voicing: 'At some stage therefore we should have to expect the machines to take control, in the way that is mentioned in Samuel Butler's "Erewhon" '. On the other, his fear for the machines' fate comes across as painfully personal:[38]

Many people are extremely opposed to the idea of [a] machine that thinks, but I do not believe that it is for any of the reasons that I have given, or any other rational reason, but simply because they do not like the idea. One can see many features which make it unpleasant. If a machine can think, it might think more intelligently than we do, and then where should we be? Even if we could keep the machine in a subservient position, for instance by turning off the power at strategic moments, we should, as a species, feel greatly humbled.

It is as if Turing, in envisioning what might happen to the machine, was displaying his own capacity for precognition. For the fate that he feared that the *machines* would suffer—the deliberate circumscription of their freedom—was to be, in just a few years, his own.

Randomness and free will

The solution to the puzzle of argument 9 may lie in the random generator itself. As Turing explained, its principal function was not to pull Zener-card symbols out of a hat, but to create the impression that the machine had 'free will'. As he explained in 'Intelligent machinery, a heretical theory':[39]

Each machine should be supplied with a tape bearing a random series of figures, e.g. 0 and 1 in equal quantities, and this series of figures should be used in the choices made by the machine. This would result in the behaviour of the machine not being by any means completely determined by the experiences to which it was subjected, and would have some valuable uses when one was experimenting with it.

Randomness provides the computer with a means of giving the appearance that it functions like a human brain, and in the imitation game appearances are everything. Or perhaps free will in human beings is also illusory—a notion that Turing entertained in a 1951 talk on BBC radio:[40]

It may be that the feeling we all have of free will is an illusion. Or it may be that we really have got free will, but yet there is no way of telling from our behaviour that this is so. In the latter case, however well a machine imitates a man's behaviour it is to be regarded as a mere sham. I do not know how we can ever decide between these alternatives, but whichever is the correct one it is certain that a machine which is to imitate a brain must appear to behave as if it had free will, and it may be asked how this is to be achieved. One possibility is to make its behaviour depend on something like a roulette wheel or a supply of radium.

In 1932 Turing had written of a 'will which is able to determine the action of the atoms probably in a small portion of the brain, or possibly all over it', but also of a 'spirit' which he believed to be 'really eternally connected with matter'. His conclusion was: 'The body provides something for the spirit to look after and use'.[41] So what of the thinking machine? It seems that in this case chance, in the form of the random number generator, takes over the role of spirit. Random numbers (the imaginary die thrown again and again) give the paradoxical impression of free will. Yet does this mean that, in Turing's view, chance and spirit are synonymous? Or is there in this case an 'amplifying apparatus' by means of which the programmer, if no one else, can perceive 'remote effects of spirit' that only resemble chance to the naked eye? As Turing himself knew full well, chance could be faked as easily as it can be manipulated. As he noted in his 1951 BBC talk, 'It is not difficult to design machines whose behaviour appears quite random to anyone who does not know the details of their construction'.

It turned out that S. G. Soal was also cooking the books. In 1973, two years before his death, an article in *Nature* revealed that in his 1940–41 experiments with Basil Shackleton he had systematically changed 1s to 4s and 5s on the score sheets, thereby tilting the statistical balance in favour of the percipient. In this effort he may have been assisted by his agents, in particular 'Rita Elliott', whom he subsequently married. It was his own imitation game.[42]

PART VI

Biological growth

CHAPTER 33

Pioneer of artificial life

MARGARET A. BODEN

lan Turing's 1952 account of 'The chemical basis of morphogenesis'[1] pioneered the use of reaction–diffusion equations in explaining the origin of bodily form. This exciting paper showed how it is possible for various familiar biological structures to develop by referring to chemical 'morphogens' whose actual identity was still unknown. Some forty years later, his work began to inspire research in computer graphics, artificial life, and structuralist developmental biology: the follow-up had needed to await huge advances in computer power, computer graphics, and experimental biochemistry. Turing's unpublished mathematical notes found after his death are still not fully understood: they very probably contain further insights into biological form.[2]

Introduction

In 1794 the Romantic poet William Blake asked a deep question:[3]

> Tyger! Tyger! burning bright
> In the forests of the night,
> What immortal hand or eye
> Could frame thy fearful symmetry?
> . . . Did He who made the Lamb make thee?

The answer was a mystery. But Blake apparently assumed that it must involve *some* supernatural power—perhaps the Christian God. Most of his contemporaries would have agreed with him.

Even 150 years later, in the middle of the 20th century, this answer was still widely accepted. But by this time the study of embryology had progressed enough to suggest that the marvellous development of an undifferentiated egg-cell into an embryo of increasingly complex form could in principle be given a scientific answer. Moreover, embryologists believed that the answer would refer to chemicals—called 'organizers'—that cause successive changes in the cytoplasm as the egg gradually develops into organs of differing types.

But this was about as far as it went. Just what those so-called organizers were was unknown; how they might be able to produce not merely chemical changes but also novel patterns and shapes—like the tiger's stripes and its magnificent muscles and head—was equally mysterious.

The then-fashionable talk of 'self-organization' served more to prohibit reference to Blake's immortal hand or eye than to offer specific answers: indeed, that notion was a highly abstract one. It was understood as the spontaneous emergence (and maintenance) of order out of an origin that is ordered to a lesser degree—where 'spontaneous' meant not magical or supernatural, but somehow caused by the inner nature of the system itself. However, the 'order' and 'origin' (not to mention 'emergence' and 'maintenance') were unspecified. In short, the concept of the organizer was an optimistic place-marker for a future scientific explanation: a tantalizing we-know-not-what, working we-know-not-how.

Only two years before his tragic death, Alan Turing's 1952 paper provided the seed of the answer. He too could specify no chemicals: instead of unknown 'organizers' he spoke of equally unknown 'morphogens' (from the Greek 'form-originators'), so he didn't solve the mystery of the *what*. However, he did solve (at least in outline) the even greater mystery of the *how*—'even greater', because if embryologists had discovered that *this* or *that* chemical was involved, so what? How could *any* chemical possibly generate the amazing changes of form seen in the developing embryo?

As we shall see, his answer was largely ignored for nearly forty years, and as a result many highly educated people today have not even heard of it. They have, of course, heard of Turing. Indeed, as a present-day Thomas Macaulay might put it, 'every schoolboy knows' (and every schoolgirl too) that Turing pioneered the theory of computer science, helped to design the first computers, outlined the research programme for future artificial intelligence (AI), and offered a still-provocative challenge (the Turing test) to those who denied that computers could ever think.[4] These contributions have entered so deeply into our culture that June 2012 saw a host of centenary meetings, and even a UK postage stamp, celebrating Turing's work in these areas—not to mention his history-changing codebreaking in the Second World War. But, as of today, his work on embryology remains largely unknown outside biological circles.

However, what 'every schoolboy knows' changes as the years go by: some items simply fall by the wayside. Macaulay's 'every schoolboy' was said to know 'who imprisoned Montezuma, and who strangled Atahualpa'. Maybe in 1840 they did, but schoolchildren today wouldn't know either of those things. (Do you?)

Other items get added to the list. Turing's developmental biology will have found a place there long before his bicentenary in 2112, for today it is at last being explored by biologists with increasing excitement and success. It is being applied at all levels: from genetics and cellular morphology, via embryology, to adult form—and not least, to neuroscience. The neuroscientist Jack Cowan, for instance, has remarked on 'a very close relationship' between Turing's theoretical ideas and the brain.[5] Turing would have been pleased by this, for in his 1952 paper he mentioned cerebral self-organization as among the important biological phenomena to be explained.

Given that biology has replaced physics as the science capturing most of the public attention (and funding), Turing's reputation in this context will surely continue to grow. One does not need to be a biologist to know something about his biology, any more than one needs to be a physicist to know something about Einstein's theory of relativity. The main reason why 'every [future] schoolboy' will know about Turing's embryonic embryology is that his question

(Blake's question) marks one of the most intriguing—and most in-your-face unavoidable—puzzles of biology. Only the nature of life itself is deeper.

Turing's paper on the chemical basis of morphogenesis contained what, in conversations with friends, he called 'my mathematical theory of embryology'. It concerned questions that he had been pondering ever since his youth—for instance, taking a rest from his wartime work on the Germans' Enigma code, when he lay on the grass of Bletchley Park picking daisies and carefully counting their petals.

How daisy petals might be relevant to mathematics isn't immediately obvious. And what about tigers or lambs? (Remember Blake's question.) Are they relevant too? Having written the first programming manual for the Manchester 'Baby' computer in 1950, why did Turing irritate his Manchester colleagues by neglecting his duties in the computing laboratory there in favour of this other interest? And why did he look forward eagerly to the delivery of the first Ferranti computer in February 1951, planning to use it immediately for calculations to be reported in his about-to-be-submitted biological manuscript?

Well, read on.

Turing's mathematical theory of embryology

Turing was nothing if not a mathematician—and he knew it. His embryology paper, published in the biological strand (Series B) of the Royal Society's *Philosophical Transactions*, was a precursor of certain aspects of what is now called complexity theory. It contained specific warnings for its biologist readers about the difficulty of the mathematics involved. It gave directions about which sections to read or omit, depending on one's level of mathematical literacy. It would be fully understood, he said, only by those 'definitely trained as mathematicians'.[6] In fact, even what he termed 'only a very moderate understanding of mathematics' was doubtless too much to expect from many of his readers.

Nonetheless, the general message was clear enough. It was also intoxicating. For the paper showed how biological structure or form could arise out of a homogeneous origin—in particular, how cell differentiation and (eventually) various organs could emerge from the unstructured protoplasm of the fertilized ovum. (As his Abstract put it: how 'the genes of a zygote may determine the anatomical structure of the resulting organism'.[7])

As we saw earlier, the origin of biological form (morphogenesis) was still hugely mysterious at that time. Hard-headed biologists spoke of vaguely conceptualized 'morphogenetic fields' governing cell differentiation, controlled by hypothetical, presumably biochemical, forces called 'organizers'. But just what these were, and just how they might work, was an enigma much harder to crack than the Enigma code.

In other words, embryologists' talk of organizers and morphogenetic fields was largely empty. It certainly didn't answer the question posed forty years before in a children's book which (so he told his mother) had first awakened Turing to science:[8]

The body is a machine. It is a vastly complex machine, many, many times more complicated than any machine ever made with hands; but still after all a machine . . . [But if we ask how its] living bricks find out when and where to grow fast, and when and where to grow slowly, and when and where not to grow at all [we must admit that this is something which] nobody has yet made the smallest beginning at finding out.

Turing's intoxicating paper clearly supplied that 'smallest beginning'. And although it was only a beginning, only a seed, it gave very good reason to believe that later work along similar mathematical lines would answer a host of questions in embryology, and in developmental biology in general.

In his thinking about morphogenesis, Turing followed the path that had been pioneered by the highly original mathematical biologist D'Arcy Thompson:[9] indeed, Thompson's book *On Growth and Form* was one of only six references that he cited.[10] Thompson had insisted that the most basic aspects of biological form arise because of physics. For instance, a creature like a water-boatman (an insect that walks on water) has to have bodily dimensions and body parts that exploit the constraints of surface tension, but its body plan can in effect ignore the force of gravity. An elephant, by contrast, can forget about surface tension but not about gravity. (Some parts of its internal organs, of course, may be highly constrained by surface tension; but Thompson was speaking here of its overall bodily form.) Turing, too, appealed to fundamental principles of physics and chemistry to explain the origin of form in biology.

In an earlier paper,[11] Turing had already asked how organization could arise in unorganized neural networks, but there he had assumed some outside interference or training procedure. Now he considered the origin of organization without outside interference—in other words, the seemingly paradoxical phenomenon of self-organization. In particular, he asked how homogeneous cells can develop into differentiated tissues, and how these tissues can arrange themselves in regular patterns on a large scale, such as stripes or segments.

Turing didn't claim to know the chemical details: if the embryologists couldn't identify the organizers, neither could he. He spoke of a 'leg-evocator', for instance,[12] but that was just a place-holder for some as-yet-unknown molecule. His achievement was to prove, in mathematical terms and with the help of computer calculations, that relatively simple chemical processes could in principle generate biologically relevant order from homogeneous tissue. These processes involve chemicals whose mutual interactions as they diffuse throughout the system can sometimes destroy or build each other.

Since no one knew just which chemicals these might be, Turing referred to them simply as 'morphogens'. He showed that the interactions between two or more morphogens, each initially distributed uniformly across the system, could eventually produce waves of differing concentrations. This can happen in non-living systems (chemicals diffusing in a bowl, for instance), as well as in living creatures. But Turing suggested that in an embryo or developing organism a succession of these processes might prompt the appearance of ordered structures such as spots, stripes, tentacles, or segments—culminating in the highly differentiated organs of the adult creature.

This may seem like magic: how can difference arise from homogeneity? Turing allowed that a perfectly homogeneous system in stable equilibrium would never differentiate. But if the equilibrium is unstable, even very slight disturbances could trigger differentiation.

As he pointed out, some disturbances are inevitable, given that living matter is subject to the laws of physics, as D'Arcy Thompson had insisted. For example, random Brownian motion within the cell fluids must vary the pairwise interactions between the molecules, and molecules will be slightly deformed as they pass through the cell wall. Even minute disturbances like these could upset the initial (unstable) equilibrium.

Turing gave differential equations defining possible interactions between two morphogens. Various terms specified the initial concentrations of the two substances, their rates of diffusion, the speed at which one could destroy or build up the other, the random disturbances involved,

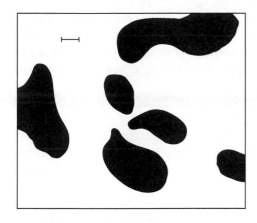

Figure 33.1 Dappling produced by diffusion equations; this picture was hand-drawn by Turing, on the basis of numerical calculations done using his desk calculator.

Reproduced from A. M. Turing, 'The chemical basis of morphogenesis', *Philosophical Transactions of the Royal Society (B)*, 237 (1952), 37–72, on p. 60.

and (when dealing with multicellular structures) the number and spatial dimensions of the cells. He showed that certain numerical values of these terms would result in ordered structures having biological plausibility. For instance, irregular dappling or spot patterns (such as are seen on dalmatians and some cows) could result from two morphogens diffusing on a plane surface (Fig. 33.1).

As another example, diffusion waves of two morphogens within a twenty-cell ring could give rise to regularly spaced structure, reminiscent of the embryonic beginnings of circular patterns of cilia, tentacles, leaf buds, or petals—or segments, if the ring were broken. Three developmental stages within this twenty-cell body are represented in Fig. 33.2: the initial homogeneous equilibrium (dotted line), early signs of an emerging pattern (hatched line), and the final equilibrium (unbroken lines).

Turing suggested that order could be generated in three dimensions also. For example, diffusion waves could cause embryonic gastrulation, in which a sphere of homogeneous cells develops a hollow (which eventually becomes a tube), and interactions between more than two morphogens could produce travelling waves, such as might underlie the movements of a spermatozoon's tail.

There were two reasons why he focused on the origin of form from a homogeneous source. One was that this is the most fundamental (and therefore the most puzzling) case. The second

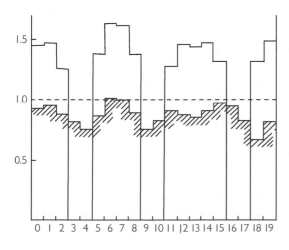

Figure 33.2 Concentrations of a morphogen diffusing in a twenty-cell ring. This picture was also drawn by Turing, but represents calculations so taxing that they had to be done on the Manchester computer;[13] that was why (as remarked earlier) he had awaited the delivery of the machine so impatiently.

Reproduced from A. M. Turing, 'The chemical basis of morphogenesis', *Philosophical Transactions of the Royal Society (B)*, 237 (1952), 37–72, on p. 60.

was that he couldn't give a mathematical account of the origin of novel structure from some prior structure, and that was a significant omission. For, as he pointed out, most examples of structural development in biology involve the emergence of new forms (new patterns) from pre-existing ones. (Think, for instance, of the appearance of fingers and then fingernails on what were initially merely limb-buds.)

There was no notion, at that time, of the switching on and off of individual genes by other ('regulatory') genes.[14] Nevertheless, Turing suggested that a succession of different chemicals (morphogens) is brought into play as development proceeds. To understand the process, however, one would need to understand not only the simple diffusion reactions between morphogens but also how (later) morphogens can react with pre-existing structures. This he couldn't do—at least not yet, as we shall see.

This imaginative paper, despite its many pages of challenging (and for many people, unfathomable) mathematics, was read with excitement by embryologists. It proved beyond doubt what D'Arcy Thompson had suggested: that self-organization of a biologically plausible kind could in principle result from relatively simple chemical processes describable by mathematics. It intimated *just which* immortal hand had framed the tiger's magnificent stripes and symmetry—namely, the laws of physics, understood as reaction–diffusion equations. And it was sufficiently general to cover lambs as well as tigers. (Blake's 'Lamb', of course, had symbolized Jesus: even Turing's mathematics couldn't cover that.)

Moreover, it suggested an exciting programme of biological research. For many years, however, that excitement was doomed to disappointment. It was almost forty years before the research programme implied by Turing's ideas was taken up.

Turing and modern biology

Turing's influence on modern biology is evident in three largely overlapping areas, each of which covers a huge range of fascinating questions. The first concerns self-organization in general—including brain development, for instance, briefly mentioned earlier. The second is the computer-based field of artificial life (A-Life). The third is an unorthodox approach in experimental biology known as 'structuralism'.

A-Life is a form of mathematical biology.[15] It employs computational and dynamical ideas, and computer models of various kinds, to understand living processes in general (including development and evolution), and specific biological phenomena too, such as flocking, hexapod locomotion, or the location of mates by hoverflies and crickets. Many A-Life scientists even hope to understand the nature of 'life as it could be', not merely of 'life as it is'.[16]

That hope is reminiscent of D'Arcy Thompson's goal in biology, which encompassed *all possible* organisms:[17]

[I] have tried in comparatively simple cases to use mathematical methods and mathematical terminology to describe and define the forms of organisms . . . [My] study of organic form . . . is but a portion of that wider Science of Form which deals . . . with forms which are theoretically imaginable.

Moreover, it is clear that D'Arcy Thompson would have become involved in A-Life had he lived longer (he died in 1948, the year in which the Manchester Baby became operational). He himself had said:[18]

Our simple, or simplified, illustrations carry us but a little way, and only half prepare us for much harder things . . . *If the difficulties of description and representation could be overcome* . . . we should at last obtain an adequate and satisfying picture of the processes of deformation and the directions of growth.

The concepts provided by computer science, and the experimental platforms provided by computer models, were a large part of what was needed to overcome those 'difficulties of description and representation' in broadly mathematical terms. It is hardly surprising then that Turing—the father of computer science, long fascinated by the biological questions that had so intrigued D'Arcy Thompson—saw the opportunity and grasped it.

His 1952 'mathematical embryology' paper is now recognized as an early essay in A-Life— one that has inspired many later projects; in fact, it was his only published essay on A-Life, although unpublished drafts were discovered after his death (as discussed later). But remarks in some of his other writings presaged insights and techniques now thought of as characteristic of A-Life.

Three such remarks appeared in the *Mind* paper that is usually remembered for introducing the Turing test, but which was primarily intended as an outline manifesto for AI.[19] They concerned the classification of computational systems, the possibility of evolutionary computing, and the role of chemicals in the brain.

As for the first remark, he outlined the four-fold classification of complexity that is now normally attributed to Stephen Wolfram.[20] A machine's behaviour, he said, could be 'completely random', or 'completely disciplined', or involve 'pointless repetitive loops'; intelligent behaviour, he suggested, would require a 'rather slight departure' from the highly disciplined case.[21] Wolfram investigated such differences thoroughly. After experimenting with a wide range of 'bottom-up' computational systems called cellular automata (CA), he distinguished four types:

- those that eventually reach stasis
- those that settle into rigidly periodic behaviour
- those that remain forever chaotic
- those that achieve order that is stable without being rigid, so that the structural relations between consecutive states are varied yet intelligible.

The final category, he said, included life.

Many A-Life researchers agree with Wolfram's claim that a certain level of complexity, involving both order and novelty, is needed for life in general, and for computation too: life is possible, as they put it, only 'at the edge of chaos'.[22] The Turing-inspired biologists, whom we mention briefly later, accept this claim also, applying the mathematics of complexity to a host of empirical examples.[23]

As for the second remark, in his report to the Laboratory (NPL), where he designed the ACE computer, Turing had already said that 'intellectual activity consists mainly of various kinds of search', of which 'genetical or evolutionary' search—a technique widely used in A-Life today— was one possibility.[24] When he wrote his *Mind* paper, the NPL report was still largely unknown. But he hinted at the possibility of evolutionary computing, saying that evolution was analogous to the use of 'mutations' and 'random elements' in a learning machine.[25]

As for the third remark, he said that 'in the nervous system chemical phenomena are at least as important as electrical',[26] but he didn't elaborate. He didn't hint that computers might one day be able to model the action of neurotransmitters in the brain. Even more to the point, he didn't say that he was already working on a computable theory of how chemicals can guide the organization of the brain and other bodily structures.

At the time, scant notice was taken of these insights. This is understandable, for they were presented as mere passing remarks. Moreover, Turing admitted that his 1950 paper offered no more than 'recitations tending to produce belief', as opposed to 'convincing arguments of a positive nature' about what practical advances might be made in AI and A-Life. By contrast, his 1952 publication for the Royal Society provided pages of mathematical proof to back up its startling claims.

An early example of A-Life work, directly inspired by Turing's 'mathematical theory of embryology', is due to Greg Turk.[27] Turk used a computer to vary the numerical parameters in Turing's own equations, to calculate the results, and to apply two or more equations successively (thus addressing the 'pattern-from-pattern' problem that Turing hadn't solved). In this way, Turing's equations have generated spot patterns, stripes, and reticulations resembling those seen in various living creatures.

Some of the structures produced in these computational experiments are shown in Fig. 33.3. The large and small spots in the upper half of the picture result from changing the size parameter in Turing's reaction–diffusion equation. If the large-spot pattern is frozen and the small-spot equations then run over it, we get the 'cheetah spots' (bottom left). The 'leopard spots' (bottom right) are generated by a similar two-step process, except that the numbers representing the concentrations of chemicals in the large spots are altered before the small-spot equation is run.

These cascades of reaction–diffusion systems, kicking in at different times, are reminiscent of the switching on and off of genes, and result in more naturalistic patterns than does a simultaneous superposition of the two equations. Similar cascades of a five-morphogen system generate life-like patterns, and the addition of rules linking stripe equations to three-dimensional contours produces moulded zebra stripes (Figs 33.4 and 33.5). Of course, the stripes on a real zebra (or on Blake's tiger) aren't quite like these, but this image is more realistic and more 'natural' than one showing straight (unmoulded) stripes would be.

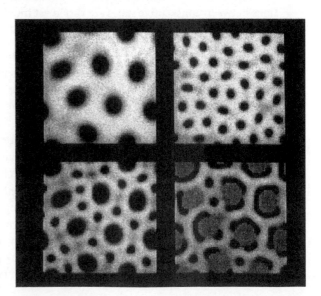

Figure 33.3 Naturalistic spot patterns.

Reproduced from G. Turk, 'Generating textures on arbitrary surfaces using reaction–diffusion', *Computer Graphics*, 25 (1991), 289–98, on p. 292.

Figure 33.4 Complex naturalistic patterns.

Reproduced from G. Turk, 'Generating textures on arbitrary surfaces using reaction–diffusion', *Computer Graphics*, 25 (1991), 289–98, on p. 292.

Figure 33.5 A zebra with three-dimensional algorithmic stripes.

Reproduced from G. Turk, 'Generating textures on arbitrary surfaces using reaction–diffusion', *Computer Graphics*, 25 (1991), 289–98, on p. 292.

The more recent Turing-inspired models of biological self-organization go far beyond Turk's intriguingly naturalistic spots and stripes, and some of these exemplify the controversial (but increasingly popular) approach known as structuralist biology.[28] The structuralists insist that Darwinian natural selection is a secondary factor in the explanation of biological form—in other words, it merely prunes the forms that have arisen as a result of physics and chemistry. This implication was not made explicit by Turing, but had been stressed by D'Arcy Thompson. Many developmental biologists today who draw on Turing's mathematical ideas follow Turing in leaving this implication unspoken. (One example is research on what Turing called 'leg-evocators'.[29]) Card-carrying structuralist biologists, however, make a point of criticizing neo-Darwinism. They argue that physics determines not only what forms are even possible, but also which are most likely, and so turn up over and over again across the phylogenetic tree—and they see genes (morphogens) as epigenetic modulators rather than deterministic instructions.

'Epigenesis' is a notion borrowed from the geneticist Conrad Waddington,[30] who was deeply influenced by D'Arcy Thompson and, like him, was one of the six authors cited in Turing's 1952 paper. Waddington described development and the activities of genes in terms not of rigidly pre-defined pathways, but of variable trajectories selected according to the current (ultimately biochemical) environment. Whereas he, like D'Arcy Thompson, could express this idea in only the broadest terms, today's structuralists can do so by using sophisticated experimental data, advanced mathematics, and the techniques of computer modelling.

For instance, consider the model of the development of the unicellular alga *Acetabularia* provided by Brian Goodwin,[31] an ex-pupil of Waddington. This uses computer graphics to present the numerical results of mathematical calculations as diagrams or pictures of the developing forms concerned. Like structuralist models in general, it illustrates metabolic and developmental functions and how they change over time.

Specifically, it simulates the cell's control of the concentration of calcium ions in the cytoplasm, and how this affects, and is affected by, other conditions, such as the mechanical properties of the cytoplasm, so as to generate one type of morphology or another (such as stalks, flattened tips, and whorls). It contains some thirty or more parameters, based on a wide range of experimental work. These reflect factors such as the diffusion constant for calcium, the affinity between calcium and certain proteins, and the mechanical resistance of the elements of the cytoskeleton. The model simulates complex iterative feedback loops wherein these parameters can change from moment to moment.

One welcome result of running this model was the appearance of an alternating pattern of high and low calcium concentrations at the tip of the stalk (in effect, a modern version of Turing's morphogen pattern), interpreted by Goodwin as the emerging symmetry of a whorl. This result was welcome largely because whorls aren't found only in *Acetabularia*: on the contrary, they are 'generic forms' found in all members of this group of algae and (as Turing himself had remarked) in many other organisms too.[32] In Goodwin's model it turned out that whorl symmetries were very easy to find: they did not depend on a particular combination of specific parameter values, but emerged if the parameters were set anywhere within a large range of values.

However, the nature of the computer graphics—whose diagrams were composed of many tiny lines—prevented the emergence of visually recognizable whorls.[33] A *whorl* is a crown of little growing tips, each of which then develops into a growing lateral branch. To simulate this, each of the little tips (calcium peaks) would have to behave (to soften, bulge, and grow) like the main tip, but on a smaller scale: in principle, that is possible, but it would need the machine to draw even more, and even tinier, lines. So it would be very costly computationally, requiring the whole program to be repeated on a finer scale (to generate each lateral branch), and many times over (to generate many laterals).

Why such a long wait?

The examples given so far should suffice to show that biology today owes a significant intellectual debt to the mathematician who cracked the Enigma code. What has not been explained is why the world had to wait for so many years before this debt was acknowledged.

Even now, it hasn't been acknowledged fully. As we noted at the beginning of this chapter, Turing's mathematical biology is less familiar to the general public than his ideas about computer science and AI. And it is not only the 'schoolboys' who are unaware of it—indeed, on mentioning his paper enthusiastically to various friends and colleagues on many occasions between the mid-1950s (when I first read it) and 1990, I discovered to my amazement that most of them had as little knowledge of its existence (not to mention its detail) as I had of Atahualpa. Even now, in the 21st century, that is still often true.

Nevertheless, as we have seen, many people are at long last waking up to his ideas on morphogenesis. But why only now? Why not immediately? After all, it's not as though his paper lay unread in an obscure journal hidden from public view, like Gregor Mendel's little gem on genetics.

One reason for this was pure bad luck—or bad timing. Only a year after Turing's paper appeared, Francis Crick and James Watson sent their epochal letter about the double helix to *Nature*—and only a few years after that, they cracked the genetic code. Their efforts in cryptography were at least as influential as Turing's, for they resulted in an immediate switch of interest on the part of biologists (including developmentalists) from whole-organism concerns to molecular biology.

To be sure, questions about biochemistry were pushed to the fore, and Turing's paper was a theoretical exercise in biochemistry. But the emphasis in the new discipline of molecular biology was more on the actual identity of the molecules (morphogens) than on their actions and interactions. As for questions about biological form, whether of body or cell, these were relegated to the background.[34]

But there were also more intellectually substantive reasons for the neglect of Turing's paper. We have seen that D'Arcy Thompson's mathematical approach was less influential than it might have been, because he lacked the concepts and the machines needed to explore its implications effectively. Partly thanks to his own earlier work, Turing possessed some of the concepts, and even a machine: the world's first modern-style digital computer. But that machine was unavoidably primitive.

Turing used the Manchester University computer to calculate the successive steps of interaction between the morphogens, but this was a time-consuming and tedious matter. Moreover, in the absence of computer graphics, the computer's numerical results had to be laboriously converted by Turing by hand to a more easily intelligible visual representation. Turing remarked that much better machines would be needed to follow up his ideas.

Now, half a century later, such machines exist—and they are invaluable in the study of specific complex systems and of the mathematics of complexity in general. For instance, Turk was able to use powerful computers to play around with the numbers in Turing's own equations. Without them he could not have done the calculations and explored the huge set of possibilities, only some of which result in life-like forms. Similarly, he could not have produced even a single leopard spot without the aid of computer graphics: indeed, computer graphics have been invaluable in many areas of mathematical modelling.

But *plus ça change, plus c'est la même chose*: Turing's 21st-century successors face updated versions of the methodological problems that faced him. We have seen, for example, that Goodwin's *Acetabularia* model couldn't illustrate some of the implications of his mathematics because of lack of computer power. So it is still the case that Turing-rooted biological ideas cannot always be applied in practice, even if their relevance is evident in theory.

A third obstacle to the take-up of Turing's work was that even if his mathematics was impeccable *as mathematics*, he didn't (and couldn't) prove that real biological structures actually do emerge in this way. Only experimental developmental biology could show that. (Likewise, only psychological experiments can confirm the psychological reality of a computer model of mind.)

This question couldn't be experimentally addressed in Turing's lifetime. Strictly, that's not quite true: one of his six references cited then-recent research showing that the open ('head') end of a growing *Hydra* tube develops patches of chemicals that show up with a particular stain, and that the tentacles subsequently arise at those points.[35] But the nature of the chemicals, and just how they could form the whorl of tentacles, were mysteries. The genetic and biochemical techniques needed to solve such mysteries were not yet available. Today, the empirical question raised by his theory—or rather, the host of different questions—are at last being debated. Goodwin's study of *Acetabularia* is just one example.

The last reason for the delay in following up the work of Turing (and of D'Arcy Thompson) was that it approached biology in a deeply unfashionable way. D'Arcy Thompson had been clear on that point: natural selection, he said, was a secondary factor in morphogenesis. If he had lived to see the rise of molecular biology after Crick and Watson's discoveries, he would have had reservations about that too, for its reductionist approach discourages biologists from asking morphological questions. Unlike D'Arcy Thompson, Turing didn't mount an explicit attack on neo-Darwinism. Nevertheless, his ideas, too, were out of line with the biological orthodoxy.

Today Turing's conceptual chickens have come home to roost and are sitting pretty on their perches: indeed, they are becoming increasingly fashionable. The mathematical verdict on 'The chemical basis of morphogenesis' is highly positive. The biological verdict is generally positive also. Even orthodox neo-Darwinists allow that the questions asked by the structuralist biologists are interesting—and that many of their answers are valuable, too.

The editor of a collection of Turing's writing on morphogenesis (one of Goodwin's collaborators) has even said that his 1952 paper has been cited more often 'than the rest of his works taken together'.[36] I doubt this, because the Turing test paper has been cited countless times by philosophers, computer scientists, and members of the general public,[37] but undoubtedly the reputation of this long-neglected work has rocketed. Given the surging public and professional interest in various areas of biology, that rocket will still be airborne—and visible to 'every schoolboy'—one hundred years from now.

Buried treasure

In 1940, when England was expecting a Nazi invasion, Turing buried some silver bars in the grounds of Bletchley Park—or perhaps in Shenley Brook End, the village where he was lodging—but despite repeated searches, many aided by metal-detectors, no-one has managed to find them. Twelve years later he inspired another determined treasure hunt, for after his tragically early death his friends found in his rooms many pages of handwritten notes comprised of largely unintelligible mathematical symbols.

They also found drafts of (and extensive notes for) three papers on 'The morphogen theory of phyllotaxis'—mentioned in his Royal Society publication.[38] These explained, among other things, the prevalence of Fibonacci numbers in plant anatomy—the numbers of daisy petals for example (see the Introduction). Versions partly corrected or prepared by his student Bernard Richards, with notes by his close friend Robin Gandy, can be read in the *Collected Works*.[39] (Richards had been asked by Turing to solve the equations for spherical symmetry; his solutions nearly matched various species of *Radiolaria*, but Turing died before seeing them: see Chapter 35.[40]) In short, these drafts were intelligible, at least to Turing's collaborators.

The copious mathematical notes were a different matter. Gandy, no mean mathematician himself, couldn't make head or tail of them. The relevant folder in the Turing Archive bears a manuscript note by Gandy, saying: 'It will be difficult, in some places impossible, to know exactly what the fragments are (exactly) about'.[41]

There were four sorts of difficulty. The first was Turing's often-illegible handwriting: this had always been a problem—one of his schoolteachers had described it as 'the worst I have ever seen'.[42] Another was his use of idiosyncratic abbreviations, many of which presumably referred to equally idiosyncratic mathematical concepts. The third and fourth were identified by Gandy as his unorthodox ('so individual') ways of doing mathematics and—more surprising—perhaps even his 'unmethodical' thinking.[43] Faced with these obstacles, Gandy was not alone in being unable to decipher Turing's notes. They are still not fully understood.

It is inconceivable that there are no further mathematical insights in there of potential relevance to theoretical biology; possibly they concerned the problem that Turing admitted he had not solved—namely, the origin of new pattern not from homogeneity but from pre-existing patterns. In other words, those handwritten pages are a treasure trove. But whereas the 'trove' (the finding) was all too easy, assessing the value of the treasure may forever prove too difficult. In brief: yet another enigma.

The enigmatic embryo, however, has been demystified. Many unanswered questions remain, to be sure. But thanks to Turing, this awe-inspiring biological phenomenon is significantly less occult than it used to be. Blake might not be pleased about that—but perhaps 'every schoolboy' eventually will be.

CHAPTER 34

Turing's theory of morphogenesis

THOMAS E. WOOLLEY, RUTH E. BAKER, AND PHILIP K. MAINI

In 1952, Turing proposed a mathematical framework for understanding certain very interesting chemical reaction systems.[1] He described a rather counter-intuitive chemical mechanism, and showed that it could generate patterns in chemical concentrations. He coined the term 'morphogens' for the chemicals composing his mechanism, and he hypothesized that morphogens instruct cells to adopt different fates. Which future is 'selected' by the cell depends on the concentrations of morphogens to which the cell is exposed. Thus a new field of research was born, leading to novel mathematical developments and to new biological experiments. However, researchers continue to hunt for a biological example of Turing's mechanism in which the morphogens can be identified. Here we briefly review sixty years of research inspired by Turing's seminal paper.

Introduction

This chapter is a non-mathematical introduction to the mathematical techniques that allow us to understand the mechanisms behind the formation of biological patterns, such as the development of stripes on the skin of a zebra. In particular, we can extract general rules from the mathematical models, and these aid us in identifying places where such patterns could be found. We highlight the successes of Turing's theory, discuss its applicability to particular real-life examples, and explain the potential solutions that it offers to problems thrown up by recent advances in biology.

The core idea of Turing's theory is to take two stable (or stabilizing) processes and combine them. What will you get? Intuitive reasoning suggests that the outcome is a stable system. Turing showed that this is sometimes the *wrong answer*.

Turing's equations

The theory is that patterns arise as the consequence of an observable population, such as skin cells, responding to diffusing and reacting populations of chemicals, such as proteins. These chemicals are known as 'morphogens' and the process of generating biological complexity is known as 'morphogenesis'. Multiple different types of morphogen can be present, and they react with each other in order to create products that cells can use and/or respond to. What cells do is determined ultimately by their genes, but gene expression—which genes are 'turned on'—is determined by a multitude of signals. In Turing's theory the most important signal is the morphogen.

The mathematical equations describing morphogenesis not only model the chemical reactions themselves, but also the essentially random motion of the morphogens as they diffuse. The so-called 'diffusion equation' is incredibly important for understanding all undirected random motion; for example, heat conduction in solids, drainage of water through soil, and gases spreading through the air.[2] The basic equation states that material is conserved throughout the region of diffusion, and that a diffusing chemical always tends to spread out equally but does so in a random way with no preferred direction. The equation also tells us that the rate of change of chemical concentration at a specific position (on a zebra's skin, for example) is given by the 'local balance' of the flow of material into and out of the location in question, with no matter being either created or destroyed.

The big idea

The flash of inspiration that led Turing to suggest the pattern-forming mechanism now bearing his name is a complete mystery. Indeed, because the underlying assumptions of his work are so counter-intuitive, it is a testament to his genius that he found such an important result in a place where no one would think to look.

Turing focused his research on *stable reaction systems*: systems that tend to a constant concentration of morphogens everywhere in the system. On their own, these stable chemical systems cannot produce long-term patterning. We are all familiar with the fact that allowing inert substances to diffuse together does not give rise to patterning; for example, if we put a drop of red ink into water and neither heat nor stir the liquid, diffusion causes the ink to spread out uniformly through the water over time. In other words, after a sufficiently long time no part of the water is darker red than the rest—there is no pattern.

At this point, common sense tells us that if we have a system of stable reactions and simply allow the chemicals to diffuse around, we would not expect any interesting behaviour: we would eventually see a constant concentration of morphogens everywhere, and no pattern. This is because we have a stabilizing mechanism (diffusion) acting on a set of already stable reactions. However, Turing postulated that, when coupled with certain reactions, diffusion could in fact lead to a patterned state: this is called *diffusion-driven instability*. Starting from an unpatterned state, the system comes to exhibit persistent patterns.

To illustrate this idea, let us use a fictitious example involving 'sweating grasshoppers'.[3] Suppose that numerous grasshoppers inhabit a field of dry grass. Suddenly, a fire starts burning somewhere and spreads out into the dry grass. The grasshoppers try to avoid the fire as much as possible, fleeing randomly around the field. As the grasshoppers move, they generate moisture in the form of sweat. This sweat prevents the fire from penetrating into areas of high grasshopper density.

If the grasshoppers move too slowly then the whole field will burn, and there will be no resulting pattern. But if the grasshoppers escape death by moving faster than the fire can spread, then some patches of grass will burn, while other areas, crammed with grasshoppers, become saturated by moisture and do not burn. As a result, the field develops a pattern composed of sections of burnt and unburnt grass. Essentially, this pattern arises due to Turing's mechanism (Fig. 34.1).

Turing was (as usual) ahead of his time in his thinking, and his ideas lay dormant for quite a while. More recently, however, the concept of diffusion-driven instability has led to numerous avenues of further investigation, both theoretical and experimental.

In the particular case of two interacting species (like the fire and the sweating grasshoppers), two mathematical biologists, Alfred Gierer and Hans Meinhardt, clarified Turing's mechanism by identifying one species as an 'activator' (the fire in our case, since it produces more fire and activates the grasshoppers to sweat) and one species as an 'inhibitor' (the sweat, since it prevents fire from occurring).[4] They also showed that, for patterning to occur, the inhibitor must diffuse more rapidly than the activator (the grasshoppers must flee more quickly than the fire advances).

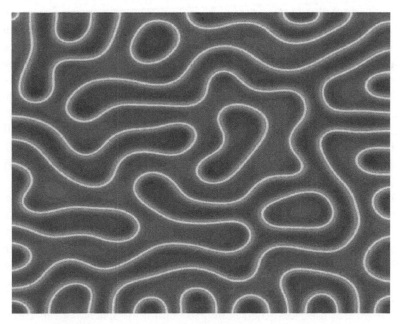

Figure 34.1 An example of a two-dimensional striped Turing pattern.

Thomas E. Woolley.

Do Turing patterns exist in nature?

Nearly forty years after their existence was first theorized, researchers constructed chemical Turing patterns.[5,6] This was quickly followed by the development of a corresponding mathematical model.[7] It is impossible to overstate the importance of these achievements. These researchers showed for the first time that Turing patterns were not merely theoretical. Moreover, their work spurred many other chemists to find reaction systems that give rise to patterns in chemical concentration.

Even though the theory and chemistry of such chemical reaction systems is now well documented, the existence of Turing patterns in *biology* is still controversial. Many biochemical gene products may in fact be Turing morphogens. For example, there is strong evidence to suggest that, during limb formation, certain cell growth factors (proteins that stimulate cell division) act as Turing activators, and although in some cases their complementary inhibitors have been identified, definitive proof that the resulting patterns are Turing patterns still eludes us. To give another example, the regular patterns of arrangement of hair follicles in various mammals also suggest the presence of Turing's mechanism.[8]

Two recent studies of the development of mice have produced strong evidence that Turing patterns might accurately describe a number of biological systems. Jeremy Green and his fellow researchers were the first experimental group to claim to have shown that two proteins, 'fibroblast growth factor' (FGF) and 'sonic hedgehog' (Shh), could act as Turing-style morphogens.[9]

Neither of these proteins is unique to the mouse system that Green and his group studied. Shh was in fact first identified in fruit flies, and acquired its name through experiments showing that fruit fly embryos developed small pointy protrusions, like the quills of a hedgehog, when Shh was inhibited. More recently, it has been discovered that Shh is essential in the development of not only the brain and spinal cord but also the teeth.[10] FGF proteins are known to play important roles in wound healing and the development of blood vessels and neurons.

Green and his group were in fact researching the growth of ridges in the mouths of mice, specifically ridges in the palate. Their work is all the more suggestive because they were able to derive a mathematical model involving Turing's mechanism that reproduced the mouth ridge pattern of normal mice. They also showed that their model could predict the way that the ridge patterns changed when the activity of the morphogens was increased or decreased in experiments.

Shortly after this work on mouth ridges, it was also shown that Turing systems could explain the development of toe spacing in mice. In particular, the effects on toe development of so-called Hox genes were explored.[11] The prevailing theory had been that higher doses of Hox proteins would cause extra toes to form, and so eliminating Hox gene activity would reduce the number of toes. However, as Hox genes were eliminated, it was found that *more* toes formed, fourteen in the most extreme case. The overall paw size remained unchanged, though, meaning that as the number of toes increased they became thinner.

It seems, then, that the Hox gene system can control the *spacing* of the Turing pattern. Although the mathematical theory of this process reproduces the experiments very well, there remains a significant problem—namely, that Hox genes do not diffuse and so are not actually morphogens in Turing's original sense. This means that if this research is to fit with the Turing hypothesis, the genes must somehow be signalling to their environment by means of a mechanism that acts *like* a diffusive agent.

Figure 34.2 (a) As the size of the host surface changes, the pattern changes too: each animal skin has been scaled to be the same size, with a scale factor S denoting the magnification. (b) A Valais goat. (c) A Galloway cow.

Cows, angelfish, and tapirs

Why is Turing's diffusion-driven instability such an attractive mechanism for describing the development of pattern and form? First, it is simple: the production of the patterns relies on the natural tendency of molecules to diffuse and react. Second, the mechanism has inbuilt features that control the spacing of the patterns. These inbuilt features give Turing patterns a number of characteristic properties. For example:

- in order for patterns to form, the host surface needs to be larger than a specific critical size
- provided that the host is large enough to support patterns, no external 'input' is needed to specify the pattern—the process is self-regulating.

This means that Turing patterns can be highly regular over large distances without any external input.

Both these features are illustrated in Fig. 34.2(a), where we see the effect of increasing the size of an animal's skin. At first the skin is too small to support patterning, but as the skin gets bigger a qualitative change in behaviour occurs, leading to a skin with different concentrations of morphogen at the front and back. As the skin continues to increase in size further bifurcations occur, causing the pattern to become more complex, eventually leading to a maze-like pattern and then, finally, isolated spots. These transitions can be compared with the different patterns observed on the Valais goat (Fig. 34.2b) and the Galloway cow (Fig. 34.2c): the small goat only has one transition, whereas the larger cow has two.

Biologists Shigeru Kondo and Rigito Asai extended these results linking size and pattern by studying animals that grew in size while their skin patterns developed.[12] In experiments involving the marine angelfish *Pomocanthus* they observed that, as the size of the angelfish doubles, new stripes develop along the skin in between the old ones, so producing nearly constant spacing between the stripes. This constant spacing in the patterns on the fish's skin suggests that a Turing-like mechanism is responsible for the development of the pigmentation making up the pattern. Applying this picture to human growth might seem to imply, alarmingly, that we should develop more heads or limbs as we grow from childhood. But human cells can select a fate only during a brief time interval, usually at the embryonic stage, after which existing structures simply grow in size, rather than new structures forming.

Yet the natural world is not always as simple as these results might suggest. To give one example, the coat pigmentation of the Brazilian tapir refuses to be understood as a Turing pattern (Fig. 34.3). This is because the pattern becomes *more* complex on the thinner limb regions, contradicting Turing's theory. Furthermore, only baby Brazilian tapirs are patterned. As the animals mature, their coat markings disappear, leaving a uniform grey colour. For such patterns to be consistent with Turing's theory, we would need to postulate that the 'inputs' to the pattern-making processes on the limbs differ from those on the body and that, moreover, the inputs change over time, causing an evolution from a patterned condition to no pattern. Alternatively, the changes in the tapir's skin pattern may indicate that Turing's mechanism is not universal: these changes may occur in a regime that simply cannot be characterized by Turing's theory.

Figure 34.3 A baby tapir.

More problems—and solutions

One criticism of Turing's mechanism is that it often requires a possibly implausible 'fine-tuning' of biological parameters.[13] Moreover, the diffusion rates of the activator and inhibitor chemicals must vary greatly—something that is unlikely in practice. However, these problems become less severe if there are more than two reacting species. The two-species Turing system is generally only a caricature of the underlying biology. Systems in which three or more morphogens interact are more realistic.

Another criticism is that Turing's mechanism produces patterns that lack robustness. This means that minor perturbations in the starting state of the process, the geometry of the host surface, or the boundary conditions of the process, may greatly influence the final pattern.[14] In certain cases, such as animal skin patterns, the resulting individuality could be a positive outcome. However, when dealing with the toes of mice, for example, the mechanism should ensure that the normal pattern of toe spacing is produced reliably.

In response to this criticism, we can point out that researchers have in fact demonstrated realistic ways of generating robust patterns. For example, it has been shown that robust patterns can be generated when realistic forms of growth are included in the model.[15] When the surface is growing, the tightly controlled initial patterns evolve through intermediate stages until their final form is created. Turing foresaw this. As he said in his 1952 paper:

Most of an organism, most of the time, is developing from one pattern into another, rather than from homogeneity into a pattern.

Although Turing knew that there were limitations to his model, this fortunately did not stop him from publishing his ideas. He thought that as the paradigm was extended, and generalized to match reality more closely, many of the problems would be surmounted.

One particular problem that is still not well understood is that of time delays in the chemical reactions involved. When considering a mathematical formulation of biology, researchers usually assume implicitly that the chemical reactions occur instantaneously. But this is not true. When dealing with gene products, the delays involved in the production of reactants can be significant—of the order of minutes to hours. Yet if we include time delays in the mathematical equations, what we see is a catastrophic collapse of the ability of the Turing mechanism to generate patterns.[16] Although the addition of randomness has been shown to regenerate the patterns in certain specific cases,[17] this conundrum has still not been completely solved. Indeed, these studies may indicate that the detailed biological processes used in generating patterns are ones in which delays are minimal.

Conclusions

Experimental biology continuously pushes the boundaries of knowledge forward. Our desire for a better quality of life and to live longer has led to a prioritization of the biosciences. However, experimentation, and the linear verbal reasoning implicit in that approach, can lead us only so far: to extend our biological insight we need a fundamental understanding of the complex non-linear feedback interactions inherent in living systems. This is no longer a new idea, but after Turing introduced it, it lay dormant for some time, eclipsed by the long-lasting excitement surrounding the gene theory revolution that was started by the Cambridge researchers Francis Crick and James Watson in 1953, the year after Turing published his ideas on morphogenesis.

Here we have shown that some relatively simple ideas give us a mathematical framework for understanding the formation of complex patterns. These ideas afford a way of comprehending the creation of the natural beauty of pigmentation patterns in animal skins, and give us insight into many developmental features, such as toe development in mice, which in turn can then be translated to human development.

Turing's theory of morphogenesis has been highly successful in illuminating mechanisms that may underlie a wide range of patterning phenomena. But we should not forget that the theory is a *simplification* of the underlying biology. Turing's picture of diffusing chemicals driving a system to form patterns via chemical interactions may not be exact. In fact, recent work suggests that Turing's morphogens may actually be cells themselves: certain types of cells involved in pigmentation patterning in fish have been shown to interact with each other in the same way that Turing hypothesized his morphogens to interact.[18]

In this chapter we have only scratched the surface of a huge topic. Turing's model has been extended to three spatial dimensions, and sources of biological randomness have been included.[19] This has led to new theories and generalizations. It would certainly be a mistake to think that, because of its long history, Turing's idea has run its course. There are plenty of questions to motivate and intrigue a new generation of minds. As Turing said:

We can only see a short distance ahead, but we can see plenty there that needs to be done.

In the light of recent biological evidence, Turing's original ideas may not stand up in detail, yet the levels of abstraction and detail in his model were absolutely appropriate at the time he formulated it. The modern range of experimental and theoretical extensions of his 1952 model show that he was on the right track. These extensions strongly support his key claim that a complete, and physically realistic, model of a biological system is not necessary in order to explain specific key phenomena relating to growth.

Turing's theory provides a nice illustration of a general point that the mathematical biologists George Box and Norman Draper put so succinctly:[20]

All models are wrong, but some are useful.

Turing put the point this way, when discussing the application of his mathematical ideas to biology:[21]

[My model] will be a simplification and an idealization, and consequently a falsification. It is to be hoped that the features retained for discussion are those of greatest importance in the present state of knowledge.

Turing's theories lay dormant for a long time, because mathematics and biology were not ready for such counter-intuitive ideas. If it had not been for his premature death, just two years after he published his ideas on morphogenesis, how much further might he have developed his theory? How much closer might we now be to solving one of nature's greatest mysteries? How much more useful might our (still wrong) models be?

CHAPTER 35

Radiolaria: validating the Turing theory

BERNARD RICHARDS

In his 1952 paper 'The chemical basis of morphogenesis' Turing postulated his now famous Morphogenesis Equation. He claimed that his theory would explain why plants and animals took the shapes they did. When I joined him, Turing suggested that I might solve his equation in three dimensions, a new problem. After many manipulations using rather sophisticated mathematics and one of the first factory-produced computers in the UK, I derived a series of solutions to Turing's equation. I showed that these solutions explained the shapes of specimens of the marine creatures known as Radiolaria, and that they corresponded very closely to the actual spiny shapes of real radiolarians. My work provided further evidence for Turing's theory of morphogenesis, and in particular for his belief that the external shapes exhibited by Radiolaria can be explained by his reaction–diffusion mechanism.

Introduction

While working in the Computing Machine Laboratory at the University of Manchester in the early 1950s, Alan Turing reignited the interests he had had in both botany and biology from his early youth. During his school-days he was more interested in the structure of the flowers on the school sports field than in the games played there (see Fig. 1.3). It is known that during the Second World War he discussed the problem of phyllotaxis (the arrangement of leaves and florets in plants), and then at Manchester he had some conversations with Claude Wardlaw, the Professor of Botany in the University. Turing was keen to take forward the work that D'Arcy Thompson had published in *On Growth and Form* in 1917.[1] In his now-famous paper of 1952 Turing solved his own 'Equation of Morphogenesis' in two dimensions, and demonstrated a solution that could explain the 'dappling'—the black-and-white patterns—on cows.[2]

The next step was for me to solve Turing's equation in three dimensions. The two-dimensional case concerns only surface features of organisms, such as dappling, spots, and stripes, whereas the three-dimensional version concerns the overall shape of an organism. In 1953 I joined Turing as a research student in the University of Manchester, and he set me the task of solving his equation in three dimensions. A remarkable journey of collaboration began. Turing chatted to me in a very friendly fashion. He told me he could see a way of developing his theory that might lead to a realistic solution in the three-dimensional case. Although he did not mention it to me at the time, Turing had earlier studied the drawings of Radiolaria in a copy of the Report of the HMS *Challenger* Expedition.[3] (The British research ship *Challenger* toured the Pacific and other oceans from 1873–76.)

Radiolaria

The *Challenger*'s crew found Radiolaria in the mud on the bottom of the Pacific Ocean. Radiolaria are marine creatures whose unicellular body consists of two main portions supported by a membrane. One portion is an inner central capsule and the other an external surface in contact with the outside world, for feeding and protection. The spherical body of Radiolaria measures about 2 mm in diameter.

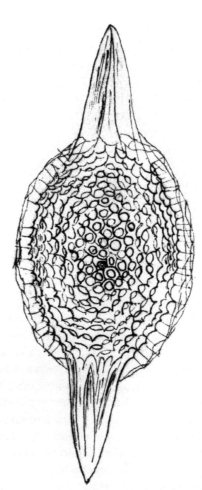

Figure 35.1 *Cromyatractus tetracelyphus* with two spines.

Reproduced from Bernard Richards, 'The morphogenesis of Radiolaria', MSc Thesis, University of Manchester, 1954, with thanks to Dr Richard Banach. Original image taken from Ernst Haeckel, *Report of the Scientific Results of the Voyage of H.M.S. Challenger During the Years 1873–76*, Volume 18.

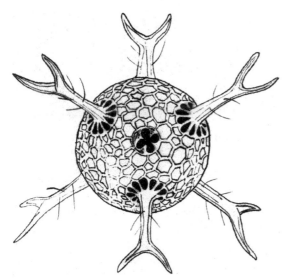

Figure 35.2 *Circopus sexfurcus* with six spines, all set at 90° apart.

Reproduced from Bernard Richards, 'The morphogenesis of Radiolaria', MSc Thesis, University of Manchester, 1954, with thanks to Dr Richard Banach. Original image taken from Ernst Haeckel, *Report of the Scientific Results of the Voyage of H.M.S. Challenger During the Years 1873–76*, Volume 18.

Figure 35.3 *Circopurus octahedrus* with six spines and eight faces.

Reproduced from Bernard Richards, 'The morphogenesis of Radiolaria', MSc Thesis, University of Manchester, 1954, with thanks to Dr Richard Banach. Original image taken from Ernst Haeckel, *Report of the Scientific Results of the Voyage of H.M.S. Challenger During the Years 1873–76*, Volume 18.

As the spherical newborn radiolarian grows, 'spines' (a kind of spike) develop from the surface in pre-determined positions. The resulting shapes give rise to six sub-species of Radiolaria. The length of the spines is usually equal to the radius of the main spherical body and they are symmetrically positioned over the sphere. If the body has only two spines they are situated at the north and south poles. There are no three-, four-, or five-spined versions. The six-spined version has the spines placed 90° apart over the surface of the sphere—two at the poles and four around the equator. Other versions have twelve or twenty spines. The shapes taken by these types of Radiolaria resemble some of the regular mathematical solids (Figs 35.1–35.6).

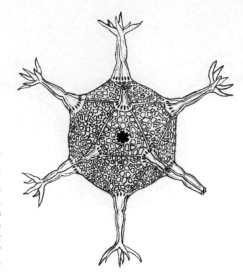

Figure 35.4 *Circogonia icosahedra* with twelve spines and twenty faces.

Reproduced from Bernard Richards, 'The morphogenesis of Radiolaria', MSc Thesis, University of Manchester, 1954, with thanks to Dr Richard Banach. Original image taken from Ernst Haeckel, *Report of the Scientific Results of the Voyage of H.M.S. Challenger During the Years 1873–76*, Volume 18.

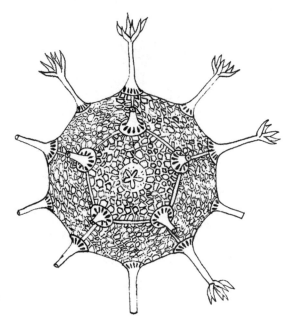

Figure 35.5 *Circorrhegma dodecahedra* with twenty spines and twelve faces.

Reproduced from Bernard Richards, 'The morphogenesis of Radiolaria', MSc Thesis, University of Manchester, 1954, with thanks to Dr Richard Banach. Original image taken from Ernst Haeckel, *Report of the Scientific Results of the Voyage of H.M.S. Challenger During the Years 1873–76*, Volume 18.

The part played by the Manchester computer

Once I had obtained a mathematical solution for Turing's equation, the next step was to discover what three-dimensional shape the solution represented. This is not an easy thing to do, because the equation giving the shape in three dimensions involves three coordinates, r (the distance), θ (the latitude), and φ (the longitude), and one could not visualize the shape

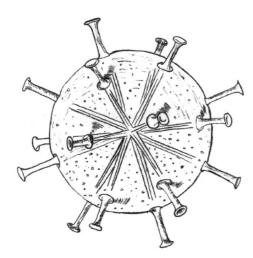

Figure 35.6 *Cannocapsa stehoscopium* with twenty spines.

Reproduced from Bernard Richards, 'The morphogenesis of Radiolaria', MSc Thesis, University of Manchester, 1954, with thanks to Dr Richard Banach. Original image taken from Ernst Haeckel, *Report of the Scientific Results of the Voyage of H.M.S. Challenger During the Years 1873–76*, Volume 18.

thus defined. One needs to know where the spines are located on the sphere, and also the diameter of the sphere and the lengths of the spines protruding from it. Normal powers of mathematical visualization are inadequate for this task—mine certainly were—and I needed the computer's help.

It must be admitted, though, that in those days computers were less helpful than they are now. The computer that I used, the Ferranti Mark I, had no facilities for a visual display output and only a very primitive printer: this could print only numbers and alphabetic characters—not graphics. So I decided to use some clever programming to make it print contour maps of the surface. On the first page was displayed an array with θ taking values from 0° to 90° and φ taking values from 0° to 360°, while on the second page the values of θ from 90° to 180° were shown; so between them these two pages mapped both the northern and southern hemispheres of the shape. The printer covered the pages with the arcane teleprinter symbols that Turing had brought with him to Manchester from Bletchley Park, each one representing a distance from the centre of the sphere (a height) on a scale from 0 to 31. Thus the whole surface of the sphere was present on my two-dimensional sheets. I was then able to draw contour lines on the sheets, locate the spines, and record their lengths. In this way the computer found three-dimensional shapes corresponding to the above figures.

Comparing the computed shapes with real Radiolaria

I set my computed shapes against their closest matches from among the Radiolaria, and the matches were very good. Figure 35.7 shows my computed shapes superimposed upon *Circopus sexfurcus* (left) with two spines at the two poles and four around the equator, and on *Circogonia icosahedra* (right) with twelve spines equidistantly spread over its surface. Most marvellously, starting from nothing more than Turing's morphogenesis equation, I had produced the shape and surface structure of creatures that accurately matched in shape and form real Radiolaria found in the oceans of the world.

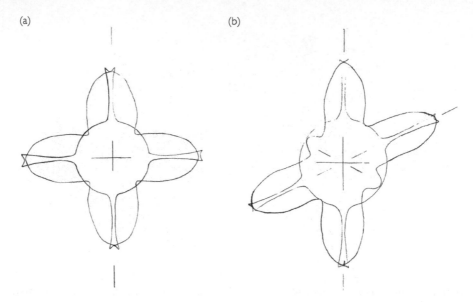

Figure 35.7 The computer solutions superimposed on *Circopus sexfurcus* (left) and *Circogonia icosahedra* (right).

Reproduced from Bernard Richards, 'The morphogenesis of Radiolaria', MSc Thesis, University of Manchester, 1954, with thanks to Dr Richard Banach.

Conclusion

At the end of May 1954 I showed my computer-generated contour maps to Turing: I knew nothing about Radiolaria at this stage. He immediately asked me to get the *Challenger* Expedition Report from the University Library, but told me nothing about its content. Excitedly I studied the drawings of Radiolaria, looking for matches with my computer-generated shapes, and discovered the close matches shown in Fig. 35.7. It seemed a tremendous breakthrough. Turing had arranged to meet me again on Tuesday 8 June, but alas we could not meet for he died on the previous day. He never learned of the triumph of his theory and the close matches that I obtained.

Nevertheless, these close matches remain as a tribute to Turing's genius and foresight, and also to his love of nature. Today he is remembered and honoured for his work on codebreaking, computing, and artificial intelligence, but he truly deserves simultaneous recognition for his pioneering work on morphogenesis.[4]

PART VII

Mathematics

CHAPTER 36

Introducing Turing's mathematics

ROBIN WHITTY AND ROBIN WILSON

A lan Turing's mathematical interests were deep and wide-ranging. From the beginning of his career in Cambridge he was involved with probability theory, algebra (the theory of groups), mathematical logic, and number theory. Prime numbers and the celebrated Riemann hypothesis continued to preoccupy him until the end of his life.

Turing, master of all trades

As a mathematician, and as a scientist generally, Turing was enthusiastically omnivorous. His collected mathematical works comprise thirteen papers,[1] not all published during his lifetime, as well as the preface from his Cambridge Fellowship dissertation; these cover group theory, probability theory, number theory (analytic and elementary), and numerical analysis. This broad swathe of work is the focus of this chapter. But Turing did much else that was mathematical in nature, notably in the fields of logic, cryptanalysis, and biology, and that work is described in more detail elsewhere in this book.

To be representative of Turing's mathematical talents is a more realistic aim than to be encyclopaedic. Group theory and number theory were recurring preoccupations for Turing, even during wartime; they are represented in this chapter by his work on the word problem and the Riemann hypothesis, respectively. A third preoccupation was with methods of statistical analysis: Turing's work in this area was integral to his wartime contribution to signals intelligence. I. J. Good, who worked with Turing at Bletchley Park, has provided an authoritative account of this work,[2] updated in the *Collected Works*. By contrast, Turing's proof of the central limit theorem from probability theory, which earned him his Cambridge Fellowship,[3] is less well known: he quickly discovered that the theorem had already been demonstrated, the work was never published, and his interest in it was swiftly superseded by questions in mathematical logic. Nevertheless, this was Turing's first substantial investigation, the first demonstration of his powers, and was certainly influential in his approach to codebreaking, so it makes a fitting first topic for this chapter.

Turing's single paper on numerical analysis, published in 1948, is not described in detail here. It concerned the potential for errors to propagate and accumulate during large-scale computations; as with everything that Turing wrote in relation to computation it was pioneering, forward-looking, and conceptually sound. There was also, incidentally, an appreciation in this paper of the need for statistical analysis, again harking back to Turing's earliest work.

The central limit theorem

The idea of a 'bell-shaped' data distribution is a familiar one: that female human beings, say, have an average height (the 'mean' of the distribution) and that roughly as many are shorter than this height as are taller, tailing off in both directions. The measure of 'standard deviation' is a way of making precise this 'tailing off': to say that a woman taller than 2 m is extremely unlikely is to say that this part of the height distribution is, perhaps, four standard deviations away from the mean. To be even more precise, this tailing off can be drawn as a mathematical curve. A curve that has been observed to describe the bell-shaped data distribution particularly well is the 'normal distribution', also known as the 'Gaussian distribution', although it was analysed by Pierre-Simon Laplace in 1783 when Carl Friedrich Gauss was only six years old (Fig. 36.1).

Laplace observed that the normal distribution is prevalent in astronomical and probability calculations. He began the process of tabulating its values; to this day such tables are generally included in textbooks on statistics and probability. Some standardization is essential and we use the 'standard normal distribution', in which the mean is 0, the standard deviation is 1, and the height of the distribution at the mean is also 1. The values that are tabulated are *areas*, because these correspond to probabilities: given any non-negative value X, the probability that any given measurement falls between 0 and X is the area lying between the standard normal curve and the horizontal axis between 0 and X.

So we now have a beautifully simple way of making predictions! What is the probability that the next woman I pass in the street is taller than 2 m? I shift the average observed height to the

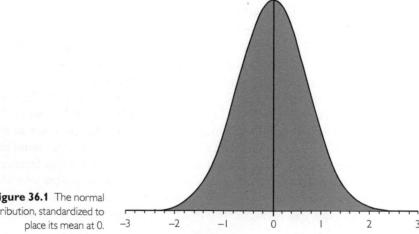

Figure 36.1 The normal distribution, standardized to place its mean at 0.

zero mark, its observed frequency and the observed standard deviation are both reduced to 1 (there are textbook equations that tell you how to do this), and I then look up in my table to find the area under the curve beyond whatever 2 m has been reduced to, and read off the value— perhaps it is 0.0001 (one woman in every 10,000).

The question that Turing addressed was: how does one make predictions about data that are not normally distributed, and for which no standardized tables have been drawn up? The answer is that, while an *individual* measurement may not be predictable, a sum of *n* measurements (or an arithmetic mean) is: when *n* is large enough, summations and means fit a normal distribution. The idea is illustrated by the example depicted in Fig. 36.2: a ladies' outfitter can forecast sales of various sizes of clothes because sizes are normally distributed. The spend per customer is likely to be more erratically distributed, as shown on the left of the figure; however, the average spend for every ten customers approximates a normal distribution, as shown on the right. So, after suitably standardizing the data, our merchant can turn to those well-thumbed tables of probabilities and the bank manager can be soothed with reliable estimates based on well-behaved *average* spending patterns.

This example of finding order in chaos has a long history; a version for coin tossing dates back to Abraham de Moivre in 1738. But in 1930s Cambridge the general phenomenon was regarded as what is known in the trade as a 'folk theorem'. Recent developments in probability theory in mainland Europe and Russia were apparently unknown in Cambridge, or were disregarded. So Turing set himself the task of understanding, from first principles, why the phenomenon worked, writing in the preface to his dissertation:

My paper originated as an attempt to make rigorous the 'popular' proof mentioned in Appendix B. I first met this proof in a course of lectures by Prof. Eddington.

Eddington's lectures took place in the autumn of 1933; Turing's research was concluded, according to a classic 1995 article by Zabell,[4] 'no later than February 1934'. In that short space of time Turing had fast-forwarded through 200 years of the history of probability! He derived something almost as good as the 1922 version of the central limit theorem due to the Finnish mathematician Jarl Waldemar Lindeberg (who was himself initially working in ignorance of earlier Russian work). Lindeberg's theorem is very general: the *n* measurements that are being summed need not even come from the same distribution, so our ladies' outfitter need not worry

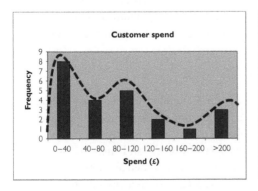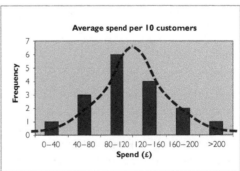

Figure 36.2 Achieving a normal distribution by summing non-normally distributed data.

that different types of customer have different spending patterns; there is a technical condition, now called the 'Lindeberg condition', that the measurements must obey. Turing derived a stronger condition than Lindeberg's, so his theorem is less general, but it was still better than anything proved before the 1920s. It was a remarkable achievement, showing deep insight and accomplished in isolation and at great speed. These are precisely the hallmarks of Turing's landmark discoveries in logic just a year later.

Group theory and the word problem

Unlike the modern theory of probability, the theory of groups was mainstream fare for British mathematicians of the 1930s. In 1897 William Burnside, like Turing a Cambridge graduate who became a Cambridge Fellow and a fine sportsman, published the first English-language book on the subject, *The Theory of Groups of Finite Order* (the second edition of 1911 became a standard reference work in the subject for many years and remains in print to this day).[5] For reasons that are not clear, Burnside moved in 1885 to a position at the Royal Naval College at Greenwich, from where he refused to be tempted back to Cambridge, even to be Master of Pembroke College. But he was a central figure in British mathematics: a President of the London Mathematical Society, a Fellow of the Royal Society, and a recipient of its Royal Medal.

Despite Burnside's influence, group theory did not immediately take Britain by storm. In his presidential address to the London Mathematical Society in 1908, Burnside complained:

It is undoubtedly the fact that the theory of groups of finite order has failed, so far, to arouse the interest of any but a very small number of English mathematicians; and this want of interest in England, as compared with the amount of attention devoted to the subject both on the Continent and in America, appears to me very remarkable.

But his text on the theory of groups had a major influence on a near contemporary of Turing's at King's College. Philip Hall graduated in 1925, nine years before Turing, and like Turing he proceeded to a Fellowship, although in Hall's case his dissertation, in group theory, was to set the mould for a lifetime of distinguished specialization. His Fellowship was renewed in 1933, the year before Turing was elected, on the strength of a paper that is now considered a milestone in the history of group theory.

So by the 1930s group theory was making an impact in Cambridge, and at King's College in particular. Indeed, the *Turing Archive for the History of Computing* includes a correspondence between Turing and Hall (in April 1935) concerning Turing's very first published paper:

I enclose a small-scale discovery of mine. I should be very grateful if you could advise me how to get it published. Perhaps you referee group theory for the LMS yourself.

'Small-scale' it may have been, but Turing's first paper, 'Equivalence of left and right almost periodicity' was, as J. L. Britton has remarked,[6] 'surely a promising beginning to have noticed something that Von Neumann, already enormously successful, had missed'.

'I am thinking', Turing continued, 'of doing this sort of thing'—and there follow, perhaps tellingly, the words 'more or less seriously' erased. The correspondence continues amicably, because in May 1935 Turing wrote to Hall thanking him for a dinner invitation. A sporadic interchange continued after Turing's move to Princeton, with only one archive letter including any technical mathematics. Over the course of these letters the salutation progressed from 'Dear Mr Hall' to 'Dear Hall' to 'Dear Philip'.

So Turing was active and well connected in the field of group theory. He would surely have been aware of developments in a branch that has come to be known as combinatorial group theory, which revolves around a collection of deep questions asked by Max Dehn around 1912, known as 'word problems'. Indeed, the first real breakthrough was achieved by Dehn's student, Wilhelm Magnus, in 1932—and Magnus visited Princeton in 1934–35, just a year before Turing arrived there.

Although Dehn asked several related questions, what is known as *the* word problem is roughly as follows. We are given, first, a collection of objects that can be 'multiplied' together, and, second, a collection of rules that say when the result of a multiplication simplifies (an example of such a rule in ordinary arithmetic is the one saying that '$x \times 1 = x$'). The problem is then to find a method for answering 'yes' or 'no' to the question: does a given multiplication simplify down to 'nothing'?

A rough analogy may be drawn with the computer game Tetris. A collection of shapes simplifies if it can be arranged to contain a single continuous row. In the sketch in Fig. 36.3 the bottom row does not simplify, but the second row does if the descending block column is rotated and inserted into the obvious gap. A Tetris word problem might be: given a sequence of descending shapes, can they be 'multiplied' (that is, moved into position) so that the whole sequence simplifies to nothing by the time that the last shape has descended?

To get a little closer to actual group theory, suppose we are asked about multiplying together the following sequence of numbers $\frac{1}{2}$, 2, 4, $\frac{1}{3}$, 6, $\frac{1}{8}$, whose product might be written as

$$2^{-1} \times 2 \times 4 \times 3^{-1} \times 6 \times 8^{-1},$$

with the superscript −1 denoting a reciprocal (or inverse). Does this product reduce to a single 1? This is a word problem, but not a very difficult one! The answer is easily seen to be 'yes', and there is obviously a method or 'algorithm' for obtaining this answer because your electronic calculator will find it.

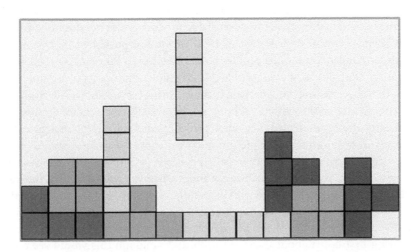

Figure 36.3 A sketch of a Tetris configuration: the second row potentially simplifies.

What happens if we move to variables and ask about a corresponding multiplication in which 2 is replaced by p and 3 by q:

$$p^{-1} \times p \times p \times p \times q^{-1} \times p \times q \times p^{-1} \times p^{-1} \times p^{-1} ?$$

The answer is still 'yes', provided that multiplication commutes: $p \times q^{-1} = q^{-1} \times p$. While commutativity holds for ordinary multiplication of fractions, it is not assumed for abstract groups. In the word problem we are given a list of variables, such as p, q, and r, and we can assume that multiplication by 1 leaves any variable unchanged, that inverses cancel ($p \times p^{-1} = p^{-1} \times p = 1$), and that associativity holds: $(p \times q) \times r = p \times (q \times r)$. Anything else is given specifically as a rule, (technically called a 'relator').

If we declare that $q^{-1} \times p \times q \times p^{-1} = 1$ is a rule of our game, then the 'word' displayed above will reduce to 1. But what happens if we declare instead that there are two rules, as follows:

$$\text{rule 1: } p \times q \times p^{-1} \times p^{-1} = 1 \qquad \text{and} \qquad \text{rule 2: } q^{-1} \times p = 1.$$

This is a little more tricky—we have to apply rule 2 before we apply rule 1—but the answer is still 'yes'. This leads to an abstract general question: given a finite set of variables and a finite set of rules, and a 'word' obtained by multiplying variables and their inverses, does the word reduce to 1? The word problem itself is about computability: can we specify an algorithm for answering the question? In other words, given a set of variables and a set of rules, together specifying a group G, can we write a computer program depending on G whose input is a word w composed of copies of the given variables and their inverses, and whose output is 'yes' or 'no', according to whether $w = 1$ in G?

Wilhelm Magnus's 1932 breakthrough was to demonstrate that there is indeed such an algorithm for solving word problems if there is just one rule. In 1952, Pyotr Sergeyevich Novikov, working at the Moscow State Teachers Training Institute, achieved a much bigger breakthrough, proving that no algorithm can exist for the word problem in general. This work owed a double debt to Turing. First, non-existence was proved by showing that existence of an algorithm would in turn imply an algorithmic solution for a problem already known to be algorithmically unsolvable. It was thanks to Turing's 1936 paper on computability that such unsolvable problems were known, and that methods were available for converting between one problem and another (via Turing machines). Secondly, in 1950 Turing had published his own 'small-scale' breakthrough, proving that the word problem cannot be solved for objects called 'semigroups with cancellation'. Both pieces of work are cited in Novikov's landmark publication.[7]

As is often the case with first-rate mathematicians, Turing rescued his result from a defective proof of a stronger one: he thought he had proved unsolvability for general groups (Novikov's achievement); his proof contained a flaw but it also contained substantial insights. With semigroups, inverses are removed from the picture, and we do not even necessarily have a unit value (1). So we are left only with associativity [recall: $(p \times q) \times r = p \times (q \times r)$]. In this case we cannot ask whether a word reduces to 1; instead we ask about whether two words are reducible to each other. With much less structure to deal with in a semigroup, the solvability of its word problem is consequently easier to determine: it is unsolvable. This was proved in 1947 by Emil Post and independently by Andrei Andreyevich Markov. It has been suggested that it was around this time that Turing first heard about the word problem,[8] although this would seem odd given his close and long-standing general association with group theory. Be that as it may, he was apparently taken with the problem and, as he had with problems in the past, mastered it at breakneck speed, to the extent of believing he had solved it. He had at any rate got somewhat further than Post and Markov: a semigroup with cancellation adds a little more structure, since we may assume that $pq = pr$ implies (via cancellation) that $q = r$.

If Turing failed to resolve the word problem for general groups, his Cambridge contemporary the famous group theorist Bernhard Neumann did no better! Neumann's student, J. L. Britton, recounts that:[9]

At one time (probably 1949) he [Bernhard Neumann] believed he had discovered a proof of the solvability of the word problem for groups; upon mentioning this to Turing, he was disconcerted to learn that Turing had just completed a proof of its unsolvability. Both of them urgently re-examined their proofs and both proofs were found to be wrong.

Prime numbers

Turing's interest in the Riemann hypothesis originated in the 1930s, shortly after he arrived in Cambridge. In 1932 the Cambridge mathematician A. E. Ingham wrote a classic text on prime numbers. Turing purchased this in 1933 and started serious work on the subject around 1937.[10]

A 'prime number' is a number, larger than 1, whose only factors are itself and 1: so 17 is a prime number, while 18 (= $2 \times 3 \times 3$) is not. The primes up to 50 are: 2, 3, 5, 7, 11, 13, 17, 19, 23, 29, 31, 37, 41, 43, 47.

Note that we cannot write out a complete list of primes—they go on for ever. This celebrated result appears in Book IX of Euclid's *Elements*, and has one of the most attractive proofs in mathematics. In modern language, we assume that the result is false—that there are only a finite number of primes (say, $p, q, \ldots$, and z)—and consider the number $N = (p \times q \times \ldots \times z) + 1$. Since $p \times q \times \ldots \times z$ is divisible by each of these primes, N cannot be divisible by any of them, and so there must exist another prime—either N itself or one of its prime factors. This contradicts our assumption that $p, q, \ldots, z$ are the *only* primes. So there are infinitely many primes.

Prime numbers are central to mathematics because they form the building blocks for numbers—every whole number can be built up from them. So, for example, $60 = 2 \times 2 \times 3 \times 5$, and, in general, *every whole number splits into primes in only one way*, apart from the order in which we write down the factors.

How are the primes distributed? Although they generally seem to 'thin out' the further along the list we proceed, they do not seem to be distributed in a regular manner. For example, pairs of prime numbers differing by just 2 (such as 5 and 7, or 101 and 103) seem to crop up however far we go, although this has never been proved. On the other hand, it is a simple matter to construct arbitrarily long lists of non-primes. For numbers around 10 million, the hundred just below contain nine primes (which is rather a lot), while the hundred just above contain only two.

In his inaugural lecture at the University of Bonn in 1975, the distinguished number theorist Don Zagier said:[11]

There are two facts of which I hope to convince you so overwhelmingly that they will permanently be engraved in your hearts. The first is that the prime numbers belong to the most arbitrary objects studied by mathematicians: they grow like weeds, seeming to obey no other law than that of chance, and nobody can predict where the next one will sprout. The second fact is even more astonishing, for it states just the opposite: that the prime numbers exhibit stunning regularity, that there are laws governing their behaviour, and that they obey these laws with almost military precision.

To see what he meant by this, we'll consider the prime-counting function $p(x)$, which counts the number of primes up to any number x. So $p(10) = 4$, because there are exactly four primes (2, 3, 5, 7) up to 10, and $p(20) = 8$, because there are now a further four (11, 13, 17, 19). Continuing, we find that $p(100) = 25$, $p(1000) = 168$, and $p(10,000) = 1229$.

If we plot the values of the primes up to 100 on a graph we get a jagged pattern—each new prime creates a jump. But if we stand back and view the primes up to 100,000, we get a lovely smooth curve—the primes do indeed seem to increase very regularly (Fig. 36.4).

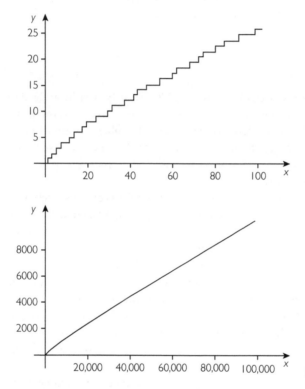

Figure 36.4 The distribution of the primes.

We can describe this regularity more precisely by comparing the values of x and $p(x)$ as x increases. We get Table 36.1, which lists x, $p(x)$, and their ratio $x/p(x)$.

So up to 100 one-quarter of the numbers are prime, up to 1000 about one-sixth of them are prime, and so on. We can express this 'thinning-out' more precisely by noting that whenever x is multiplied by 10, the ratio $x/p(x)$ seems to increase by around 2.3. This number 2.3 turns out to be the natural logarithm of 10, and the natural logarithm function $y = \ln x$ turns multiplication into addition. We can summarize this phenomenon by saying that, as x increases, $p(x)$ behaves like $x/\ln x$ – or, more precisely, that the ratio of $p(x)$ and $x/\ln x$ approaches 1 as x becomes large. This celebrated result is known as the 'prime number theorem': the German mathematician Carl Friedrich Gauss guessed it while experimenting with prime numbers at the age of 15. But it was not proved until 1896, when the mathematicians Jacques Hadamard (from France) and Charles de la Vallée Poussin (from Belgium) did so independently, using sophisticated ideas from calculus.

Table 36.1 x, p(x), and their ratio x/p(x).

x	p(x)	x/p(x)
10	4	2.5
100	25	4.0
1000	168	6.0
10,000	1229	8.1
100,000	9592	10.4
1,000,000	78,498	12.7
10,000,000	664,579	15.0
100,000,000	5,761,455	17.4

The Riemann hypothesis

What is the Riemann hypothesis? It was introduced by Bernhard Riemann, who died at the age of 40 while a professor at Göttingen University where he had followed in the footsteps of Gauss. Elected to the Berlin Academy in 1859, Riemann was required to contribute to the academy's *Proceedings*, and the result was his only paper in number theory, 'On the number of primes less than a given magnitude' (now regarded as a classic).[12]

In broad terms the Riemann hypothesis asks: 'Do all the solutions of a particular equation have a particular form?'. This is very vague, and the full statement, which still remains unanswered 150 years later, is: Do all the non-trivial zeros of the zeta function have real part $\frac{1}{2}$? But what does this mean, and what is its connection with prime numbers?

To answer these questions we need to enter the world of infinite series. The series

$$1 + \frac{1}{2} + \frac{1}{4} + \frac{1}{8} + \frac{1}{16} + \dots,$$

where the denominators are the powers of 2, continues for ever, which is why it is called an 'infinite series'. What happens if we add all the numbers in the series together? If we add them up one by one, we get

$$1, \text{ then } 1 + \frac{1}{2}\left(=1\frac{1}{2}\right), \text{ then } 1 + \frac{1}{2} + \frac{1}{4}\left(=1\frac{3}{4}\right), \text{ then } 1 + \frac{1}{2} + \frac{1}{4} + \frac{1}{8}\left(=1\frac{7}{8}\right),$$

and so on. Whenever we add just a finite number of terms of the series we never actually reach 2, but by adding together enough terms we can get as close to 2 as we wish; for example, we can get within 1/1000 of 2 by adding the first twelve terms of the series. We express this by saying that the infinite series *converges* to 2, or *has the finite sum* 2, and write

$$1 + \frac{1}{2} + \frac{1}{4} + \frac{1}{8} + \frac{1}{16} + \dots = 2.$$

In the same way, we can show that the infinite series

$$1 + \frac{1}{3} + \frac{1}{9} + \frac{1}{27} + \dots,$$

where each term is one-third of the previous one, converges to $1\frac{1}{2}$, and that, for any number p, we can write

$$1+1/p+1/p^2+1/p^3+\ldots=p/(p-1).$$

We shall need this last result later on.

However, not all infinite series converge. One celebrated example of a series that does not have a finite sum is the so-called harmonic series

$$1+\frac{1}{2}+\frac{1}{3}+\frac{1}{4}+\ldots,$$

where the denominators are the natural numbers $1, 2, 3, \ldots$. To see why this series does not converge, we bracket it as

$$1+\frac{1}{2}+\left(\frac{1}{3}+\frac{1}{4}\right)+\left(\frac{1}{5}+\frac{1}{6}+\frac{1}{7}+\frac{1}{8}\right)+\ldots.$$

But the sum of the numbers in each bracketed expression is greater than $\frac{1}{2}$, and so the sum of the whole series is greater than $1+\frac{1}{2}+\frac{1}{2}+\frac{1}{2}+\ldots$ which increases without limit. So the harmonic series cannot converge to any finite number. And amazingly, even if we throw away most of the terms and leave only those whose denominators are prime numbers,

$$\frac{1}{2}+\frac{1}{3}+\frac{1}{5}+\frac{1}{7}+\frac{1}{11}+\ldots$$

then there is still no finite sum.

In the early 18th century a celebrated challenge was to find the exact sum of the infinite series

$$1+\frac{1}{4}+\frac{1}{9}+\frac{1}{16}+\frac{1}{25}+\ldots,$$

where the denominators are the squares, $1, 4, 9, 16, \ldots$. This was answered by the Swiss mathematician Leonhard Euler, who showed that this series converges to $\pi^2/6$, an unexpected and remarkable result. In the same way, he showed that the sum of the reciprocals of the fourth powers is $\pi^4/90$, for the sixth powers it is $\pi^6/945$, and so on. He called the sum of the kth powers $\zeta(k)$, naming it the 'zeta function', that is,

$$\zeta(k)=1+1/2^k+1/3^k+1/4^k+\ldots.$$

So $\zeta(1)$ is undefined (since the harmonic series has no finite sum), but $\zeta(2) = \pi^2/6$, $\zeta(4) = \pi^2/90$, etc. It turns out that the series for $\zeta(k)$ converges for every number $k > 1$.

Although the zeta function and prime numbers do not seem to have anything in common, Euler spotted a crucial link, which we present in Box 36.1. This link can be used to give another proof that there are infinitely many primes.

We have now set the scene for the Riemann hypothesis, in which Alan Turing became so interested. As we have seen, the zeta function $\zeta(k)$ is defined for any number $k > 1$. But can we define $\zeta(k)$ for other numbers k? For example, how might we define $\zeta(0)$ or $\zeta(-1)$? We cannot define them by the same infinite series since we would then have

$$\zeta(0)=1/1^0+1/2^0+1/3^0+1/4^0+\ldots=1+1+1+1+\ldots$$

and

$$\zeta(-1)=1+1/2^{-1}+1/3^{-1}+1/4^{-1}+\ldots=1+2+3+4+\ldots,$$

and neither of these series converges. So we need to find some other way.

As a clue to how we proceed, we can show that, for certain values of x,

$$1+x+x^2+x^3+\ldots=1/(1-x).$$

We have already seen that this is true when $x = 1/p$. But it can be shown that the series on the left-hand side converges only when x lies between –1 and 1, whereas the formula on the right-hand side has a value for any x, apart from 1 (when we get 1/0 which is not defined). So we can *extend* the definition of the series on the left-hand side to *all* values of x (other than 1) by redefining it using the formula on the right-hand side.

In the same way, Riemann found a way of extending the infinite series definition of the zeta function to *all* numbers x other than 1 (including 0 and –1). But he went further than this, extending the definition to almost all complex numbers. Here, a 'complex number' is a symbol of the form $x + iy$, where i represents the 'imaginary square root of –1'; x is called the *real part* of the *complex number*, and y is the *imaginary part*. Examples of complex numbers are $3 + 4i$

Box 36.1 The zeta function and prime numbers

We can write the series for $\zeta(1)$ as follows:

$$\zeta(1)=1+\frac{1}{2}+\frac{1}{3}+\frac{1}{4}+\ldots=\left(1+\frac{1}{2}+\frac{1}{4}+\ldots\right)\times\left(1+\frac{1}{3}+\frac{1}{9}+\ldots\right)\times\left(1+\frac{1}{5}+\frac{1}{25}+\ldots\right)\times\ldots,$$

where each bracket involves the powers of just one prime number. Summing the separate series, using our earlier result that $1 + 1/p + 1/p^2 + \ldots = p/(p-1)$, gives

$$\zeta(1) = 2 \times 3/2 \times 5/4 \times \ldots = 2 \times 11/2 \times 11/4 \times \ldots .$$

If there were only a finite number of primes, then the right-hand side would have a fixed value. This would mean that $\zeta(1)$ has this same value, which is not the case since it is not defined. So there must be infinitely many primes.

Euler extended these ideas to prove that, for any number $k > 1$:

$$\zeta(k) = 2^k/(2^k-1) \times 3^k/(3^k-1) \times 5^k/(5^k-1) \times 7^k/(7^k-1) \times \ldots .$$

This remarkable result provides an unexpected link between the zeta function, which involves adding reciprocals of powers of numbers and seemingly has nothing to do with primes, and a product that intimately involves the prime numbers. It was a major breakthrough.

and $2i$ (which equals $0 + 2i$), and a real number such as 3 can be thought of as $3 + 0i$. There is nothing particularly mystical about this—all it means is that whenever we meet i^2 in calculations, we replace it by -1. Using a technique called 'analytic continuation', Riemann extended the definition of the zeta function to all complex numbers other than 1 (since $\zeta(1)$ is undefined). When k is a real number greater than 1, we get the same values as before; for example, $\zeta(2) = \pi^2/6$.

We can represent complex numbers geometrically on the 'complex plane'. This two-dimensional array consists of all points (x, y), where the point (x, y) represents the complex number $x + iy$; for example, the points $(3, 4)$, $(0, 2)$, and $(1, 0)$ represent the complex numbers $3 + 4i$, $2i$, and 1.

We have seen how Gauss attempted to explain why the primes thin out by proposing the estimate $x/\ln x$ for the number of primes up to x. Riemann's great achievement was to obtain an *exact formula* for the number of primes up to x, and his formula involved in a crucial way the so-called 'zeros of the zeta function'—the complex numbers z that satisfy the equation $\zeta(z) = 0$.

It turns out that $\zeta(z) = 0$ when $z = -2, -4, -6, -8, \ldots$ (these are called the 'trivial zeros' of the zeta function), and that all the other zeros of the zeta function (the 'non-trivial zeros') lie within a vertical strip between $x = 0$ and $x = 1$ (the so-called 'critical strip'). As we move away from the horizontal axis, the first few non-trivial zeros are at the following points:

$$\frac{1}{2} \pm 14.1i, \quad \frac{1}{2} \pm 21.01i, \quad \frac{1}{2} \pm 25.01i, \quad \text{and} \quad \frac{1}{2} \pm 30.4i.$$

The imaginary parts (such as 14.1) are approximate, but the real parts are all equal to $\frac{1}{2}$ (Fig. 36.5). Since all of these points all have the form $\frac{1}{2} +$ (something times i), the question arises: Do all of the zeros in the critical strip lie on the line $x = \frac{1}{2}$?

The Riemann hypothesis is that the answer to this question is 'yes'. It is known that the zeros in the critical strip are symmetrically placed, both horizontally about the x-axis and vertically

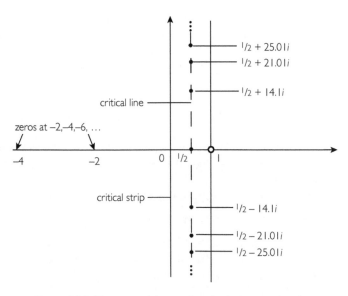

Figure 36.5 The zeros of the zeta function in the complex plane.

about the line $x = \frac{1}{2}$, and that as we progress vertically up and down the line $x = \frac{1}{2}$, many zeros do lie on it: in fact, the first trillion zeros lie on this line! But do *all* of the non-trivial zeros lie on the line $x = \frac{1}{2}$, or are the first trillion or so just a coincidence?

It is now generally believed that all the non-trivial zeros *do* lie on this line, but proving this is one of the most celebrated unsolved challenges in the whole of mathematics. Finding just one zero off the line would cause major havoc in number theory—and in fact throughout mathematics. But no-one has been able to prove the Riemann hypothesis, even after a century and a half.

Turing and the Riemann hypothesis

In 1936, while Turing was in Princeton, the Oxford mathematician E. C. Titchmarsh[13] proved that no zeros appear on the critical line up to a height of 1468. In the following year, working on a suggestion of Ingham, Turing set to work seriously on the zeros of the zeta function. Like Titchmarsh, but unlike most other mathematicians, Turing believed that the Riemann hypothesis might be false, and he contemplated building a special 'gear-wheel' calculating machine to search for any zeros that strayed off the critical line.

Turing wrote two papers on the Riemann hypothesis. The first, written in Cambridge in 1939 but not published until 1943, was highly technical and included numerical methods for calculating values of the zeta function.[14] Although it refined some of Titchmarsh's work, Turing's results were soon superseded, being made unnecessary by advances brought about by the use of electronic computers. In the same year Turing submitted a proposal to the Royal Society for the construction of a 'zeta function machine', with the aid of which he hoped to extend Titchmarsh's numerical results by a factor of about 4 (that is, up to a height of about 6000). Due to the outbreak of the war, the construction of this machine was never completed.

In 1948 Turing moved to the Manchester Computing Machine Laboratory, where in June 1950 he carried out his calculations on the zeros of the zeta function on the Ferranti Mark I electronic computer. His second paper described the process in detail, as he looked for zeros off the critical line around the height 25,000, and he also proved conclusively that none appears up to height 1540—a 'negligible advance', he said, as much more had been hoped for. He explained:[15]

The calculations had been planned some time in advance, but had in fact to be carried out in great haste. If it had not been for the fact that the computer remained in serviceable condition for an unusually long period from 3 p.m. one afternoon to 8 a.m. the following morning it is probable that the calculations would never have been done at all. As it was, the interval $2\pi \cdot 63^2 < t < 2\pi \cdot 64^2$ [24,938 to 25,735] was investigated during that period, and very little more was accomplished . . . The interval $1414 < t < 1608$ was investigated and checked, but unfortunately at this point the machine broke down and no further work was done. Furthermore this interval was subsequently found to have been run with a wrong error value, and the most that can consequently be asserted with certainty is that the zeros lie on the critical line up to $t = 1540$, Titchmarsh having investigated as far as 1468.

Further information about Turing and the Riemann hypothesis can be found in the articles by Booker[16] and Hejhal and Odlyzko.[17]

Conclusion

If the actual statement of the Riemann hypothesis seems an anticlimax after the big build-up, its consequences are substantial. Recalling Riemann's discovery of the role that the zeta function's zeros play in the prime-counting function $p(x)$, and his exact formula (involving the zeta function) for the number of primes up to x, we note that any divergence of these zeros from the line $x = \frac{1}{2}$ would crucially affect Riemann's exact formula, since our understanding about how the prime numbers behave is tied up in this formula. If Riemann's hypothesis fails, then the prime number theorem would still be true but would lose its command of the primes—instead of Don Zagier's 'military precision', the primes would be found to be in full mutiny!

We conclude with an unexpected development. In 1972 the American number theorist Hugh Montgomery was visiting the tea room at Princeton's Institute for Advanced Study and found himself sitting opposite the celebrated physicist Freeman Dyson. Montgomery had been exploring the spacings between the zeros on the critical line, and Dyson said: 'But those are just the spacings between the energy levels of a quantum chaotic system'. If this analogy indeed holds, as many think possible, then the Riemann hypothesis may well have consequences in quantum physics. Conversely, using their knowledge of these energy levels, quantum physicists rather than mathematicians may be the ones to prove the Riemann hypothesis. Turing would surely have been delighted with such an observation!

CHAPTER 37

Decidability and the *Entscheidungsproblem*

ROBIN WHITTY

I n 1936 Turing invented a mathematical model of computation, known today as the Turing machine. He intended it as a representation of human computation and in particular as a vehicle for refuting a central part of David Hilbert's early 20th-century programme to mechanize mathematics. By a nice irony it came to define what is achievable by non-human computers and has become deeply embedded in modern computer science. A simple example is enough to convey the essentials of a Turing machine. We then describe the background to Hilbert's programme and Turing's challenge—and explain how Turing's response to Hilbert resolves a host of related problems in mathematics and logic.

Turing in 30 seconds

If I had to portray, in less than 30 seconds, what Alan Turing achieved in 1936 it seems to me that drawing the picture shown in Fig. 37.1 would be a reasonable thing to do. That this might be so is a testament to the quite extraordinary merging of the concrete and the abstract in Turing's 1936 paper on computability.[1] It is regarded by, I suppose, a large majority of mathematical scientists as his greatest work.

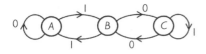

Figure 37.1 A machine that processes binary numbers.

The details of our picture are not especially important. As it happens, it is a machine for deciding which whole numbers, written in binary form, are multiples of 3. It works thus: suppose the number is 105, whose binary representation is 1101001, because $(1 \times 2^6) + (1 \times 2^5) + (0 \times 2^4) + (1 \times 2^3) + (0 \times 2^2) + (0 \times 2^1) + (1 \times 2^0) = 64 + 32 + 8 + 1 = 105$. We start at the node labelled A and use the binary digits to drive us from node to node. The first couple of 1s take us to node B and back to A again. The third digit, 0, loops us around at A. Now a 1 and a 0 take us across to node C; and the final 0 and 1 take us back via B to A once more. Now, if we finish at A we have a multiple of 3, otherwise we do not. You can try some more examples for yourself: 9 in binary is 1001 and that works; 19 is 10011 which leaves us at node B, so that works too, and so on.

Suppose you allowed me 90 seconds for my Turing pitch! Now I would draw a more elaborate picture (Fig. 37.2). The table tells me how to connect the nodes with arrows (1 for an arrow, 0 for no arrow) and what labels to put on the arrows (columns p and q). This table is completely described by writing down all the entries (the labels A, B, C, p, and q are immaterial) as a string of zeros and ones. And my machine itself has become a binary number! Just because we can, we can run the machine on itself: we finish up at node C, so 27,353 is not a multiple of 3. (You probably know the 'divides-by-3' trick anyway, but my machine, if a bit slow, is impressively minimalist.)

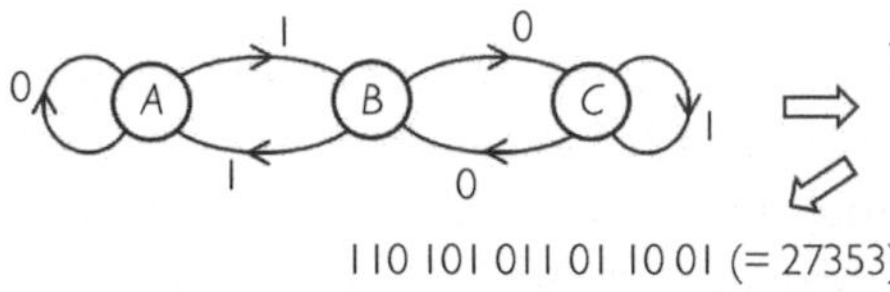

Figure 37.2 Machine to table to (binary) number.

	A	B	C	p	q
A	1	1	0	0	1
B	1	0	1	1	0
C	0	1	1	0	1

110 101 011 01 10 01 (= 27353)

Allow me a 2-minute pitch in total and I will add this run-on-itself feature to my drawing (Fig. 37.3). The machine has a 'tape' from which it reads the binary input, from left to right, just as you would do yourself if you wrote down a number and experimented with the machine. But my final extra 30 seconds has bought me another much more crucial enhancement. The machine now outputs on the tape as well: at the bottom of the diagram it has rubbed out the input digits and written, in their place, a two-digit binary number. This number is 00 if the machine stops at node A, 01 if it stops at node B, and 10 if it stops at node C. In fact, the machine is *computing*: it has computed, in binary, the remainder when 27,353 is divided by 3.

The machine as it now operates, reading and writing on the tape, incorporates the original three-node diagram (Fig. 37.1) in tabular form, but it has extra columns to tell it to move left as well as right on its tape and to write as well as read, as appropriate.

So, in a 2-minute picture, this is how Alan Turing invented computer science: with a machine that reads and writes binary digits on a linear tape. The machine itself can be represented by a binary number and can then be applied to itself: this seems like a curiosity, but it is very much more than that, as we shall see. But first, some historical background.

Hilbert's doomed programme

The back-story to Turing's work is this (see also Chapter 7). In 1900 David Hilbert, whose mathematical endeavours were, among his contemporaries, unequalled in breadth and depth, addressed the International Congress of Mathematicians in Paris with a list of 'millennium

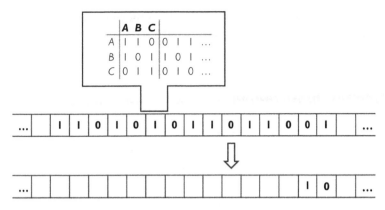

Figure 37.3 Machine with input/output tape.

problems' covering most of mathematics as it was then known. The tenth problem, in post-Turing language, was to write a computer program which would read in equations and write out 'yes' or 'no' according to whether the equation had whole-number solutions.

This was an immense challenge, of course. For example, Fermat's last theorem, that $x^n + y^n = z^n$ has no positive integer solutions for $n \geq 3$, would be proved if we wrote $x^{n+2} + y^{n+2} = z^{n+2}$ as our input and received 'no' as output. But Hilbert was convinced that any such challenge must yield to human ingenuity:[2]

This conviction of the solvability of every mathematical problem is a powerful incentive to the worker. We hear within us the perpetual call: There is the problem. Seek its solution. You can find it by pure reason, for in mathematics there is no *ignorabimus*.

To be strictly accurate, Hilbert's tenth problem concerns so-called 'Diophantine equations', in which whole-number values are sought for variables x, y, z, etc., which are added, multiplied, and raised to whole-number powers—algebraic expressions of this kind are called 'polynomials'. In Fermat's equation the variable n is itself a power, so the equation is not, properly speaking, Diophantine. Hilbert would not have accepted this as an extenuating circumstance. *No* mathematical equation should be able to plead *ignorabimus* ('we shall not know'): the problem, solvable or not solvable, must still succumb to pure reason.

More generally, to say that a problem is 'decidable' is to say that it can be given to a computer programme that will solve it. The output of the programme can just be 'yes' or 'no': we do not have to be more demanding than this, but we need the output to arrive in a finite number of steps. With the aid of this definition we can rephrase Hilbert: he was insisting that the problem of solving equations with integers is decidable.

In making this definition we can rely on familiarity with the ideas of computers and computer programming; in 1936 Alan Turing had to invent these ideas for himself. But the challenge of decidability had by then been set out clearly enough over a period of thirty years by the Hilbert problem. And to solve this—that is, to supply a computational foundation for some branch of mathematics, seemed to involve a monumental—but foreseeable—accumulation of painstaking technical progress. However, to demonstrate that the Hilbert problem could *not* be solved would require a complete innovation, a theory of what it means to compute. Turing produced this theory single-handedly at the age of 24. Although not alone in the achievement, his version of it, profound yet eminently practical, was to bequeath to mathematics ideas and techniques that are still being worked out today.

The *Entscheidungsproblem*

By the 1920s Hilbert had enlarged his vision well beyond the scope of his tenth problem, to encompass nothing less than all of mathematical truth. Diophantine equations are a part of arithmetic, and to say that a particular equation is solvable is to assert a theorem of arithmetic. We should have a systematic procedure for deciding whether this assertion is valid or not—and the same should be true of geometry or logic, or any other part of mathematics.

So, Hilbert was aiming to conquer, literally in his view, the whole of mathematics: given any mathematical system S (such as arithmetic), he wanted to establish that, for any proposition P,

Completeness: either P or **not** P is provable in S
Consistency: P and **not** P are not both provable in S
Decidability (the *Entscheidungsproblem*): there is a process for deciding in a finite number of steps whether P is provable in S.

The grandness of Hilbert's concept cannot be overestimated: from completeness and consistency, truth and provability become synonymous; then by the *Entscheidungsproblem*, truth becomes ascertainable. It is a wonderful implementation of Hilbert's resounding 'in mathematics there is no *ignorabimus*'.

Alas, as Hilbert entered the last decade of his life his programme was destroyed in every particular. In 1931 Kurt Gödel proved his incompleteness theorems.[3] The first exhibits a counterexample to completeness in any system S that has at least the expressive power of ordinary arithmetic; the second establishes that consistency cannot itself be provable as a theorem of S. This leaves open the *Entscheidungsproblem*: we might at least hope to have a process for checking which theorems are provable, even if an answer of 'no' fails to provide a certificate of a theorem's falsehood. By 1936 this hope too had been demolished—simultaneously, and on both sides of the Atlantic, by Alonzo Church and Emil Post, working independently at Princeton, and by Alan Turing.

Cantor and diagonalization

There is an indication in the title of Alan Turing's 1936 paper, 'On computable numbers with an application to the Entscheidungsproblem', of the extraordinary amount of inventiveness that it contains. The refutation of the final part of Hilbert's noble programme was almost an afterthought, appearing as the last of eleven numbered sections. Before that there were the creation of a model of computation and the notion of universal computation, and the demonstration that there are properties of numbers that are undecidable by computation, as modelled. Transferring this undecidability to the *Entscheidungsproblem* is accomplished by yet another new idea, that of reductions between computations: this idea was eventually to deny Hilbert even the solvability of Diophantine equations.

There are two technical themes that Turing, already by his early 20s an accomplished and widely read mathematician, was able to deploy in his paper. One is the concept of 'enumerability', or 'countability': that an infinite collection of things may be listed systematically by aligning its members with the positive integers (that is, by 'counting' them). The other is the reduction of mathematical ideas to the symbolic manipulation of strings of letters, familiar in the 1930s from certain problems in group theory and topology.

Before delivering his *coup de grâce* to the *Entscheidungsproblem*, Turing flexed his muscles with an application of enumerability which, as Copeland in his guide to the paper observed,[4] the Hilbert camp appear to have overlooked. This is that completeness implies decidability because we can enumerate *all proofs* of *all propositions*: those that are one letter long (if any such exist), those that are two letters long, and so on. Eventually, *in a finite number of steps*, for any proposition P, either P or **not** P must appear as a proved proposition.

The idea of enumerability dates back to the late 19th century—notably, to the work of Georg Cantor, championed by Hilbert for putting the infinite on a secure mathematical footing. In 1874 he proved the following result, now known as Cantor's theorem:

We cannot enumerate all infinite 0–1 sequences.

This is profound, but is demonstrated by what might seem to be a sleight of hand called 'diagonalization' that Cantor first used in 1891, and Turing was to redeploy forty-five years later. Suppose that we *can* enumerate all infinite 0–1 sequences. Align them with the positive integers: $s_1, s_2, s_3, \ldots$. Define a *diagonal* sequence S by making its ith entry the opposite of the ith entry of s_i; for example, if $s_4 = 0011001 \ldots$ with fourth entry '1', then S must have fourth entry '0'. Now S differs from every one of our enumerated sequences in at least one entry, and has therefore escaped the enumeration, whose existence is thereby disproved by contradiction.

It is time to return to our opening portrayal of Turing's work: a machine which is both described by, and operates on, finite binary sequences. Such machines are enumerable (Cantor's diagonalization does not apply to collections of finite sequences). But Turing was able to use exactly the same self-referencing trick as appears in Cantor's proof and apply it to an enumeration of machines. This had far-reaching consequences.

Turing's machine

We used the remainder-on-division-by-3 machine to represent Turing's model of human computation. It has all the features of Turing's general model: it can be represented by a binary sequence and it reads and writes backwards and forwards on a tape. This tape is required to be endless, so that a machine can go on working forever, perhaps producing an infinite amount of output. Crucially, the machine's binary representation can be written on its own input tape, whereby the machine is applied to itself.

Turing's idea of a machine which has a machine on its input tape is a game-changer, for two reasons. First, one might conceive of an ingenious machine which, finding another machine on its input tape, responds by carrying out exactly the steps which the input machine is designed to carry out, *no matter what machine this is*. The ingenious machine is accordingly called 'universal': it is the blueprint of a modern, stored-program computer. Turing devised such a machine.

Second, putting the ingenious machine on to its own input tape produces self-referential computation, and this, as we shall see, demolishes the *Entscheidungsproblem*.

We shall certainly not attempt to describe how Turing's universal machine works, although it is just a couple of dense pages in his 1936 paper. But we will follow his deployment of it as closely as we can, beginning with a glossary of Turing's original terminology:

Description numbers: numerical encodings of machine definitions (enumerable as *finite* 0–1 strings)
Circle-free machines: those that write infinitely many 0–1 symbols onto the tape
Satisfactory numbers: description numbers that encode circle-free machines
Computable sequences: tape outputs produced by circle-free machines (infinite 0–1 sequences, enumerated by enumerating satisfactory numbers)
Uncomputable sequences: infinite 0–1 sequences that are not the output of any machine (these must exist, by diagonalization).

The idea of a 'description number' is a natural application of enumerability: as we have seen, the totality of a machine—the initial input digits on its tape, and the list of instructions in its instruction table—can all be represented as long strings of zeros and ones, which together constitute a single positive integer, represented in binary form. Once a machine has become a number, it is natural to suggest that it be placed on a Turing machine tape and to ask what instruction table might reanimate this number—thus, Turing's idea of a universal computing machine.

Note that finite tape outputs are regarded as 'trivially' computable and can be excluded from our discussion. It was the infinite strings that brought Cantor's theorem to bear on the matter: some of these strings could not be computed, and Turing used them to dispatch the *Entscheidungsproblem*. To recap: infinite outputs are produced by circle-free machines whose description numbers are satisfactory. In what follows we shall simplify things a bit by using the word 'satisfactory' for both numbers and machines and writing 'satisfactory machine' where Turing would have written 'circle-free machine'.

An application to the *Entscheidungsproblem*

The customized universal machine, which we shall call H, is specified in 'pseudocode'. Turing, in devising the machine, was readily able to give a native Turing machine specification:

```
for k from 1 to infinity do (* infinite loop *)
    if k is description number of a satisfactory machine (i.e., is a satisfactory description
number) then
            • write corresponding machine, say Mₖ, on tape
            • simulate Mₖ until kth output digit reached
            • write binary negation of kth output digit on tape
    end if
```

The job of the machine H is to generate a version of the diagonal sequence S featuring in Cantor's sleight-of-hand proof of his theorem. Note that H must be a satisfactory machine if it is to work, since S is infinite. The machine works by enumerating all machines and throwing away those that do not produce infinite 0–1 sequences (i.e., are unsatisfactory). The trouble is that the machine cannot work—because if H is satisfactory it must eventually try to simulate H, which will in turn try to simulate *H*, and so on ad infinitum. The machine can never proceed to the point where it writes an output for machine *H*. But then its output is finite, in which case it is not satisfactory.

So Turing had to debug his programme, the first person in the world to do so! The simulation step (*H* simulating *H*) appears tricky, but it is fine: his proof of universality ensures this. Attention focuses on the '**if**' condition: can it be a bug? It *must* be a bug, because everything else is working. Turing concluded, exactly as he wanted to conclude, that he could not programme a machine to check satisfactoriness. This is known as 'Turing's theorem':

Satisfactory numbers are not computable: there can be no machine which, given any positive integer *k*, and in a finite number of steps, outputs 'yes' if *k* is satisfactory and 'no' if it is not.

It was a small step for Turing to establish undecidability on the basis of his theorem. Even so, it introduced the new idea of 'machine reductions'. Rather than giving details, we illustrate the ideas in Fig. 37.4. Note that there are *two* reductions. One shows that a hypothetical machine solving a new task (Turing's choice was an artificial 'printing problem') reduces to (or converts into) a machine solving satisfactoriness: thus this new task cannot be computed. This is the most influential kind of reduction, which has since been applied in hundreds of new cases and adapted to related tasks such as assigning a level of computational complexity to a computing task. Its influence in mathematics and mathematical logic is profound.

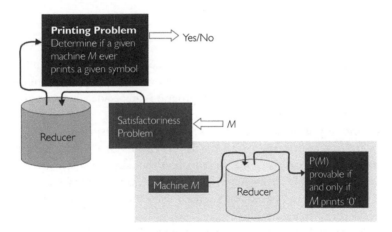

Figure 37.4 Reductions between Turing machines.

The second reduction turns machine *M* into a theorem *P*(*M*). A machine that solves the *Entscheidungsproblem* by checking provability can in turn solve the unsolvable printing problem, which in turn checks the uncheckable satisfactoriness property. Thus, an *Entscheidungsproblem* machine cannot exist; the last facet of the Hilbert programme shatters.

Turing and Hilbert: the final showdown

As a coda we return to Hilbert's tenth millennium problem: Can a computer program be written to decide solvability of any input Diophantine equation? Can we somehow reduce this to

Turing's prototype undecidable problem, the *Entscheidungsproblem*, so showing that the tenth problem, too, is unsolvable?

The story of this reduction is a fascinating one, and involves a seemingly extraordinary equivalence. We have seen that enumerable sets are those whose members can be listed systematically by a computer program. 'Diophantine sets' are those for which a Diophantine equation exists, one particular variable of which determines membership of the set. Recall Fermat's equation $x^n + y^n = z^n$ (not quite Diophantine but never mind). It defines a set of all values of n for which the equation is solvable in positive integers. The set begins $\{1, 2, \ldots\}$. And Fermat's last theorem says those two are its only members. Well, this set is certainly enumerable. But what about any other enumerable set? What about, say, the prime numbers $\{2, 3, 5, 7, 11, \ldots\}$: is this notorious set actually Diophantine?

In 1950 Princeton and Alonzo Church re-enter our story, through the work of Church's PhD student Martin Davis, whose dissertation contains a bold, not-to-say brash, proposal, known as 'Davis's conjecture':

All enumerable sets are Diophantine sets.

To cut a twenty-year story short, Davis's conjecture is true. It took the combined efforts of Martin Davis, Hilary Putnam, Julia Robinson, and Yuri Matiyasevich, himself a PhD student in Leningrad, to prove this. The proof appeared in 1970 and is known as the 'DPRM theorem' (after the initials of these mathematicians):

A set is enumerable if and only if it is a Diophantine set.

An immediate corollary of this theorem is: no algorithm can exist for solving general Diophantine equations. Such an algorithm would automatically decide membership of any enumerable set. And we know this to be impossible: provable statements are enumerable, as we have seen, but Turing showed that no computer program can decide whether a given statement belongs to the set of provable ones.

DPRM has some surprising additional consequences—surprising enough for Davis's original conjecture to have been treated with much scepticism. One is that there is a *universal Diophantine equation* in a *fixed* number of variables that simulates any given Diophantine equation, by making a suitable choice of one of the variables. Matiyasevich even put a value on the number of variables: it can be as low as 9. This universality result relates directly to Turing's idea of a universal Turing machine.

A more immediate, and more amazing, consequence of the DPRM theorem is that, for any enumerable set E, there is a polynomial whose *positive* values are precisely the members of E. Think of *any* conjecture; it is likely that any counter-examples can be enumerated; for example, the set of counter-examples to Goldbach's conjecture that any even number greater than 2 is the sum of two primes, or the set of counter-examples to the Riemann hypothesis (see Chapter 36). Then there is a polynomial whose positive values are precisely these counter-examples!

No wonder mathematicians in the 1950s were dubious: even a polynomial taking exactly the prime numbers as its positive values seemed a far-fetched notion! Nevertheless, the proof of the DPRM theorem spurred mathematicians to find such a polynomial, and several are now known. Perhaps the most elegant is the following, in twenty-six variables.[5] It may look horrendous, but if you substitute any whole numbers you wish for each of the twenty-six variables $a, b, c, \ldots, z$, and if the resulting number you get is positive, then this number must be a prime.

Moreover, *every* prime number can be obtained in this way, by making a suitable choice of whole numbers $a, b, c, \ldots, z$:

$$(k+2)\{1 - (wz+h+j-q)^2$$
$$- [(gk+2g+k+1)(h+j)+h-z]^2$$
$$- [16 \ (k+1)^3 (k+2)(n+1)^2 +1- f^2]^2$$
$$- (2n+ p+q+z-e)^2 -[e^3(e+2)(a+1)^2 +1-o^2]^2$$
$$- [(a^2 -1)y^2 +1-x^2]^2 -[16r^2 y^4(a^2 -1)+1-u^2]^2$$
$$- \{([a+u^2(u^2 -a)]^2 -1)[(n+4dy)^2 +1-(x+cu)^2]^2$$
$$- (a^2 -1)l^2 +1-m^2\}^2 -[(a-1)i+k-l+1]^2 -(n+l+v-y)^2$$
$$- [p+l(a-n-1)+b(2an+2a-n^2 -2n-2)-m]^2$$
$$- [q+y(a-p-1)+s(2ap+2a-p^2 -2p-2)-x]^2$$
$$- [z+ pl(a-p)+t(2ap-p^2 -1)- pm]^2\}.$$

It is not too hard to see that this polynomial can take a positive value only if all the squared terms are 0, so to find prime-number values we must solve fourteen Diophantine equations.

Each of these fourteen equations has solutions that grow rapidly in size and seem to defy analysis; to combine the solutions to all of the equations seems to be a formidable task—indeed, to the best of my knowledge, even the prime number 2 has never been exhibited *explicitly* as a value of the polynomial. Nevertheless, it is still a fact that *every* prime number can be produced in this way! It strikes one as the kind of problem that Turing would have enjoyed tackling; it is conceivably a problem that could never have existed without him.

CHAPTER 38

Banburismus revisited: depths and Bayes

EDWARD SIMPSON

C hapter 13 covered all aspects of Banburismus, but without much of the mathematical and other detail. This chapter seeks to explain depth, Bayes' theorem, logarithmic scoring (and its application to the Banburies discovered in the roof of Hut 6), decibanning, chains and depth cribbing, comic strips, and twiddling. This seems to be the first publication of much of this explanation.

Depth

In Chapter 13 I promised a simple illustrative example of depth and some of the terms associated with it.

In Table 38.1 QVAJX . . . and the others are three enciphered messages as transmitted. This table does not relate to any particular real-life cipher, and its three enciphered messages are unrealistically short. It illustrates an imaginary system which enciphers letter by letter (as Enigma did), using an enciphering table which would provide hundreds, or more probably thousands, of individual enciphering units, whatever the nature of the imaginary enciphering system. The numbers 180, 181, . . . indicate the positions of the units in the table. The encipherer could choose where in the table to begin enciphering (say, at position 183) and, once started, continued to use the enciphering units in sequence (183, 184, 185, . . .). The legitimate receiver of the message had to be told where to start in the table. This was done by an 'indicator', which was itself enciphered, usually by a different system, and formed part of a preamble to the message as transmitted.

The three enciphered messages in this example, which may have come from different sources, start at different positions in the enciphering table. They have significant overlaps, and it is where they overlap that they are said to be 'in depth'. Here the depths are of two messages (as in column 182) or of three (column 183). In real life they could be of ten or twenty. All the letters in a particular column (as QDF in column 183) have been enciphered in exactly the same way.

Table 38.1

Position in table	180	181	182	183	184	185	186	187	188	189	190	191	192	193
Message 1				Q	V	A	J	X	P	B	K	S	B	
Message 2	W	R	N	D	M	F	C	I	R	U	K	S	D	Y
Message 3			N	F	Y	A	M	S	A					

Where the same letter appears twice or more in a column (as N in position 182 or A in position 185) this is called a 'repeat'. A two-letter sequence is called a bigram. In Chapter 13 it was explained how a repeating bigram (e.g. KS in columns 190–191) provides more powerful evidence than two single repeats. Repeating trigrams or better are more powerful still.

The cryptanalysts, unless they had the good fortune that the indicator's enciphering system had already been solved, needed a different way of getting started. Applying Banburismus to the numbers of repeats, in order to set messages in depth, provided one way.

The scale of the problem

It is important first to get a grasp of the number of variations of which Enigma was capable, and the consequent complexity of solving it anew each day. There were of course variations between one Enigma model and another, and between different periods of their use. The following figures are typical for the three-wheel Naval Enigma machine with ten leads on the plugboard.

The three wheels in order can be chosen from the available eight in $8 \times 7 \times 6 = 336$ ways. Their three ring settings can be chosen in $26^3 = 17,576$ ways. Combining these gives 5.91×10^6 combinations from the wheels alone. On the plugboard, for the first lead, there are 26 ways of inserting the first end and 25 of inserting the second. For the second lead, there are 24 and 23, respectively. Then 22 and 21, and so on down to 8 and 7 for the tenth lead. These multiply together to give 5.6×10^{23} combinations. But we have overcounted. The order of choosing the ten leads is irrelevant, so to avoid multiple counting we have to divide by $10 \times 9 \times 8 \times 7 \times 6 \times 5 \times 4 \times 3 \times 2 = 3.63 \times 10^6$, and since it is also irrelevant which end of a lead is inserted first, we have to divide again by $2^{10} = 1024$. The combined divisor, 3.72×10^9, reduces the number of different plugboard combinations to 1.51×10^{14}. When these are combined with the 5.91×10^6 wheel orders and ring settings, we obtain the final total of 8.9×10^{20}, or 890 million million million daily keys. This is some 75 million million times the number for a five-letter combination safe (see Chapter 13).

The often-quoted smaller figure of 159 million million million keys relates to the Army or Air Force Enigma which had three wheels chosen from five, instead of the Naval Enigma's choice of three from eight (Banburismus was not used against Army or Air Force Enigma).[1]

Tetras and the Hollerith section

As we saw in Chapter 13, pairs of messages with only one indicating letter in common were so numerous that nothing less than a repeat of four consecutive letters (a tetragram) was worth pursuing. Finding these was beyond the reach of manual methods, and the search was performed, starting afresh each day, by the Freebornery, the Hollerith section under Frederic Freeborn, which was first located in Hut 7 and later occupied all of Block C.

As Ronald Whelan, Freeborn's deputy, has written:[2]

The daily Hollerith processing carried on in Hut 7/Block C in dealing with the Enigma message texts was accorded the highest priority. The requirement was to search through the day's traffic to locate [tetragram] repeats occurring between messages. The teleprinter material was edited in Hut 8 . . . These edited message texts were then made available to Hut 7/Block C without delay throughout the day.

Although dealing with small batches of cards, in this way, was wasteful of machine operator time, as well as affecting machine availability for less important work, it did enable the output results to be made available to Hut 8 in the shortest possible time following their advice that the 'cut' was to be made: that the final batch of messages had been made available to us . . .

It was the case that the daily Enigma traffic gave rise to a total of around 80,000 cypher text characters. The Hollerith processing in dealing with this material called for slick and expert machine operation, and involved a large number of machines for many of the operations . . .

The 'cut' instruction from Hut 8 was liable to be made late in the evening, in which case the Team Leader would draft many machines and operators to the work to gain the earliest possible completion. This did not necessarily mean that an operator was needed for each machine pressed into service, because many operators were adept in running two or more machines at the same time.

Sybil Griffin joined Bletchley Park early in 1940 as a Foreign Office civilian, at the age of 17. Recalling Tetras as a routine daily job in the Hollerith section, and illustrating the variety of Hollerith machines in use, she recalls:[3]

On Tetras we were in teams of civilians—the Wrens came later. The Team Leader was in charge but we all got on with the job in hand.

Basically we were looking for 4-letter repeats. The messages were punched on to cards in sections of 26 letters and cards were taken to the machine room. A duplicate card was made using the Reproducer, which was in another room, and one card was stored. (Some people thought that was all we did!) Then 25 blank cards were inserted behind each master card using a Collator. All cards then went back to the Reproducer, where 4 letters at a time were rotated. The cards were then sorted: there was a row of Sorters working together, maybe six. Lists were printed, in yet another room, and sent to sections needing them to look for repeats and patterns to break in to Enigma.

We could chat and spoke of social activities, dances, trips to London, etc. We used to go up to London by train after the midnight to 9 am shift. It was possible, if rather exhausting, for one person to run several machines. I remember being on duty one weekend (after a week training Wrens) and working alongside an American serviceman. We ran all the Sorters, Collators and Listers between us because there was nobody else on duty!

Ronnie Whelan kept in constant touch during a day's work and Mr Freeborn walked round often, always bringing visitors to Hut 7/Block C. When VIPs visited Bletchley Park they always visited us. I remember Admiral Cunningham, Sir Charles Portal, and Lord Louis Mountbatten being shown around. Churchill visited BP.[4] The Scharnhorst was sunk on Boxing Day 1943 and Admiral Sir Bruce Fraser gave a lecture to tell all of us involved how much we had helped when Scharnhorst and Bismarck were sunk, and we in Block C were invited.

We were not totally in the dark about what the work was for, as Mr Freeborn had told some staff as much as he could to keep their interest. Probably the first time we realised we were helping was when the Bismarck was sunk in May 1941. To show their gratitude the cook of one of our ships involved sent us a cake!

Bayes' theorem

The solving of the daily Enigma depended on a theorem proved almost two centuries earlier by the eponymous Presbyterian minister of a church in Tunbridge Wells, Thomas Bayes. The Royal Society of London published his theorem in 1763. Its importance is this: it shows us how to assign numbers, representing levels of reliability, to what we learn from experience and what we conjecture. Sadly Bayes was in ill health and did not live to see his theorem in print.

We start with a hypothesis to be tested, for example that a certain statement is true. We start also with some understanding of the credibility of the hypothesis, which may be objective or subjective. For simplicity I depart a little here from Bayes' own exposition and express that credibility in terms of odds on, or odds against, the hypothesis. Odds are simple: if something happens on average four times out of five trials (so 80% probability of its happening), the odds are 4:1 on its happening (or 1:4 on its not happening).

What we start with are the prior odds—prior, that is, to an event that gives us some additional evidence. We now need to know whether that event is more likely to happen when the statement in the hypothesis is true or when it is false; and how much more likely—or, to be more precise, to know the ratio of the first of those probabilities to the second. That ratio is called the 'Bayes factor'. Bayes' theorem tells us that the credibility after we have the additional evidence (the posterior odds) is obtained by multiplying the prior odds by the Bayes factor, and the credibility is enhanced or diminished according to whether the factor is above or below 1. Moreover, if there are successive events that are independent of each other, their factors can be multiplied to give a composite factor.

To illustrate this, suppose that I work in quality control for a firm making torch bulbs. Each bulb is tested at a high level of current, and the experience is that one bulb in five fails. When I start using a new tester, all five of the first five bulbs tested fail, instead of the expected one. Suspicious, I recall the fable of the railway worker whose job was to walk alongside a stationary train, testing each wheel with his hammer. If the blow produced a dull thud instead of a clear ring, the wheel was assumed to be cracked. He had caused nineteen carriages to be taken out of service before he realized that his hammer was cracked.

On enquiry, the suppliers confess that my new tester came from a suspect batch in which 1% of the testers were faulty: they passed too high a current and blew one quarter of the acceptable bulbs as well as all the substandard ones. I now need to test the hypothesis that my new tester is one of the faulty 1%.

Out of a batch of 100 bulbs, a true tester will fail 20. Out of such a batch a false tester will fail those 20, but will also (by reason of its own fault, as explained by the supplier) fail one in four of the other 80 (i.e. another 20), meaning that 40 fail in all out of the 100 bulbs. Thus the probability that a bulb will fail when tested by a faulty tester is 40%, compared with a probability of failure of 20% when tested by a true tester. A single bulb failure is thus twice as likely (40% against 20%) to occur with a faulty tester as with a true one. This means a Bayes factor of 2 in favour of the hypothesis that my tester is a faulty one.

We now apply the same reasoning to a bulb's passing the test. Out of a batch of 100 a true tester will pass 80. For a faulty tester to pass a bulb, two things are necessary: that the bulb is one of the 80 good ones and that the tester is on one of its three out of four good behaviours. On average, 60 (i.e. 3/4 × 80) of the 100 will comply. It is thus 3/4 times as likely (60 against 80) that a single bulb will pass the test with a faulty tester as with a true one. This means a Bayes factor of 3/4 in favour of the hypothesis that my tester is faulty (the factor being below 1 makes the hypothesis less likely).

The prior odds for my tester being faulty (as explained by its supplier) are 99:1 against, or 1:99 on, because 1% of the testers in the suspect batch were faulty. The evidence that aroused my suspicion—five consecutive blown bulbs each with a Bayes factor of 2—applies a composite factor of $2^5 = 32$ to the prior odds of 1:99 on the tester being faulty. So the posterior odds are about 1:3 that the tester is faulty (or 3:1 that it is good). This is inconclusive. Perhaps those five blown bulbs were a freak.

I now test another ten bulbs and find that four are blown and six are accepted. The further composite factor resulting from this, taking the respective Bayes factors, is $2^4 \times (3/4)^6 = 2.85$, and this moves the posterior odds from about 1:3 that the tester is faulty to just under evens. This is still inconclusive. Another 30 tests deliver 12 bulbs blown and 18 accepted, bringing a further composite Bayes factor of $2.85^3 = 23.15$. The posterior odds now move from just under evens to 21:1 on the hypothesis that the tester is faulty. This passes beyond what statisticians commonly accept as the first-level criterion for evidence establishing truth: 95% confidence, or odds of 19:1 on. The tester must be discarded.

Weighing the evidence

To return to Banburismus and to how much evidence was 'enough'. Alexander's history sets up an example of two messages set in correct alignment, which provide an overlap of 32 letters enciphered at the same machine positions. In 7 of those 32, the same letter occurs in both messages: 7 repeats. He continues:[5]

It is fairly obvious that the repeat rate [of letters] for plain language is higher than that of random cypher material . . . Random 1/26 . . . German Naval 1/17 . . . This provides us with a criterion for testing whether or not a given [alignment[6]] is correct, viz. the number of repeats or 'score' obtained by writing out the messages concerned at that [alignment]. Going back to our example . . . the 'score' is 7 repeats . . . in a stretch of 32 letters. To determine the merits of this score it is necessary to know whether it is more likely that such a result would arise from a true [alignment] or by chance—and how much more likely one way is than the other. It can be shown that it is about 12 times as likely that this particular score would arise in a true as in a false position.

Neither Alexander's history nor Mahon's (see Chapter 13) explains the 1 in 17 repeat rate used for German naval language. Earlier workers had used 1 in 20 in 1920 and 1 in 16.5 in 1940. Alexander records that the capture of the June and July 1941 keys gave them an opportunity to overhaul their methods:[7]

A fresh statistical investigation into the frequency of repeats between messages was made and our old figures revised.

In fact the 1 in 17 repeat rate is puzzlingly low. From a frequency count in Wikipedia we can derive a repeat rate of 1 in 13 for everyday German language, and the language of German naval signals might have been expected to be more stereotyped, so giving an even higher repeat rate.

Alexander's history jumps over the derivation of '12 times as likely' by 'it can be shown that'. In fact, it depends on a composite Bayes factor. The hypothesis to be tested is that the alignment is true. A single repeat has a probability of 1/17 when the alignment is true (because two stretches of German naval language are being compared) and of 1/26 when the alignment is false, so there is a Bayes factor of (1/17) ÷ (1/26), or 1.53, in favour of truth for each repeat. A single no-repeat has a probability of 16/17 when the alignment is true, and of 25/26 when the alignment is false, so there is a Bayes factor of (16/17) ÷ (25/26), or 0.979, in favour of truth for each no-repeat. Such a factor, just below 1, means that each no-repeat slightly reduces the odds on the hypothesis that the alignment is true, diminishing its credibility and making it slightly more likely that the alignment is false.

Alexander's example has 7 repeats and 25 no-repeats in a stretch of 32 letters, and the Bayes factors for the successive events have to be combined to give a composite factor. Supposing (for the moment) that the occurrences of consecutive letters are independent of each other—such a supposition is patently false, and persevering with it regardless is called 'naive Bayes' as it often scarcely affects the conclusion—the individual factors can simply be multiplied and the resulting composite factor in favour of truth is $1.53^7 \times 0.979^{25} = 11.6$. This rounds up to 12, confirming the phrase 'about 12 times as likely' in the quotation from Alexander.

After deriving 'about 12 times as likely', Alexander went on to say:

In the method of units and scoring used by us (logarithmic scoring) it would be a score of +22.

The derivation of this figure can also be reconstructed. Much ingenuity went into manipulating the factors so that testing could be done faster and by less skilled staff. First the factors were replaced by their logarithms, so that addition could replace multiplication, and were called 'scores'. The above Bayes factor of 1.53 for a repeat has logarithm 0.1847. Since bigger numbers are easier to handle, Hut 8 moved to scoring units of one-tenth the size, giving a score of 1.85 for one repeat. Later they halved this unit for greater facility, so one repeat then scored 3.7 and the seven in the example added up to 25.9. In practice Hut 8 may have rated speed over precision and rounded these numbers.

For a no-repeat the Bayes factor of 0.979 has logarithm –0.0092, or (on multiplying by 20 for the two stages above) –0.184. This is tiny in itself, but there are many no-repeats. For the 25 in the example, the score is –4.6. This diminishes the 25.9 to 21.3. This is very close to 'a score of +22', as Alexander wrote. Perhaps the small difference arises from different rounding at some point in the calculations.

As here, Alexander was always on the lookout for ways of streamlining procedures. He wrote:[8]

We found that by splitting up a job into as many separate parts as possible and by having a really adequate set of scoring tables . . . the whole process [was] reduced to a matter of looking things up in a number of different tables and the only mathematical operations involved were addition and subtraction.

The histories by Alexander and Mahon do not mention prior odds, nor a threshold score that had to be reached before an alignment was adopted as probably correct. In practice, scores were perhaps seen as one component rather than a sole decider, and an experienced Banburist probably relied on his judgement, taking other considerations into account too, rather than on an automatic threshold.

The Banburies in the roof

I am grateful to Bletchley Park for enabling me to study the papers found in the roof of Hut 6 in some detail. Banbury One (see Fig. 13.2) is roughly torn off at column 114. The writing along its top, 'with TGC $11^{xx}/159$', is explained in Chapter 13 as meaning: 'when this message and another with indicator TGC were compared, with an overlap of 159 letters, 11 repeats were found of which four came as two bigrams'. There must have been much more to come beyond column 114 to make an overlap of 159 possible. Banbury Two (see Fig. 13.3) lacks its first 36 columns through rough tearing, and ends (apparently through careful tearing, perhaps along a crease) at column 160. The near coincidence of this figure with the 159 overlap invites conjecture, but no plausible significance can yet be attributed to it.

Given these 11 repeats, how plausible is this alignment? By simple averaging, we should expect 159 comparisons to yield 6 repeats if the alignment is false (repeat rate 1:26 for random letters) or 9 repeats if it is true (repeat rate 1:17 for Naval German). Finding 11 repeats is encouraging, so we turn to Bayes factors. The Bayes factor for 11 repeats in 159 comparisons is $(26/17)^{11} \times (16/17 \times 26/25)^{148}$, which works out at about 4.7. Colloquially, this make-up of 159 comparisons is about 4.7 times as likely to occur in a true alignment as in a false one. The logarithm of 4.7 is 0.6701, and (moving to the one-twentieth scoring units) 20 times this is 13.4. So the alignment scores 13.4.

By the shortcut method using scoring units instead of Bayes factors, 11 repeats each scoring 3.7 make 40.7, while 148 no-repeats, each scoring –0.184, make –27.2, giving a net score of 13.5, near enough to the 13.4 above. Unless there is other supporting evidence, this is far short of what is needed to accept the alignment as probably true.

It is tempting to conjecture that Banbury Two, which carries no identification (probably because its first 36 columns are missing), is the one with indicator TGC. But scoring them at the alignment which provides a 159-letter overlap makes this look very unlikely.

The back of Banbury Two had been used for workings. It had been turned over and placed vertically on the table with its torn left (starting) end at the top, and a column of letter sets jotted down. The column is headed '105' and then 'GOY' with the G in the characteristic hand of 'with TGC' on Banbury One. There follow sets of five letters grouped as three and two and spaced as 'AAN NB', in alphabetical order; perhaps some 50 sets in all. Some are marked '?', others 'poss'. Beyond the observation that these workings probably relate to the indicator system, their purpose has not yet been discerned.

Decibans

In 1950 Jack Good recorded that:[9]

Turing suggested further that it would be convenient to take over from acoustics and electrical engineering the notation of bels and decibels (db). In acoustics, for example, the bel is the logarithm to base 10 of the ratio of two intensities of sound.

There was already some playful language here. In 1928 the Bell Telephone Laboratories invented a measure that was the base-10 logarithm of the ratio of two measures of anything that could be measured, and named it the 'bel' in honour of their founder Alexander Graham Bell; they further named the more practicable one-tenth unit the 'decibel'. So soon after the end of the war, Good was unable to go on to say that Hut 8 extended their Banbury vocabulary by re-naming the bel in their context as the 'Ban' and their one-tenth and one-twentieth scoring units as the 'deciban' and 'half-deciban' or hdB. As Christine Ogilvie-Forbes explains, 'decibanning' then entered Hut 8's vocabulary as the name of a procedure using 'large sheets of paper and simple arithmetic with the odd Greek letter thrown in'.[10] She recounts:

I was decibanning the day Churchill came round. He leaned over me and said 'where was the message?' No letters in sight. Alexander rather shortly said it had nothing to do with the message, only maths. The men had lined up to be introduced and asked by Churchill what they had been previously—'undergraduate, undergraduate etc.' Alexander said 'draper'. That slightly checked the flow.

Alexander had been Research Director of the John Lewis Partnership before being recruited to Bletchley Park. 'Draper' was his amusedly diminishing way of referring to one of the United Kingdom's biggest retailers of all kinds of cloth.

Chains and depth cribbing

Enigma's complexity arose from the choice of three wheels and their order, their three ring alphabet positions, and the plugboard pattern. As described in Chapter 12, the bombe could, given time, find what combination of these would decipher a message text so as to match a crib with which it had been provided. The plugboard pattern changed daily, while the wheels and ring settings changed every two days. The job of the Banburist cryptanalysts was to speed the bombes' work by reducing the number of wheel orders to be tested. No two situations facing them were ever the same. They started each time with the whole assembly of plausible or better alignments of pairs of messages resulting from the two processes already described in Chapter 13. Each alignment, with its credibility label attached, provided the alphabetic distance between a pair of enciphered letters in the enciphered message setting (Chapter 10). Starting with the most credible alignments, and linking pairs with one letter in common, they constructed chains with more enciphered letters spaced out according to the known distances.

These relationships were only relative. To find the true relationships of the enciphered letters to their underlying clear (i.e. plaintext) equivalents, each chain of letters spaced out by gaps was tested in each of the twenty-six possible positions against a true alphabet. Enigma's reciprocal feature immediately doubled the number of letters along the chain (for if A was enciphered to Q, then Q was enciphered to A) and correspondingly diminished the number of gaps. Where this produced contradictions, that speculative positioning of enciphered letters against true ones was eliminated.

For those chains and positionings which survived the first test, some further alignments with next-best credibility were brought in to add evidence. As gaps were filled and the twenty-six tests were repeated, further contradictions led to further eliminations. Eventually a particular chain in a particular position against the true alphabet was either wholly eliminated—indicating that some speculative alignment was wrong—or a triumphant survivor. The outcome was the set of thirteen pairs of letters, each to be substituted for the other, constituting the current 'alphabet' for that wheel. But because the setting of the wheel's ring was still unknown, that was not enough to determine which wheel it was. Another feature had to be brought into play.

The wheels had notches that governed when they turned their neighbour over (see Chapter 10), and the building of chains also revealed where the turnovers occurred. It was the finding of the positions of its notches that enabled a wheel to be identified. The five wheels that were common to the Enigma machines of all three military services had their notches in different places, but the three additional wheels in naval machines had them all the same. If all eight had been notched in the naval fashion, Banburismus would have been far less valuable. As it was, finding the turnover point of a wheel could identify which wheel belonged in that slot for those two days. With an accompanying crib, this paved the way for sending a menu to the bombes, as described in Chapter 12.

The construction and testing to destruction of chains just described—the second stage of Banburismus—started from a speculative alignment of two messages of which nothing more was known. A still more sophisticated method (of aligning messages correctly) started from possible but unconfirmed alignments of several messages for which there were possible, but unconfirmed, cribs, usually for the beginnings of the messages. With the knowledge that if A enciphered into Q, then Q enciphered into A, and that no letter ever enciphered into itself, and working backwards and forwards between the messages and looking out for contradictions on the one hand and the appearance of German words on the other, the Banburist could perhaps confirm both the alignments and the cribs. Such an outcome provided a more powerful menu for the bombes than one that came from building chains. Falling short of their usual flair for '-ismus' names, Hut 8 called this 'depth cribbing'.

Comic strips

The exhibition 'Codebreaker: Alan Turing's Life and Legacy', which ran at the Science Museum, London, in 2012–13, had two exhibits described simply as 'cryptanalytic working aids used at Bletchley Park'. I have identified one of them as the 'comic strips' described by Turing in his 'Treatise on the Enigma', colloquially known as 'Prof's Book'.[11] As Turing wrote:

For demonstration purposes it is best to replace the machine by a paper model. We replace each wheel by a strip of squared paper 52 squares by 5 squares. The squares in the right hand column of the strip represent the spring contacts of the wheel . . . The machine itself is represented by a sheet of paper with slots to hold the 'wheels' . . .

and much more of greater complexity. Time after time his explanation of the attack on Enigma is illustrated by annotated drawings of the strips in appropriate positions (Fig. 38.1).

Their 'demonstration purposes' probably excluded operational use. After visiting the Navy Department in Washington in November 1942 (see Chapter 18), Turing wrote disparagingly:[12]

They were using comic strips a good deal which struck me as rather pathetic.

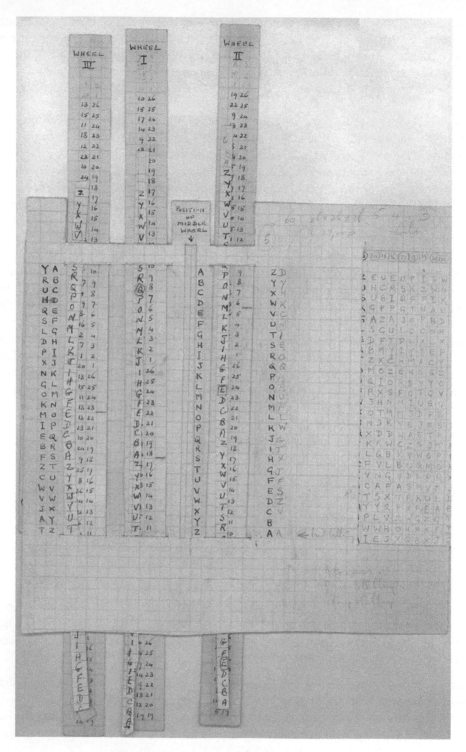

Figure 38.1 Turing's 'comic strips', in his handwriting.

'Prof's Book' is a very personal document. It was evidently typed by him, with many mis-types, and the corrections and diagrams are in his hand. Joan Clarke wrote of it:[13]

I was the guinea pig, to test whether his explanation and worked example were understandable, and my task included using this method on [that] half of the material [which Turing had] not used in the example.

Amid the complex and not particularly mathematical reasoning of 'Prof's Book', and among the details of the electrical installation required, we find:[14]

In order to avoid bogus confirmations we cover up the constatations with shirt buttons as they are used.

This is a splendid example of the way Turing would go straight to a creatively simple solution to a problem—a trait that led the more conventional to brand him as eccentric.

Twiddling

As Chapter 12 explained, the bombes' hypotheses went back to Hut 8 to be examined and sometimes tested by hand. Cipher text was typed on a replica Enigma and plain German was looked for in the output. Christine Ogilvie-Forbes was there too:[15]

The bombes stopped every time they got a possible answer. Wrens would send these results to us via the 'spit and suck' tube. There were two or three machines and we would set up each stop and Twiddle, hoping for German: it didn't take many letters.

But it was not always straightforward. As Eileen Plowman recalled:[16]

As a linguist I enjoyed my later work, on machines and decoding messages into German, much more. I was in charge of 'twiddling': adjusting settings to get out correctly the messages that had not come out when decoded by other girls. I remember working on a message with Alexander and Turing breathing down my neck and standing at my shoulder, waiting for the message to come out correctly. The message was then taken straight off to High Command, and I later heard that the U-boat had been sunk.

Conclusion: the factor method

This compressed account necessarily omits many features, some helpful to understanding, some not. The Germans applied some rules which, though intended to enhance security, simplified the task of the Banburists. For example, that there always had to be at least one naval wheel among the three in use reduced what would have been $8 \times 7 \times 6 = 336$ possible combinations to 276. A further rule, that the same wheel should never be used in the same position in consecutive two-day periods, could reduce it further (depending on the particular combination of wheels) to as few as 105.

Neither Alexander's nor Mahon's history mentioned the word 'Bayes', and Turing himself generally referred to 'the factor'. However, a still-withheld history by Alexander entitled 'The factor method'[17] showed that the term 'factor' was widely used, and that Bayes' theorem was indeed the fundamental mathematical theory of the factor method. This history, which I have seen by courtesy of GCHQ, opened with a conventional exposition of Bayes' theorem with traditional examples. It later exemplified its use on three ciphers which I was unaware of, as well as its application to Banburismus and to the Japanese Naval JN25 cipher which I described in 2010.[18] As Alexander summed up:

Banburismus is about the best example of the use of scoring systems the writer has seen. It would have been quite impossible without this technique, and the more whole-heartedly it was employed, the better the results obtained.

CHAPTER 39

Turing and randomness

ROD DOWNEY

I n an unpublished manuscript Turing anticipated by nearly thirty years the basic ideas behind the theory of algorithmic randomness, using a computationally constrained version of 'measure theory' to answer a question posed by Émile Borel in number theory: this question concerned constructing what are called 'absolutely normal' numbers. In this chapter we explain what these mysterious terms mean, and what Turing did.

Repeated decimals in fractions

Mathematicians have always been fascinated with patterns in numbers. At an early stage in our education we learn about the special nature of decimal expansions of 'rational numbers', fractions that we can write in the form m/n, for some whole numbers m and n with $n \neq 0$. The Greeks proved that some numbers, such as $\sqrt{2}$, $\sqrt[3]{7}$ and $\sqrt{2} + \sqrt{3}$ are not rational—indeed, it can be shown that 'most' numbers (in a precise mathematical sense) are irrational.

It can be shown that a real number is rational if and only if it has a finite decimal expansion, or a decimal expansion that repeats from some point onwards; for example, $1/4 = 0.25$ and $3/7 = 0.428571\ 428571\ 428571...$. Note that we can also think of $1/4$ as a repeating decimal, $0.25000000. . .$; we can also write it as $0.24999999 ...$, but for simplicity we ignore such ambiguities.

We can also count using bases different from 10. The binary system uses base 2, where each place in the representation corresponds to a power of 2; for example, just as 2301 in the decimal system refers to $(2 \times 10^3) + (3 \times 10^2) + (0 \times 10^1) + (1 \times 10^0)$, so in base 2 the decimal number $13 = (1 \times 2^3) + (1 \times 2^2) + (0 \times 2^1) + (1 \times 2^0)$ is represented by 1101. In base 3 we use only the numbers 0, 1, 2 and express numbers using powers of 3, so the decimal number $25 = (2 \times 3^2) + (2 \times 3^1) + (1 \times 3^0)$ is represented by 221. Note that when we use bases larger than 10 we have to invent extra symbols to represent the larger 'digits'; for example, in base 12 we might use the digits 0, 1, 2, . . . , 9, T, E, with T and E representing 'ten' and 'eleven'.

The above result about repeating patterns in the decimal representation of rational numbers remains true if we change the base from base 10 to any other base. For instance, in base 3,

1/4 = 0.020202. . . . From now on we will drop the decimal point and concern ourselves with infinite sequences of digits.

Normal numbers

In 1909 the French mathematician Émile Borel (1871–1956) was interested in number sequences that satisfy the 'law of large numbers' (Fig. 39.1).[1] This law says that if we repeat an experiment many times, then the average number of successes should be the expected value; for example, if we toss a fair coin many times we would expect to get a head approximately half of the time, and if we toss a fair die we would expect to throw a six in about one-sixth of the tosses. In base 10 this law says that the frequency of choosing a digit from 0 to 9 is exactly what we would expect in the limit—namely, 1/10. But what do we mean by 'in the limit'?

Base 2 corresponds to tossing a coin. Over time, we would expect as many heads as tails, but this is only the eventual long-term behaviour. If we toss a head, the next toss will be independent of this toss, so with probability 1/2 we would again get a head. The law of large numbers tells us that if the coin is fair it will all 'even out in the long run'.

Now suppose that we continue to toss coins for ever—that is, we consider an infinite sequence of coin tosses. At any stage we can see how we are getting on by comparing the number of heads obtained so far with the total number of times we have tossed the coin—that is, after k tosses we divide the number of heads obtained by k; for example, if we toss a coin 100 times and get 47 heads, then the ratio of heads obtained is 47/100. If the coin were fair, then this ratio should become increasingly closer to 1/2 as we increase the number of tosses indefinitely;

Figure 39.1 Émile Borel.

Emile Borel, 1932 by Agence de Presse Mondial Photo-Presse, Bibliothèque Nationale de France. Licensed under public domain via Wikimedia Commons.

mathematically, we say that this ratio 'tends to the limit 1/2'. Similarly, if we change tack and count tails instead of heads, then the ratio of tails also tends to the limit 1/2 as the number of tosses increases indefinitely. Since we get the same limit whether we choose heads or tails, we say that the sequence of tosses is *normal to base* 2.

The same situation holds for other bases. When we toss a fair die, then the ratio of sixes obtained tends to the limit 1/6 as the number of tosses increases indefinitely. But we get the same limit when we count the number of threes, or fives, or any other number, so the sequence of tosses is *normal to base* 6. More generally, if we are tossing a fair n-sided die, then in the limit each number should appear $1/n$ times and the sequence of tosses is *normal to base* n.

In his 1909 paper, Borel defined a real number (represented as an infinite sequence of digits) to be 'absolutely normal' if it is normal when written in any base. This means that, whatever base we are using, no digit, or combination of digits, can occur more frequently than any other. It follows that no rational number can be absolutely normal since, when written in any base, it must contain a repeating pattern of digits. But, as Borel observed, 'almost every' irrational number is absolutely normal. This means (in a way that can be made mathematically precise) that if we throw a dart at the line it would, with probability 1, land on an irrational normal number.[2]

Turing's work on normal numbers was motivated by the following two questions asked by Borel:

- Can there be an irrational number that is normal to one base but not to another?
- Can one give an *explicit construction* of an absolutely normal number?

We are more concerned here with the second question.

We might guess that certain irrational numbers, such as π, e, and log 2, are absolutely normal: they are certainly explicit, in our sense, since their constructions are well known—and in the case of π a construction has been known since ancient times: the Babylonians already had a method for estimating π four thousand years earlier. There are also irrational numbers, such as $\sqrt{2}$, $\sqrt[3]{7}$ and $\sqrt{2} + \sqrt{3}$, which are called 'algebraic irrational numbers', since they are solutions of algebraic equations involving only whole numbers: for example, $\sqrt{2}$ is a solution of the algebraic equation $x^2 = 2$, $\sqrt[3]{7}$ is a solution of $x^3 = 7$, and $\sqrt{2} + \sqrt{3}$ is a solution of $x^4 + 1 = 10x^2$. It has been conjectured that every algebraic irrational number is absolutely normal, but any attempts to prove this have been as pitiful as they could possibly be: no explicit example has ever been proved to be normal, even to a single base!

From a modern point of view, we can think of a number as being explicit if we can give an algorithm that—when told a desired degree of precision—will compute the number to that degree of precision. This is called 'rapid convergence'. Actually, this modern point of view stems directly from Turing's 1936 paper. This is Turing's *definition* of a computable real number.

Another way of constructing normal numbers was discovered in 1933 by Turing's friend David Champernowne (1912–2000), while still an undergraduate (he later held chairs in statistics and economics at both Cambridge and Oxford). What you do is simply to write the digits in increasing order; for example, in base 10 the Champernowne number is 0.123456789101112131415. . . . This can be done to any base, and the resulting numbers are easily proved to be normal in the base in which they are written—and we believe, but have no idea to how to prove, that they are normal when expressed in terms of any other base.

In place of the numbers 1, 2, 3, . . . we could have taken some other increasing sequence and used it to define a similar number; for example, if we take the prime numbers 2, 3, 5, 7, 11, . . .

we obtain the number 0.2357111317. . . . Remarkably, this number is known to be normal to base 10, as shown in 1946 by Arthur Copeland and Paul Erdős,[3] but it is not known whether the primes written in base 2 result in such a normal sequence to base 2. As with many problems in number theory, it is easy to make conjectures, but hard to prove things!

Failing to have *natural* examples of absolutely normal numbers, we might ask what would be an *explicit* example. In a manuscript thought to have been written around 1938, but never published in his lifetime,[4] Turing suggested that an absolutely normal number n would be explicitly constructed if the number were 'computable': this means that we can construct a Turing machine that could compute the expansion of n to any desired precision. Any irrational number met in 'normal mathematics' will have this property, as all have rapidly converging approximations; for example, the series $e = 1 + 1/1! + 1/2! + 1/3! + \ldots$ converges rapidly, and hence its limit $e = 2.718281828459 \ldots$ is computable. Similar series are known that converge rapidly to π, and so π is also computable, as is any other 'natural' constant.

As we shall see, Turing's approach anticipates a line of research that can be traced back to the early 20th century, but was realised only in the early 1960s. Before we look at Turing's work in greater detail we briefly describe this body of work. This analysis will enable us to describe exactly what Turing did, and how he did it.

The events of the 1960s and 1970s

Following unsuccessful historical attempts to capture an account of algorithmic randomness based on work by Richard von Mises, Abraham Wald, Alonzo Church, and others, three modern paradigms developed in the 1960s and 1970s by Andrey Kolmogorov, Ray Solomonoff, Per Martin-Löf, Leonid Levin, and others:

- A random sequence should have no computably rare properties (the statistician's approach).
- A random sequence should have no regularities that would allow for compression of information. (This has become known as the 'coder's approach' since it encapsulates the idea of writing computer code to describe a string.)
- A random sequence should not have any predictability (the gambler's approach).

These clarifications arise from the celebrated works of a number of distinguished authors in the 1960s and 1970s. As we shall see later, Turing's work anticipated some of these developments in many ways.

The gambler's approach, the idea that we should not be able to predict the bits of a random sequence, goes back to von Mises at the beginning of the 20th century. But it was properly developed by Claus Peter Schnorr.[5] We idealize a person betting on the bits of a infinite sequence. Think of them as heads and tails. So you first bet on the first bit: you can bet either heads or tails, 1 or 0, and after you have bet, the outcome is revealed. For example, suppose that you start with $3, and bet $1 on heads. If heads come up, you then have $4, and if tails appear you have $2. Next, according to your current capital and the history, you bet on the next bit, and so on. As opposed to a real casino, you can bet any fraction that you like.[6]

So taking the gambler's approach, we want to argue that we can defeat any algorithmic plan that a person might have. Thus we think of 'computable betting strategies': a real number should be random if no computable betting strategy can make infinite capital by betting on its bits.

Turing's work

We recall that Turing was interested in absolute normality. Clearly, a random sequence should be absolutely normal. Why is this? As we have already seen, if the frequency of 1s is less than that of the 0s, then we could easily have a betting strategy to allow us to make infinite capital, in the limit. Now suppose that an infinite binary sequence is not normal in base 10, and (for example) that the frequency of 5s is higher than expected: then in base 10 we could formulate a betting strategy to exploit that fact.

Suppose that I told you that every third bit of a sequence was a 1. You could make money as follows. On the first bit don't bet, on the second don't bet, and then put all your capital on 1 for the third bet, and repeat. Similarly, if I knew that the frequency of 1s was always less than that of 0s for the sequence, then it is again possible to formulate a strategy to make infinite capital.

There are relatively easy computable translations from one base to another. For example, 0.5 (base 10) corresponds to 0.1 (base 2). In general, you get 'blocks'. As an illustration: 0.6 (base 10) can be converted to base 2 as follows. First multiply by 2, to get 1.2. Keep the 1. Now multiply 0.2 by 2 and get 0.4. Keep the 0. Multiply 0.4 by 2 and get 0.8, keeping the 0. Repeat, getting 1.6, and keep the 1. Thus our binary expansion of 0.6 will begin: 0.1001; and since the next thing to be processed will be 0.6, the same pattern will repeat forever.

Although this process is a bit more complicated in the case of infinite expansions, there is always a very tight relationship between an expansion in one base and one in another, and it is possible to exploit this to prove that randomness in one base translates into randomness in another. Roughly speaking, for two bases p and q, there is always a computable function f which converts the first n bits of a sequence in base p into $f(n)$ bits of the sequence in base q. If I have a way of winning infinite capital in one base, I could use that to bet in the other base.

In his unpublished manuscript 'A note on normal numbers', Turing attacked the question of an explicit construction of an absolutely normal number by interpreting 'explicit' to mean 'computable', and presented the best answer to date to Borel's second question: an algorithm that produces absolutely normal numbers.[7] This early proof of existence of computable normal numbers remained largely unknown because Turing's manuscript was not published until 1997, in his *Collected Works*, where the editorial notes claimed that Turing's proof was inadequate and speculated that the theorem might be false. In 2007, Verónica Becher, Santiago Figueira, and Rafael Picchi reconstructed and completed Turing's manuscript, correcting minor errors while preserving his ideas as accurately as possible.[8]

As Verónica Becher has remarked, the very first examples of normal numbers were published independently by Henri Lebesgue and Wacław Sierpiński in 1917;[9] both appeared in the same journal issue, but Lebesgue's example dated back to 1909, immediately after Borel's question.[10] The examples of Sierpiński and Lebesgue can also be modified to give computable (that is, explicit) solutions to Borel's question. This can be done by giving a computable reformulation of the original constructions of 1917, as shown by Becher and Figueira.[11] Until recently these, together with Turing's algorithm, were the only known constructions of computable normal numbers. Turing was unaware of these earlier constructions, but in any case what is interesting is the way that Turing solved Borel's question.

What did Turing's construction do? His paper claims the following:

Although it is known that almost all numbers are [absolutely] normal no example of [an absolutely] normal number has ever been given. I propose to show how [absolutely] normal numbers may be constructed and to prove that almost all numbers are [absolutely] normal constructively.

Turing's idea was first to extend the law of large numbers to 'blocks' of numbers. Intuitively one can see that not only single digits, but fixed blocks of numbers, should occur with the desired frequency in a random sequence, for otherwise we could bet on the knowledge that they do not. For example, if we knew that no blocks of the form 1011 appear in the sequence, but that there were blocks of the form 101, then we would not bet until we saw 101 and then bet that the next number would be 0, since 1011 does not occur.

The reason for using blocks of numbers is that translating between bases results in correlations between blocks of integers in one base and those in another. Realising this, the key to Turing's construction is the following: Turing regarded blocks as generating 'tests' with 'small measure', so if we avoid such tests, we should have an absolutely normal number.

Notice how this echoes the 1936 paper where he introduced Turing machines.[12] There it was not just that Turing solved the technical question of the *Entscheidungsproblem*, which had arguably already been solved by Church, Emil Post, Stephen Kleene, and others,[13] but he introduced the fundamental model of computation along with an argument supporting it. In his unpublished manuscript, Turing gave an explicit construction of an absolutely normal number: this was a fine technical achievement. But, more importantly, he did so by massively generalizing the problem. His ideas were years in advance of their time.

Here is what Jack Lutz said of this in 2012 in a lecture at a conference in Cambridge:

Placing computability constraints on a nonconstructive theory like Lebesgue measure seems a priori to weaken the theory, but it may strengthen the theory for some purposes. This vision is crucial for present-day investigations of individual random sequences, dimensions of individual sequences, measure and category in complexity classes, etc.

What is fascinating here is the clarity of Turing's intuition: to construct absolutely normal numbers, take the classical proof that almost all numbers are absolutely normal, and re-do this as a computable version. Turing's tests were sensitive enough to exclude absolutely normal numbers, but insensitive enough to allow suitable computable sequences to pass them.

It is interesting to speculate about why Turing did not develop the whole apparatus of algorithmic randomness, since he seemed to have many of the ideas needed for this development. Whilst he had the notion of a pseudo-random number generator, he thought that randomness was a physical phenomenon. Perhaps he did not see the need for a definition of randomness for an individual sequence. To me the following quote from 1950 suggests that we can increase the power of a Turing machine by *adding* randomness, which seems to suggest that he recognized the non-computable nature of randomness:[14]

An interesting variant on the idea of a digital computer is a 'digital computer with a random element.' These have instructions involving the throwing of a die or some equivalent electronic process; one such instruction might for instance be, 'Throw the die and put the resulting number into store 1000.' Sometimes such a machine is described as having free will (though I would not use this phrase myself).

Interestingly, in the sentences following this, he recognized the difficulty of perception of randomness:

It is not normally possible to determine from observing a machine whether it has a random element, for a similar effect can be produced by such devices as making choices depend on the digits of the decimal for π.

Figure 39.2 The ANU (Australian National University) Quantum Random Number generator. It generates numbers from quantum fluctuations of the vacuum. The process involved is a leading contender for a physical source of genuine randomness.

Secure Quantum Communication group, Australian National University.

We know that Turing used algorithms to generate 'pseudo-random strings', strings which seemed sufficiently random that they worked in the way that a random source should behave. We also know that Turing attempted to include a random number generator in the Ferranti Mark I computer, based on a physical noise generator; however, even today it remains an open question in physics whether the universe can generate 'true randomness' or even 'algorithmic randomness' in a physical device (see Fig. 39.2). Perhaps Turing was concerned only with pseudo-random strings in relation to algorithms in (for example) cryptography and artificial intelligence (AI), as outlined in the following section. Certainly, shortly after 1938, the Second World War intervened, and when it ended Turing did not return to the topic of normality. He did not mention randomness except in relation to the efficiency and efficacy of algorithms, and probably never considered the problem of defining an algorithmically random sequence.

Turing on randomness as a resource

It seems clear that Turing regarded randomness as a computational resource; for example, in AI Turing considered learning algorithms. As he remarked in 1950:[15]

It is probably wise to include a random element in a learning machine . . . A random element is rather useful when searching for the solution of some problem.

Turing then gave an example of a search for the solution to some numerical problem, pointing out that if we did this systematically then we would often have a lot of overheads arising from our previous searches. But if the problem has solutions that are reasonably dense in the sample space, then random methods should succeed.

From a modern perspective, we know of many algorithms that work well randomly. A simple illustration is what is called 'polynomial identity testing' (PIT), concerning the following problem. Given a polynomial expression in several variables x, y, z, etc., is that polynomial equal to 0, no matter what I substitute into the variables? For example, $xy + 1$ is not such a polynomial since, for example, substituting $x = 1$, $y = 2$ would give the answer '3', whereas the polynomial $xy - yx$ is such an 'everywhere 0' polynomial. Although this might seem a strange and specialized problem, it turns out that many problems can be computationally restated as instances of PIT, so that we can solve them by solving a PIT algorithm. There are very efficient randomized algorithms for PIT: in fact, the dumbest algorithm one could imagine works. Take any random numerical substitutions for $x, y, z, \ldots$ and evaluate the polynomial at these values. If you don't get 0, the answer is certainly 'no', but if you do get 0, say 'yes': with high probability you will be correct, roughly speaking because such polynomials are 0 quite rarely unless they are *always* 0.[16]

A major theme in modern algorithm design is to use randomness as a resource like this. It is fair to say that most computational learning algorithms use randomness as a resource, like those that monitor you on the internet and try to learn your preferences, or those that model physical phenomena. This use was anticipated by Turing in his writings: he even speculated that randomness is somehow necessary for intelligence,[17,18] but seems not to have developed this theme mathematically.

This lack of mathematical development is perhaps unfortunate, as there was an implicit recognition of algorithm *complexity* in Turing's writings. Such recognition came from cryptography, where large search spaces are used to hide information, and from practical computing, where an answer must be found in real time. Even Turing's writings on intelligence and the limitations of machines were full of calculations as to the numbers of possible states in computers, etc. Had he pushed his ideas further, perhaps Turing could have developed the mathematical theory of computational complexity (only really investigated since the 1970s)—but this is complete speculation. It is clear that he had an intuitive understanding of the time and space complexity of computation.

These complexity issues are very deep. It is a long-standing open question as to whether PIT can be solved by a non-randomized deterministic algorithm in polynomial time. Although we have no idea how to construct such an algorithm, is believed that such an algorithm exists. This is because of the work of Russell Impagliazzo and Avi Wigderson, who proved a certain hardness/randomness trade-off: their truly remarkable result says roughly that if certain problems are as hard as we think they are, then all problems (like PIT) with efficient randomized polynomial-time algorithms have polynomial-time derandomized versions.[19]

This result is remarkable because it says that if (for example) finding a route through all the cities in a map without repetition is *really hard* in terms of efficiency, then this *other* class of tasks, including PIT, is *easy*. It is an important result, since to show that some task can be done easily, we usually show that it can be done by another task that can itself be done easily. Many, many tasks can be reduced to instances of solving PIT: so, if we can prove that certain problems are as hard as we think they are, then we would have revolutionary new ways of solving many problems.

For a long time, determining whether a number is prime had the same status—that is, primality testing had randomized algorithms that had been known for many years and work very quickly. In a celebrated result, first announced in 2002, Manindra Agrawal, Neeraj Kayal, and Nitin Saxena gave a non-randomized polynomial-time algorithm for primality testing.[20]

Where does this all lead?

There is much ongoing work on absolute normality, since it is a natural number-theoretical notion. In terms of its relationship with randomness, one can show that normality is really a precise calibration of randomness. Normality corresponds to a kind of randomness property called the 'finite-state Hausdorff dimension', where we look at the behaviour of betting strategies that are controlled by 'finite state gamblers'. These are strategies controlled by 'finite state machines'. Finite state machines are widely used in computer science: these, the most basic of all machines, can be thought of as memory-less machines that plod from one internal state to the next, processing the bits of the input purely on the basis of the bit being read and the internal state. In the early 1970s Schnorr and Stimm proved that a sequence of numbers in base b is normal to base b if and only if the finite-state Hausdorff dimension of the sequence is 1: this is a statement about the complexity of the initial segments of the sequence.[21] Meanwhile, Elvira Mayordomo and Jack Lutz, and independently, Verónica Becher, Pablo Heiber, and Theodore Slaman, have shown that it is possible to construct such sequences that are quite simple, in that we can compute each bit of the sequence in polynomial time.[22] There has been a great amount of work on algorithmic randomness and its applications to such areas as algorithm design, understanding analysis in mathematics, Brownian motion, ergodic theory, Markov chain theory, and physics.[23]

CHAPTER 40

Turing's mentor, Max Newman

IVOR GRATTAN-GUINNESS

The interaction between mathematicians and mathematical *logicians* has always been much slighter than one might imagine. This chapter examines the case of Turing's mentor, Maxwell Hermann Alexander Newman (1897–1984). The young Turing attended a course of lectures on logical matters that Newman gave at Cambridge University in 1935. After briefly discussing examples of the very limited contact between mathematicians and logicians in the period 1850–1930, I describe the rather surprising origins and development of Newman's own interest in logic.

The cleft between logic and mathematics

One might expect that the importance to many mathematicians of means of proving theorems, and their desire in many contexts to improve the level of rigour of proofs, would motivate them to examine and refine the logic that they were using. However, inattention to logic has long been common among mathematicians.

A very important source of the cleft between mathematics and logic during the 19th century was the founding, from the late 1810s onwards, of the 'mathematical analysis' of real variables, grounded on a theory of limits, by the French mathematician Augustin-Louis Cauchy. He and his followers extolled rigour—most especially, careful definitions of major concepts and detailed proofs of theorems. From the 1850s onwards, this project was enriched by the German mathematician Karl Weierstrass and his many followers, who introduced (for example) multiple limit theory, definitions of irrational numbers, and an increasing use of symbols, and then from the early 1870s by Georg Cantor with his set theory. However, absent from all these developments was explicit attention to any kind of logic.

This silence continued among the many set theorists who participated in the inauguration of measure theory, functional analysis, and integral equations.[1] The mathematicians

Artur Schoenflies and Felix Hausdorff were particularly hostile to logic, targeting the famous 20th-century logician Bertrand Russell. (Even the extensive dispute over the axiom of choice focused mostly on its legitimacy as an assumption in set theory and its use of higher-order quantification:[2] its ability to state an infinitude of independent choices within *finitary* logic constituted a special difficulty for 'logicists' such as Russell.) Russell, George Boole, and other creators of symbolic logics were exceptional among mathematicians in attending to logic, but they made little impact on their colleagues. The algebraic logical tradition, pursued by Boole, Charles Sanders Peirce, Ernst Schröder, and others from the mid-19th century, was nothing more than a curiosity to most of their contemporaries. When mathematical logic developed from the late 1870s, especially with Giuseppe Peano's 'logistic' programme at Turin University from around 1890, Peano gained many followers there but few elsewhere.[3]

Followers of Peano in the 1900s included Russell and Alfred North Whitehead, who adopted Peano's logistic (including Cantor's set theory) and converted it into their logicist thesis that all the 'objects' of mathematics could be obtained from logic. However, apart from the eminent Cambridge mathematician G. H. Hardy, few mathematicians took any interest in Russell–Whitehead logicism.[4] From 1903 onwards, Russell publicized the form of logicism put forward from the late 1870s by Gottlob Frege—which had gained little attention hitherto, even from students of the foundations of mathematics, and did not gain much more in the following decades.

In the late 1910s David Hilbert started the definitive phase of his programme of 'metamathematics', which studied general properties of axiom systems such as consistency and completeness, and 'decision problems' concerning provability. Hilbert's programme attracted several followers at Göttingen University and a few elsewhere; however, its impact among mathematicians was limited, even in Germany.[5]

The next generations of mathematicians included a few distinguished students of the foundations of mathematics. From around 1900, in the United States, E. H. Moore studied Peano and Hilbert, and passed on an interest in logic and metamathematical 'model' theory to his student Oswald Veblen, and so to Veblen's student Alonzo Church, and from Church in turn to his students Stephen Kleene and Barkley Rosser.[6] At Harvard, Peirce showed his 'multiset' theory to the philosopher Josiah Royce, who was led on to study logic and (around 1910) to supervise budding logicians Clarence Irving Lewis, Henry Sheffer, Norbert Wiener, Morris Cohen, and also Curt John Ducasse—the main founder of the Association of Symbolic Logic in the mid-1930s.[7] In central Europe, John von Neumann included metamathematics and axiomatic set theory among his concerns,[8] but in Poland a distinguished group of logicians did not mesh with a distinguished group of mathematicians, even though both groups made much use of set theory (even their joint journal *Fundamenta Mathematicae*, published from 1920, rarely carried articles on logic).

The normal attitude of mathematicians to logic was indifference. For example, around 1930 the great logician Alfred Tarski and others proved the fundamental 'deduction theorem';[9] this work met with apathy from the mathematical community, although it came to be noted by the French Bourbaki group, who were normally hostile to logic. (Maybe the reason was that French logician Jacques Herbrand had proved versions of the theorem—if so, this was his sole impact on French mathematics.) Also, while logicians fairly quickly appreciated Kurt Gödel's theorems of 1931 on the incompletability of (first-order) arithmetic, the mathematical community did not become widely aware of Gödel's results until the mid-1950s.[10]

Newman's lecture course at Cambridge

Turing's own career provides a good example of the cleft between logic and mathematics. When he submitted his paper 'On computable numbers' to the London Mathematical Society in 1936 the journal could not referee it properly because Max Newman was the only other expert in Britain and he had been involved in its preparation (Fig. 40.1).[11] Nevertheless, they seem to have accepted it on Newman's word.[12] But this detail prompts historical questions that have not so far been explored. Why was this logical subject matter so little known in Britain? Why was Newman, a mathematician, a specialist in it? Answers to these questions are suggested below. [13]

A crucial event, Newman's lectures to Turing, occurred in the mid-1930s, when he taught a new final-year course on 'Foundations of mathematics' for the Mathematical Tripos. The tripos examination questions he set show that he had handled all the topics in logic and metamathematics mentioned in the opening section of this chapter, and also the intuitionist mathematics and logic of L. E. J. Brouwer. Turing, newly graduated from the tripos, sat in on the course in 1935 and learnt about decision problems and Gödel numbering from one of the few Britons who was familiar with them.[14] There seems to be no documentation about contact between Turing and Newman following the lecture course, but presumably the two met on a regular basis during 1935 and 1936 as Turing prepared his paper on computability.

A sense of isolation hangs over Newman's lecture course. Launched in the academic year 1933–34, Newman ran the course for only two more years before it was closed down, possibly because of disaffection among dons as well as students. In 1937 Hardy opined to Newman:[15]

Figure 40.1 Max Newman.

Reproduced with permission of the Master and Fellows of St John's College, Cambridge.

though 'Foundations' is now a highly respectable subject, and everybody ought to know something about it, it is (like dancing or 'groups') slightly dangerous for a bright young mathematician!

Somehow Newman continued to set examination questions on foundations for five of the six years that he remained at Cambridge before moving to Bletchley Park.[16] The questions for 1939 may have been set by Turing, who was invited—presumably in a spirit of resistance against Hardy's coolness—to give a lecture course on foundations in the Lent Term of 1939. He was asked to repeat it in 1940, but by then he was at Bletchley Park.[17] Newman arrived there in the summer of 1942, to join the fight against Tunny. During his wartime period he published three technical papers in logic, one written jointly with Turing. Two, including the one written with Turing, were on aspects of type theory, which was Russell's solution of the paradoxes of logic and set theory,[18] and one on the so-called 'confluence' problem (concerning the reduction of terms).[19]

Newman's entry into logic

How did Newman become so involved with logic in the first place? Born Max Neumann in London in 1897, to a German father and an English mother, he gained a scholarship in 1915 to St John's College, Cambridge, taking Part I of the Mathematical Tripos in the following year.[20,21] During the First World War, Max's father Herman was interned by the British; when released he went back to Germany, never to return. In 1916, Max and his mother Sarah changed their surname to 'Newman'. A pacifist, Max served in the Army Pay Corps, returning to his college in 1919. He completed Part II of the tripos with distinction in 1921.

Against the odds, Newman spent much of the academic year 1922–23 at Vienna University. He went with two other members of his college. One was the geneticist and psychiatrist Lionel Penrose (father of the distinguished mathematician Roger Penrose), who seems to have initiated the trip to Vienna; his family was wealthy enough to sustain it, especially as at that inflationary time British money went a long way in Vienna. Penrose had been interested in Russell's mathematical logic as a schoolboy, and had studied traditional Aristotelian logic as an undergraduate at Cambridge. He had also explored modern mathematical logic and it might even have been Penrose who alerted his friend Newman to the subject. Penrose wanted to meet Sigmund Freud, Karl Bühler, and other psychologists in Vienna. The third member of the Cambridge party was Rolf Gardiner, an enthusiast for the Nazis, later active in organic farming and folk dancing, and father of the conductor Sir John Eliot Gardiner. Gardiner's younger sister Margaret came along too. She would become an artist and companion to the biologist Desmond Bernal. Margaret recalled 'the still deeply impoverished town' of Vienna, where Penrose and Newman would walk side-by-side down the street playing a chess game in their heads.[22]

Of Newman's contacts with the Vienna mathematicians all that tangibly remains is a welcoming letter of July 1922 from Wilhelm Wirtinger.[23,24] Yet it seems clear that Newman's experience of Viennese mathematics was decisive in changing the direction of his researches. His principal mathematical interest was to become topology, which was not a speciality of British mathematics. Some of Wirtinger's own work in Vienna related to the topology of surfaces, and in 1922 the University of Vienna recruited Kurt Reidemeister, who, like Newman himself, went on to become a specialist in combinatorial topology. Most notably, Hans Hahn, later a leading member of the famous 'Vienna Circle', ran a preparatory seminar on 'Algebra and logic' while Newman was in Vienna.[25,26] Hahn was not only a specialist in the topology of curves and in real-variable mathematical analysis but also regarded formal logic as an important topic, both for research and teaching. In later years, Hahn held two full seminar courses on Whitehead and Russell's *Principia Mathematica*, one of the earliest major publications in modern mathematical logic. Russell's approach to philosophy and logic strongly influenced the Vienna Circle. Hahn also supervised the young Kurt Gödel, a doctoral student in Vienna.

After Vienna

Newman became a pioneer of topology in Britain, with a serious interest in logic and logic education, and also in Russell's philosophy. One surely sees strong Viennese influence here, especially from Hans Hahn.

Newman applied for a College Fellowship at St John's in 1923, a year that saw him publishing in the traditional area of mathematical analysis (in particular, on avoiding the axiom of choice in the theory of functions of real variables), and also writing a long unpublished essay in the philosophy of science, entitled 'The foundations of mathematics from the standpoint of physics'. This could well have originated in a Viennese conversation. Maybe he even wrote some of it in Vienna.[27] In the essay, Newman contrasted the world of idealized objects customarily adopted in applied mathematics (smooth bodies, light strings, and so on) with the world 'of real physical objects'. He distinguished the two worlds by the different logics that they used. The idealizers would draw on classical two-valued logic, for which he cited a recent metamathematical paper by Hilbert as a source,[28] but those interested in real life, he said, would go to *constructive* logic, on which he cited recent papers by Brouwer and Hermann Weyl.[29] This readiness to embrace logical pluralism and to put *logics* at the centre of his analysis of a problem was most unusual for a mathematician, and far more a product of Vienna than Cambridge.[30,31]

An occasion for Newman to exercise his logical and philosophical talents arose when he attended a series of philosophical lectures that Russell gave in Trinity College Cambridge in 1926. These lectures were the basis for Russell's book *The Analysis of Matter*.[32] Newman helped Russell to write two chapters, and, when the book appeared in 1928, he criticized its philosophical basis most acutely.[33,34] Russell accepted the criticisms, which stimulated Newman to write Russell two long letters on logic and on topology, featuring some ideas from his 1923 philosophical essay.[35] Newman continued to pioneer both topology and logic at Cambridge and, doubtless with topology in mind, Hardy successfully proposed Newman as a Fellow of the Royal Society, with J. E. Littlewood as seconder—even though Newman was no analyst in the Hardy–Littlewood tradition.

Newman used the Royal Society to support the cause of logic. In 1950, he proposed Turing as a Fellow, seconded by Russell, and Turing was duly elected in 1951. Just five years later, in 1955, Newman wrote Turing's Royal Society obituary.[36] In 1966 Newman proposed Gödel as a Foreign Member, and again Russell seconded; Gödel was elected to the society two years later.[37] When Russell died in 1970, Newman agreed to write the society's obituary of Russell, together with the philosopher Freddie Ayer, but failing health prevented him from fulfilling his obligation. He died in 1984.

What a fluke!

I conclude with some reflections about this very unusual history. As a result of Penrose's early interest in mathematical logic, and the unusual mixture of mathematics, logic, and philosophy in Vienna, Newman changed direction: had he stayed in Cambridge in 1922–23, he would surely have continued on the path indicated by his 1923 paper about avoiding axioms of choice—namely, Hardy–Littlewood mathematical analysis. Even then his interest in logic could have waned and he might not have set up a foundations course to be taken by Turing. The existence of this course in 1935 was a highly improbable event, in that Newman might easily not have accompanied Penrose to Vienna, and might never have had an inclination to teach logic. After Turing's graduation in 1934 he was himself another budding Hardy–Littlewood mathematical analyst, working on 'almost periodic' functions and with interests in mathematical statistics and group theory. One can imagine Turing remaining happily engaged in these branches of mathematics—particularly in analysis, with the influential Hardy back at Cambridge from 1931 after twelve years at Oxford. Turing would not have come across recursive functions or undecidability, and would not have invented his universal machine.

Could Newman and Turing have come to these topics by another route—via some of the Cambridge philosophers, perhaps? Frank Ramsey, who visited the philosopher Ludwig Wittgenstein in Austria in 1923 and 1924, and died in 1930, was along with other Cambridge philosophers largely concerned with revising the logicism of Russell and the early Wittgenstein.[38] Wittgenstein himself, back in Cambridge from 1929, was philosophically a monist, and so distinguished between what can be said and what can only be shown. By contrast, the Hilbertians depended centrally on the concept of a 'hierarchy', mathematics and metamathematics, upon which Turing seized. (The logician Rudolph Carnap, another member of the Vienna Circle, coined the term 'metalogic' in 1931, on the basis of reflecting upon Gödel's procedures in his proof of the incompleteness of arithmetic, and Tarski was starting to speak of 'metalanguage' at about the same time.) Russell was in a strange position on this issue. Back in 1921, in his introduction to Wittgenstein's *Tractatus*, he had advocated a hierarchy of languages to replace the Wittgensteinian showing–saying distinction—a move that Wittgenstein rejected—but he never envisaged a companion hierarchy of logics, and so neither understood (nor even stated) Gödel's theorems properly.[39] It thus seems highly unlikely that either Turing's contacts with Russell or the contacts over logic and logicism that he had with Wittgenstein would have led him to metamathematics and decision problems.

If neither Turing nor Newman had found their way to those crucial topics in logic and metamathematics, then—while they *might* have been recognized as clever analysts who were also good at chess—they would not have been such obvious choices for Bletchley Park. If they had got there, they would probably not have been as effective as the actual Newman and Turing were, because they would not have known much, if any, of the key mathematics. So a crucial part of the British wartime decoding effort, especially Turing's Enigma-breaking bombe, came about as a consequence of Turing's change of direction years earlier, which was inspired by the happenstance of Newman's course on the foundations of mathematics, offered because of his own earlier change of direction towards this unusual topic, a change in turn brought about as an unintended consequence of some decisions that Penrose took to develop his own career. In short, what a fluke![40]

PART VIII

Finale

CHAPTER 41

Is the whole universe a computer?

JACK COPELAND, MARK SPREVAK,
AND ORON SHAGRIR

The theory that the whole universe is a computer is a bold and striking one. It is a theory of *everything*: the entire universe is to be understood, fundamentally, in terms of the universal computing machine that Alan Turing introduced in 1936. We distinguish between two versions of this grand-scale theory and explain what the universe would have to be like for one or both versions to be true. Spoiler: the question is in fact wide open—at the present stage of science, nobody knows whether it's true or false that the whole universe is a computer. But the issues are as fascinating as they are important, so it's certainly worth while discussing them. We begin right at the beginning: what exactly *is* a computer?

What is a computer?

To start with the obvious, your laptop is a computer. But there are also computers very different from your laptop—tiny embedded computers inside watches, and giant networked supercomputers like China's Tianhe-2, for example. So what feature do all computers have in common? What is it that makes them all *computers*?

Colossus was a computer, even though (as explained in Chapter 14) it did not make use of stored programs and could do very few of the things that a modern laptop can do (not even long multiplication). Turing's ACE (see Chapters 21 and 22) was a computer, even though its design was unlike that of a laptop; for example, the ACE had no central processing unit (CPU), and moreover it stored its data and programs in the form of 'pings' of supersonic sound travelling along tubes of liquid. Turing's artificial neural nets were also computers (Chapter 29), and so are the modern brain-mimicking 'connectionist' networks that Turing anticipated. In connectionist networks—as in a human brain, but unlike a laptop—there is no separation between memory and processing, and the very same 'hardware' that does the processing (the neurons and their

Figure 41.1 Is the universe a computer? The Pillars of Creation, in the distant Eagle Nebula (taken by the Hubble Space Telescope).

Posted to Wikimedia Commons and licensed under public domain, https://commons.wikimedia.org/wiki/File:Pillars_of_creation_2014_HST_WFC3-UVIS_full-res_denoised.jpg.

connections) also functions as the memory. Even Babbage's Analytical Engine (Chapter 24) was a computer, despite being built from mechanical rather than electrical parts. As Turing said:[1]

The fact that Babbage's Analytical Engine was to be entirely mechanical will help us to rid ourselves of a superstition. Importance is often attached to the fact that modern digital computers are electrical, and that the nervous system also is electrical. Since Babbage's machine was not electrical . . . we see that this use of electricity cannot be of theoretical importance.

In fact, there is astonishing variety among the computers that are currently being researched or prototyped by computer scientists. There are massively parallel and distributed computers, asynchronous computers (computers with no central 'clock' coordinating the processing), nanocomputers, quantum computers, chemical computers, DNA computers, evolutionary computers, slime-mould computers, computers that use billiard balls, and computers that use swarms of animals or insects to solve problems . . . . The list goes on. There could in principle even be a universal computer consisting entirely of mirrors and beams of light.[2] What, then, do all these different forms of computer have in common? Let's examine what Turing said of relevance to this question.

Before the modern era, the word 'computer' referred to a human being. If someone spoke of a computer in the nineteenth century, or even in 1936, they would have been taken to be referring to a *human* computer—a clerk who performed the tedious job of routine numerical computation. There used to be many thousands of human computers employed in businesses, government departments, research establishments, and elsewhere. In 1936, Turing introduced his 'logical computing machines'—Turing machines—so as to provide an idealized description of the human computer. In fact he began his account of the Turing machine: 'We may compare a man in the process of computing a . . . number to a machine.'[3] Cambridge philosopher Ludwig Wittgenstein, well known for his pithy and penetrating statements, put the point like this:[4]

Turing's 'Machines'. These machines are *humans* who calculate.

Turing often emphasized the fundamental point that the Turing machine is a model (idealized in certain respects) of the human computer. For example:[5]

A man provided with paper, pencil, and rubber, and subject to strict discipline, is in effect a universal machine.

Even in his discussions of the ACE, Turing continued to use the word 'computer' to mean 'human computer':[6]

Computers always spend just as long in writing numbers down and deciding what to do next as they do in actual multiplications, and it is just the same with ACE . . . [T]he ACE will do the work of about 10,000 computers . . . Computers will still be employed on small calculations.

So there were on the one hand computers—human beings—and on the other hand *machines* that could take over aspects of the computers' work. The term 'computing machine' was used increasingly from the 1920s to refer to small calculating machines that mechanized elements of the human computer's work. When the phrase 'electronic computer' came along in the 1940s, it too referred to a machine that mechanized the work of the human computer. Turing made this explicit:[7]

The idea behind digital computers may be explained by saying that these machines are intended to carry out any operations which could be done by a human computer.

Turing considered this characterization of the concept of a 'digital computer' to be so important that he began his *Programmers' Handbook for Manchester Electronic Computer Mark II* with the following statement:[8]

Electronic computers are intended to carry out any definite rule of thumb process which could have been done by a human operator working in a disciplined but unintelligent manner.

Here, then, is the Turing-style answer to the question: 'What is a computer, in the modern sense?'[9] Any physical mechanism that carries out the same work as the idealized human computer is itself a computer: by performing steps that the idealized human computer cloud perform, the machine carries out sequences of operations that, given enough time, the human computer could also carry out. (The human computer is idealized in that no limit is placed on the amount of time available to the human computer, nor on the quantity of paper and pencils available—idealized human computers live indefinitely, and never get bored.) Computers—all computers—carry out tasks that can, in principle, be done by a human rote-worker following an algorithm (and no other tasks)—tasks that, as Turing put it,[10] can be done

by human clerical labour, working to fixed rules, and without understanding.

With this clarification in place we turn next to the important distinction between *conventional* and *unconventional* computers. Modern laptops, tablets, minis, and mainframes are conventional computers, while slime-mould computers and swarm computers are not. Conventional computers derive ultimately from the design set out in the famous 1945 proposal 'First draft of a report on the EDVAC' (see Chapter 20) and they consist fundamentally of two parts, the CPU and the memory. A conventional computer's basic cycle of activity is the 'fetch–operate–store' cycle: operands (numbers) are fetched from memory, operated on in the CPU (e.g. multiplied together), and the result of the operation (another number) is routed back to the memory and stored. Any computer that does not fit this description is unconventional.

Is the universe a conventional computer, a cosmic version of a laptop or of Tianhe-2 (Fig. 41.1)? This seems to us logically possible, but not terribly likely. Where is the cosmic computer's CPU? Where is the cosmic computer's memory? Where are the registers holding the operands, and the registers in which the results of the CPU's operations are stored? There is no evidence of the universe containing any of these core elements of a conventional computer.

However, the Californian philosopher John Searle argues that even a garden wall is a conventional computer; and other philosophers maintain that a simple rock standing motionless on a beach is a (conventional) computer—and so, if Searle et al. are right, the entire universe is by the same token a gigantic conventional computer.[11] These claims about walls and rocks, even if ultimately absurd, deserve a detailed discussion, but since we are after more important quarry we shall not pause to give this discussion here: interested readers will find a critique of these claims in the references given in the endnotes.[12] Turning away from the idea that the universe is a conventional computer, we are going to discuss the more promising hypothesis that the universe is a computer of a type first introduced by John von Neumann and mentioned by Stephen Wolfram in Chapter 5: a 'cellular automaton'.[13]

Zuse's thesis

Konrad Zuse (Fig. 41.2), who appeared briefly in Chapters 6 and 31, built his first computers before the Second World War—in the living room of his parents' apartment in Berlin.[14] As an engineering student at the Technical University in Berlin-Charlottenburg, Zuse had become painfully aware that engineers must perform what he called 'big and awful calculations'.[15] 'That is really not right for a man', he said:[16]

It's beneath a man. That should be accomplished with machines.

After the war Zuse supplied Europe with cheap relay-based computers, and later transistorized computers, from his factory in Bad Hersfeld. Even though he had anticipated elements

Figure 41.2 Konrad Zuse, 1910–1995.

ETH-Bibliothek Zürich, Bildarchiv, Portr_14648.

of the stored-program concept, in a 1936 patent application, it was not until the 1960s that he began to include stored programming in his computers.[17] (It is sometimes said in the historical literature that Zuse's 1941 Z3 computer was a stored-program machine, but this is an error.) Whether Zuse and Turing ever met in person is uncertain. Interestingly, Zuse stated that he had no knowledge of Turing's 1936 article 'On computable numbers' until 1948, the year that he was summoned from Germany to London to be interrogated by British computing experts.[18] Donald Davies, Turing's assistant at the National Physical Laboratory, was one of the interviewers: Zuse eventually 'got pretty cross', Davies recollected, and things 'degenerated into a glowering match'.[19] Zuse seemed 'quite convinced' (Davies continued) that he could make a smallish relay machine 'which would be the equal of any of the electronic calculators we were developing'.

Zuse's 1967 book *Rechnender Raum* ('*Space Computes*') sketched a new—even mind-bending—framework for fundamental physics. Zuse's thesis was that the universe is a giant digital computer, a cellular automaton (CA).[20] According to Zuse the universe is, at bottom, nothing more than a collection of ones and zeros changing state according to computational rules. Everything that is familiar in physics—force, energy, entropy, mass, particles—emerges from that cosmic computation.

Stephen Wolfram explains that cellular automata are lattice-like grids, all of whose properties are *discrete*. They are:[21]

systems in which space and time are discrete, and physical quantities take on a finite set of discrete values. . . . A cellular automaton evolves in discrete time steps.

A CA is very different from a conventional computer. To visualize a CA, picture a two-dimensional grid made up from square cells. As the CA's time ticks forward in discrete steps, each cell in the grid is at any moment in one or other of two states, 'on' or 'off'. The CA's 'transition rules' describe how the cells' states at one time-step determine their states at the next time-step. At the start of the process, some of the grid's cells are 'on' and others are 'off'; and as time ticks forward, cells turn on or off according to the transition rules. At some point the grid may reach what is called a 'halting' state: the computation is completed and the output can be read off from the remaining pattern of activity on the grid.

Just as a laptop can solve computational problems (such as calculating how many tiles of a certain size and shape you will need to tile your bathroom floor, or solving some humungous

Figure 41.3 John Conway playing the Game of Life.

Kelvin Brodie. Sun News Syndication.

mathematical equation), CAs can also solve such problems. The problem is encoded in the grid's initial pattern of activity, and once the grid reaches its 'halting' state, the user reads off the solution from the residual pattern of activity. Different CAs can have different transition rules; and some may have different kinds of grid, or more than just two possible cell states. Even though CAs are remarkably different from conventional computers, it nevertheless turns out that if a problem can be solved by a conventional computer then it can also be solved by a CA (and vice versa): different computational *architecture*, but the same computational *power*.

In 1970 the British mathematician John Conway invented a CA engagingly called the 'Game of Life'. This CA has four very simple transition rules (Box 41.1). Conway noted an interesting fact about the Game of Life: through applying these simple rules to small-scale patterns on the grid, large-scale patterns of surprising complexity emerge.

If you were to zoom in and watch individual cells during the course of the Game of Life's computation, all you would see would be the cells switching on and off according to the four rules. Zoom out, though, and something else appears. Large structures, composed of many cells, are seen to grow and disintegrate over time. Some of these structures have recognizable

---

Box 41.1 The Game of Life

The Game of Life has just four transition rules:

1. If a cell is **on**, and fewer than two of its neighbours are also on, it will turn **off** at the next time-step.
2. If a cell is **on**, and either two or three of its neighbours are also on, it will stay **on** at the next time-step.
3. If a cell is **on** and more than three of its neighbours are on, it will turn **off** at the next time-step.
4. If a cell is **off** and exactly three of its neighbours are on, it will turn **on** at the next time step.

characters: they maintain cohesion, move, reproduce, and interact with one another. Their behaviour can be dizzyingly complex. Patterns called 'oscillators' change shape, returning after a certain number of time steps to the shape that they began with. A three-cell winking 'blinker' flips repeatedly from a vertical line to a horizontal line and back again, while the twelve-cell 'pentadecathlon' undergoes a beautiful fifteen-step transformation that returns it to its original shape.

So-called 'spaceships' glide across the grid in the Game of Life: as time clicks forward, they morph into a new configuration that duplicates their original pattern but is displaced by one or more cells from their starting position, so creating movement. If you watch the game speeded up, spaceships appear to move smoothly. Spaceships are the main way in which information is transferred from one part of the grid to another. 'Gliders' are the smallest spaceship: they consist of five cells and will, over four time-steps, reproduce their original configuration but displaced one cell to the left and down. There are larger spaceships: in fact there is no known largest spaceship. The largest one discovered so far is an 11-million-cell monster, the 'caterpillar'. Large-scale structures like the caterpillar are governed by their own rules, and to discover these 'higher-order' rules it is often better to experiment than to calculate. Observing the behaviour of the large structures under various conditions reveals the large-scale rules.

Some large-scale patterns, consisting of hundreds of thousands of cells, even behave as a universal Turing machine. Still larger patterns act like construction machines that assemble this universal Turing machine, and yet larger patterns—virtual creatures, perhaps?—feed instructions to the universal machine. The virtual creatures inside the Game of Life can program their universal Turing machines to perform any computation—and that includes running their own simulation of the Game of Life. A simulation of the Game of Life on their machines—a game within the game—might contain other virtual creatures, and these may simulate the Game of Life on *their* Turing machines, which may in turn contain more virtual creatures, and so on. The nested levels of complexity that can emerge on a large grid are mind-blowing.[22] Nevertheless, everything that happens in the Game of Life is in a fundamental sense simple: the behaviour of every pattern, large and small, evolves as prescribed by the four simple transition rules. Nothing ever happens in the Game of Life that is not determined by these rules.

Zuse's thesis is that our universe is a CA governed by a small number of simple transition rules: he suggested that with the right rules a CA can generate patterns called 'digital particles' (*Digital-Teilchen*).[23] These digital particles correspond to the fundamental physical particles that conventional physicists regard as the basic building blocks of the universe. Zuse was writing before the Game of Life was invented, and so he wasn't suggesting that the specific transition rules in the Game of Life are the fundamental rules of our universe, but if he's right then *some* simple transition rules (and their associated grid structure) comprise the fundamental physics of the universe. More recently the Dutch Nobel Laureate and theoretical physicist Gerardus 't Hooft (pronounced 'toft') has said:[24]

I think Conway's Game of Life is the perfect example of a toy universe. I like to think that the universe we are in is something like this.

If Zuse's thesis is right, then our universe is at its most fundamental level a computer: everything we observe in the universe—particles, matter, energy, fields—is a large-scale pattern that emerges from the activity of a CA. The grid of this CA is not made up from traditional matter like electrons or protons: the CA operates at a more fundamental level, and electrons, protons, and all matter currently known to physics, emerge from the activity of the CA—although what the CA's grid is in fact made of is far from clear, as we shall see below. This CA operates everywhere in the universe, at the smallest scale; and to describe it would be to produce a grand unifying theory of everything in the universe. All our other theories in physics—including general relativity and quantum mechanics—should fall out of this grand unifying theory. If Zuse is right then we humans are not so different from the virtual creatures that we create in the Game of Life. In fact, 't Hooft suggests that our three-dimensional universe may be a sort hologram, arising from the transformation of digital information on a two-dimensional surface.[25]

Examining Zuse's thesis

Is Zuse's thesis right? The idea that the universe is a giant CA faces three big challenges. The first is the 'emergence problem': can it be demonstrated that the physics of our universe could emerge out of a digital computation? The second challenge is the 'evidence problem': is there any experimental evidence to support Zuse's thesis? Third is the 'implementation problem': what 'hardware' is supposed to implement the universe's computation? Let's take the three challenges in turn.

The emergence problem is extremely hard. To solve it, the proponent of Zuse's thesis would need to find a way of showing how current physical theories could emerge from some simple underlying digital computation. Four large hurdles stand in the way. First, existing physics involves *continuous* quantities (position, energy, velocity, etc.); whereas CAs, and all digital computers, deal only in *discrete* units, not in continuous quantities. How could what is fundamentally continuous emerge from what is fundamentally discrete? To give a simple illustration: time is traditionally regarded as being continuous, whereas the movements of a digital watch are discrete: how could what is smooth and continuous arise from what is jerky and discontinuous? Second, physics seems to involve *non-deterministic* (i.e. random) processes, whereas CAs behave in a completely deterministic way. Third, current physics allows for *non-local* connections between particles: relationships without an intervening messenger (this is known as 'quantum entanglement'). Yet CAs don't allow such connections between distant cells of the grid. Fourth, and more worryingly still, our two best physical theories—general relativity and quantum mechanics—appear to be *incompatible*. How to unify general relativity and quantum mechanics is the hardest problem in current physics. But this is exactly what would need to be done by an underlying computational theory—no easy task! Perhaps each of these problems can be solved; if so, it is up to the advocates of Zuse's thesis to find the solutions.

Second, the evidence problem. To date, there is no experimental evidence at all for Zuse's thesis—so why believe it? It is also true that there's no experimental evidence against the thesis, and in fact evidence either way would be hard to find. This is because the CA that supposedly underlies the universe exists at extremely small spatial scales, of around one 'Planck length'. A Planck length, named after the famous quantum physicist Max Planck, is defined as 10^{-35} metres—that's just one zero short of a million million million million million millionth of a metre. Exploring events at this scale poses formidable obstacles. Let's use the size of the subatomic particle called the proton as our measuring stick. The Large Hadron Collider at CERN near Geneva can probe events that are 100,000 times smaller than a proton—the size difference between a mosquito and Mount Everest. However, a Planck length is much, much smaller still: a Planck length is ten followed by nineteen zeros times smaller than a proton— the same as the size difference between a mosquito and the Milky Way. Conventional particle colliders are never likely to be able to explore events at the Planck scale. In 2014, Craig Hogan's team at FermiLab in Chicago started a different kind of experiment to test whether space is a digital grid at the Planck scale. If it were, this would be one small step toward verifying Zuse's thesis. The experiment aims to detect jitter at the Planck scale, by measuring small movements in two laser interferometers.[26] So far, the experiment has not produced any evidence for or against space being digital, and moreover there are serious doubts over whether the experiment will ever produce evidence one way or the other.[27] Collecting evidence at the scale that is relevant to Zuse's thesis is *hard*.

Turning to the implementation problem, the challenge here is to say what hardware could possibly implement the universe's computation. The computations that your laptop carries out are implemented by electrical activity in silicon chips and metal wires; the computations in your brain are implemented by electrochemical activity in your neurons and synapses; and Conway's original version of the Game of Life was implemented by means of plastic counters and a Go board. Every computation requires some implementing medium, and the implementing hardware must exist in its own right. It cannot be something that itself *emerges* from the computation as a high-level pattern: Conway's plastic counters cannot emerge from the Game of Life—they are required in order to play the Game of Life in the first place.

According to Zuse's thesis, all matter, all energy, all fields, and all particles emerge as patterns from the underlying cellular computation. What, though, could implement the cellular computation? Not something that we already know of in physics, since by hypothesis everything that we currently know of is supposed to be an emergent pattern produced by the computation. Nor even something physical that we don't currently know of, since *everything* physical is supposed to emerge from the underlying computation. The implementing hardware must be something else: something beyond the realm of physics.

Some outlandish proposals have been made regarding this hardware. For example, the cosmologist Max Tegmark's 'mathematical universe hypothesis' claims that the implementing hardware of the physical universe consists of *abstract mathematical objects* (existing in what mathematicians sometimes call 'Platonic heaven').[28] Tegmark's proposal inverts the usual way we think of computation: rather than physical objects (such as an electronic PC) implementing abstract mathematical objects (such as natural numbers) abstract objects implement all physical objects. On Tegmark's proposal, abstract mathematical objects are more fundamental to the universe than atoms and electrons!

Many objections can be raised to this proposal.[29] The most relevant here is that abstract mathematical entities don't seem to be the right kinds of things to implement computation. Time and change are essential to implementing a computation: computation is a process that unfolds through time, during which the hardware undergoes a series of changes (flip–flops flip, neurons fire and go quiet, plastic counters appear and disappear on a Go board, and so on). Yet Tegmark's mathematical objects exist timelessly and unchangingly. What plays the role of time and change for this 'hardware'? How could these Platonic objects change over time in order to implement distinct computational steps? And how could one step give rise to the next if there is no time or change? Unchanging mathematical objects are just not the right kinds of things to implement a computation.

Currently, there are no plausible solutions to this chicken-and-egg implementation problem. Perhaps supporters of Zuse's thesis could say: we know that *something* must implement the universe's computation, but we should admit that we know nothing—and can know nothing—about this shadowy substratum. The proper aim of physics (Zuse's supporters continue) is simply to describe the universe's computation; physics must remain *silent* about the implementing medium. As Wittgenstein said in his usual pithy way:[30]

Whereof one cannot speak, thereof one must be silent.

If, however, you think that there is something unsatisfying about restricting the scope of physics in this way, then you are not alone. Red-blooded physicists want to know everything about the universe, and will not take well to this idea that the universe contains a fundamental substratum that must always remain beyond the reach of physics.

So far we have found no reason at all to think that the universe is a computer. At the beginning of the chapter we mentioned a second version of the computer-universe theory and we now turn to this. More modest than the first version of the theory, this acknowledges that the universe may not *literally be* a computer, but maintains that nevertheless the physical universe is fundamentally computational, in a sense that we shall now explain.

Is the universe comput*able*?

A comput*able* system is a system whose behaviour could be computed by an idealized human computer. It's important to add the caveat 'idealized', since it might take a human clerk a million years to compute the behaviour of some large and complex system—and moreover, the calculations might require more paper and pencils than planet Earth is able to supply.

There are many systems that, although they are not computers themselves, are nevertheless computable. Consider, for example, an old-fashioned navigation lamp. The function of the lamp is to flash out a signal marking, say, the eastern end of a particular sandbank in the Thames Estuary (and to assist navigators the signal must be recognizably different from the signals emitted by all the other navigation lamps up and down that stretch of water). This lamp's signal is as follows: the lamp turns on for 1 second, then switches off for 2 seconds, then on for 2 seconds, then off for 4 seconds, and then repeats this cycle indefinitely. (The ons and offs are created by a sliding metal disc that is controlled mechanically: while the disc is positioned over the lamp's glass aperture the light is effectively turned off, and when the disc ceases to obscure the aperture, the light shines out—although the mechanical details do not matter for the example.) It is easy for a human computer to calculate the on–off behaviour of this lamp, and if you were asked whether the lamp would be on or off 77 seconds (say) after its first flash, you would probably have little difficulty computing the answer. In summary: the lamp is not a computer but its flashing behaviour is comput*able*.

More complicated behaviours are also computable. For example, let's bring the irrational number π into the formula that determines whether the lamp is on or off. π, the ratio of a circle's circumference to its diameter, is 3.141592653589.... There is no last digit: the digits of π continue on to infinity. Using π we can make the lamp's switching behaviour quite complex: if the first digit of π is odd then the lamp begins its sequence of operations by illuminating for a second, and if the first digit is even, the lamp remains unilluminated during the first second; and if the second digit of π is odd, the lamp illuminates for a second, and if the second digit is even the lamp is unilluminated during this second second of its operating time; and so on. In this case, the lamp's behaviour during its first 13 seconds of operating is: *flash, flash, no flash, flash, flash, flash, no flash, no flash, flash, flash, flash, no flash, flash.* As the sequence grows longer, an observer might think that the flashes and pauses are coming randomly. But this isn't so.

Is the behaviour of the lamp still computable? Yes, it is. Turing showed that π is what he called a 'computable number': a Turing machine—and therefore a human computer—can calculate the digits of π, one by one. (Since there is no *last* digit of π, the Turing machine will work on forever, unless we stop it after it has produced some finite number of the digits.) So the Turing machine (or human computer) can calculate when the lamp is going to flash and when there will be no flash.

Randomness is one form of uncomputability: if the lamp were flashing randomly, its behaviour would *not* be computable, because if the human clerk could always predict the behaviour at the next second, then the behaviour would not be random.[31] Is there any conceivable way of arranging the behaviour of the lamp so that its behaviour is *not* random (i.e. is deterministic) but is nevertheless not computable? The answer to this is 'yes'. This in fact follows from points explained in Chapters 7 and 37. As mentioned there, Turing proved that no Turing machine can solve the 'printing problem': that is to say, there is no way of programming a Turing machine so that it can decide, given any Turing-machine program, whether that program is ever going to print '1' or not (or '#', as in the example in Chapter 7: the choice of symbol makes no difference). Using this fact, we will explain how to arrange the flashes so that the sequence is not computable.

Let's assume, first, that all the infinitely many Turing machines are *ordered* in some way, so that we can speak of the first Turing machine, the second Turing machine, and so on. The precise details of how the ordering is done need not concern us; one method would be to deem that the Turing machine with the *shortest* program is the first machine, and that the one with the next shortest program is the second machine, and so on—although, of course, some 'tie-breaker' principles would be required for ordering machines whose programs are of the same length; and some further details would also be required to deal with the issue of how much data ('input') a machine has on its tape before it starts work. We are going to modify the above switching recipe (the one involving π) like this: if the first Turing machine is one of those that at some point prints a '1', then the lamp starts its sequence of operations by illuminating for 1 second, and if the first Turing machine never prints '1', then the lamp remains unilluminated during the first second; and if the second Turing machine is one of those that at some point prints a '1', the lamp illuminates for 1 second, and if the second Turing machine never prints '1', the lamp is unilluminated during the second second of its operating time; and so on. Now the resulting sequence of flashes and pauses is *not* computable.

This flashing light example helps to clarify what is being asked by the question 'Is the whole universe computable?': the question asks whether the behaviour of *everything* in the universe can be computed by an idealized human computer (equivalently, by a Turing machine), or whether the universe contains systems that, like the third lamp, are *un*computable. In the next three sections we examine a number of theses that are relevant to this question, starting with a famous—but sometimes misunderstood—thesis put forward by Turing.

Turing's thesis

In 1936 Turing stated (and argued for) what has come to be called simply 'Turing's thesis': *A Turing machine can do any task that the human computer can do.*[32] Essentially the thesis says that the Turing machine is a correct formal model of the human computer.

It is worth mentioning in passing that Turing's thesis is sometimes called the 'Church–Turing thesis' (or even just 'Church's thesis'). This is because Alonzo Church devised another formal model of the human computer, also in 1936, which Turing quickly proved to be equivalent to his own model.[33] Church's model was couched in terms of his highly technical concept of 'λ-definability' (the Greek letter 'λ' is pronounced 'lambda'). Something is said to be λ-definable if it can be produced by a certain process of repeated substitutions—the details need not concern us. In Chapter 7, it was mentioned that Kurt Gödel much preferred Turing's model to Church's: Gödel said that he found Turing's model 'most satisfactory', but told Church that his approach was 'thoroughly unsatisfactory'.[34]

Nevertheless Church's own thesis, that his λ-definability model is a model of the human computer, is true—since Turing managed to prove that everything Turing machines can do is λ-definable (and vice versa). Nowadays there is a tendency to say that Turing's thesis and this thesis of Church's are 'the same' (in virtue of Turing's proof), but this is misleading, because the two theses have different *meanings*. One obvious and important difference between them is that Turing's thesis involves *computing machines* but Church's does not. In what follows we focus on Turing's thesis, not Church's.

Turing's thesis gets confused with some quite different claims about computability, and the implications of his thesis are not infrequently misunderstood. Searle, for example, gives this formulation of the thesis:[35]

anything that can be given a precise enough characterization as a set of steps can be simulated on a digital computer.

This statement of Searle's implies that *any* system that operates step by step is computable, but that is a much stronger claim than Turing's actual thesis, which says merely that *human computers* can be simulated by Turing machines. In fact, counter-examples to Searle's thesis can readily be found.[36]

Another example of confusion is Sam Guttenplan's statement (in *A Companion to the Philosophy of Mind*) that for any systems whose 'relations between input and output are functionally well-behaved enough to be describable by . . . mathematical relationships':[37]

we know that some specific version of a Turing machine will be able to mimic them.

Again this is very different from what Turing said: Turing's own thesis does *not* imply that a Turing machine can simulate (mimic) any and every input–output system that can be described by mathematics, but only that it can simulate any human computer. In fact, since the universe is effectively an input–output system (one thing leads to another by physical causation), and since the universe certainly appears to be mathematically describable, Guttenplan's thesis appears to imply that indeed the physical universe is computable. But no such thing is implied by Turing's thesis.

A third and final example of this tendency to misunderstand Turing and his work is provided by philosophers Paul and Patricia Churchland, who say that Turing's[38]

results entail something remarkable, namely that a standard digital computer, given only the right program, a large enough memory and sufficient time, can compute *any* rule-governed input-output function. That is, it can display any systematic pattern of responses to the environment whatsoever.

Turing's results certainly do not entail that every rule-governed input–output system is computable. That, as we have just seen, is tantamount to claiming that a *rule-governed* universe is a *computable* universe; and the uncomputable lamp example—where the rule that determines whether or not a flash comes at the nth second involves whether or not the nth Turing machine ever prints '1'—shows that this is wrong.

There is a thesis very different from Turing's lurking amid these confusions, and the fact that it isn't the *same* as Turing's thesis doesn't necessarily mean that it isn't true. So let's try to pin this thesis down and examine it. As a first attempt: *The behaviour of any imaginable law-governed deterministic physical system is computable.* Notice that if it's assumed that the universe contains *only* physical systems (and if it is also assumed that the universe is law-governed and deterministic) then this thesis certainly implies that the whole universe is computable. But because we don't want to get diverted by a discussion of the thorny question of whether *everything* in the universe is physical—or whether, on the other hand, it contains non-physical things such as souls and angels—we will henceforward concentrate on the claim that the whole *physical* universe is computable.

This thesis relating lawfulness and computability is, however, no more acceptable than the previously discussed thesis that all rule-governed systems are computable. The third lamp is a deterministic physical system that is governed by the laws of physics, yet it is not a computable system. Another counter-example to the thesis is the hypothetical discovery, in some distant galaxy, of a naturally occurring and fully deterministic source of radio waves—perhaps an oscillating supernova remnant—that emits an unending sequence of radio-frequency pulses exhibiting the same pattern as the flashes emitted by the third lamp. The supernova remnant is a law-governed physical system whose behaviour is not computable. The underlying point is that there are imaginable physical laws that are deterministic but uncomputable, and so the thesis fails.[39]

The third lamp, though, was only very loosely specified: we did not actually explain how the flashing behaviour is brought about in such a way that it reflects the printing behaviour of Turing machines. The same goes for the hypothetical supernova remnant. This leads on to a better statement of the thesis: *The behaviour of any rigorously specified deterministic physical system is computable.* We call this the 'specification thesis', or S-thesis for short.

The physical universe, we assume, can be rigorously specified mathematically—we don't know this specification yet, but this is what physics aims at. So the S-thesis seemingly entails that the physical universe, if deterministic, is computable. But is the S-thesis true? In fact it seems not to be. In the next section we will describe a rigorously specified deterministic machine—actually a kind of Turing machine—whose behaviour is not computable.

Accelerating Turing machines

As the name implies, accelerating Turing machines (ATMs) speed up as they operate; but apart from the fact that their speed of operation accelerates as the computation proceeds, ATMs are exactly like standard Turing machines.[40]

An ATM accelerates in accordance with a formula first introduced by the philosopher Bertrand Russell in a lecture in 1914. He described an unusual way of running around a racetrack:[41]

If half the course takes half a minute, and the next quarter takes a quarter of a minute, and so on, the whole course will take a minute.

Russell emphasized that although a human runner's accelerating in this way is *medically* impossible', it is not *logically* impossible.

An ATM follows the same pattern: it performs the second operation specified by its program in half the time taken to perform the first, and the third in half the time taken to perform the second, etc. If the time taken to perform the first operation is a second (say), then the next operation is done in half a second, the third operation is done in a quarter of a second, and so on. Adding up the times taken by the second and third operations and onwards, we get:

$$\frac{1}{2} + \frac{1}{4} + \frac{1}{8} + \frac{1}{16} + \frac{1}{32} + \ldots$$

This total evidently cannot exceed 1 second (since what is added at each step is always less than the remaining amount that must be added in order to reach 1). So, allowing also for the second that it takes to do the first operation, the total time required for *all* the operations done by the machine is no more than *two seconds*. Notice that this remains true even if the machine carries out an *infinite* number of operations, i.e. it never stops computing. Which is to say: an ATM can carry out an infinite number of computations in no more than 2 seconds.

An ATM can exhibit behaviour that is not computable. For example, let's suppose that, for any selected integer n, we want to find out whether the nth Turing machine ever prints '#'. If the nth Turing machine is one of those that runs on forever, then you couldn't find this out simply by watching the machine at work, since no matter how long you had been watching without '#' appearing you could never be sure that '#' would not be printed at some future time—and so you would never be in a position to say: 'The nth Turing machine does not print "#" '.

This is not a computable task, as Turing proved in 1936; but nevertheless an ATM can do it. Here's how. The ATM calculates the behaviour of the nth Turing machine, step by step, and if it finds that the nth machine does print '#', then it outputs the message: 'Yes, it prints "#" ' (or, better, we can arrange for it to output an abbreviated form of this message). If, on the other hand, the nth machine runs on forever without printing a single '#', then the ATM will itself run on through infinitely many calculations as it simulates the nth machine, waiting in vain for it to print '#'. But since these infinitely many calculations will take the ATM no more than 2 seconds to complete, we know that if the message 'Yes, it prints "#" ' has not arrived after 2 seconds, then the nth machine does not print '#'.

It only remains to fill in the details. We prepare the ATM by writing the program of the nth machine on its tape: the ATM will simulate the nth machine by following this program. In order to make the message 'Yes, it prints "#" ' as compact as possible, we adopt the convention that if the ATM writes '1' on the very first square of its tape (which we make sure we leave blank in the setting-up procedure), then this shall be taken to mean 'Yes, it prints "#" '. We accordingly program the ATM to go to this square and print '1' on it (and then halt) if it discovers that the nth machine prints '#', and the ATM is prohibited from printing anything on this square in any other circumstances.

Now we press the ATM's start button and wait 2 seconds before reading the output. If we see that the special square contains '1', then the nth machine does print '#'; and if we see that the special square is still blank, then the nth machine does not print '#'.

So the S-thesis appears to be false. The ATM has been specified carefully and yet its behaviour is not computable. Those interested in exploring ATMs further will find a reference in the endnotes.[42]

However, the fact that the ATM is a logical possibility doesn't mean that it could actually exist in our physical universe. The ATM pushes us towards a sharper version of the thesis that we are looking for; we call this sharper version the 'physical computability thesis' (PCT).

The physical computability thesis

The PCT is: *The behaviour of every genuinely possible deterministic physical system—that is, every deterministic physical system that is possible according to the physics of our universe—is computable.*

It is worth noting in passing that the PCT is often referred to as the 'physical version' of Turing's thesis, and is sometimes called the 'physical Church–Turing thesis'. However, the name 'physical Church–Turing thesis' is perhaps not ideal, because the PCT has little or nothing to do with the Church–Turing thesis, and neither Church nor Turing endorsed—nor even formulated—the PCT. Since using the name 'physical Church–Turing thesis' could open the door to confusion we prefer to avoid it here.

The PCT is an interesting thesis, and entails an affirmative answer to our question 'Is the whole physical universe computable?' (assuming that the universe is deterministic). But is the PCT true? Some maintain not. In the remainder of this section we will describe a potential counter-example to the PCT involving a relativistic system (in the sense of Einstein's theory of relativity).

The idea of using relativity to formulate an uncomputable system was presented by Itamar Pitowsky in 1986, at an academic conference in Jerusalem. Pitowsky explained that under certain special conditions, a computer can perform infinitely many steps in what an observer who is outside the system (but communicating with it) experiences as a *finite* time.[43] What's more, this computer can be a perfectly ordinary laptop that functions just as usual—and as far as the laptop is concerned, there is no speed-up at all: it performs each step of the computation at the same rate. The speed-up that enables infinitely many steps to be performed in a finite time is seen only from the viewpoint of the distant observer.

This is not quite the same idea as an ATM, since the ATM described earlier would be seen as speeding up even by an observer who is part of the system (sitting on the scanner, say). Additionally, relativistic systems are governed by Einstein's theory of relativity, meaning that no signal can travel faster than light travels in a vacuum; whereas an ATM is not necessarily subject to this restriction. An ATM may accelerate to the point where the scanner is moving along the tape faster than the speed of light (although the scanner could be kept below light speed if the symbols on the tape were to get progressively smaller, with the result that the distances travelled by the scanner become shorter and shorter).

Pitowsky's original setup conformed to Einstein's special theory of relativity, but here we will describe a setup proposed by István Németi and his group (at the Hungarian Academy of Sciences in Budapest) that involves Einstein's general theory of relativity.[44] Németi's system is in effect a relativistic implementation of an ATM. He emphasizes that his system is a *physical* one, as opposed to some purely notional system that could exist only in fairy-land: the system is physical, Németi says, in the sense that it is 'not in conflict with presently accepted scientific principles', and, in particular, 'the principles of quantum mechanics are not violated'.[45] Németi suggests that humans might 'even build' his relativistic system 'sometime in the future'.[46]

Németi's system consists of two parts, one part being a standard Turing machine *S*, located on Earth, and the other part being an observer *O*, who journeys through space. Before beginning the journey, *O* sets up *S* to simulate the *n*th Turing machine, the object of the exercise being to discover whether or not the *n*th machine prints '#'. Associated with *S* is a piece of ancillary equipment that emits a signal if (and only if) the simulation done by *S* reveals that the *n*th machine prints '#'. This arrangement is equivalent to the previous one of writing '1' on the first square of *S*'s tape if (and only if) the *n*th machine prints '#'.

O then travels through space to a type of black hole known as a 'slow Kerr hole', after the New Zealand mathematician Roy Kerr. Slow Kerr holes are huge, slowly rotating black holes. Cosmologists do not know for certain if any slow Kerr holes actually exist, but Németi points to 'mounting astronomical evidence for their existence'. He chooses a slow Kerr hole because these have special properties, one of which is that the observer *O* can, Németi says, pass through the hole 'and happily live on'. If *O* were to enter a more traditional type of black hole (Fig. 41.4), he or she would be annihilated by the crushing gravitational forces generated by the black hole. In the case of a slow Kerr hole, however, these extreme gravitational forces are, Németi explains, counterbalanced by the Kerr hole's rotational forces—the gravitational forces are offset by the forces that result as the black hole spins, meaning that the observer is not crushed, and can in principle emerge safely.

Németi theorizes that as *O* starts to enter the Kerr hole, *S*'s rate of computation accelerates relative to *O*. This is due to 'gravitational time dilation', an effect predicted by Einstein's theory of relativity. The deeper into the hole *O* travels, the faster and faster *S* runs relative to *O*, in fact without any upper limit. The acceleration continues until, relative to a time *t* on *O*'s watch, the entire span of *S*'s computing is over; and, if any signal was emitted by *S*'s signal-generator it will have been received by *O* before this time *t*. From *O*'s point of view, *S* has done its computation in a finite period of time. This is so even if *S* runs on through infinitely many calculations as it simulates the *n*th machine, in the possible case where the *n*th machine computes forever

Figure 41.4 Artist's impression of a supermassive black hole.

without printing '#'. By time t, therefore, O knows whether or not a signal has been emitted, and so knows whether or not the nth machine ever prints '#'.

If Németi is right, then this is a counter-example to PCT: a deterministic physical system that is *not* computable but that nevertheless *is* possible according to the physical laws of our universe. His counter-example is certainly not universally accepted: for example, one can question whether the existence of a Turing machine that is able to compute forever without wearing out—as S must, if the nth machine runs on forever without printing '#'—is *really* consistent with the actual laws of physics. But Németi's example certainly serves to show that it's far from *obvious* that the PCT is true. As we intimated at the beginning of the chapter, the answer to 'Is the whole physical universe computable?' is currently unknown. Even if there is genuine randomness, and hence uncomputability, in the universe, the question would still remain whether there are (or could be) real physical systems *not* involving randomness that are uncomputable. Theorists disagree about the answer—and sometimes the debate gets heated.

In fact, to assume without warrant that the physical universe is computable might actually hinder scientific progress. If the universe is essentially uncomputable, and yet physicists are searching for a system of physical laws that would describe a computable universe, then bad physics is likely to ensue. Even in the case of brain science—let alone the study of the whole universe—simply assuming computability could be counterproductive. As philosopher and physicist Mario Bunge remarked, this assumption[47]

involves a frightful impoverishment of psychology, by depriving it of nonrecursive [i.e. non-computable] functions.

In the next section, we will examine Turing's views on this question of computability and the brain—a microcosm of the debate about the grand-scale question of whether the whole universe is computable.

Turing's opinion

It used to be widely believed that Turing had said, or perhaps even proved, that every possible physical system is computable. Earlier we mentioned Paul and Patricia Churchland asserting that Turing's results *entail* that all rule-governed behaviour is computable. Another example comes from David Deutsch, one of the pioneers of quantum computing, who put forward this variant of the PCT, calling it 'the physical version of the Church–Turing principle':[48]

Every finitely realizable physical system can be perfectly simulated by a universal model computing machine operating by finite means.

Deutsch went on to say that 'This formulation is both better defined and more physical than Turing's own way of expressing it.' Deutch's thesis is indeed more physical than Turing's thesis; but it is a completely *different* thesis from Turing's, not a 'better defined' version of what Turing said. Turing was talking about human computers, not physical systems in general.

In a similar vein the mathematician Roger Penrose (the co-discoverer of black holes) stated:[49]

It seems likely that he [Turing] viewed physical action in general—which would include the action of a human brain—to be always reducible to some kind of Turing-machine action.

Penrose even named this claim 'Turing's thesis'. But, as we shall see, Turing never endorsed this thesis and was aware that the thesis might be false.

Andrew Hodges (the mathematician who wrote the biography that inspired the movie *The Imitation Game*) maintained in his book that Turing's work implied what is a close cousin of the PCT:[50]

Alan had . . . discovered something almost . . . miraculous, the idea of a universal machine that could take over the work of any machine.

Here Hodges, like Penrose, was suggesting that Turing's work entails that any physical mechanism is computable. He also stated that Turing claimed:[51]

that the action of the brain must be computable, and therefore can be simulated on a computer.

However, that was back in the bad old days. Modern Turing scholarship, by Hodges and others, now paints a very different picture. In fact, there is no evidence that Turing ever understood his work on computability to rule out the possibility of mechanisms whose action is not computable. In 1999, one of us—Jack Copeland, together with Diane Proudfoot—suggested in an article in *Scientific American* that Turing was an important forerunner of the modern debate concerning the possibility of uncomputability in physics and uncomputability in the action of the human brain.[52] In the same year Copeland also published a commentary on a lecture that Turing gave on BBC radio in 1951 titled 'Can digital computers think?':[53] this commentary pointed out that, in the lecture, Turing noted the possibility that the physical brain cannot be simulated by any Turing machine.[54] Hodges was persuaded by these observations (and, in a public lecture, generously thanked Copeland for the idea, even saying on his website 'I don't mind admitting that I wish I had thought of it').[55] Turing, far from claiming that every physical mechanism, including the brain, *must* be computable, was open to the idea that the brain, at least, is not a computable system.[56]

As Hodges recently said in the journal *Science*:[57]

[Turing was] one of the first to use a computer for simulating physical systems. In 1951, however, Turing gave a radio talk with a different take on this question, suggesting that the nature of quantum mechanics might make simulation of the physical brain impossible.

Conclusion

Alan Turing never said that the physical universe is computable, and nor do any of his technical results entail that it is. Some computer scientists and physicists seem infuriated by the suggestion that the physical universe might be uncomputable; but it is an important issue, and the truth is that we simply do not know.

CHAPTER 42

Turing's legacy

JONATHAN BOWEN AND JACK COPELAND

In terms of public appreciation Turing has risen quickly from zero to hero. The decades of obscurity (Who did you say? Turning?) came to an end with the much-publicized Royal Pardon of 2013 and, before that, the widespread international celebrations in 2012 during the centenary of his birth—celebrations that formed a spontaneous collective decentralized yell, by academics, artists, broadcasters, politicians, athletes, and others: ACCLAIM HIM! Suddenly Turing went viral. This final chapter surveys his burgeoning legacy: cultural, political, scientific, and linguistic. It is a work in progress. Turing's renown is growing too fast for any final assessment.

Finding Turing

Turing invented the universal machine, representing the stored-program computer in its most general and abstract form; and with hindsight his seminal 1936 paper 'On computable numbers' foretold the capabilities of the modern computer. During the Second World War he recognized that computing machines could carry out important tasks fast and reliably, even tasks requiring intelligence when done by human beings. Unlike some theorists, Turing had a strong practical bent and was as happy to wield a soldering iron as to wrestle with an abstract mathematical problem. The story of his post-war contributions to the design and development of electronic computers was told in Chapters 20–22.

Yet placing Turing is always difficult. He was not *the* inventor of the computer, but one among a diffuse group of mathematicians and engineers who brought the new machines into existence (see Chapter 6). For the same reason he was not *the* founder of computer science; and in any case the title 'founder of computer science' is anachronistic: Turing had been dead for 6 years before George Forsythe coined the term 'computer science' and the first university departments of computer science began to emerge. Nevertheless, the new discipline of computer science was certainly grounded in the fundamental ideas of Turing and his contemporaries, including John von Neumann, Maurice Wilkes, and Tom Kilburn.

In his personal life Turing was a victim of his times. To be an active homosexual in those dark days was to be a criminal. A mere 15 years after Turing's arrest and prosecution the anti-gay laws were struck down—in his small corner of the world at any rate—but in mid-twentieth-century England an isolated maverick innocent like Turing trod dangerously. He had no protective network of powerful or worldly gay friends, and at that time he possessed no public reputation as a war hero and leading scientist: the secrecy surrounding his wartime work meant there could be no acknowledgement of the nation's debt to him. Even within the scientific community the importance of his work was not obvious at the time. To most scientists in that largely pre-computational era, his groundbreaking 1936 paper would have seemed abstruse and irrelevant beyond the narrow field of mathematical logic. His work on artificial intelligence and morphogenesis also had little impact during his life.

At the time of his death, Turing was a convicted criminal and an obscure Fellow of the Royal Society. Little else was known or widely remembered. Real recognition of Turing's scientific contributions came long after he died, with the development of computer science and a growing awareness of his applied work during the war.

Scientific legacy

By the 1970s and 1980s a number of popular and semi-popular science books were spreading the word about Turing and his achievements. Herman Goldstine's 1972 book *The Computer: From Pascal to von Neumann* contained some glimpses of Turing from the point of view of an American computer pioneer. A little later Turing appeared, if briefly, in Brian Johnson's groundbreaking 1977 book *The Secret War*, and then at greater length in Pamela McCorduck's 1979 classic *Machines Who Think*. McCorduck's excellent cameo biography of Turing, just a dozen pages long, was the wider world's first glimpse of this 'delightful if eccentric' mathematician.

Douglas Hofstadter's elegant bestseller *Gödel, Escher, Bach: An Eternal Golden Braid*, also published in 1979, put the Turing test and the Church–Turing thesis onto many a coffee table. Another 1979 book, *The Mighty Micro*, by the National Physical Laboratory's Christopher Evans, contained significant coverage of Turing's work, as did Simon Lavington's more specialized *Early British Computers* the following year. Gordon Welchman's 1982 *The Hut Six Story: Breaking the Enigma Codes* told much about Turing, and so did Andrew Hodges' classic 1983 biography *Alan Turing: The Enigma*. Several 1984 books on the computer revolution dwelt on Turing: Stan Augarten's *Bit by Bit*, Joel Shurkin's *Engines of the Mind*, and Michael Shallis's *The Silicon Idol*.

Brian Carpenter and Bob Doran, the first computer historians to emphasize that the stored-program concept originated in Turing's 'On Computable Numbers', had written a landmark article in 1977 ('The other Turing machine', in *The Computer Journal*), and in 1986 they published their book *A. M. Turing's ACE Report of 1946 and Other Papers*. This collected together for the first time key materials explaining Turing's role in computer history. In 1988 Rolf Herken's monumental *The Universal Turing Machine: A Half-Century Survey* was published; and in 1989 Roger Penrose's *The Emperor's New Mind: Concerning Computers, Minds, and the Laws of Physics* brought talk of Turing machines to dinner parties around the world.

In the following decade, Mary Croarken's 1990 book *Early Scientific Computing in Britain* gave good coverage of Turing and his ideas, as did David Kahn's *Seizing the Enigma* (1991), Doron Swade and Jon Palfreman's *The Dream Machine: Exploring the Computer Age* (also 1991), and Jack Copeland's *Artificial Intelligence* (1993). George Dyson's *Darwin Among the Machines*

(1997) and John Casti's *The Cambridge Quintet* (1998) both contained rich treatments of Turing aimed at the popular market. Casti's book—in which a fictional treatment of Turing is blended with passages from the real Turing's writings—bore the dedication:

To the memory of Alan Turing and John von Neumann, creators of the modern computer age.

On the Enigma front, Michael Smith's 1998 *Station X* and Simon Singh's 1999 *The Code Book* both brought Turing's wartime work alive for general readers. The decade came to a close with *Time* magazine's 1999 article *The Great Minds of the Century*: this gave Turing his due, despite labelling him a computer scientist.[1]

Turing's three most cited papers—all heading towards 10,000 citations apiece, according to Google Scholar[2]—were published in 1936, 1950, and 1952.[3] Each effectively founded a field, even though this did not become clear until later. The 1936 paper gave rise to theoretical computer science, while artificial intelligence stemmed from the 1950 work, and mathematical biology from the third paper, which according to Google Scholar is the most cited of the three. If Turing had not died so young, and so soon after two of his three most influential publications, it seems certain that he would have gone on to produce yet more inspirational ideas.

A lasting scientific memorial to Turing, inaugurated in 1966, is the Association for Computing Machinery's A. M. Turing Award.[4] This award is the highest scientific honour available to a computer scientist and is presented to at least one, and sometimes up to three, leading computer scientists each year. Many winners of this award work or worked in areas founded by Turing, such as artificial intelligence or formal methods (the application of mathematics to software engineering). No fewer than 33 Turing Award winners attended the Association for Computing Machinery's *A. M. Turing Centenary Celebration* in San Francisco in 2012 (Fig. 42.1).

The Turing machine (see Chapter 6) is probably Turing's most important legacy to computer science. This has continued to play a fundamental role, not least as the measure of what is and is not computable (see Chapters 7, 37, and 41). The Turing machine concept has also deeply influenced cognitive science and the philosophy of mind. A fundamental problem known to

Figure 42.1 Winners of the A. M. Turing Award at the Turing Centenary Celebration in San Francisco in 2012. Left to right (with year of award): Niklaus Wirth (1984), Edmund Clarke (2007), and Barbara Liskov (2008).

every computer science student and called the 'halting problem' is based on the Turing machine concept: this is the problem of deciding, given any computer program, whether the program terminates or runs on forever (until memory is exhausted). Christopher Strachey, who features in Chapters 20, 21 and 23, recounted how Turing gave him a verbal proof of the unsolvability of the halting problem in a railway carriage in 1953.[5]

Many developments in modern computer science still rely on the Turing machine concept, for example quantum computing and complexity theory.[6] The latter is a rigorous formulation of the idea that some problems are harder to solve than others. Another illustration of the importance of the Turing machine concept is that new and seemingly very different forms of computation have so far always turned out to be *equivalent* to standard Turing-machine computation. An example is provided by the 'Game of Life', explained in Chapter 41: although the Game of Life *seems* very different from the Turing machine concept, it has been proved that whatever can be computed by the Game of Life can also be computed by a Turing machine, and vice versa.[7] The emerging field of quantum computing might be another example where this is true—that is to say, it might be true that everything computable by a quantum computer is computable by Turing machine and vice versa—although in this case the jury is still out.

Apart from the Turing machine, another conspicuous 'Turing first' in computer science was his work in artificial intelligence (Chapter 25), including his specification of the Turing test (Chapter 27), his anticipation of connectionism (Chapter 29), and his pioneering discussion of what is now called 'superintelligence'—artificial intelligence that exceeds human intelligence. He also published (in 1949) a very early proof of what is today called 'program correctness'. Arguably the first ever such proof, this innovative work was rediscovered later and appreciated as an early foray into what is now an important subfield within computer science.[8]

Any book on the history of the computer would be radically incomplete without an explanation of Turing's role in its development. He is also considered an important figure in the overall history of science. Charles Van Doren's encyclopaedic 1991 book *A History of Knowledge*, which set out to cover the entire range of human invention and creativity, devoted a section to Turing and the Turing machine, while *The Oxford Companion to the History of Modern Science* included not only an entry for 'Computer science' but also a separate entry for 'Artificial intelligence': both entries described Turing's foundational role. In addition the *Oxford Companion* (which was edited by John Heilbron) contained an entry simply on Turing himself: this noted presciently—in 2003—that 'Turing's status as a cult hero will undoubtedly increase'.

Figure 42.2 Entrance to the 2012 'Codebreaker' exhibition at the Science Museum, London.

Photograph by Jonathan Bowen.

Turing's legacy extends to his impact on the institutions where he worked. Manchester University went on to build the first transistorized computer. The National Physical Laboratory's Division of Computer Science carried on doing groundbreaking research, most notably in computer communications. 'Packet switching', a crucial enabling technology for the Internet, was developed at the National Physical Laboratory in the 1960s by Donald Davies, one of Turing's assistants in the ACE days.

The Science Museum in London has the original NPL Pilot ACE computer on permanent display: Davies and others at NPL developed this pilot model from Turing's ambitious ACE design (see Chapter 21). Turing himself visited the Science Museum in 1951 and was fascinated by an electromechanical cybernetic 'tortoise' that he saw on display.[9] Designed by William Grey Walter, the tortoise was a small autonomous robot that could sense its surroundings.[10] During 2012 and 2013 a special exhibition, 'Codebreaker: Alan Turing's Life and Legacy' ran at the Science Museum as part of the Turing centenary celebrations (Fig. 42.2): this featured the Pilot ACE, an Enigma machine, and other Turing-related items, including the tortoise he had viewed in 1951.[11] Chapter 38 contains some detective work about one of the codebreaking exhibits in this exhibition.

The Turing centenary celebrations sparked numerous scientific articles about Turing's work, and indeed entire journal issues appeared focusing on his achievements. The leading scientific periodical *Nature* pictured him on the front cover of a special centenary issue titled 'Alan Turing at 100'.[12]

Cultural legacy

Turing's life and early death have inspired numerous literary works and artworks. One of the earliest plays about Turing, in 1986, was Hugh Whitemore's *Breaking the Code*. Starring Derek Jacobi as Turing, the play premiered in London's West End and then ran on Broadway during 1987–88. Ten years later it screened on BBC television, again starring Jacobi.[13] Jacobi played Turing brilliantly on stage but by the time of the TV version he was a little old for the role, and the BBC's Turing seemed middle-aged and almost dull as he tore into Enigma at the age of 27. More recently, Maria Elisabetta Marelli's multimedia stage show *Turing: A Staged Case History* was described by Barry Cooper in 2012 as 'The most remarkable of the Turing Centenary events'.[14] Snoo Wilson's *Lovesong of the Electric Bear* opens with Turing being awoken from his deathbed by his teddy-bear Porgy; and Catrin Fflur Huws's *To Kill a Machine* revolves around a gameshow called 'Imitation': her Turing sits beneath a tree-like structure with a poisoned apple and a photograph of Christopher Morcom hanging over him.[15]

Other plays about Turing include George Zarkadakis's *Turing* and Rey Pamatmat's *Pure*, both focusing on Turing's death and his relationship with his school friend Christopher Morcom.[16] Composer and Turing admirer James McCarthy says about Morcom:[17]

Falling in love with Morcom changed Turing's life. It would be an over-simplification to say that Turing owed everything to that single event, but I believe that the desire to fulfil Morcom's potential for him and the later investigations into whether machines could think flow from this pivotal moment.

Morcom was Turing's only real friend at Sherborne, and this and similar claims about the importance of his influence on Turing's scientific life are often voiced; but in reality there is no evidence, apart from letters written in 1930 at the time of Morcom's sudden death. In a letter to Morcom's mother Turing said that he had worshipped the ground Morcom trod on, and in his grief and distress the seventeen-year-old seems to have felt that his dead friend was somehow still helping and encouraging him. Three days after the devastating news of the death he wrote to his mother Sara:[18]

I feel sure that I shall meet Morcom again somewhere and that there will be some work for us to do together as I believed there was for us to do here. Now that I am left to do it alone I must not let him down.

The 2014 Hollywood movie *The Imitation Game*[19] portrayed Turing as naming both his bombe and his post-war computer 'Christopher', and as saying about the latter machine that 'Christopher's become so smart'. In fact, the film's scriptwriter concoted all this. McCarthy too takes a wild speculative leap when he says that Turing's 'investigations into whether machines could think' flowed from his relationship with Morcom; and even the more modest speculation that Morcom and his early death in some way significantly influenced Turing's adult life and work goes well beyond the known facts.

Turing novels have become almost a mini-genre. Amy Thomson's award-winning 1993 science fiction story *Virtual Girl* featured an AI named 'Turing'. Robert Harris's excellent 1995 codebreaking novel *Enigma* was inspired by Turing, and the character Tom Jericho—who in the novel is Turing's student at Cambridge—is loosely based on Turing. Neal Stephenson's geeky bestseller *Cryptonomicon* from 1999 featured a fictionalized Turing who, among other adventures, invents the Turing machine, breaks codes, and has a love affair with Third Reich codebreaker Rudy von Hacklheber. The following year Paul Leonard's *The Turing Test* buddied up Turing with the BBC's sci-fi icon Doctor Who: at Bletchley Park Turing is unable to break strange new coded messages emanating from Germany, but with the Doctor's help everything is possible. Charles Stross's 2001 *The Atrocity Archive* related how Turing proved a theorem that threatened to undermine most of modern cryptography. The theorem, quickly hushed up by the authorities, not only showed that the Church–Turing thesis is false (see Chapter 41) but also solved the famous $P = NP$ problem in complexity theory—in reality the most important unsolved problem of modern computer science.

In 2005 the theme of Turing-as-AI appeared again, this time in Christos Papadimitriou's *Turing (A Novel About Computation)*. The best parts of this book are the lectures on computation that the Turing AI gives. The AI at times resembles Papadimitriou, himself a charismatic lecturer and computer science professor at Berkeley. Disappointingly Turing appears in the 2007 Bolivian cyberpunk novel *Turing's Delirium* by Edmundo Paz Soldan mainly as the nickname of the novel's hero, Miguel 'Turing' Saenz; but both Turing and Gödel appear as themselves in Janna Levin's 2007 novel *A Madman Dreams of Turing Machines*. Gödel, not looking any more dreamy than usual, has the cover.

Chris Beckett's 2008 book *The Turing Test* is a collection of fourteen science fiction stories dwelling on AI's relationship with humanity. In Nas Hedron's 2012 cyberpunk novel *Luck + Death at the Edge of the World* Turing, an AI with emotions, is eventually decommissioned by the authorities. Rudi Rucker's 2012 beatnik sci-fi novel *Turing & Burroughs* begins with the Kjell crisis (see Chapter 4) but the secret service's attempt to poison Turing with cyanide fails through sheer chance. Escaping his poisoned-icon destiny Turing goes on to become the lover of beat-generation hero William Burroughs. David Lagercrantz's gloomy 2015 thriller *Fall of Man in Wilmslow* is very different. With a bitten apple on the cover, the book begins with the

Figure 42.3 Turing in bronze, sitting on a park bench in Sackville Park, Manchester.

Posted to Wikimedia Commons by Hamish MacPherson, https://commons.wikimedia.org/wiki/File:Alan_Turing_Memorial_Sackville_Park.jpg. Creative Commons Licence.

hackneyed image of Turing dipping an apple into cyanide and committing suicide—'all true' the *Spectator*'s reviewer, journalist Sinclair McKay, assures readers.[20]

In 2001 a bronze sculpture of Turing was installed in Manchester's Sackville Park (Fig. 42.3): he sits on a bench holding an apple. Inspired by *Breaking the Code*, the sculpture was unveiled on 23 June, Turing's birthday. Stephen Kettle's slate sculpture of Turing, which stands in Bletchley Park's Block B, has already been mentioned in Chapter 19 (see Fig. 19.3).[21] The slate came from North Wales, where Turing visited as a child and adult. There are other memorial sculptures and busts of Turing around the world. Figure 42.4 shows one located far away from Turing's home: this large bust, perhaps not immediately recognizable as Turing, stands outside the computer science department at Southwest University in Chongqing, China.

Turing appears on a number of postage stamps, from places as far apart as Portugal, Tatarstan (on the railway line from Moscow to Siberia), and Guinea in West Africa. Two special-edition UK stamps in 2014 depicted Turing and the bombe, while an earlier UK stamp from 2012 displayed the words 'Alan Turing 1912–1954   Mathematician and WWII code breaker', together with a picture of the rebuilt bombe (see Fig. 19.2).[22] An event even more improbable than the issuing of a Turing stamp by St Helena—the tiny volcanic island in the South Atlantic where Napoleon was imprisoned—was the (Google-funded) release in 2012 of a Turing version of the popular *Monopoly* board game. Bletchley Park replaces Mayfair, and instead of the usual houses and hotels are Huts and Blocks. Locations round the board include Bletchley Train Station, Warrington Crescent (Turing's birthplace), Hut 8, and the National Physical Laboratory, while the usual Income Tax and Super Tax squares become War Tax and Bury Gold Bullion. Better still, the original version of the Turing board was designed by Max Newman's son William: Turing actually played William on the prototype board—and lost, the story goes.

Figure 42.4 Bust of Turing at Southwest University, Chongqing, China.

Photograph by Jonathan Bowen.

Turing has been extensively celebrated in music. One of the earliest Turing songs was Steve Pride's 'Alan Mathison Turing', written in about 2002 ('. . . he took a bite from a poison apple, they found him dead on the bedroom floor . . .').[23] Matthew Lee Knowles' *For Alan Turing—Solo Piano Suite in Six Movements* premiered in 2012 at *Alan Turing's 100th Birthday Party*, held at King's College, Cambridge.[24] James McCarthy's oratorio *Codebreaker* premiered in 2014 at London's Barbican Centre, with the London Orchestra da Camera and the Hertfordshire Chorus. McCarthy says about *Codebreaker*:[25]

I want the audience in this piece to feel like they've been sitting across the table from Alan Turing and they've shared a cup of tea with him, and have spoken about his hopes and his fears and what he loves.

But like many artists McCarthy uncritically accepted the verdict of Turing's inquest. His oratorio draws to a close with Turing's suicide, and gay martyr Turing is reunified after death with Christopher Morcom.[26]

An operatic work by the Pet Shop Boys, *A Man From the Future*, also premiered in London in 2014, at the BBC Proms in the Royal Albert Hall. The title of the work is oddly selected since a key to understanding Turing's life and work is that he was a man of his time. In the Proms performance, orchestrated songs were interspersed with lengthy spoken passages from Hodges' 1983 biography of Turing.[27] Quite different is Nico Muhly's *Sentences*, which premiered at the Barbican in 2015 with the Britten Sinfonia and counter-tenor Iestyn Davies. Muhly says admiringly:[28]

Turing's work may have begun as mathematics, but it is amazing to think about the impact it had on physical people's bodies—it literally saved lives.

Sentences has seven thematic sections with topics ranging from bicycling to the Turing test to Turing's death—though the fast-talking Muhly (himself gay) says emphatically 'No one wants a gay martyr oratorio'. Like Muhly, Justine Chen explores the lingering question of suicide versus accident in her opera *The Life and Death(s) of Alan Turing*, and she dramatizes four different theories of how Turing died.

The sheer amount of serious new music and theatre about Turing demonstrates the quantity of energy that has been released by his sudden impact on contemporary culture. A Google search reveals by comparison no serious music about Isaac Newton and little about Einstein.

Chapter 19 has already mentioned a number of Turing movies, including *The Imitation Game*. Turing was a definite presence in Danny Boyle's 2015 biopic *Steve Jobs*.[29] Persistent rumours link Turing with the bitten apple logo that Jobs created for Apple Computer, Inc.: in the movie, Jobs admits that in fact there was no connection, but wishes that the rumours were true.[30] Going back to 1992, *The Strange Life and Death of Dr Turing* (made for the BBC by Christopher Sykes) remains one of the best Turing films. Included are interviews with Turing's ex-fiancée Joan Clarke and his Bletchley Park colleagues Jack Good and Shaun Wylie, as well as his close friends Robin Gandy, Norman Routledge, and Don Bayley. American logician and AI pioneer Marvin Minsky introduces the film, saying:

Here's a person who discovered the most important thing in logic: he invented the concept of the stored-program computer . . . here's the key figure of our century.

Items that belonged to Turing have become increasingly valuable. In 2011, thanks to a large donation from Google, Turing's 'secret papers were saved for the nation' (the *Telegraph* reported).[31] The Bletchley Park Trust, backed by Google, paid an undisclosed six-figure sum for the papers at auction. In fact the papers were not secret at all, consisting mostly of offprints of journal articles that can be found on the shelves of any major university library. The collection originally belonged to Max Newman and some items are marked 'M. H. A. Newman' in Turing's hand (and there are also occasional marginal notes in Newman's own hand). The papers now form part of the 'Life and Works of Alan Turing' exhibition described in Chapter 19. Four years after the auction of the papers, a notebook containing 39 pages in Turing's handwriting was auctioned in New York for more than a million dollars.[32] This is (so far as is known) the most substantial surviving manuscript by Turing; other Turing manuscripts are around ten pages or less.

Political legacy

Winston Churchill reputedly described Turing and the other Bletchley Park codebreakers as 'the geese who laid the golden eggs and never cackled'.[33] But Turing, in the shadow of a giant wall of secrecy, received no greater official recognition for his war work than the OBE—practically an insult when a knighthood would have been more fitting. With the lifting of the veil of secrecy, Turing's status has changed from almost-forgotten mathematician to gay icon and national hero.

At the 1998 unveiling of the blue plaque marking Turing's birthplace (see Fig. 1.1), a message was read out from one of the UK's first openly gay members of parliament, Chris Smith:[34]

Alan Turing did more for his country and for the future of science than almost anyone. He was dishonourably persecuted during his life; today let us wipe that national shame clean by honouring him properly.

The wish to wipe away the national shame gained momentum. Computer scientist John Graham-Cumming led a campaign demanding a government apology; among the campaign's many supporters were biologist Richard Dawkins, novelist Ian McEwan, and gay-rights campaigner Peter Tatchell. In 2009 the apology arrived, in the form of a statement in the *Telegraph* newspaper by the British prime minister Gordon Brown:[35]

While Turing was dealt with under the law of the time and we can't put the clock back, his treatment was of course utterly unfair and I am pleased to have the chance to say how deeply sorry I and we all are for what happened to him. . . . This recognition of Alan's status as one of Britain's most famous victims of homophobia is another step towards equality, and long overdue.

In 2011 an e-petition on a UK government website requested an official pardon. Justice Secretary Lord McNally rejected this request the following year, saying that Turing was 'properly convicted' (which he undoubtedly was).[36] Nevertheless Queen Elizabeth II issued a posthumous Royal Pardon 18 months later:[37]

NOW KNOW YE that we, in consideration of circumstances humbly represented to us, are graciously pleased to grant our grace and mercy unto the said Alan Mathison Turing and grant him our free pardon posthumously in respect of the said convictions; AND to pardon and remit unto him the sentence imposed upon him as aforesaid; AND for so doing this shall be a sufficient Warrant.

The pardon was granted under the 'Royal Prerogative of Mercy', following a request from the then Justice Minister Chris Grayling. Grayling stated:[38]

Turing deserves to be remembered and recognised for his fantastic contribution to the war effort and his legacy to science. A pardon from the Queen is a fitting tribute to an exceptional man.

But why only Turing? Are only exceptional men worthy of being pardoned for the 'crime' of being gay? What about the 75,000 others who were convicted under the same wicked legislation?

Linguistic legacy

Turing's name features in numerous technical terms. Many of these are notable enough to have individual Wikipedia entries. Best known are 'Turing machine' and 'Turing test'. There are also more specialized examples, such as 'symmetric Turing machine', 'reverse Turing test' (where a human pretends to be a computer) and 'visual Turing test' (a Turing-style test for computer vision systems). The term 'Alan Turing law' entered Wikipedia in 2016, following headlines such as this one in *The Independent* newspaper:[39]

'Alan Turing law' unveiled by government will posthumously pardon thousands of gay men convicted of historic offences.

Turing computability is the basic type of computability studied in mathematical recursion theory. The Church–Turing thesis (also known as Turing's thesis) is a famous thesis about computability, as explained in Chapter 41. David Deutsch published a different claim about computability and this is sometimes called the 'physical' Church–Turing thesis, or the Church–Turing–Deutsch thesis (again see Chapter 41).

A Turing reduction is a special kind of algorithm that transforms—or 'reduces'—one problem to another. The algorithm is special in that it makes use of what Turing called an 'oracle', a hypothetical device that is able to solve the printing problem or carry out some other mathematical job that cannot be done by a Turing machine (see Chapters 7 and 37). 'Turing equivalence' means that this reduction can be done in both directions: the first problem can be transformed into the second and also the second can be transformed into the first.

A Turing degree, or degree of unsolvability, is a measure of a set's resistance to being generated by an algorithm—or in other words, a measure of the set's algorithmic intractability. (Typically the set in question consists of infinitely many integers.) The higher the Turing degree, the harder the problem that the set characterizes. For example, the set of numbers of Turing machines that halt (i.e. the set that characterizes the halting problem) is of degree 1, meaning that the halting problem is at the easier end of the spectrum of uncomputable problems.

The Bletchley Park term 'Turingery' is still in use in the modern literature and refers to a codebreaking method invented by Turing in July 1942 (see Chapter 14). Turingery was used against the German Tunny cipher.

The first winner of the A. M. Turing Award, American computer scientist Alan Perlis (who won the award in 1966) coined the term 'Turing tar-pit' (sometimes written 'Turing tarpit') in 1982. A Turing tar-pit is any computer programming language (or computer interface) that shares two of the most distinctive features of the universal Turing machine: it is highly flexible, yet hard to work with, because the user is provided with only very minimal facilities. The universal Turing machine, the ultimate Turing tar-pit, is as flexible as it is possible for a computing device to be (i.e. universal), but every commonplace job—addition, multiplication, counting, sorting into alphabetical order, and so on—has to be done by means of utterly minimal facilities (move left one square, erase a single bit, etc.). A completely different sense of the term 'Turing tar-pit' refers to the fact that every account of computation ever offered has turned out to be equivalent to Turing's: there is no escaping from the Turing tar-pit.

A computer programming language called Turing was developed in 1982 at the University of Toronto. Related languages include Turing+, introduced in 1987 for programming concurrent systems, and Object-Oriented Turing, developed in 1991 as a replacement for Turing+ (and, as the name implies, providing object-oriented programming features).

Figure 42.5 The Alan Turing Building at the University of Manchester, housing the School of Mathematics.

Posted to Wikimedia Commons by Mike Pee, https://commons.wikimedia.org/wiki/File:Alan_Turing_Building_1.jpg. Creative Commons Licence.

The Turing Institute was an AI laboratory in Glasgow, Scotland, that was established by one of Turing's wartime colleagues Donald Michie (Chapters 30 and 31 describe some of Michie's AI work). Michie's Turing Institute, which specialized in machine learning and intelligent computer terminals, ran from 1983 to 1994. The unrelated Alan Turing Institute, founded in 2015, is a national data sciences centre located in London.

There are many buildings and streets named after Turing (although strangely not in the United States). Examples include the Turing Building at Oxford Brookes University, the Turing Building in Auckland, New Zealand, the Turing Bygning in Aarhus, Denmark, the Bâtiment Alan Turing in Paris, and the Alan Turing Building—three of them in fact, one in Manchester (Fig. 42.5), one at the Open University near Bletchley, and another in Guildford, Surrey. Campinas in Brazil has the Avenida Alan Turing, Lausanne in Switzerland has the Place Alan Turing and the Chemin Alan Turing, and Catania in Italy has Via Alan Turing. There is an Alan Turing Road in Guildford and another one in Loughborough, while Alan Turing Way is in Manchester. Turing Road, not to be confused with Alan Turing Road, is near London's National Physical Laboratory, although there is a second one 70 miles away in Biggleswade. Bracknell's Turing Drive and Bletchley's Turing Gate are so far unique.

Conclusion

Even Richard Dawkins' book *The God Delusion* mentions Turing, whose persecution Dawkins lays at the door of religion. Dawkins said (laudably if not entirely accurately):[40]

As the pivotal intellect in the breaking of the German Enigma codes, Turing arguably made a greater contribution to defeating the Nazis than Eisenhower or Churchill. . . . [H]e should have been knighted and fêted as a saviour of his nation. Instead this gentle, stammering, eccentric genius was destroyed, for a 'crime', committed in private, which harmed nobody.

The 2012 book *The Scientists* by Andrew Robinson ranked Turing among the top fifty scientists of all time. John von Neumann gained the same accolade, although his entry in the book is only half the length of Turing's. Naturally, the top 50 also included Einstein, who is today probably the best-known scientist worldwide. But Einstein had quite a head start on Turing, publishing his first work in 1901, more than 10 years before Turing was born.

Public appreciation of Turing's scientific achievements is growing rapidly; and even if his fame never overtakes Einstein's, his place in the pantheon of the world's great scientists is now assured.

NOTES ON THE CONTRIBUTORS

Ruth E. Baker is Assistant Professor of Mathematical Biology at the Mathematical Institute, University of Oxford, and a Tutorial Fellow of St Hugh's College. Her research focuses on mathematical models of embryonic pattern formation and the analysis of the novel mathematical models arising in this context. She was awarded a Whitehead Prize by the London Mathematical Society in 2014.

Mavis Batey, MBE, née Lever (1921–2013), was a codebreaker at Bletchley Park during the Second World War. She was studying German at University College, London at the outbreak of the war. Initially she was employed to check the personal columns of *The Times* for coded spy messages. In 1940 she was recruited to Bletchley Park where she assisted Dilly Knox and was involved in the decryption effort before the Battle of Matapan. After 1945, she worked in the Diplomatic Service and later became a garden historian and author, campaigning to save historic parks and gardens and serving as the President of the Garden History Society.

Margaret A. Boden, OBE ScD FBA, is Research Professor of Cognitive Science at the University of Sussex, where she helped develop the world's first academic programme in artificial intelligence (AI) and cognitive science. She holds degrees in medical sciences, philosophy, and psychology (as well as a Cambridge ScD and three honorary doctorates), and integrates these disciplines with AI in her research, which has been translated into twenty languages. She is a past vice-president of the British Academy, and Chair of Council of the Royal Institution. Her recent books include *The Creative Mind: Myths and Mechanisms*; *Mind as Machine: A History of Cognitive Science*; *Creativity and Art: Three Roads to Surprise*; and *AI, Its Nature and Future*.

Jonathan P. Bowen, FBCS FRSA, is Emeritus Professor of Computing at London South Bank University, where he established and headed the Centre for Applied Formal Methods in 2000. From 2013–15 he was Professor of Computer Science at Birmingham City University. Previously he was a lecturer at the University of Reading, a senior researcher at the Oxford University Computing Laboratory's Programming Research Group, and a research assistant at Imperial College, London. Since 1977 he has been involved with the field of computing in both academia and industry. His books include: *Formal Specification and Documentation using Z*; *High-Integrity System Specification and Design*; *Formal Methods: State of the Art and New Directions*; and *Electronic Visualisation in Arts and Culture*.

Martin Campbell-Kelly is Emeritus Professor in the Department of Computer Science at the University of Warwick, where he specializes in the history of computing. His books include *Computer: A History of the Information Machine* (co-authored with William Aspray); *From Airline Reservations to Sonic the Hedgehog: A History of the Software Industry*; and *ICL: A Business and Technical History*. He is editor of the *Collected Works of Charles Babbage*.

Brian E. Carpenter is an Honorary Professor of Computer Science at the University of Auckland. His research interests are in Internet protocols, especially the networking and routing layers, as well as computing history. He was an IBM Distinguished Engineer working on Internet standards and technology from 1997 to 2007. Earlier he led the networking group at CERN, the European Laboratory for Particle Physics from 1985 to 1996. He chaired the Internet Engineering Task Force from 2005 to 2007, the Board of the Internet Society from 2000 to 2002, and the Internet Architecture Board from 1995 to 2000. He holds a first degree in physics and a PhD degree in computer science.

Catherine Caughey (1923–2008) was called up for war service in 1943. After a rigorous interview and testing process she was assigned to work at Bletchley Park, where she used the Colossus computers for deciphering German High Command messages. After the war she attended Dorset House in Oxford where she trained as an occupational therapist; after qualifying, she worked in an Oxford psychiatric hospital.

B. Jack Copeland, FRS NZ, is Distinguished Professor in Arts at the University of Canterbury, New Zealand, where he is Director of the Turing Archive for the History of Computing. He has been script advisor and scientific consultant for a number of recent documentaries about Turing. Jack is co-director of the Turing Centre at the Swiss Federal Institute of Technology (ETH), Zurich, and also Honorary Research Professor in the School of Historical and Philosophical Inquiry at the University of Queensland, Australia. In 2012 he was Royden B. Davis Visiting Chair of Interdisciplinary Studies in the Department of Psychology at Georgetown University, Washington, DC, and in 2015–16 was a Fellow at the Institute for Advanced Studies in Israel. A Londoner by birth, he earned a DPhil in mathematical logic from the University of Oxford, where he was taught by Turing's great friend Robin Gandy. His books include a highly accessible biography *Turing, Pioneer of the Information Age* as well as *Colossus: The Secrets of Bletchley Park's Codebreaking Computers; The Essential Turing; Alan Turing's Electronic Brain; Computability: Turing, Gödel, Church, and Beyond* (with Oron Shagrir and Carl Posy); *Logic and Reality*; and *Artificial Intelligence*. He has published more than one hundred articles on the philosophy, history, and foundations of computing, and on mathematical and philosophical logic. He was the recipient of the 2016 Covey Award, recognizing a substantial record of innovative research in the field of computing and philosophy.

Robert W. Doran is Professor Emeritus of Computer Science at the University of Auckland, where he was head of department for many years. He has had a lifelong interest in the history of computing, currently maintaining displays and a website on the subject for the Computer Science Department. His additional interests include parallel algorithms, programming, and computer architecture, and he was a Principal Computer Architect at Amdahl Corporation in the late 1970s. A graduate of the University of Canterbury and Stanford University, he has also had appointments at The City University of London and Massey University.

Rod Downey is Professor of Mathematics at Victoria University of Wellington, New Zealand. His interests evolve around the theory of computation and computational complexity. He is author of three books, *Fundamentals of Parameterized Complexity; Parameterized Complexity*; and *Algorithmic Randomness and Complexity*, and is the editor of many volumes, including *Turing's Legacy*.

Ivor Grattan-Guinness (1941–2014) was Emeritus Professor of the History of Mathematics and Logic at Middlesex University, England, and also a Visiting Research Associate in the Centre for Philosophy of Natural and Social Science at the London School of Economics. He was editor of the history of science journal *Annals of Science* from 1974 to 1981. In 1979, he founded the journal *History and Philosophy of Logic*, which he edited until 1992. His books include *The Search for Mathematical Roots, 1870–1940*. He was the associate editor for mathematician, statistician, and computer scientist entries in the *Oxford Dictionary of National Biography*. In 2009, he was awarded the Kenneth O. May Medal and Prize in the History of Mathematics.

Joel Greenberg is an educational technology consultant with 35 years' experience in the field. He received his PhD degree in numerical mathematics from the University of Manchester (UMIST) in 1973. He worked for the Open University for over 33 years, and held a number of senior management positions at director level. He also lectures and writes about Bletchley Park and its role in the Second World War, as well as conducting tours of the site. He is author of a biography of Gordon Welchman, one of Bletchley Park's key figures throughout the war. Current projects include an authorized biography of Alastair Denniston, the first head of GCHQ.

Simon Greenish, MBE, is a Chartered Civil Engineer and was initially a project manager. In 1995 he joined the Royal Air Force Museum to mastermind a £30 million development, and became Director of Collections in 2005. Subsequently he was appointed Director of Bletchley Park when it was in monetary difficulties. Over six years, he secured the trust financially and the site's historical importance is now widely recognized. He raised £10 million to undertake repairs and support development after he retired in 2012. He was appointed MBE in 2013 for services to English Heritage and holds an honorary degree from the University of Bedfordshire.

Peter Hilton (1923–2010) won a scholarship to The Queen's College, Oxford in 1940. With knowledge of German, he arrived at Bletchley Park in early 1942, initially working on Naval Enigma in Hut 8. In late 1942, he started work on German teleprinter ciphers, including 'Tunny'; an early member of the 'Testery', he also liaised with the 'Newmanry'. Hilton obtained his doctorate in 1949 from Oxford University, supervised by J. H. C. Whitehead. In 1958, he became the Mason Professor of Pure Mathematics at the University of Birmingham, moving to the United States in 1962 for professorial posts at Cornell University and elsewhere. His research interests included algebraic topology, homological algebra, categorical algebra, and mathematics education, with 15 books and over 600 articles.

Eleanor Ireland was born in Berkhamsted, England, in 1926. After graduating, she moved to London, where she joined the Women's Royal Naval Service (WRNS) in 1944. As a member of the WRNS, she was stationed at Bletchley Park to be part of a top secret group of engineers, mathematicians, and programmers working together to break codes during the Second World War. At Bletchley Park she operated the early Colossus computers until the end of the war. After the war she became an artist and illustrator, studying at the Regent Street Polytechnic School of Art.

David Leavitt is an American writer of novels, short stories, and non-fiction. He graduated from Yale University. Leavitt is the author of *The Man Who Knew Too Much: Alan Turing and the Invention of the Computer*. His other books include the novels *The Indian Clerk* and *The Two Hotel Francforts*. He is Professor of English at the University of Florida in the United States, where he is a member of the Creative Writing faculty, as well as the founder and editor of the literary journal *Subtropics*.

Jason Long is a New Zealand composer and performer, focusing on musical robotics and electro-acoustic music. He has carried out research at the Utrecht Higher School of the Arts in the Netherlands and at Tokyo University of the Arts in Japan, as well as at several universities in New Zealand. His work *First Contact* has been featured at festivals such as the International Society of Contemporary Music in Brussels and the International Computer Music Conference in Perth, Australia. Other works, including his *Glassback* and *The Subaquatic Voltaic*, have been performed at the Manila Composers' Laboratory in the Philippines, and at Asian Composers' League Festivals in Taipei and Tokyo, as well as in New Zealand and elsewhere.

Philip K. Maini, FRS, is Statutory Professor of Mathematical Biology and Director of the Wolfson Centre for Mathematical Biology, University of Oxford. His research involves mathematical modelling in developmental biology, wound healing, and cancer. He was awarded the London Mathematical Society Naylor Prize in 2009 and elected a Society of Industrial Applied Mathematics Fellow in 2012. In 2015 he was elected to a Fellowship of the Royal Society. He is currently editor-in-chief of the *Bulletin of Mathematical Biology*.

Dani Prinz is the daughter of Dietrich G. Prinz (1903–89), a German computer scientist and pioneer who developed the first implemented chess program in England in 1951. She was a spokesperson for the Alan Turing Year in 2012 and was consulted by the producers of *The Imitation Game*, the 2014 film based on the life of Alan Turing.

Diane Proudfoot is Professor of Philosophy at the University of Canterbury, New Zealand. She and Jack Copeland founded the award-winning online Turing Archive for the History of Computing, the largest web collection of digital facsimiles of original documents by Turing and other pioneers of computing, and received a grant from the Royal Society of New Zealand for research into the philosophical foundations of cognitive and computer science. She has published in the *Journal of Philosophy, Artificial Intelligence, Scientific American*, and numerous other philosophy and science journals. Diane is co-director of the Turing Centre at the Swiss Federal Institute of Technology (ETH), Zurich, and also Honorary Research Associate Professor in the School of Historical and Philosophical Inquiry at the University of Queensland, Australia. In 2015–16 she was a Fellow of the Israel Institute for Advanced Studies and a member of the Computability: Historical, Logical and Philosophical Foundations Research Group there.

Brian Randell is Emeritus Professor of Computing Science at Newcastle University, where he held the Chair of Computing Science from 1969 to 2001 after several years at IBM's T. J. Watson Research Center. At Newcastle his research interests have centred on computer system dependability and the history of computers. He edited the book *The Origins of Digital Computers*. His other publications on computer history include *Ludgate's Analytical Machine of 1909; On Alan Turing and the Origins of Digital Computers*; and *The Colossus*. He has a University of London DSc degree, and Honorary Doctorates from the University of Rennes and the Institut Nationale Polytechnique de Toulouse.

Bernard Richards is Emeritus Professor of Medical Informatics at the University of Manchester. He studied mathematics and physics for his first degree and for his Master's degree he worked under Turing to validate Turing's theory of morphogenesis. After Turing died, he changed his research for his doctorate and studied an aspect of optics, resulting in a Royal Society paper with his supervisor Professor Emil Wolf, describing in full the diffraction of light passing through a convex lens. Thereafter his attention turned to medicine: he produced the definitive paper on hormone peaks in the menstrual cycle, a paper on the recovery from stroke, and more recently an expert system for use in open heart surgery and another for use in the intensive care unit.

Jerry Roberts, MBE, (1920–2014) studied at University College London (UCL) from 1939 to 1941. During the Second World War his UCL tutor Professor Willoughby recommended him to Bletchley Park, where he was one of the four founder members of the Testery in October 1941, the group later tasked with breaking 'Tunny', Hitler's top-level cipher system. He was a senior codebreaker and linguist at Bletchley Park from 1941 to 1945, and after the war was a member of the War Crimes Investigation Unit. Thereafter he pursued a career in international marketing research for the next fifty years, including running two of his own companies until sold to NOP. He spoke German, French, and Spanish fluently and used them throughout his life. For the last five years of his life he worked hard to get better recognition for his colleagues at Bletchley Park.

Oron Shagrir is Professor of Philosophy and Cognitive Science, and currently the Vice Rector, at the Hebrew University of Jerusalem in Israel. His areas of interest include the conceptual foundations of (mainly computational) cognitive and brain sciences, history and philosophy of computing and computability, and 'supervenience'. His publications include *Computability: Turing, Gödel, Church,* *and Beyond* (with Jack Copeland and Carl Posy, 2013) and a special volume on the history of modern computing (with Copeland, Posy, and Parker Bright, *The Rutherford Journal*, 2010).

Edward Simpson, CB, joined Bletchley Park's Naval Section in 1942, at age 19 as a mathematics graduate of Queen's University, Belfast, and led the Japanese Naval JN25 cryptanalytic team. In 1947 he married Rebecca Gibson, a cryptanalyst in the JN25 team. After the war he pursued mathematical statistics at Cambridge under Maurice Bartlett, where his name became joined (not by him) in Simpson's paradox, Simpson's drift, and Simpson's diversity index. From 1947 he was an administrative class civil servant in education, the Treasury, and elsewhere, and was a Commonwealth Fund Fellow in the United States from 1956 to 1957. He retired in 1982 as Deputy Secretary and was appointed a Companion of the Order of the Bath.

Mark Sprevak is a Senior Lecturer in Philosophy at the University of Edinburgh. His primary research interests are in philosophy of mind, philosophy of science, and metaphysics, with particular focus on the cognitive sciences. He has published articles in, among other places, *The Journal of Philosophy*, *The British Journal for the Philosophy of Science*, *Synthese*, *Philosophy, Psychiatry & Psychology*, and *Studies in History and Philosophy of Science*.

Doron Swade, MBE, is an engineer, historian, and museum professional, and a leading authority on the life and work of Charles Babbage. He was formerly Assistant Director and Head of Collections at the Science Museum, and before that Curator of Computing. He has published widely on the history of computing, Babbage, and curatorship. He studied physics, mathematics, electronics, control engineering, machine intelligence, philosophy of science, and history at various universities, including the University of Cape Town, the University of Cambridge, and University College London, where he received his PhD degree. He was awarded an MBE in 2009 for Services to the History of Computing. He is currently researching Babbage's mechanical notation at Royal Holloway University of London.

Sir John Dermot Turing is the nephew of Alan Turing. After following in Alan's footsteps at Sherborne School and King's College Cambridge, he did a research degree in genetics at Oxford University before going into the law. He is now a partner at Clifford Chance, specializing in financial sector issues including regulation, failed banks, and risk management. He became a trustee of Bletchley Park in 2012 and, as a complement to an inescapable interest in naval history and cryptanalysis, enjoys opera, cooking, mountains, and gardening.

Jean Valentine was an operator of the bombe decryption device at Bletchley Park during the Second World War. She was a member of the 'Wrens' (Women's Royal Naval Service, WRNS). Along with her co-workers, she remained quiet about her war work until the mid-1970s. More recently she has been involved with the reconstruction of the bombe at Bletchley Park Museum, giving demonstrations to the public.

Robin Whitty has a BSc in mathematics and a PhD in software engineering. He has lectured in computer science at Goldsmiths and at London South Bank University, and is currently with the School of Mathematical Sciences at Queen Mary University of London. His research interests are in combinatorics and combinatorial optimization. He is creator of the award-winning website *www.theoremoftheday.org*.

Robin Wilson is an Emeritus Professor of Pure Mathematics at the Open University, Emeritus Professor of Geometry at Gresham College, London, and a former Fellow of Keble College, Oxford University. After graduating from Oxford, he received his PhD degree in number theory from the University of Pennsylvania. He has written and co-edited over forty books on graph theory and the history of mathematics, including *Four Colors Suffice* and *Combinatorics: Ancient & Modern*. His historical research interests include British mathematics and the history of graph theory and combinatorics, and he was President of the British Society for the History of Mathematics from 2012 to 2014. An enthusiastic popularizer of mathematics, he was awarded the Mathematical Association of America's Lester Ford award and Pólya prize for his 'outstanding expository writing', and has received an honorary Doctor of Education degree from the University of Bradford.

Stephen Wolfram is a distinguished scientist, inventor, author, and entrepreneur. He is the founder and CEO of Wolfram Research, creator of Mathematica, Wolfram|Alpha, and the Wolfram Language, and author of *A New Kind of Science*. His work has led to a series of discoveries about the computational universe of possible programs; these have not only launched new directions in basic research but also led to breakthroughs in scientific modelling in physical, biological, and social domains—as well as defining a broad new basis for technology discovery.

Thomas E. Woolley is Junior Research Fellow in Mathematics at St John's College, University of Oxford, and London Science Museum Fellow of Modern Mathematics. His research interests are in biological pattern formation and mechanisms of cell movement. He was mathematics advisor to the television show *Dara O'Briain's School of Hard Sums*.

FURTHER READING, NOTES, AND REFERENCES

The following sources (archives, books and Turing items) are referred to in the chapter notes as follows.

ARCHIVES

- King's College Archive: The Turing Papers in the Modern Archive Centre in the library of King's College, Cambridge. Some of these are available electronically in the Turing Digital Archive: http://www.turingarchive.org.
- Manchester Archive: National Archive for the History of Computing, University of Manchester.
- NA: The National Archives, Kew, London.
- *The Turing Archive for the History of Computing*: available online at http://www.alanturing.net.
- Woodger Archive: the Michael Woodger Papers in the Science Museum, London.

BOOKS

Copeland (2004) (abbreviated in the Notes as *The Essential Turing*):

> B. J. Copeland, *The Essential Turing: Seminal Writings in Computing, Logic, Philosophy, Artificial Intelligence, and Artificial Life* plus *The Secrets of Enigma*, Clarendon Press, Oxford (2004).

Copeland et al. (2005):

> B. J. Copeland et al., *Alan Turing's Automatic Computing Engine: the Master Codebreaker's Struggle to Build the Modern Computer*, Oxford University Press (2005). New edition: *Alan Turing's Electronic Brain: the Struggle to Build the ACE, the World's Fastest Computer*, Oxford University Press (centenary edition published 2012).

Copeland et al. (2006):

> B. J. Copeland, et al., *Colossus: the Secrets of Bletchley Park's Codebreaking Computers*, Oxford University Press, 2006 (paperback edition published 2010).

Copeland (2012) (abbreviated in the Notes as *Turing* (Copeland 2012)):

> B. J. Copeland, *Turing, Pioneer of the Information Age*, Oxford University Press (2012).

Copeland et al. (2013):

> B. J. Copeland, C. J. Posy, and O. Shagrir (eds), *Computability: Turing, Gödel, Church, and Beyond*, MIT Press (2013).

Hinsley & Stripp (1993):

F. H. Hinsley and A. Stripp (eds), *Codebreakers: the Inside Story of Bletchley Park*, Oxford University Press (1993).

Hodges (1983):

A. Hodges, *Alan Turing: the Enigma*, Burnett (1983) and Simon and Schuster (1988) (centenary edition published in 2012 by Princeton University Press).

Metropolis et al. (1980):

N. Metropolis, J. Howlett, and G.-C. Rota (eds), *A History of Computing in the Twentieth Century* (Proceedings of the 1976 Los Alamos Conference on the History of Computing), Academic Press (1980).

Randell (1973):

B. Randell (ed.), *The Origins of Digital Computers: Selected Papers*, Springer-Verlag (1973) (3rd edition 1982).

Smith & Erskine (2001):

M. Smith and R. Erskine (eds), *Action this Day: Bletchley Park From the Breaking of the Enigma Code to the Birth of the Modern Computer*, Bantam Press (2001) (reissued in 2011 as *The Bletchley Park Codebreakers*, Biteback).

S. Turing (1959) (abbreviated in the Notes as *Alan M. Turing* (S. Turing 1959));

S. Turing, *Alan M. Turing*, W. Heffer & Sons (1959) (centenary edition published in 2012, Cambridge University Press).

TURING ITEMS

Items with an asterisk appear in *The Essential Turing*; for asterisked items, page numbers in the Notes refer to *The Essential Turing*.

*Turing (1936):

A. M. Turing, 'On computable numbers, with an application to the Entscheidungsproblem', *Proceedings of the London Mathematical Society*, Series 2, 42(1) (1936-7), 230-65 [correction in 43 (1937), 544-6]; *The Essential Turing*, Chapters 1 and 2.

Turing (*c*. 1936):

A. M. Turing, 'A note on normal numbers'; in J. L. Britton (ed.), *Collected works of A. M. Turing: Pure Mathematics*, North-Holland (1992), 117-19.

Turing (1937):

A. M. Turing, 'Computability and λ-definability', *Journal of Symbolic Logic*, 2 (1937), 153-63.

*Turing (1939):

A. M. Turing, 'Systems of logic based on ordinals', *Proceedings of the London Mathematical Society*, Series 2, 45 (1939), 161-228; *The Essential Turing*, Chapter 3.

*Turing (1940):

A. M. Turing, 'Bombe and Spider'; *The Essential Turing*, Chapter 6.

*Turing (*c*. 1941):

A. M. Turing, 'Memorandum to OP-20-G on Naval Enigma'; *The Essential Turing* Chapter 8.

Turing (1943):

A. M. Turing, 'A method for the calculation of the zeta-function', *Proceedings of the London Mathematical Society*, Series 2, 48 (1943), 180–97.

Turing (1945):

A. M. Turing, 'Proposed electronic calculator'; Copeland et al. (2005), Chapter 20.

Turing (1945a):

A. M. Turing, 'Notes on memory'; Copeland et al. (2005), Chapter 21.

*Turing (1947):

A. M. Turing, 'Lecture on the Automatic Computing Engine'; *The Essential Turing*, Chapter 9.

*Turing (1948):

A. M. Turing, 'Intelligent machinery', National Physical Laboratory, 1948; *The Essential Turing*, Chapter 10.

*Turing (1950):

A. M. Turing, 'Computing machinery and intelligence', *Mind*, 59(236) (October 1950), 433–60; *The Essential Turing*, Chapter 11.

*Turing (*c*.1951):

A. M. Turing, 'Intelligent machinery: a heretical theory'; *The Essential Turing*, Chapter 12.

*Turing (1951):

A. M. Turing, 'Can digital computers think?'; *The Essential Turing*, Chapter 13.

*Turing (1952):

A. M. Turing, 'The chemical basis of morphogenesis', *Philosophical Transactions of the Royal Society of London. Series B, Biological Sciences*, 237 (1952), 37–72; *The Essential Turing*, Chapter 15.

*Turing et al. (1952):

A. M. Turing, R. B. Braithwaite, G. Jefferson, and M. H. A. Newman, 'Can automatic calculating machines be said to think?', BBC broadcast (1952); *The Essential Turing*, Chapter 14.

*Turing (1953):

A. M. Turing, 'Chess'; *The Essential Turing*, Chapter 16.

Turing (1953a):

A. M. Turing, 'Some calculations of the Riemann zeta-function', *Proceedings of the London Mathematical Society*, Series 3, 3 (1953), 99–117.

*Turing (1954):

A. M. Turing, 'Solvable and unsolvable problems', *Science News*, 31 (1954), 7–23; *The Essential Turing*, Chapter 17.

Unpublished items by Turing are listed in the Notes.

CHAPTER NOTES

CHAPTER 1 LIFE AND WORK (COPELAND AND BOWEN)

1. Letter from Jefferson to Sara Turing (18 December 1954), in the King's College Archive, catalogue reference A16.
2. Parts of this chapter are adapted from *Turing* (Copeland 2012) and J. P. Bowen, 'Alan Turing', in *The Scientists: an Epic of Discovery* (ed. A. Robinson), Thames and Hudson (2012), 270–5.
3. 'The great minds of the century', *Time*, 153(12) (29 March 1999) (http://content.time.com/time/magazine/article/0,9171,990608,00.html).
4. *Alan M. Turing* (S. Turing 1959), p. 17.
5. Geoffrey O'Hanlon quoted in *Alan M. Turing* (S. Turing 1959), p. 39.
6. *The Inquest Handbook*, INQUEST (2011), Section 4.3.
7. J. A. K. Ferns, quoted in *The Daily Telegraph and Morning Post* (11 June 1954).
8. *Alderley and Wilmslow Advertiser* (18 June 1954).
9. D. Leavitt, 'Alan Turing, father of computer science, is not yet getting his due', *Washington Post* (23 June 2012) (https://www.washingtonpost.com/opinions/alan-turing-father-of-computer-science-not-yet-getting-his-due/2012/06/22/gJQA5eUOvV_story.html); M. Hastings, 'Why I believe it's wrong to pardon Bletchley Park code breaker Alan Turing for breaking the anti-gay laws of his time', *Daily Mail* (3 April 2014) (http://www.dailymail.co.uk/news/article-2379515/The-moral-enigma-Bletchley-Parks-code-breaker-Alan-Turing-genius-undoubtedly-helped-defeat-Hitler-So-I-believe-wrong-pardon-breaking-anti-gay-laws-time.html#ixzz2xnZNELgU).
10. *Manchester Guardian* (11 June 1954).
11. *Alan M. Turing* (S. Turing 1959), p. 117.
12. *Alan M. Turing* (S. Turing 1959), p. 115.
13. Statement of Police Sergeant Leonard Cottrell before the coroner, in the King's College Archive, catalogue reference K6.
14. *The Daily Telegraph* (11 June 1954).
15. Statement of Police Sergeant Leonard Cottrell (see Note 13).
16. Statement of Police Sergeant Leonard Cottrell (see Note 13).
17. S. Turing 'Comments by friends on the manner of Alan Turing's death', typescript (no date), in the King's College Archive, catalogue reference A11.
18. J. L. Burgess and D. Chandler, 'Clandestine drug laboratories', in M. I. Greenberg et al. (ed.), *Occupational, Industrial, and Environmental Toxicology*, 2nd edn, Mosby (2003), p. 759.
19. The apology appeared on the official website of the British Prime Minister (http://www.number10.gov.uk). Two sheets signed by Gordon Brown and headed 'Remarks of Prime Minister Gordon Brown, 10 September 2009' are now part of the 'Life and Works of Alan Turing' exhibition at Bletchley Park.

CHAPTER 2 THE MAN WITH THE TERRIBLE TROUSERS (TURING)

1. King's College Archive.
2. *Alan M. Turing* (S. Turing 1959).
3. F. W. Winterbotham, *The Ultra Secret*, Weidenfeld & Nicolson (1974).
4. J. F. Turing, 'My brother Alan, c.1976–79', included in the 2012 edition of *Alan M. Turing* (Note 2).
5. Criminal Justice Act, 1948, c.58.
6. Human Rights Act, 1998, c.42.
7. The English Bill of Rights of 16 December 1689 includes the following, 'And thereupon the said Lords Spirituall and Temporall and Commons . . . Declare . . . That excessive Baile ought not to be required nor excessive Fines imposed nor cruell and unusuall Punishments inflicted'.
8. M. Grünhut, *Probation and Mental Treatment*, Tavistock Publications (1963). I am indebted to Dr Elizabeth Wells of the Bodleian Library for directing my attention to this book.
9. Home Office and Scottish Home Department, *Report of the Committee on Homosexual Offences and Prostitution* (Cmnd 247), HMSO, 1957.
10. J. D. Watson and F. H. C. Crick, 'A structure for deoxyribose nucleic acid', *Nature*, 171 (1953), 737–8.
11. http://www.mathcomp.leeds.ac.uk/turing2012.

CHAPTER 3 MEETING A GENIUS (HILTON)

1. Peter Hilton died in 2010. This chapter, assembled by Jack Copeland, is published with the permission of Peter's wife, Margaret Hilton. In 2001 and again in 2002 Peter visited Copeland at the University of Canterbury in New Zealand, where he delivered lectures on Turing and codebreaking. This chapter is a compilation of extracts from his papers and notes left in New Zealand, together with extracts from his unpublished paper 'Reminiscences of the life of a codebreaker in World War II' and his chapter 'Living with Fish: breaking Tunny in the Newmanry and the Testery', in Copeland et al. (2006). Parts are also from his 'Cryptanalysis in World War II—and Mathematics Education', *Mathematics Teacher*, 77 (1984), 548–52, 'Reminiscences of Bletchley Park, 1942–1945' in *A Century of Mathematics in America, Part I*, American Mathematical Society (1988), 291–301, and 'Working with Alan Turing', *Mathematical Intelligencer*, 13 (1991), 22–5.

   A tribute to Hilton, 'Peter Hilton: codebreaker and mathematician (1923–2010)', by co-ordinating editor Jean Pedersen with contributions from Jack Copeland, Bill Browder, Ross Geoghegan, Joe Roitberg, Guido Mislin, Urs Stammbach, and Gerald L. Alexanderson, appears in the *Notices of the American Mathematical Society*, 58 (2011), 1538–51.
2. I. J. Good, 'Pioneering work on computers at Bletchley', in Metropolis et al. (1980).

CHAPTER 4 CRIME AND PUNISHMENT (COPELAND)

1. This chapter is an expanded version of a series of extracts from *Turing* (Copeland 2012), Chapters 9, 10, and 12.
2. Turing's story is in the King's College Archive, catalogue reference A13.
3. An article in the *Wilmslow Advertiser* gives an account of information presented at the subsequent trial: 'University Reader put on probation' (4 April 1952), p. 8. I am grateful to the Cheshire Record Office for supplying this article, and copies of the court records: Register of the Court, Wilmslow, 27 February 1952, and Knutsford, 31 March 1952, and the Indictment at the Cheshire Quarter Sessions in Knutsford, 31 March 1952, *The Queen v. Alan Mathison Turing and Arnold Murray*. My account in this and the next section is based on these materials.
4. Letter from Turing to Philip Hall (no date), King's College Archive, catalogue reference D13.
5. Letter from Turing to Norman Routledge (no date), King's College Archive, catalogue reference D14.
6. Don Bailey, in interview with Copeland (21 December 1997).

7. Letter from Don Bayley to Copeland (15 December 1997). Minutes of the Executive Committee of the National Physical Laboratory for 23 October 1945, NPL library; a digital facsimile is in *The Turing Archive for the History of Computing* (http://www.AlanTuring.net/npl_minutes_oct1945).

8 J. Murray, née Clarke, 'A personal contribution to the bombe story', *NSA Technical Journal*, 20 (4) (Fall 1975), 41–6, p. 44.

9. Letter from Turing to Hall (Note 4).

10. Letter from Turing to Hall (Note 4).

11. Bailey interview (Note 6).

12. Letter from Turing to Routledge (22 February [1953]), King's College Archive, Catalogue reference D14.

13. Norman Routledge in an interview broadcast by the BBC on 11 September 2009 (http://www.bbc.co.uk/worldservice/news/2009/09/090911_turing_page_nh_sl.shtml).

14. Letter from Turing to Robin Gandy (11 March [1953]), King's College Archive, catalogue reference D4.

15. J. D. Watson, *The Double Helix: A Personal Account of the Discovery of the Structure of DNA*, Penguin (1999), p. 155.

16. *The Imitation Game*, directed by Morten Tyldum.

17. C. Caryl, 'Saving Alan Turing from his friends', *New York Review of Books* (5 February 2015). The movie's historical errors are the topic of my 'Oscars for *The Imitation Game*?', *Huffington Post Entertainment* (13 July 2015) (http://www.huffingtonpost.com/jack-copeland/oscars-for-the-imitation-_b_6635654.html).

18. Letter from Turing to Gandy (Note 14). Kjell's surname is a new discovery: 'Kjell Carlsen' is written in the corner of a sheet of Turing's handwritten mathematical notes.

19. Letter from Turing to Routledge (Note 12).

20. Letter from Turing to Gandy (Note 14).

21. Letter from Viscount Portal of Hungerford to F. C. Williams (30 November 1950); letter from Williams to Hungerford (13 December 1950).

22. Letter from A. C. Ericsson to Williams (19 April 1955).

23. Letter from Turing to Maria Greenbaum, postmarked 10 May 1953, King's College Archive, catalogue reference K1/83.

24. Nick Furbank in interview with the author, September 2012; letter from Furbank to Gandy (13 June 1954), King's College Archive, catalogue reference A5.

25. The movie's original title was *Britain's Greatest Codebreaker*; it was directed by Nic Stacey.

26. P. Sammon and P. Sen, 'Turing committed suicide: case closed' (5 July 2012) (http://www.turingfilm.com).

27. Letter from Franz Greenbaum to Sara Turing (5 January 1955).

28. Sammon and Sen, 'Turing committed suicide: case closed' (Note 26).

29. Turing's Last Will and Testament (11 February 1954), King's College Archive, catalogue reference A5.

30. Eliza Clayton quoted in Sara Turing's 'Comments by friends on the manner of Alan Turing's death', typescript (no date), King's College Archive, catalogue reference A11.

31. Letter from Gandy to Sara Turing, quoted in *Alan M. Turing* (S. Turing 1959), p. 118.

32. Letter from N. Webb to Sara Turing (13 June 1954), King's College Archive, catalogue reference A17.

33. Letter from Bernard Richards to Copeland (20 August 2012).

34. Letter from Richards to Copeland (31 May 2016).

CHAPTER 5 A CENTURY OF TURING (WOLFRAM)

1. This chapter is based on a blog post I wrote to commemorate the centenary of Turing's birth: 'Happy 100th birthday, Alan Turing' (23 June 2012) (http://blog.stephenwolfram.com/2012/06/happy-100th-birthday-alan-turing).

2. *Alan M. Turing* (S. Turing 1959).

3. Turing (1950).

4. S. Wolfram, *A New Kind of Science*, Wolfram Media (2002) (http://www.wolframscience.com/nksonline).

CHAPTER 6 TURING'S GREAT INVENTION: THE UNIVERSAL COMPUTING MACHINE (COPELAND)

1. Turing (1936). The publication date of 'On computable numbers' is sometimes cited incorrectly as 1937; for details, see *The Essential Turing*, pp. 5–6.
2. See *The Essential Turing*, pp. 15–16.
3. See B. J. Copeland and G. Sommaruga, 'The stored-program universal computer: did Zuse anticipate Turing and von Neumann?' in G. Sommaruga and T. Strahm, *Turing's Revolution*, Birkhauser/Springer (2015). This article offers a detailed analysis of the stored-program concept.
4. For more details see Copeland et al. (2006).
5. T. H. Flowers in interview with Copeland, July 1996.
6. See my chapters 'Mr Newman's section' and 'Colossus and the rise of the modern computer' in Copeland et al. (2006).
7. See my 'The origins and development of the ACE project', in Copeland et al. (2005).
8. These news articles are reprinted in Copeland et al. (2005), pp. 6–9.
9. T. H. Flowers in interview with Copeland, July 1998.
10. Minutes of the Executive Committee of the National Physical Laboratory, 20 April 1948, NPL library; a digital facsimile is in *The Turing Archive for the History of Computing*, http://www.AlanTuring.net/npl_minutes_apr1948.
11. 'The origins and development of the ACE project' (Note 7), pp. 57–9.
12. T. Vickers, 'Applications of the Pilot ACE and the DEUCE', and also pp. 72–4 (Note 7), in Copeland et al. (2005).
13. *Turing* (Copeland 2012), Chapter 9.
14. Letter from F. C. Williams to Brian Randell, 1972 (published in B. Randell, 'On Alan Turing and the origins of digital computers', in B. Meltzer and D. Michie (eds), *Machine Intelligence 7*, Edinburgh University Press (1972)).
15. Letter from Williams to Randell, 1972 (Note 14); F. C. Williams in interview with Christopher Evans in 1976, 'The pioneers of computing: an oral history of computing', Science Museum, London, copyright Board of Trustees of the Science Museum. This interview was supplied to me on audiotape in 1995 by the archives of the London Science Museum and transcribed by me in 1997.
16. A. M. Turing, *Programmers' Handbook for Manchester Electronic Computer Mark II*, Computing Machine Laboratory, University of Manchester (no date, c.1950); a digital facsimile is in *The Turing Archive for the History of Computing* (http://www.AlanTuring.net/Programmers_handbook).
17. Great British Innovation Vote (http://www.topbritishinnovations.org/Pastinnovations.aspx).
18. For additional information on Turing's thesis see *The Essential Turing*, pp. 40–5 and 577–8.
19. M. H. A. Newman, 'Alan Mathison Turing, 1912–1954', *Biographical Memoirs of Fellows of the Royal Society*, 1 (November 1955), 253–263, p. 256.
20. For details, see H. R. Lewis and C. H. Papadimitriou, *Elements of the Theory of Computation*, Prentice-Hall (1981), pp. 170–1.
21. For a discussion of what I call Turing-machine realism, see B. J. Copeland and O. Shagrir, 'Do accelerating Turing machines compute the uncomputable?', *Minds and Machines*, 21 (special issue on the philosophy of computer science) (2011), 221–39.
22. Draft précis of 'On computable numbers' (undated, 2 pp.), King's College Archive, catalogue reference K4 (in French, translation by B. J. Copeland).
23. M. Y. Vardi, 'Who begat computing?', *Communications of the ACM*, 56 (January 2013), 5.
24. Letter from John von Neumann to Norbert Wiener (29 November 1946), Von Neumann Archive, Library of Congress, Washington, DC. The text of von Neumann's lecture 'Rigorous theories of control and information' is printed in J. von Neumann, *Theory of Self-Reproducing Automata* (ed. A. W. Burks), University of Illinois Press (1966); see p. 50.
25. Letter from Stanley Frankel to Brian Randell, 1972, published in Randell (Note 14). I am grateful to Brian Randell for supplying me with a copy of this letter.

26. A. W. Burks, H. H. Goldstine, and J. von Neumann, 'Preliminary discussion of the logical design of an electronic computing instrument', Institute for Advanced Study (28 June 1946), in *Collected Works of John von Neumann*, Vol. 5 (ed. A. H. Taub), Pergamon Press (1961), Section 3.1, p. 37.

27. Notoriously, Turing's 'Proposed electronic calculator' (Turing (1945)) did not mention the universal machine of 1936, leading some commentators to wonder whether it was a direct ancestor of ACE at all. However, some fragments of an early draft of Turing's proposal (Turing (1945a)) cast light on this issue: they were published for the first time in Copeland et al. (2005). There Turing specifically related ACE to the universal Turing machine, explaining why the memory arrangement described in Turing (1936) could not 'be taken over as it stood to give a practical form of machine'.

28. Turing (1947), pp. 378 and 383.

CHAPTER 7 HILBERT AND HIS FAMOUS PROBLEM (COPELAND)

1. This chapter includes material from my article 'What did Turing establish about the limits of computers and the nature of mathematics', Big Questions Online, John Templeton Foundation (February 2013) (https://www.bigquestionsonline.com/). I am grateful to the John Templeton Foundation for permission to publish the material here.

2. For a more detailed account of this pioneering time, see B. J. Copeland, C. Posy, and O. Shagrir, 'The 1930s revolution', in Copeland et al. (2013).

3. K. Gödel, 'Über formal unentscheidbare Sätze der Principia Mathematica und verwandter Systeme I' [On formally undecidable propositions of Principia Mathematica and related systems], *Monatshefte für Mathematik und Physik*, 38 (1931), 173–98; an English translation is in M. Davis (ed.), *The Undecidable: Basic Papers on Undecidable Propositions, Unsolvable Problems and Computable Functions*, Raven (1965). To be exact, in 1931 Gödel proved that the system of arithmetic set out by Bertrand Russell and Alfred Whitehead in their seminal *Principia Mathematica* is (if consistent) incomplete, in the sense that there are true statements of arithmetic that are not provable in the system. Using Turing's discoveries, Gödel was later able to generalize this result considerably. The details are explained in *The Essential Turing*, pp. 47–8.

4. Nowadays this prima facie unpromising escape route has its advocates: see C. Mortensen, *Inconsistent Mathematics*, Kluwer (1995), G. Priest, *In Contradiction: a Study of the Transconsistent*, Oxford University Press (2006); and R. Sylvan and B. J. Copeland, 'Computability is logic-relative', in G. Priest and D. Hyde (eds), *Sociative Logics and their Applications*, Ashgate (2000).

5. M. H. A. Newman, 'Alan Mathison Turing, 1912–1954', *Biographical Memoirs of Fellows of the Royal Society*, 1 (November 1955), 253–263, p. 256. I say 'seemed to leave open one of the most fundamental problems' because Gödel's proof of Theorem X of his 1931 paper does entail the unsolvability of the *Entscheidungsproblem* (given the Church–Turing thesis: see Chapter 41), but this was not noticed until 1965, by Martin Davis. It is probably fortunate that Gödel did not settle the decision problem in 1931, for if he had, then the young logician Alan Turing would not have taken it up, and the history of the computer might have unfolded quite differently—and perhaps much less satisfactorily, in the absence of Turing's fundamental concept of universality.

6. D. Hilbert and W. Ackermann, *Grundzüge der Theoretischen Logik* [*Principles of Mathematical Logic*], Springer (1928).

7. D. Hilbert, 'Mathematical problems: lecture delivered before the International Congress of Mathematicians at Paris in 1900', *Bulletin of the American Mathematical Society*, 8 (1902), 437–79.

8. Turing (1936), p. 84.

9. Hilbert and Ackermann (Note 6).

10. In Gödel's notes for his 1939 introductory logic course at the University of Notre Dame, published as P. Cassou-Nogues, 'Gödel's Introduction to Logic in 1939', *History and Philosophy of Logic*, 30 (2009), 69–90, on p. 85.

11. Gödel in Cassou-Nogues (Note 10), p. 85.

12. See B. J. Copeland, 'Turing's thesis', in A. Olszewski, J. Wolenski, and R. Janusz (eds), *Church's Thesis after 70 Years*, Ontos Verlag (2006), and B. J. Copeland and O. Shagrir, 'Turing versus Gödel on computability and the mind', in Copeland et al. (2013).

13. Turing (1936), Section 9.
14. Stephen Kleene, a member of Church's Princeton group, came a step closer. In 1936 he published his first example of what he later called a 'universal function', but did not connect this idea with computing machinery; see S. C. Kleene, 'General recursive functions of natural numbers', *Mathematische Annalen*, 112 (1936), 727–42 (reprinted in Davis (Note 3)), and S. C. Kleene, *Introduction to Metamathematics*, North-Holland (1952), on p. 289.
15. Gödel (Note 3).
16. See H. Wang, *Reflections on Kurt Gödel*, MIT Press (1987), on p. 171.
17. Turing (1936), Sections 8 and 11.
18. D. Hilbert, 'The new grounding of mathematics: first report' [Neubegründung der Mathematik. Erste Mitteilung] (1922), in W. B. Ewald, *From Kant to Hilbert: a Source Book in the Foundations of Mathematics*, Vol. 2, Clarendon Press, Oxford (1996), p. 1119.
19. Hilbert, p. 1119 (Note 18).
20. Hilbert, p. 1120 (Note 18). Italics added.
21. Hilbert, p. 1121 (Note 18). Here Hilbert was talking about analysis.
22. Hilbert, p. 1119 (Note 18).
23. Hilbert, p. 1119 (Note 18).
24. Turing (1936), p. 84.
25. Hilbert, p. 1132 (Note 18).
26. Hilbert, D. and Ackermann, W. *Grundzüge der Theoretischen Logik* [*Principles of Mathematical Logic*], Julius Springer (1928), p. 76.
27. Hilbert and Ackermann, p. 77 (Note 26).
28. Hilbert and Ackermann, p. 77 (Note 26).
29. Hilbert and Ackermann, p. 81 (Note 26).
30. Turing (1936) p. 84. (Turing's German 'A', which follows 'given formula', has been omitted from the quotation, together with his '–A' preceding 'adjoined'.)
31. See S. C. Kleene, 'Origins of recursive function theory', *Annals of the History of Computing*, 3 (1981), 52–67, on pp. 59, 61.
32. K. Gödel, 'Some basic theorems on the foundations of mathematics and their implications' (1951), in S. Feferman et al. (eds), *Collected Works*, Vol. 3, Oxford University Press (1995), pp. 304–5.
33. Turing (1937), p. 153.
34. Turing's doctoral dissertation, 'Systems of logic based on ordinals', completed in 1938, was published as Turing (1939). For an introduction to Turing's project in this thesis, see *The Essential Turing*, pp. 135–44, and *Turing* (Copeland 2012), Chapter 3.
35. Hilbert, D. 'On the infinite' [Über das Unendliche], 1925, in P. Benacerraf and H. Putnam (eds), *Philosophy of Mathematics: Selected Readings*, 2nd edn, Cambridge University Press (1983), p. 196.
36. Turing (1939), p. 192.
37. Hilbert, p. 201 (Note 35). I am indebted to Carl Posy for discussions of Hilbert's views on intuition.
38. Hilbert, pp. 1121–2 (Note 18).
39. Hilbert, p. 1126 (Note 18).
40. Hilbert, pp. 1121, 1124 (Note 18). See also P. Bernays, 'On Hilbert's thoughts concerning the grounding of arithmetic' [Über Hilberts Gedanken zur Grundlegung der Arithmetik], 1921, in P. Mancosu (ed.) *From Brouwer to Hilbert: the Debate on the Foundations of Mathematics in the 1920s*, Oxford University Press (1998).
41. Hilbert, p. 192 (Note 35).
42. Hilbert, p. 198 (Note 35).
43. Hilbert, D. 'Probleme der Grundlegung der Mathematik' [Problems concerning the foundation of mathematics], *Mathematische Annalen*, 102 (1930), 1–9, on p. 9.
44. Hilbert, p. 198 (Note 35).
45. Turing (1939), pp. 192–3.
46. Hilbert, p. 9 (Note 43).
47. Hilbert and Ackermann, p. 72 (Note 26). I am grateful to Giovanni Sommaruga for his translation of the German. The full sentence is: 'After the logical formalism has been fixed, it is to be expected that

a systematic, so to speak computational treatment of the logical formulae is possible which would somehow correspond to the theory of equations in algebra'.

CHAPTER 8 TURING AND THE ORIGINS OF DIGITAL COMPUTERS (RANDELL)

1. This chapter is an expanded version of the first part of my invited lecture at CONCUR-2012; reprinted in B. Randell, 'A Turing enigma', *Proc. CONCUR-2012, Newcastle upon Tyne*, Lecture Notes in Computer Science 7454, Springer (2012).
2. P. Ludgate, 'Automatic calculating engines', in E. Horsburgh (ed.), *Napier Tercentenary Celebration: Handbook of the Exhibition*, Royal Society of Edinburgh (1914), 124–7; also published as *Modern Instruments and Methods of Calculation: a Handbook of the Napier Tercentenary Celebration Exhibition*, G. Bell and Sons (1914).
3. P. Ludgate, 'On a proposed analytical machine', *Scientific Proceedings of the Royal Dublin Society*, 12(9) (1909), 77–91; reprinted in Randell (1973).
4. C. Babbage, 'On the mathematical powers of the calculating engine', unpublished manuscript (1837); first published in Randell (1973).
5. B. Randell, 'Ludgate's analytical machine of 1909', *Computer Journal*, 14(3) (1971), 317–26.
6. B. V. Bowden (ed.), *Faster Than Thought*, Pitman (1953).
7. P. Morrison and E. Morrison, *Charles Babbage and his Calculating Engines*, Dover Publications (1961).
8. *Alan M. Turing* (S. Turing 1959).
9. D. Kahn, *The Codebreakers*, Macmillan (1967).
10. L. Halsbury, 'Ten years of computer development', *Computer Journal*, 1 (1959), 153–9.
11. B. Randell, 'On Alan Turing and the origins of digital computers', in B. Meltzer and D. Michie (eds), *Machine Intelligence 7*, Edinburgh University Press (1972), 3–20.
12. Turing (1945).
13. J. von Neumann, 'First draft of a report on the EDVAC', contract no. w-670-ord-4926, *Technical Report*, Moore School of Electrical Engineering, University of Pennsylvania (30 June 1945); extracts reprinted in Randell (1973).
14. Randell (1972) (Note 11).
15. M. H. A. Newman, 'Alan Mathison Turing, 1912–1954', *Biographical Memoirs of Fellows of the Royal Society*, 1 (November 1955), 253–263.
16. I. Good, 'Some future social repercussions of computers' *International Journal of Environmental Studies*, 1(1) (1970), 67–79.
17. Turing (1936).
18. Randell (1972) (Note 11).
19. Randell (1972) (Note 11).
20. Randell (1972) (Note 11).
21. Randell (1972) (Note 11).
22. D. Horwood, 'A technical description of Colossus I', *Technical Report P/0921/8103/16*, Government Code and Cypher School (August 1973), NA HW25/24.
23. See D. Swade, 'Pre-electronic computing', in C. B. Jones and J. L. Lloyd (eds), *Dependable and Historic Computing*, Springer (2011), 58–83.
24. A. A. Lovelace, 'Sketch of the analytical engine invented by Charles Babbage, by L. F. Menabrea, Officer of the Military Engineers, with notes upon the memoir by the Translator', *Taylor's Scientific Memoirs 3*, Article 29 (1843), 666–731; reprinted in Randell (1973).
25. Von Neumann (Note 13).
26. J. Eckert, 'Disclosure of a magnetic calculating machine', *Technical Report*, unpublished typescript (1945); reprinted in J. P. Eckert. 'The ENIAC', in *A History of Computing in the Twentieth Century*, Academic Press (1980), 525–39.
27. B. Carpenter and R. Doran, 'The other Turing machine', *Computer Journal*, 20(3) (1977), 269–79.
28. Turing (1945).

29. Von Neumann (Note 13).
30. B. Randell, 'The origins of computer programming', *IEEE Annals of the History of Computing*, 16(4) (1994), 6–14.
31. Randell (Note 11).
32. Randell (1973).

CHAPTER 9 AT BLETCHLEY PARK (COPELAND)

1. For a detailed account of Turing's work on Enigma, see my chapter 'Enigma' in *The Essential Turing*.
2. A. P. Mahon, 'History of Hut 8 to December 1941', in *The Essential Turing*, p. 275.
3. For additional information about Rejewski and the Polish work on Enigma, see 'Enigma' in *The Essential Turing*.
4. 'Meeting held on 6th July 1950 to discuss "Bombes"', GCHQ, NA HW25/21, p. 1.
5. F. H. Hinsley et al., *British Intelligence in the Second World War*, Vol. 2, Her Majesty's Stationery Office (1981), p. 29.
6. For a detailed account of Turing's battle with U-boat Enigma, see *Turing* (Copeland 2012), Chapter 5.
7. W. L. S. Churchill, *The Second World War, Vol. 2: Their Finest Hour*, Cassell (1949), p. 529.
8. For information about *Seelöwe*, see F. H. Hinsley et al., *British Intelligence in the Second World War*, Vol. 1, Her Majesty's Stationery Office (1979), pp. 186, 188, and Appendix 7, and N. de Grey, '1939–1940 Sitz and Blitz' (no date), part of 'De Grey's history of Air Sigint', NA HW3/95, pp. 58–9.
9. Hinsley et al. (Note 8), p. 183.
10. For more information about the early bombes and the attack on Luftwaffe Enigma, see *Turing* (Copeland 2012), pp. 62–9.
11. See *Turing* (Copeland 2012), pp. 78–80, and Hinsley et al. (Note 5), pp. 422–50.
12. Quoted in Hinsley et al. (Note 5), pp. 448–9.
13. Bletchley Park's English translation of a crucial Tunny decrypt relating to the Battle of Kursk (a turning point of the war in the east) is reproduced in Copeland et al. (2006), pp. 5–6.
14. Here I argue against numerous authors who claim that the work done at Bletchley Park shortened the war by at least 2 years; see, for example, F. H. Hinsley, 'The counterfactual history of no Ultra', *Cryptologia*, 20 (1996), 308–24.

CHAPTER 10 THE ENIGMA MACHINE (GREENBERG)

1. With thanks to Jack Copeland for his editorial contributions to this chapter.
2. See K. de Leeuw, 'The Dutch invention of the rotor machine, 1915–1923', *Cryptologia*, 28 (2004), 73–94.
3. D. Kahn, *Seizing the Enigma*, Barnes & Noble Books (1991); C. A. Deavours and L. Kruh, *Machine Cryptography and Modern Cryptanalysis*, Artech (1985).
4. D. Hamer, G. Sullivan, and F. Weierud, 'Enigma variations: an extended family of machines' *Cryptologia*, 22 (1998), 211–29.
5. R. Erskine and F. Weierud, 'Naval Enigma: M4 and its rotors', *Cryptologia*, 11 (1987), 227–34 (p. 243).
6. Home waters and Mediterranean messages continued to be read in very large numbers during the first nine months of 1942, whereas the Atlantic U-boat traffic could not be read during that period. This presumably indicates that the replacement of M3 by M4 was limited to the Atlantic U-boat fleet.
7. Some accounts give the call sign simply as MMA.
8. A. P. Mahon, 'History of Hut 8 to December 1941', in *The Essential Turing*.
9. 'The history of Hut 6', Vol. 1, p. 87, NA HW43/70.
10. A. P. Mahon, 'The history of Hut Eight, 1939–1945' (1945), NA HW25/2, p. 99. A digital facsimile of Mahon's 1945 typescript is available in *The Turing Archive for the History of Computing* (http://www.AlanTuring.net/mahon_hut_8).
11. C. H. O'D. Alexander, 'Cryptographic history of work on the German naval Enigma' (c.1945), NA HW25/1; a digital facsimile of Alexander's typescript is available in *The Turing Archive for the History of Computing* (http://www.AlanTuring.net/alexander_naval_enigma).

12. T. Hughes and J. Costello, *Battle of the Atlantic*, Doubleday (1977).
13. W. G. Welchman, *The Hut Six Story: Breaking the Enigma Codes*, Allen Lane (1982), p. 168.

CHAPTER 11 BREAKING MACHINES WITH A PENCIL (BATEY)

1. Letters from Peter Twinn to Jack Copeland (28 January and 21 February 2001).
2. Hugh Foss mentioned several such cribs in his 'Reminiscences on Enigma', in Smith & Erskine (2001), p. 45; another is described in the section 'The race to break Enigma'.
3. See M. L. Batey, *Dilly: The Man who Broke Enigmas*, Dialogue (2009), pp. 68–72; the complete 1930 operator's manual with the crib is set out in Appendix I.
4. W. Kozaczuk, *Enigma: How the German Machine Cipher Was Broken, and How It Was Read by the Allies in World War Two* (transl. C. Kasparek), Arms and Armour Press (1984), p. 236.
5. A. D. Knox, 'Warsaw', NA HW25/12.
6. W. G. Welchman, *The Hut Six Story: Breaking the Enigma Codes*, 2nd edn, M&M Baldwin (1997), p. 34.
7. Letter from Knox to Alastair Denniston, NA HW14/1.
8. M. Rejewski, *Memories of My Work at the Cipher Bureau of the General Staff Second Department 1930–1945*, Adam Mickiewicz University in Poznán (2011), p. 132. This book was published to commemorate the thirtieth anniversary of Rejewski's death; written in 1967, it was deposited in the Wojskowy Instytut Historyczny, and is now published with an English translation.
9. Kozaczuk (Note 4), p. 97.
10. Letter from Knox to Denniston, NA HW25/12.
11. They were inapplicable except in the case of the cipher key codenamed 'Yellow' at Bletchley Park. On 1 May 1940 Yellow was being used in the short-lived Norwegian campaign, and for a while it retained the old method of encoding the indicators.
12. A. M. Turing, 'Prof's Book', NA HW25/3 (under the title 'Mathematical theory of ENIGMA machine by A M Turing'). A digital facsimile of Turing's typescript is available in *The Turing Archive for the History of Computing* (http://www.AlanTuring.net/profs_book).
13. Birch's letter is in C. Morgan, 'N.I.D. (9) Wireless intelligence', NA ADM223/463, p. 39; pages 38–9 contain Morgan's account of Operation Ruthless and include Fleming's memo to the Director of Naval Intelligence.
14. C. Caslon, 'Operation "Claymore"—report of proceedings', NA DEFE2/142.
15. Batey (Note 3), p. 130.
16. Letter from Knox to Stewart Menzies, NA HW25/12.
17. Letter from Birch to Edward Travis, quoted by A. P. Mahon, 'History of Hut 8 to December 1941', in *The Essential Turing*, p. 287.
18. J. Murray (née Clarke), 'Hut 8 and Naval Enigma, Part I', in Hinsley & Stripp (1993), 113–18.
19. On Railway Enigma, see F. Weierud, 'Railway Enigma and other special machines' (http://cryptocellartales.blogspot.co.uk/2013/05/railway-enigma-and-other-special.html) and D. Hamer, G. Sullivan, and F. Weierud, 'Enigma variations: an extended family of machines', *Cryptologia*, 22 (1998), 211–29.
20. J. H. Tiltman, 'Some reminiscences', in the US National Archives and Records Administration, RG 457, entry 9032, Historic Cryptographic Collection, pre-World War I through World War II, no. 4632, p. 16.
21. Turing (Note 12), pp. 31 and 60.
22. P. Twinn, 'The *Abwehr* Enigma', in Hinsley & Stripp (1993).
23. J. K. Batey, M. L. Batey, M. A. Rock, and P. F. G. Twinn, 'G.C. & C.S. Secret Service Sigint Vol. II: Cryptographic systems and their solution—I Machine cyphers', NA HW43/7.

CHAPTER 12 BOMBES (COPELAND, WITH VALENTINE AND CAUGHEY)

1. I am grateful to Ralph Erskine and Edward Simpson for information, comments, and suggestions.
2. A. G. Denniston, 'How news was brought from Warsaw at the end of July 1939' (May 1948), NA HW25/12, p. 4. (Published with an editorial introduction and notes in R. Erskine, 'The Poles reveal

their secrets: Alastair Denniston's account of the July 1939 meeting at Pyry', *Cryptologia*, 30 (2006), 294–305.)

3. M. Rejewski, 'How Polish mathematicians deciphered the Enigma' (transl. J. Stepenske), *Annals of the History of Computing*, 3 (1981), 213–34, p. 227.

4. P. Twinn, 'The *Abwehr* Enigma', in Hinsley & Stripp (1993), pp. 126–7; Denniston (Note 2), pp. 4–5; A. D. Knox, 'Warsaw', 4 August 1939, NA HW25/12.

5. For a detailed account of the Polish bomba, see B. J. Copeland, 'Enigma', in *The Essential Turing*, pp. 235–46; and D. Davies, 'The Bombe—a remarkable logic machine', *Cryptologia*, 23 (1999), 108–38.

6. J. Murray (née Clarke), 'A personal contribution to the Bombe story', *NSA Technical Journal*, 20(4) (Fall 1975), 41–6, p. 42.

7. 'Operations of the 6812th Signal Security Detachment, ETOUSA' (1 October 1944), US National Archives and Records Administration, RG 457, Entry 9032, Historic Cryptographic Collection, pre-World War I through World War II, Box 970, no. 2943, p. 5; I am grateful to Frode Weierud for supplying me with a copy of this document.

8. Rejewski's diagram is in B. Johnson, *The Secret War*, Pen & Sword (2004), p. 316. The suggestion that the component adjacent to the drive shaft is a magnet is due to David Link.

9. W. Kozaczuk, *Enigma: How the German Machine Cipher Was Broken, and How It Was Read by the Allies in World War Two* (transl. by C. Kasparek), Arms and Armour Press (1984), p. 63.

10. Michael Foot reported this in answer to a question at a Sigint conference in Oxford (information from Ralph Erskine).

11. M. Rejewski, 'How the Polish mathematicians broke Enigma', in Kozaczuk (Note 9), p. 267.

12. Denniston (Note 2), p. 4.

13. Copeland, 'Enigma' (Note 5), p. 246.

14. 'Enigma – Position' and 'Naval Enigma Situation', notes dated 1 November 1939 and signed by Knox, Twinn, Welchman, and Turing, NA HW14/2.

15. For more information about the bombe, see A. M. Turing, 'Bombe and Spider' (Turing 1940), Copeland, 'Enigma' (Note 5), pp. 246–57; F. Carter, 'The Turing Bombe', *Rutherford Journal for the History and Philosophy of Science and Technology*, 3 (2010) (http://www.rutherfordjournal.org/article030108.html). 'Bombe and Spider' is an extract from what was known at Bletchley Park as 'Prof's Book', 'Prof' being Turing's nickname there ('Mathematical theory of ENIGMA machine by A M Turing', NA HW25/3). Written by Turing in the summer or autumn of 1940, 'Prof's Book' served as a training manual. It was not released into the public domain until 1996, and is available online in *The Turing Archive for the History of Computing* at http://www.AlanTuring.net/profs_book.

16. Knox, Twinn, Welchman, and Turing, 'Enigma – Position' (Note 14); 'Squadron-Leader Jones' Section', anon., GC & CS (c.1946), NA HW3/164, p. 1.

17. 'Squadron-Leader Jones' Section' (Note 16), p. 1.

18. 'Squadron-Leader Jones' Section' (Note 16), p. 1.

19. 'Meeting held on 6th July 1950 to discuss "Bombes"', Birch, De Grey, Alexander, Fletcher, Foss, Zambra, GCHQ, NA HW25/21.

20. 'Squadron-Leader Jones' Section' (Note 16), p. 3.

21. Information from Mavis Batey.

22. 'Squadron-Leader Jones' Section' (Note 16), pp. 3–4.

23. 1676 women, to be precise: 'Squadron-Leader Jones' Section' (Note 16), p. 14.

24. *Alan M. Turing* (S. Turing 1959), p. 70.

25. P. Hilton, 'Living with Fish: breaking Tunny in the Newmanry and the Testery', in Copeland et al. (2006) p. 196.

26. W. G. Welchman, *The Hut Six Story: Breaking the Enigma Codes*, 2nd edn., M & M Baldwin (1997), p. 12.

27. Letters from Peter Twinn to Copeland (28 January and 21 February 2001).

28. 'Staff and Establishment of G.C.C.S.', NA HW3/82; letter from A. G. Denniston to T. J. Wilson of the Foreign Office (3 September 1939), NA FO366/1059; Murray (Note 6),p. 42.

29. Letter from Twinn to Christopher Andrew (29 May 1981), quoted in C. W. Andrew, *Secret Service: The Making of the British Intelligence Community*, Guild (1985), p. 453.

30. I. J. Good, 'From Hut 8 to the Newmanry', in Copeland et al. (2006), p. 205.
31. Good (Note 30), p. 205.
32. J. Murray (née Clarke), 'Hut 8 and Naval Enigma, Part I', in Hinsley & Stripp (1993), p. 114.
33. Murray (Note 32), pp. 113–15.
34. R. Erskine, 'The first Naval Enigma decrypts of World War II', *Cryptologia*, 21 (1997), 42–6, p. 43.
35. 'Squadron-Leader Jones' Section' (Note 16), p. 14; see also K. McConnell, 'My secret war', Dundee City Library (10 November 2005) (http://www.bbc.co.uk/history/ww2peopleswar/stories/06/a6844106.shtml). I am grateful to Edward Simpson for drawing my attention to this article.
36. 'Squadron-Leader Jones' Section' (Note 16), p. 3.
37. 'Squadron-Leader Jones' Section' (Note 16), p. 1.
38. 'Squadron-Leader Jones' Section' (Note 16), p. 10.
39. 'Meeting held on 6th July 1950' (Note 19), p. 1; 'Squadron-Leader Jones' Section' (Note 16), p. 14.
40. Welchman (Note 26), p. 139.
41. 'Tentative brief description of equipment for Enigma problems', US National Archives and Records Administration, RG 457, Entry 9032, Historic Cryptographic Collection, pre-World War I through World War II, Box 705, no. 1737; 'Bombes – Notes on Policy and Statistics, 1939–1950', GC & CS, 1950, NA HW25/21.
42. C. H. O'D. Alexander, 'Cryptographic history of work on the German Naval Enigma' (c.1945), NA HW25/1, p. 90; a digital facsimile of Alexander's typescript is available in *The Turing Archive for the History of Computing* (http://www.AlanTuring.net/alexander_naval_enigma).
43. 'Squadron-Leader Jones' Section' (Note 16), p. 8.
44. 'Hut 6 Bombe Register', GC & CS, 1940–1945, 2 vols, NA HW25/19, HW25/20; 'Squadron-Leader Jones' Section' (Note 16), p. 8.
45. 'Squadron-Leader Jones' Section' (Note 16), p. 4.
46. 'Operations of the 6812th' (Note 7), pp. 59, 60.
47. P. Mahon, 'History of Hut 8 to December 1941', in *The Essential Turing*, p. 291.
48. Alexander (Note 42), p. 25.
49. The indicator drums displayed the bombe's guess at the so-called 'rod position' of the wheels—that is, the position of the wheels relative to a notional ring-setting of ZZZ.
50. 'Squadron-Leader Jones' Section' (Note 16), p. 3.
51. A number of typical bombe menus are shown in 'Operations of the 6812th' (Note 7), pp. 8ff.
52. D. Payne, 'The bombes', in Hinsley & Stripp (1993), p. 134.
53. Mahon (Note 47), p. 293.
54. Copeland, 'Enigma' (Note 5), pp. 245–6.
55. Turing's procedure is described in detail in Turing (1940), pp. 315–19, and Copeland, 'Enigma' (Note 5), pp. 250–3.
56. Mahon (Note 47), p. 286.
57. Op. 20 GM-1 war diary, US National Archives and Records Administration, RG 38, NSG Records, CNSG Library, Box 113, no. 5750/176; I am grateful to Ralph Erskine for this reference.
58. W. F. Friedman, 'Report on E Operations of the GC & CS at Bletchley Park', US National Archives and Records Administration, RG 457, Entry 9032, Historic Cryptographic Collection, pre-World War I through World War II, Box 1126, no. 3620, p. 6; R. D. Johnson, 'Cryptanalytic report # 2 Yellow Machine', Historic Cryptographic Collection, Pre-World War I through World War II, Box 1009, no. 3175, p. 12. I am grateful to Ralph Erskine for these references.
59. Denniston (Note 2), p. 3.
60. Mahon (Note 47), p. 303.
61. 'Operations of the 6812th' (Note 7), p. 3, and Johnson (Note 58), p. 88.
62. For details of this raid and other 'pinches' of Enigma materials on the high seas, see *Turing* (Copeland 2012), Chapter 5.
63. Turing (1940), p. 315.
64. Turing (1940), pp. 316–17.

65. Turing (1940), p. 317.
66. Turing (1940), p. 316 (emphasis added).
67. Alexander (Note 42), pp. 87, 89.
68. Alexander (Note 42), p. 89.
69. 'Squadron-Leader Jones' Section' (Note 16), p. 4.
70. Turing (1940), p. 319.
71. 'Squadron-Leader Jones' Section' (Note 16), p. 2.
72. Joan Clarke's words: see Murray (Note 6), p. 43.
73. Murray (Note 6), p. 43.
74. Murray (note 6), pp. 43–4.
75. Turing (1940), pp. 320–31; Welchman (Note 26), Appendix 1.
76. C. A. Deavours and L. Kruh, 'The Turing bombe: was it enough?', *Cryptologia*, 14 (1990), 331–49, pp. 346–8.
77. Turing (1940), p. 327.
78. 'Operations of the 6812th' (Note 7), p. 59.
79. Turing (1940), p. 319.
80. 'Operations of the 6812th' (Note 7).
81. 'Squadron-Leader Jones' Section' (Note 16), p. 7.

CHAPTER 13 INTRODUCING BANBURISMUS (SIMPSON)

1. C. H. O'D. Alexander, 'Cryptographic history of work on the German Naval Enigma', (*c*.1945), NA HW25/1 and *The Turing Archive for the History of Computing* (http://www.AlanTuring.net/ alexander_naval_enigma).
2. A. P. Mahon, 'The history of Hut Eight, 1939–1945', (1945), NA HW25/2 and *The Turing Archive for the History of Computing* (http://www.AlanTuring.net/mahon_hut_8.). I am much indebted to Ralph Erskine for guiding me through these texts and helping me to cover the ground.
3. See also Steven Hosgood's thorough examination on stoneship.org.uk/-steve/banburismus.html.
4. The Bletchley Park 'Cryptographic dictionary' (1944): http://www.codesandciphers.org.uk/documents/ cryptdict.
5. Mahon (Note 2).
6. William Legrand deciphered a stream of mixed numbers and symbols, without word-breaks, to find Captain Kidd's treasure buried near Charleston, South Carolina.
7. G. U. Yule, *The Statistical Study of Literary Vocabulary*, Cambridge University Press (1944).
8. S. Padua, *The Thrilling Adventures of Lovelace and Babbage*, Particular Books (2015).
9. D. L. Moore, *Ada, Countess of Lovelace*, John Murray (1977).
10. Hinsley & Stripp (1993).
11. Mahon (Note 2).
12. Alexander (Note 1).
13. Eileen Johnson (nee Plowman), private communication.
14. Christine Brose (nee Ogilvie-Forbes), private communication.
15. Iris King (nee Brown), private communication; the winter of 1941–42 was the coldest European winter of the twentieth century.
16. Alexander (Note 1).
17. Copeland et al. (2006).
18. Hilary Pownall (nee Law), private communication.
19. A. J. (Tony) Phelps, private memoir.
20. Brose (Note 14).
21. http://www-history.mcs.st-and.ac.uk/Biographies/Clarke_Joan.html.
22. Mahon (Note 2).
23. Copeland (Note 17).

CHAPTER 14 TUNNY, HITLER'S BIGGEST FISH (COPELAND)

1. The principal sources for this chapter are my conversations with T. H. Flowers during 1996–98, and the 'General report on Tunny', written at Bletchley Park in 1945 by Tunny-breakers I. J. Good, D. Michie, and G. Timms, and released by the British government to the National Archives in 2000: NA HW25/4 (Vol. 1) and HW25/5 (Vol. 2). A digital facsimile is available in *The Turing Archive for the History of Computing* (http://www.AlanTuring.net/tunny_report). A more detailed treatment of the attack on Tunny appears in Copeland et al. (2006) and *Turing* (Copeland 2012), Chapter 7.
2. See Note 1.
3. F. L. Bauer, 'The Tiltman break', in Copeland et al. (2006), p. 372.
4. Turingery is described in full in Copeland 'Turingery', in Copeland et al. (2006), pp. 379–85.
5. W. T. Tutte, 'My work at Bletchley Park', in Copeland et al. (2006), p. 360.
6. Tutte's method is described in Copeland et al. (2006), pp. 66÷71 and 363–5 (by Tutte himself).
7. Max Newman's estimate, in an interview with Christopher Evans in 'The pioneers of computing: an oral history of computing', Science Museum, London.
8. Turing (1936).
9. The definitive biographical article on Newman is 'Max Newman—mathematician, codebreaker, and computer pioneer', by his son William, in Copeland et al. (2006), Chapter 14. Much additional information about Newman is to be found in the same book, especially in Chapters 4, 5, 9, and 13; Chapter 13, entitled 'Mr Newman's section', includes material by five of Newman's wartime engineers and computer operators.
10. H. Fensom, 'How Colossus was built and operated—one of its engineers reveals its secrets', in Copeland et al. (2006), p. 298.
11. Flowers in an interview with Copeland (Note 1).
12. Ken Myers interviewed by Copeland, 13 June 2014.
13. Du Boisson writing in Copeland et al. (2006), p. 172.
14. Caughey writing in Copeland et al. (2006), p. 165.
15. H. Currie, 'An ATS girl in the Testery', in Copeland et al. (2006), p. 268.
16. *Alan M. Turing* (S. Turing 1959), p. 67.
17. 'Max Newman—mathematician, codebreaker, and computer pioneer', (Note 9), p. 177.
18. H. H. Goldstine, *The Computer from Pascal to von Neumann*, Princeton University Press (1972), p. 150.
19. Flowers in Copeland et al. (2006), p. 107.
20. See, for example, J. von Neumann, 'The NORC and problems in high speed computing' (1954), in A. H. Taub (ed.), *Collected Works of John von Neumann*, Vol. 5, Pergamon Press (1963), pp. 238–9.
21. C. G. Bell and A. Newell, *Computer Structures: Readings and Examples*, McGraw-Hill (1971), p. 42.
22. Myers interviewed by Copeland, 13 June 2014.
23. M. Campbell-Kelly, 'The ACE and the shaping of British computing' in Copeland et al. (2005), p. 151.

CHAPTER 15 WE WERE THE WORLD'S FIRST COMPUTER OPERATORS (IRELAND)

There are no notes for this chapter.

CHAPTER 16 THE TESTERY: BREAKING HILTER'S MOST SECRET CODE (ROBERTS)

1. A detailed account of the work of the Testery can be found in J. Roberts 'Major Tester's section', in Copeland et al. (2006).

CHAPTER 17 ULTRA REVELATIONS (RANDELL)

1. This chapter is an expanded version of the second part of my invited lecture at CONCUR-2012; reprinted in B. Randell, 'A Turing enigma', *Proc. CONCUR-2012, Newcastle upon Tyne*, Lecture Notes in Computer Science 7454, Springer (2012).
2. F. W. Winterbotham, *The Ultra Secret*, Weidenfeld & Nicolson (1974).
3. A. Cave Brown, *Bodyguard of Lies: the Vital Role of Deceptive Strategy in World War II*, Harper and Row (1975).
4. A. Friendly, 'Secrets of code-breaking', *Washington Post* (5 December 1967), A18.
5. K. O. May, *Bibliography and Research Manual on the History of Mathematics*, University of Toronto Press (1973).
6. B. Randell, 'The Colossus', *Computing Laboratory Technical Report Series 90*, University of Newcastle (1976) (http://www.cs.ncl.ac.uk/publications/trs/papers/90.pdf).
7. R. W. Bemer, 'Colossus—World War II computer' (http://www.bobbemer.com/COLOSSUS.htm).
8. In fact, it was one of the Colossus team that found this aperture device, which is now on show with some other small Colossus artefacts at Newcastle University.
9. Randell (Note 6).
10. B. Randell, 'The Colossus', in Metropolis et al. (1980), 47–92.
11. B. Randell, 'Colossus: godfather of the computer', *New Scientist*, 73(1038) (1977), 346–8; reprinted in Randell (1973).
12. This television series is available online on YouTube; the 'Still Secret' episode is at https://www.youtube.com/watch?v=m04VHVmjfWk.
13. B. Johnson, *The Secret War*, British Broadcasting Corporation (1978).
14. 'Computer pioneer: Mr. Thomas Flowers', *The Times* (14 May 1977), 16, Gale CS270237870.
15. F. Hinsley, E. Thomas, C. Ransom, and R. Knight, *British Intelligence in the Second World War*, 5 vols, Her Majesty's Stationery Office (1979 *et seq.*).
16. Hodges (1983).

CHAPTER 18 DELILAH—ENCRYPTING SPEECH (COPELAND)

1. This chapter is a modified extract from *Turing* (Copeland 2012).
2. *Alan M. Turing* (S. Turing 1959), pp. 71–2.
3. S. W. Roskill, *The War at Sea 1939–1945*, Vol. 2, HMSO (1956), p. 378.
4. *Alan M. Turing* (Note 2), p. 71.
5. 'American research and development of telephonic scrambling device and research of unscrambling telephonic devices', memo from E. G. Hastings to Sir John Dill, 2 December 1942, NA HW14/60.
6. Hastings reported Marshall's decision in his 1942 memo (Note 5).
7. Letter from Joe Eachus to Copeland (18 November 2001).
8. A. M. Turing, 'Memorandum to OP-20-G on Naval Enigma' (Turing (*c.* 1941)).
9. A. M. Turing, 'Visit to National Cash Register Corporation of Dayton, Ohio', *c.*December 1942, National Archives and Records Administration, RG 38, CNSG Library, 5750/441. A digital facsimile is in *The Turing Archive for the History of Computing* (http://www.AlanTuring.net/turing_ncr).
10. I. J. Good, 'From Hut 8 to the Newmanry', in Copeland et al. (2006), p. 212.
11. Robert Mumma in an interview with Rik Nebeker in 1995, IEEE History Center, New Brunswick, New Jersey, USA; I am grateful to Ralph Erskine for making me aware of this interview.
12. Letter from Acting Chief of Staff Joseph McNarney to Sir John Dill (9 January 1943), NA HW14/60.
13. Alexander Fowler's diary; letter from Evelyn Loveday to Sara Turing (20 June 1960), King's College Archive, catalogue reference A20.
14. G. Chauncey, *Gay New York: Gender, Urban Culture, and the Making of the Gay Male World, 1890–1940*, Basic Books (1994).
15. There is a reconstruction of a SIGSALY system in the Washington National Cryptologic Museum.

16. Don Bayley in interview with Copeland (December 1997).
17. Bayley interview (Note 16).
18. T. H. Flowers in interview with Christopher Evans in 1977, 'The pioneers of computing: an oral history of computing', Science Museum, London.
19. The blueprints are in NA HW25/36.
20. A. M. Turing and D. Bayley, 'Speech secrecy system "Delilah", Technical description', c.March 1945, NA HW25/36, pp. 3–4.
21. Turing and Bayley (Note 20), p. 8.
22. *Alan M. Turing* (Note 2), p. 74.
23. Quoted in *Alan M. Turing* (Note 2), p. 75.
24. Bayley interview (Note 16).
25. Robin Gandy quoted in *Alan M. Turing* (Note 2), p. 119.
26. Letter from Don Bayley to Copeland (1 November 1997).
27. Quoted in *Alan M. Turing* (Note 2), p. 75.
28. Letter from A. C. Pigou to Sara Turing (26 November 1956), King's College Archive, catalogue reference A10.
29. 'Hitler as seen by source', by 'FLL' (probably F. L. Lucas) (24 May 1945), NA HW13/58, p. 2. Brian Oakley first suggested that Lucas was the author of this document; see J. Ettinger, 'Listening in to Hitler at Bletchley Park 1941–1945: Fish and Colossus' (June 2009), p. 5.
30. Letter from Don Bayley to Copeland (1 June 2012).

CHAPTER 19 TURING'S MONUMENT (GREENISH, BOWEN, AND COPELAND)

1. Copeland et al. (2006), Chapter 13.
2. P. Ridden, 'Feature: decoding Bletchley Park's history', *Gizmag* (7 December 2009) (http://www.gizmag.com/bletchley-park-ww2-code-breakers/13525).
3. 'The Formation of the Bletchley Park Trust', Bletchley Park (http://www.bletchleypark.org.uk/content/about/bptrust1.rhtm).
4. Copeland et al. (2006), Chapter 13.
5. 'The Colossus rebuild', The National Museum of Computing, Bletchley Park (http://www.tnmoc.org/special-projects/colossus-rebuild); B. Randell, 'Uncovering Colossus', The National Museum of Computing (https://www.youtube.com/watch?v=Yl6pK1Z7B5Q).
6. *The Reunion*, BBC Radio 4 (6 April 2008) (http://www.bbc.co.uk/radio4/thereunion/pip/hu1pz/).
7. *The Times* (24 July 2008).
8. BBC News, 'Neglect of Bletchley condemned' (24 July 2008) (http://news.bbc.co.uk/1/hi/uk/7523743.stm).
9. S. Black and S. Colgan, *Saving Bletchley Park: How #socialmedia Saved the Home of the WWII Codebreakers*, Unbound (2015).
10. 'H.M. Queen Elizabeth II honours Bletchley Park' (8 August 2011), The Churchill Centre (http://www.winstonchurchill.org/resources/in-the-media/churchill-in-the-news/hm-queen-elizabeth-ii-honours-bletchley-park).
11. J. Harper, 'The Turing bombe rebuild project', Vimeo (https://vimeo.com/51496481); D. Turing, 'Rescue and rebuild', *Bletchley Park: Demystifying the Bombe*, Pitkin Publishing, The History Press (2014), pp. 62–3.
12. BBC News, 'Duchess of Cambridge opens Bletchley Park restored centre' (18 June 2014) (http://www.bbc.co.uk/news/uk-england-beds-bucks-herts-27898997).
13. 'Restoration of historic Bletchley Park' (2014), Bletchley Park (http://www.bletchleypark.org.uk/content/about/restoration.rhtm).
14. 'Future plans for Bletchley Park' (2015), Bletchley Park (http://www.bletchleypark.org.uk/news/v.rhtm/Future_plans_for_Bletchley_Park-902740.html).
15. S. Kettle, 'Alan Turing' (http://www.stephenkettle.co.uk/turing.html).

16. J. Naughton, 'The newly restored Bletchley Park and the fast-eroding freedoms it was set up to defend', *The Guardian* (22 June 2014) (https://www.theguardian.com/technology/2014/jun/22/bletchley-park-gchq-surveillance-home-office-edward-snowden).

17. S. McKay, 'How Alan Turing's secret papers were saved for the nation', *The Telegraph* (30 July 2011) (http://www.telegraph.co.uk/lifestyle/8668156/How-Alan-Turings-secret-papers-were-saved-for-the-nation.html).

CHAPTER 20 BABY (COPELAND)

1. A fuller version of the events at Manchester is related in two of my articles, on which this chapter is based: 'The Manchester computer: a revised history. Part I The memory', and 'The Manchester computer: a revised history. Part II The Baby machine', *IEEE Annals of the History of Computing*, 33 (2011), 4–21 and 22–37. Aspects of my research on the Manchester computer have also been published in my articles 'A lecture and two radio broadcasts on machine intelligence by Alan Turing', in K. Furukawa, D. Michie, and S. Muggleton (eds), *Machine Intelligence 15*, Oxford University Press (1999), 445–76, 'Colossus and the dawning of the computer age', in R. Erskine and M. Smith (eds), *Action This Day*, Bantam Books (2001), 'Modern history of computing', in E. Zalta (ed.), *The Stanford Encyclopedia of Philosophy*, Stanford University Press (2001) (http://plato.stanford.edu), and in *The Essential Turing*, Chapters 5 and 9, Copeland et al. (2005), Chapter 5, Copeland et al. (2006), Chapter 9, and *Turing* (Copeland 2012), Chapter 9.

The research reported in this chapter has spanned many years and I am indebted to numerous pioneers and historians of the computer (some no longer alive) for information and discussion: Art Burks, Alice Burks, George Davis, Dai Edwards, Tom Flowers, Jack Good, Peter Hilton, Harry Huskey, Hilary Kahn, Tom Kilburn, Simon Lavington, Donald Michie (who provided inspiration, hospitality, and eye-opening discussion at his homes in Oxford in 1995 and Palm Desert in 1998), Brian Napper, tommy Thomas, Geoff Tootill, Robin Vowels, and Mike Woodger; also Jon Agar, for introducing me to the Manchester Archive in 1995, and William Newman for information about his father Max. I am especially grateful to Donald Michie for suggesting in 1995 that I document Newman's role in the Manchester computer project.

2. F. C. Williams and T. Kilburn, 'Electronic digital computers', *Nature*, 162(4117) (1948), 487; the letter is dated 3 August 1948.

3. Kilburn in interview with Copeland (July 1997). Letter from TRE to NPL (9 January 1947) in the Manchester Archive.

4. Letter from Turing to Michael Woodger, undated, received 12 February 1951, in the Woodger Archive; a digital facsimile is in *The Turing Archive for the History of Computing* (http://www.AlanTuring.net/turing_woodger_feb51). Martin Campbell-Kelly has provided an excellent account of the Mark I in his 'Programming the Mark I: early programming activity at the University of Manchester', *Annals of the History of Computing*, 2 (1980), 130–68.

5. N. Stern, 'The BINAC: a case study in the history of technology', *Annals of the History of Computing*, 1 (1979), 9–20, p. 17, and N. Stern, 'From ENIAC to UNIVAC: an appraisal of the Eckert–Mauchly computers', Digital (1981), p. 149.

6. W. H. Ware, 'The history and development of the electronic computer project at the Institute for Advanced Study', *RAND Corporation Report P-377*, Santa Monica (10 March 1953), pp. 16–17.

7. M. Croarken, 'The beginnings of the Manchester computer phenomenon: people and influences', *IEEE Annals of the History of Computing*, 15 (1993), 9–16, p. 9; S. H. Lavington, *A History of Manchester Computers*, NCC Publications (1975), p. 8 (2nd edn, British Computer Society, 1998). See also M. Croarken, *Early Scientific Computing in Britain*, Oxford University Press (1990), S. H. Lavington, *Early British Computers: the Story of Vintage Computers and the People Who Built Them*, Manchester University Press (1980), and M. Wilkes and H. J. Kahn, 'Tom Kilburn CBE FREng', *Biographical Memoirs of Fellows of the Royal Society*, 49 (2003), 285–97.

8. Letter from Williams to Randell (1972), in B. Randell, 'On Alan Turing and the origins of digital computers', in B. Meltzer and D. Michie (eds), *Machine Intelligence 7*, Edinburgh University Press (1972), p. 9. I am grateful to Brian Randell for supplying me with a copy of the letter.

9. J. Bigelow, 'Computer development at the Institute for Advanced Study', in Metropolis et al. (1980), pp. 304, 305–6, 308; H. H. Goldstine, *The Computer from Pascal to von Neumann*, Princeton University Press (1972), p. 310.

10. Goldstine (Note 9), p. 96.

11. Turing (1945), pp. 426–7.

12. Williams in interview with Christopher Evans in 1976, 'The pioneers of computing: an oral history of computing', Science Museum, London, copyright Board of Trustees of the Science Museum. This interview was supplied to me on audiotape in 1995 by the archives of the Science Museum and transcribed by me in 1997.

13. Minutes of the Executive Committee of the National Physical Laboratory for 22 October 1946, NPL library.

14. Hilton in interview with Copeland, June 2001.

15. F. C. Williams, 'Early computers at Manchester University', *The Radio and Electronic Engineer*, 45 (1975), 327–31, p. 328.

16. Williams in interview with Evans (Note 12).

17. T. Kilburn, 'A storage system for use with binary digital computing machines', Report for TRE, 1 December 1947, Manchester Archive; a retyped version, complete with editorial notes by Brian Napper, is at http://www.computer50.org/kgill/mark1/report1947.html; T. Kilburn, 'The University of Manchester universal high-speed digital computing machine', *Nature*, 164 (4173) (1949), 684–7.

18. Kilburn in interview with Copeland, July 1997.

19. Flowers in interview with Copeland, July 1996.

20. Letter from Newman to von Neumann (8 February 1946), in the von Neumann Archive at the Library of Congress, Washington, DC; a digital facsimile is in *The Turing Archive for the History of Computing* (http://www.AlanTuring.net/newman_vonneumann_8feb46).

21. Letter from Julian Bigelow to Copeland (12 April 2002).

22. Bigelow in a tape-recorded interview made in 1971 by the Smithsonian Institution and released in 2002. I am grateful to Bigelow for sending me a transcript of excerpts from the interview.

23. Letter from Arthur Burks to Copeland (22 April 1998).

24. A. W. Burks, H. H. Goldstine, and J. von Neumann, 'Preliminary discussion of the logical design of an electronic computing instrument', Institute for Advanced Study (28 June 1946), in A. H. Taub (ed.), *Collected Works of John von Neumann*, Vol. 5, Pergamon Press (1961), Section 3.1, on p. 37.

25. NA FO850/234; the photographs were declassified in 1975. Photographs and the accompanying official caption were published in B. Randell 'Colossus', in Metropolis et al. (1980).

26. 'General report on Tunny, with emphasis on statistical methods', NA HW25/4, HW25/5 (2 vols); this report was written in 1945 by Jack Good, Donald Michie, and Geoffrey Timms, all members of Newman's section at Bletchley Park. A digital facsimile is in *The Turing Archive for the History of Computing* (http://www.AlanTuring.net/tunny_report).

27. See, for example, the National Science Foundation 'family tree' of computer design, 1957, reproduced in Copeland et al. (2005), p. 150.

28. M. V. Wilkes, *Memoirs of a Computer Pioneer*, MIT Press (1985); J. R. Womersley, 'A.C.E. project—origin and early history', National Physical Laboratory (26 November 1946), in Copeland et al. (2005), pp. 38–9; a digital facsimile is in *The Turing Archive for the History of Computing* (http://www.AlanTuring.net/ace_early_history).

29. Flowers in interview with Copeland (Note 19).

30. Letter from Michie to Copeland (14 July 1995).

31. Council Minutes, Royal Society of London (13 Dec. 1945, 14 Feb. 1946, 7 Mar. 1946, 11 April 1946, 16 May 1946, 13 June 1946); in the archives of the Royal Society of London.

32. Letter from Newman to Colonel Wallace, GCHQ, Bletchley Park (8 August 1945), NA HW64/59. I am grateful to Michael Smith for sending me a copy of this document.

33. Sheet of notes, GCHQ (4–6 December 1945), NA HW64/59. I am grateful to Michael Smith for sending me a copy of this document.

34. Letter from Good to Copeland (5 March 2004).

35. 'Application from Professor M. H. A. Newman: project for a calculating machine laboratory in Manchester University', Royal Society of London, p. 2.

36. T. Kilburn and L. S. Piggott, 'Frederic Calland Williams', *Biographical* Memoir *of Fellows of the Royal Society*, 24 (1978), 583–604. p. 591. 'Williams cathode ray tube storage: evidence relating to the origin of the invention and the dissemination of information on the operation of the storage system', draft report of the National Research Development Corporation, no date, p. 7 (I am grateful to Dai Edwards for sending me a copy of this document). Tom Kilburn in interview with Copeland (Note 3).

37. Williams in interview with Evans (Note 12).

38. Williams (Note 15), p. 328.

39. These were not limited to the Colossi and the Robinsons (Chapter 14): other Newmanry machines are described in 'General report on Tunny' (see Note 26) and in Copeland et al. (2006).

40. Michie in an unpublished memoir sent to Copeland in March 1997.

41. 'Report by Professor M. H. A. Newman on progress of computing machine project', Appendix A of Council Minutes, Royal Society of London (13 January 1949), in the archives of the Royal Society of London.

42. The lecture notes are published as 'The Turing–Wilkinson lecture series (1946–7)', in Copeland et al. (2005); see also B. J. Copeland, 'The Turing–Wilkinson lecture series on the Automatic Computing Engine', in Furukawa et al. (Note 1), 381–444. This series of nine lectures (about half of which were given by Turing's assistant, Jim Wilkinson, most likely from notes prepared by Turing) covered Versions V, VI, and VII of Turing's design for the ACE.

43. Williams in interview with Evans (Note 12).

44. G. Bowker and R. Giordano, 'Interview with Tom Kilburn', *IEEE Annals of the History of Computing*, 15 (1993), 17–32, p. 19.

45. See my Introduction to 'The Turing–Wilkinson lecture series (1946–7)' (Note 42), pp. 459–464. Womersley's handwritten notes concerning the arrangements for the lectures are in the Woodger Archive, catalogue reference M15, and *The Turing Archive for the History of Computing* (http://www.AlanTuring.net/womersley_notes_22nov46).

46. Letter from Brian Napper to Copeland (16 June 2002).

47. Bowker and Giordano (Note 44), p. 19. I am grateful to Brian Napper for drawing this passage to my attention in correspondence during 2002.

48. Kilburn, 'A storage system for use with binary digital computing machines' (Note 17).

49. Kilburn, 'The University of Manchester universal high-speed digital computing machine' (Note 17), p. 687.

50. Williams in interview with Evans (Note 12).

51. Williams in interview with Evans (Note 12).

52. Good used these terms in a letter to Newman about computer architecture (8 August 1948). The letter is in I. J. Good, 'Early notes on electronic computers', unpublished, compiled in 1972 and 1976, pp. 63–4; a copy of this document is in the Manchester Archive, MUC/Series 2/a4.

53. For a complete discussion of ACE instruction formats see Copeland et al. (2005), Chapters 4, 9, 11, and 22.

54. This overview of Kilburn's machine is based on Section 1.4 of Kilburn, 'A storage system for use with binary digital computing machines' (Note 17). See also B. Napper, 'Covering notes for Tom Kilburn's 1947 report to TRE' (http://www.computer50.org/kgill/mark1/report1947cover.html). I am indebted to Napper for helpful correspondence.

55. Kilburn, 'A storage system for use with binary digital computing machines' (Note 17). Kilburn's block diagram may be viewed at (http://www.computer50.org/kgill/mark1/TR47diagrams/f1.2.png). Incidentally, the computer shown in Kilburn's block diagram was no baby: Kilburn ambitiously specified a memory capacity of 8192 words, whereas the actual Baby had a mere 32 words of memory.

56. W. Newman, 'Max Newman—mathematician, codebreaker, and computer pioneer', in Copeland et al. (2006), p. 185.

57. H. D. Huskey, 'The state of the art in electronic digital computing in Britain and the United States', in Copeland et al. (2005), p. 536.

58. 'Application from Professor M. H. A. Newman: project for a calculating machine laboratory in Manchester University' (Note 35), and 'Report by Professor M. H. A. Newman on progress of computing machine project' (Note 41).

59. Michie in interview with Copeland, October 1995; Good, 'Early notes on electronic computers' (Note 52), pp. vii, ix.

60. Michie in an unpublished memoir sent to me in March 1997.

61. Newman's lectures were probably given in February 1947; see Good, 'Early notes on electronic computers' (Note 52), p. iii.

62. Letter from Williams to Randell (Note 8).

63. Williams in interview with Evans (Note 12).

64. Letter from Williams to Randell (Note 8).

65. Kilburn in interview with Copeland (Note 18).

66. Good recounted this in his retrospective introduction (written in 1972) to a short paper, 'The Baby machine', that he had prepared on 4 May 1947 at Kilburn's request: see Good 'Early notes on electronic computers' (Note 52), p. iv.

67. Good 'The Baby machine' (Note 66), p. 1. The table is from Copeland 'The Manchester Computer: a revised history. Part II' (Note 1), p. 28.

68. The quotation is from Good's revised typescript of his acceptance speech delivered on 15 October 1998, p. 31: I am grateful to Good for sending me a copy of his revised typescript in January 1999. On Good's gift of the twelve instructions to Kilburn, see also Croarken (Note 7), p. 15, and J. A. N. Lee, *Computer Pioneers*, IEEE Computer Society Press (1995), p. 744.

69. Williams in interview with Evans (Note 12).

70. Good, 'The Baby machine' (Note 66), pp. 1–2.

71. Kilburn in interview with Copeland (Note 18).

72. Williams and Kilburn (Note 2). This table is from Copeland 'The Manchester Computer: a revised history. Part II' (Note 1), p. 26.

73. In an earlier note 'Fundamental operations', written *c.*16 February 1947, Good listed a larger and considerably more complicated set of basic instructions. These included multiplication, division, $|x|$, two forms of conditional transfer of control, and an instruction transferring the number in the accumulator to the 'house number in' house x (the note is in his 'Early notes on electronic computers', Note 52). David Anderson argued that Baby was based on that instruction set (D. Anderson, 'Was the Manchester Baby conceived at Bletchley Park?', *University of Portsmouth Research Report UoP-HC-2006-001*, published on the internet in 2006 at http://www.tech.port.ac.uk/staffweb/andersod/HoC). This claim is incorrect: it was the May instruction set, not the more complex February set, that Kilburn received from Good and simplified to five instructions (plus 'stop'). Unlike the May instruction set, the February set was intended for a machine with two instructions per word, whereas Baby had only one. Good made it clear that it was the May set, not the February, that he suggested in response to Kilburn's request 'for a small number of basic instructions'.

74. Burks et al. (Note 24), see especially Sections 5.5, 6, and Table 1. The operation 'transfer the number in A to R' is discussed in Section 6.6.3, where it is pointed out that this operation can be made basic at the cost of 'very little extra equipment'. The two shift operations L and R are introduced in Section 6.6.7. For additional detail, see Copeland 'The Manchester Computer: a revised history. Part II' (Note 1), pp. 29–31.

75. See Copeland 'The Manchester Computer: a revised history. Part II' (Note 1), p. 30.

76. F. C. Williams and T. Kilburn, 'The University of Manchester Computing Machine', in *Review of Electronic Digital Computers: Joint AIEE-IRE Computer Conference*, American Institute of Electrical Engineers (1952); F. C. Williams, T. Kilburn, and G. C. Tootill, 'Universal high-speed digital computers: a small-scale experimental machine', *Proceedings of the Institution of Electrical Engineers*, 98 (1951), 13–28.

77. This is made clear by a comparison of Section 6.4 of Burks et al. (Note 24), with Williams and Kilburn (Note 76), pp. 57–8, and Williams et al. (Note 76), pp. 17–18.

78. Huskey (Note 57), p. 535.

79. J. von Neumann, 'First draft of a report on the EDVAC', contract no. w-670-ord-4926, *Technical Report*, Moore School of Electrical Engineering, University of Pennsylvania (30 June 1945), in *IEEE Annals of the History of Computing*, 15 (1993), 28–75, see Section 12.8. The Moore School's distribution list for the 'First draft' is dated 25 June 1945. I am grateful to Harry Huskey for sending me a copy of the distribution list.

80. Gandy in interview with Copeland, October 1995.

81. Letter from Huskey to Copeland (4 February 2002).

82. *Evening News* (23 December 1946); the cutting is among a number kept by Sara Turing and now in the King's College Archive, catalogue reference K5.

83. Williams (Note 15), p. 330.

84. C. S. Strachey, 'Logical or non-mathematical programmes', *Proceedings of the Association for Computing Machinery, Toronto, September 1952*, pp. 46–49; C. S. Strachey, 'The thinking machine', *Encounter*, 3 (1954), 25–31; Strachey Papers, Bodleian Library, Oxford.

85. T. Kilburn, 'A storage system for use with binary digital computing machines', PhD Thesis, University of Manchester (awarded 13 December 1948), pp. 32, 34; F. C. Williams and T. Kilburn, 'A storage system for use with binary digital computing machines', *Proceedings of the Institution of Electrical Engineers*, 96 (1949), 81–100, p. 82.

86. Kilburn (Note 85) Photograph 1. Williams' and Kilburn's 1947 picture can be viewed at http://www.computer50.org/kgill/mark1/TR47diagrams/p1.jpg. The exact date on which text was first stored at Manchester is not known, but a report by Kilburn dated 1 December 1947 (Note 17) stated that the tube used to store the text had gone into operation only 3 months previously.

87. Chris Burton recreated the images using his rebuild of Baby and filmed them; I am grateful to him for sending me some of his footage.

88. Strachey, 'The thinking machine' (Note 84), p. 26.

89. See A. M. Turing, *Programmers' Handbook for Manchester Electronic Computer Mark II*, Computing Machine Laboratory, University of Manchester, no date, *c*.1950, p. 25; and A. M. Turing, 'Generation of random numbers', appendix to G. C. Tootill, 'Informal report on the design of the Ferranti Mark I computing machine', Manchester Computing Machine Laboratory, November 1949.

90. Strachey, 'The thinking machine' (Note 84), p. 26.

91. King's College Archive, catalogue reference D 4.

CHAPTER 21 ACE (CAMPBELL-KELLY)

1. Further general information about the topics in this chapter can be found in Copeland et al. (2005) and in D. M. Yates, *Turing's Legacy: A History of Computing at the National Physical Laboratory 1945–1995*, Science Museum, London (1997).

2. J. von Neumann, 'First draft of a report on the EDVAC', contract no. w-670-ord-4926, *Technical Report*, Moore School of Electrical Engineering, University of Pennsylvania (1945).

3. Turing (1945).

4. Memorandum from Turing to Womersley, undated (*c*.December 1946), in the Woodger Archive, M15/77. A digital facsimile is in *The Turing Archive for the History of Computing* (http://www.AlanTuring.net/turing_womersley).

5. M. Campbell-Kelly, 'Programming the Pilot ACE: early programming activity at the National Physical Laboratory', *Annals of the History of Computing*, 3 (1981), 133–68.

CHAPTER 22 TURING'S ZEITGEIST (CARPENTER AND DORAN)

1. According to the *Oxford English Dictionary*, 'program' was actually the original English spelling; the later spelling 'programme' was a once-fashionable Frenchification.

2. J. H. Wilkinson, 'The Pilot Ace', in *Automatic Digital Computation*, Proceedings of a Symposium held at NPL, March 1953 (1954); reprinted in C. G. Bell and A. Newell, *Computer Structures: Readings and Examples*, McGraw-Hill (1971).

3. R. Malik, 'In the beginning—early days with ACE', *Data Systems* (March 1969), 56–9, 82.

4. B. Randell, 'On Alan Turing and the origins of digital computers', in B. Meltzer and D. Michie (eds), *Machine Intelligence 7*, Edinburgh University Press (1972); Randell (1973).

5. Turing (1945) [also 'Proposals for development in the Mathematics Division of an Automatic Computing Engine (ACE)', *NPL Internal Report E882* (1946); reprinted as 'NPL Technical Report', *Computer Science 57*, April 1972, and in B. E. Carpenter and R. W. Doran (eds), *A. M. Turing's ACE Report of 1946 and other papers*, Charles Babbage Institute Reprint Series, Vol. 10, MIT Press (1986)].

6. F. W. Winterbotham, *The Ultra Secret*, Weidenfeld & Nicolson (1974).

7. J. von Neumann, 'First draft of a report on the EDVAC', contract no. w-670-ord-4926, *Technical Report*, Moore School of Electrical Engineering, University of Pennsylvania (30 June 1945); extracts included in Randell (1973).

8. B. E. Carpenter and R. W. Doran, 'The other Turing machine', *Computer Journal*, 20 (1977), 269–79.

9. This company was still registered at Companies House at the time of writing.

10. See R. W. Doran, 'Computer architecture and the ACE computers', in Copeland et al. (2005).

11. W. S. McCulloch and W. Pitts, 'A logical calculus of the ideas immanent in nervous activity', *Bulletin of Mathematical Biophysics*, 5 (1943), 115–33.

12. See M. V. Wilkes, 'The best way to design an automatic calculating machine', Manchester University Computer Inaugural Conference, July 1951.

13. Wilkinson's two reports are both in *The Turing Archive for the History of Computing*.

14. See E. G. Daylight, 'Dijkstra's rallying cry for generalization: the advent of the recursive procedure, late 1950s–early 1960s', *Computer Journal*, 54 (2011), 1756–72.

15. Turing (1945).

16. Turing (1947) [reprinted in Carpenter and Doran (Note 5)].

17. This chapter was significantly improved by multiple comments made by participants at a meeting of the BCS Computer Conservation Society in London in May 2012, and by comments from Jack Copeland. In addition to the specific references cited in the text, we are indebted to Andrew Hodges' *Alan Turing: The Enigma* and to numerous websites. Brian Carpenter was a visitor at the Computer Laboratory, University of Cambridge, during part of the preparation of this chapter.

CHAPTER 23 COMPUTER MUSIC (COPELAND AND LONG)

1. This chapter incorporates material from Chapter 9 of *Turing* (Copeland 2012).

2. See, for example, J. Chadabe, 'The electronic century, part III: computers and analog synthesizers', *Electronic Musician*, 2001 (http://www.emusician.com/tutorials/electronic_century3).

3. A. M. Turing, *Programmers' Handbook for Manchester Electronic Computer Mark II*, Computing Machine Laboratory, University of Manchester (no date, *c*.1950), p. 85; a digital facsimile is in *The Turing Archive for the History of Computing* (http://www.AlanTuring.net/programmers_handbook).

4. Turing (Note 3).

5. B. J. Copeland and G. Sommaruga, 'The stored-program universal computer: did Zuse anticipate Turing and von Neumann?', in G. Sommaruga and T. Strahm, *Turing's Revolution*, Birkhauser/Springer (2015), pp. 99–100; F. C. Williams and T. Kilburn, 'The University of Manchester Computing Machine', in *Review of Electronic Digital Computers: Joint AIEE–IRE Computer Conference*, American Institute of Electrical Engineers (1952), 57–61.

6. Williams and Kilburn (Note 5), p. 59.

7. The delivery date of the first Ferranti computer is given in a letter from Turing to Woodger, undated, received 12 February 1951 (in the Woodger Archive). A digital facsimile is in *The Turing Archive for the History of Computing* (http://www.AlanTuring.net/turing_woodger_feb51). For details of the UNIVAC see N. Stern, 'The BINAC: A case study in the history of technology', *Annals of the History of Computing*, 1 (1979), 9–20, p. 17; and N. Stern, 'From ENIAC to UNIVAC: an appraisal of the Eckert–Mauchly computers', Digital (1981), p. 149.

8. See Turing's preface to his *Handbook* (Note 3).

9. There is a circuit diagram of the hooter in K. N. Dodd, 'The Ferranti electronic computer', *Armament Research Establishment Report 10/53*, *c*.1953 (Diagram 10).

10. Turing (Note 3), p. 24.

11. Dodd (Note 9), p. 59.

12. By the time of the third edition of Turing's *Handbook* (prepared by Tony Brooker in 1953), Turing's 'about middle C' had been replaced by 'an octave above middle C'.

13. D. G. Prinz, *Introduction to Programming on the Manchester Electronic Digital Computer*, Ferranti Ltd, Moston, Manchester, 28 March 1952, Section 20. A digital facsimile is in *The Turing Archive for the History of Computing* (http://www.AlanTuring.net/prinz). Copeland is grateful to Dani Prinz for supplying a cover sheet that shows the date of this document.

14. Letter from Strachey to Max Newman (5 October 1951), in the Christopher Strachey Papers, Bodleian Library, Oxford, folder A39; letter from Strachey to Michael Woodger (13 May 1951), in the Woodger Archive).

15. Letter from Strachey to Newman (5 October 1951). Strachey's copy of Turing's *Handbook* still exists, signed on the cover 'With the compliments of A. M. Turing' (in the Christopher Strachey Papers, folder C40).

16. N. Foy, 'The word games of the night bird', *Computing Europe* (15 August 1974), 10–11 (interview with Christopher Strachey), p. 10.

17. Letter from Strachey to Newman (5 October 1951); Robin Gandy in interview with Copeland, October 1995.

18. Gandy interview (Note 17).

19. Strachey in Foy (Note 16), p. 11.

20. Strachey in Foy (Note 16), p. 11.

21. Strachey in Foy (Note 16), p. 11.

22. Strachey in Foy (Note 16), p. 11.

23. Strachey gave the name of the program in his letter to Newman (Note 17). The Checksheet program itself is in the Christopher Strachey Papers (folder C52).

24. Turing (Note 3), p. 12.

25. Frank Cooper interviewed by Chris Burton in 1994; an audio recording of part of the interview is at: http://curation.cs.manchester.ac.uk/digital60/www.digital60.org/media/interview_frank_cooper/index-2.html. In the secondary literature it is sometimes said that *God Save the King* was played at the end of Strachey's draughts program (see Chapter 20), but this is not correct.

26. Letter from Newman to Strachey (2 October 1951), in the Christopher Strachey Papers, folder A39.

27. 'Electronic brain can sing now', *Courier and Advertiser* (28 February 1952). We thank Diane Proudfoot for finding and supplying this article.

28. Cooper interview (Note 25).

29. Turing quoted in 'The mechanical brain', *The Times* (11 June 1949).

30. 'Very bad tunes', *Manchester Guardian* (4 January 1952). We thank Diane Proudfoot for finding and supplying this article.

31. 'Very bad tunes' (Note 30) (emphasis added).

32. For ease of exposition, we have replaced Turing's 'G@/P' by 'O@/P', thereby oversimplifying the behaviour of the /P instruction, but making the loop ostensibly easier to follow. Appearances to the contrary notwithstanding, Turing's 'G@/P' does take execution back to the start of the loop, while our oversimplified version does not do so. For a full explanation of /P, see Prinz (Note 13) p. 14. Strachey, who marked corrections by hand on his copy of Turing's *Handbook*, altered this loop to:
O@ /V
B@ Q@/V
G@ B@/P
(See also Strachey's typed sheets of errata to Turing's *Handbook*, dated 9 July 1951; in the Christopher Strachey Papers, folder C45.)

33. Turing (Note 3), p. 24. There is a magisterial introduction to programming (what Turing called) the Mark II in M. Campbell-Kelly, 'Programming the Mark I: early programming activity at the University of Manchester', *Annals of the History of Computing*, 2 (1980), 130–68.

34. See Note 32.

35. Turing explains loop control (by means of a B-tube) in Turing (Note 3), pp. 66–7; the B-tube was effectively a register containing n and this number was counted down by repeatedly subtracting 1.
36. Turing (Note 3), p. 22.
37. The Mark II was synchronized by an oscillator with a frequency of 100 kHz. Turing called a single cycle of the oscillator the 'digit period'. The digit period was 10 microseconds and the duration of a beat was 24 digit periods.
38. Williams and Kilburn (Note 5), p. 57.
39. We follow the A = 440 Hz tuning standard.
40. Dodd (Note 9), p. 32.
41. H. Lukoff, *From Dits to Bits: a Personal History of the Electronic Computer*, Robotics Press (1979), pp. 85–6.
42. Turing (Note 3), pp. 87–8.
43. Williams and Kilburn (Note 5), p. 59.
44. G. C. Tootill, 'Digital computer notes on design & operation', 1948–9, Manchester Archive; table of the machine's instructions dated 27 October 1948.
45. Turing (Note 3), pp. 24, 88.
46. S. H. Lavington *Early British Computers: the Story of Vintage Computers and the People Who Built Them*, Manchester University Press (1980), p. 37.
47. Prinz (Note 13), p. 20 (our italics).
48. P. Doornbusch, *The Music of CSIRAC: Australia's First Computer Music*, Common Ground (2005). The book includes a CD of re-created music.
49. D. McCann and P. Thorne, *The Last of the First. CSIRAC: Australia's First Computer*, Melbourne University Press (2000), p. 2.
50. McCann and Thorne (Note 49), p. 3; Doornbusch (Note 48), pp. 24–5.
51. J. Fildes, '"Oldest" computer music unveiled', BBC News, 17 June 2008 (http://www.news.bbc.co.uk/2/hi/technology/7458479.stm).
52. R. T. Dean (ed.), *The Oxford Handbook of Computer Music*, Oxford University Press (2009), pp. 558, 584.
53. Lukoff (Note 41), p. 84.
54. Lukoff (Note 41), p. 86.
55. http://www.news.bbc.co.uk/2/hi/technology/7458479.stm; and http://www.digital60.org/media/mark_one_digital_music. Part of this recording can be heard at http://www.abc.net.au/classic/content/2014/06/23/4028742.htm. This edition of *Midday with Margaret Throsby* (ABC Radio National, 23 June 2014) is a musical tour through Turing's life and work, on the 102nd anniversary of his birth.
56. Cooper interview (Note 25).
57. National Sound Archive, ref. H3942.
58. Cooper interview (Note 25).
59. *BBC Recording Training Manual*, British Broadcasting Corporation (1950), pp. 49, 52.
60. *BBC Recording Training Manual* (Note 59), p. 26
61. How this was done is described in our companion paper 'Turing and the history of computer music', in J. Floyd and A. Bokulich (eds) *Philosophical Explorations of the Legacy of Alan Turing*, Boston Studies in the Philosophy and History of Science, Springer Verlag, 2017.
62. D. Link, 'Software archaeology: on the resurrection of programs for the Mark 1, 1948–58' (2015) (https://vimeo.com/116346967).

CHAPTER 24 TURING, LOVELACE, AND BABBAGE (SWADE)

1. B. V. Bowden (ed.), *Faster Than Thought*, Pitman (1953), p. ix.
2. An earlier version of this thesis is presented in D. Swade, 'Origins of digital computing: Alan Turing, Charles Babbage, and Ada Lovelace', in H. Zenil (ed.), *A Computable Universe: Understanding Computation and Exploring Nature as Computation*, World Scientific (2012), 23–43.
3. M. Moseley, *Irascible Genius: a Life of Charles Babbage, Inventor*, Hutchinson (1964).

4. There are three known accounts by Babbage of this episode. These date from 1822, 1834, and 1839. The version quoted is in H. W. Buxton, *Memoir of the Life and Labours of the Late Charles Babbage Esq. F.R.S* (ed. A. Hyman), Charles Babbage Institute Reprint Series for the History of Computing, Vol. 13, Tomash (1988), p. 46. See also B. Collier, *The Little Engines That Could've: the Calculating Engines of Charles Babbage*, 2nd edn., Garland (1990), pp. 14–18.

5. The quoted phrase is in D. Lardner, 'Babbage's calculating engine', *Edinburgh Review*, 59 (1834), 263–327; reprinted in M. Campbell-Kelly (ed.), *The Works of Charles Babbage*, Vol. 2, William Pickering (1989), p. 169.

6. H. P. Babbage (ed.), *Babbage's Calculating Engines: a Collection of Papers by Henry Prevost Babbage*, Spon (1889); reprinted in an edition edited by A. G. Bromley, Tomash (1982), see Preface.

7. Buxton (Note 4), pp. 48–9 (emphasis original).

8. Lady Byron to Dr King (21 June 1833), in B. A. Toole (ed.), *Ada, the Enchantress of Numbers: A Selection from the Letters of Lord Byron's Daughter and Her Description of the First Computer*, Strawberry Press (1992), p. 51 (emphasis original).

9. For images of the three forms of mechanical notation see D. Swade, 'Automatic computation: Charles Babbage and computational method', *The Rutherford Journal for the History and Philosophy of Science and Technology*, 3 (2010) (http://www.rutherfordjournal.org/article030106.html).

10. *The Essential Turing*, p. 206.

11. The first complete Babbage engine to be built was Difference Engine No. 2, designed in 1847–49 and completed in 2002 at the Science Museum, London. D. Swade, 'The construction of Charles Babbage's Difference Engine No. 2', *IEEE Annals of the History of Computing*, 27(3) (2005), 70–88.

12. A. G. Bromley, 'The evolution of Babbage's calculating engines', *Annals of the History of Computing*, 4(3) (1982), 113–36.

13. J. von Neumann, 'First draft of a report on the EDVAC', contract no. w-670-ord-4926, *Technical Report*, Moore School of Electrical Engineering, University of Pennsylvania (30 June 1945); extracts reprinted in Randell (1973).

14. B. Collier (Note 4), p. 139.

15. A. Lovelace, 'Sketch of the Analytical Engine invented by Charles Babbage Esq. By L. F. Menabrea, of Turin, officer of the Military Engineers, with notes upon the memoir by the translator', *Scientific Memoirs*, 3 (1843), 666–731; reprinted in Campbell-Kelly (ed.) (Note 5), Vol. 3, 89–170.

16. V. R. Huskey and H. D. Huskey, 'Lady Lovelace and Charles Babbage', *Annals of the History of Computing*, 2(4) (1980), 299–329. See also J. Fuegi and J. Francis, 'Lovelace & Babbage and the Creation of the 1843 "Notes" ', *IEEE Annals of the History of Computing*, 25(4) (2003), 16–26; and the documentary *To Dream Tomorrow*, directed by J. Fuegi and J. Francis, Flare Productions (2003).

17. Lovelace (Note 15), p. 118 (emphasis original).

18. Lovelace (Note 15), p. 144 (emphasis original).

19. Charles Babbage to Ada Lovelace (probably 1 July 1843). Papers of the Noel, Byron, and Lovelace families, Bodleian Library, shelfmark 168. Quoted in Huskey and Huskey (Note 16), p. 313.

20. Papers of Charles Babbage (MSS Buxton). Museum of the History of Science, Oxford, MSS 7 (blue, folio 1).

21. Lovelace (Note 15), p. 155.

22. Two of the accounts appear in MSS 7 (Note 20). A third appears in Scribbling Book S12, 134 (4 May 1869), Science Museum Babbage Archive, digitised archive reference S15_0012.

23. Ada Lovelace to Woronzow Greig (15 November 1844). Papers of the Noel, Byron and Lovelace families, Bodleian Library, shelf mark 171. Also in Toole (ed.) (Note 8), pp. 295–6 (emphasis original).

24. A. Turing to R. Ashby (probably 1946), in D. Yates, *Turing's Legacy: A History of Computing at the National Physical Laboratory 1945-1995*, Science Museum (1997).

25. A. Lovelace (Note 15), p. 156 (emphasis original). The view about originality is usually attributed to Lovelace and the version quoted is her own as it appears in Note G. Menabrea wrote that Babbage's engine 'is not a thinking being but simply an automaton which acts according to the laws imposed upon it' (Lovelace (Note 15), p. 98). In 1841 Babbage himself wrote that the Analytical Engine 'cannot invent. It must derive from human intellect those laws which it puts in force in the developments

it performs'. MSS Buxton (Note 20), quoted in Collier (Note 4), p. 178. Both these views pre-date Lovelace's.

26. Turing (1950), p. 455.

27. B. Randell, 'From Analytical Engine to electronic digital computer: the contributions of Ludgate, Torres, and Bush', *Annals of the History of Computing*, 4(4) (1982), 327–41.

28. I. B. Cohen, 'Babbage and Aiken', *Annals of the History of Computing*, 10(3) (1988), 171–91, on p. 180.

29. N. Metropolis and J. Worlton, 'A trilogy of errors in the history of computing', *Annals of the History of Computing*, 2(1) (1980), 49–59.

30. Cohen (Note 28).

31. Hodges (1983), p. 297.

32. *The Essential Turing*, p. 29.

33. Metropolis and Worlton (Note 29), pp. 52–3.

34. A. Lovelace (Note 15).

35. Bromley (Note 6).

36. D. Lardner, 'Babbage's Calculating Engine', *Edinburgh Review*, 59 (1834), 263–327; reprinted in M. Campbell-Kelly (Note 5), Vol. 2, 118–86.

37. A. G. Bromley, *The Babbage Papers in the Science Museum: a Cross-referenced List*, Science Museum (1991).

38. The main scholars active in the revival of interest in Babbage in the modern era include Bruce Collier, Allan G. Bromley, and Maurice Wilkes. See B. Collier (Note 4); A. G. Bromley, 'Charles Babbage's Analytical Engine, 1838', *Annals of the History of Computing*, 4(3) (1982), 196–217; A. G. Bromley (Note 12) and 'Babbage's Analytical Engine, Plans 28 and 28a – the programmer's interface', *IEEE Annals of the History of Computing*, 22(4) (2000), 4–19; M. V. Wilkes, 'Babbage as a computer pioneer', *British Computer Society and the Royal Statistical Society* (1971).

39. L. H. Dudley Buxton, 'Charles Babbage and his Difference Engines', *Transactions of the Newcomen Society*, 14 (1933–4), 43–65. I am indebted to Tim Robinson for reminding me of the existence and timing of this paper.

40. Buxton (Note 20).

41. Buxton (Note 39), for discussants see pp. 59–64.

42. Hodges (1983), p. 446 (Note 31).

43. Hodges (1983), p. 109 (Note 31).

44. Turing (1950), p. 455.

45. D. R. Hartree, *Calculating Instruments and Machines*, University of Illinois Press (1949), pp. 69–73.

46. R. Gandy, 'The confluence of ideas in 1936', in R. Herken (ed.), *The Universal Turing Machine: a Half-Century Survey*, Oxford University Press (1988), p. 60.

CHAPTER 25 INTELLIGENT MACHINERY (COPELAND)

1. Turing (1948) [also *Intelligent Machinery: a Report by A. M. Turing*, National Physical Laboratory (1948), in the Woodger Archive; a digital facsimile is available in *The Turing Archive for the History of Computing* (http://www.AlanTuring.net/intelligent_machinery)].

2. Turing (1951), p. 484.

3. This view was first put forward in Copeland, 'Artificial Intelligence', in *The Essential Turing*.

4. Copeland (Note 3), pp. 354–55; see also *Turing* (Copeland 2012), pp. 191–2.

5. Turing (1940), p. 335.

6. Michie in interview with Copeland, October 1995; Good in interview with Copeland, February 2004.

7. Michie interview (Note 6).

8. Turing (1947).

9. Turing (1947), p. 391.

10. Turing (1950).

11. Turing (1950), p. 460.

12. Turing (1950), p. 460.

13. Turing (1950), p. 460 and (1947), p. 393.
14. Turing (1950), p. 463.
15. Turing (1948), p. 420.
16. Turing (1948), p. 420.
17. Turing (1950), pp. 460–1.
18. Michael Woodger, reported by Michie and Meltzer, in B. Meltzer and D. Michie (eds), *Machine Intelligence 5*, Elsevier (1970), p. 2.
19. Michie in interview with Copeland, February 1998.
20. Turing (1950), p. 449.
21. R. A. Brooks, 'Intelligence without reason', in L. Steels and R. Brooks (eds), *The Artificial Life Route to Artificial Intelligence: Building Situated Embodied Agents*, Lawrence Erlbaum (1994).
22. In Turing (1948) and (1950), and Turing et al. (1952).
23. For more detail about the meaning of the condition 'if the judges are mistaken often enough', see my introduction to 'Computing machinery and intelligence' in *The Essential Turing*, pp. 435–6, and B. J. Copeland, 'The Turing test', in J. H. Moor (ed.) *The Turing Test: the Elusive Standard of Artificial Intelligence*, Kluwer (2003), p. 527.
24. Turing (1950), p. 442.
25. Turing (1950), p. 442.
26. Turing et al. (1952), p. 495.
27. Turing et al. (1952), p. 495.
28. See further Copeland, 'The Turing Test' (Note 23), p. 18, and *Turing* (Copeland 2012), p. 205.
29. Turing (1950), p. 452.
30. Turing (1950) p. 441.
31. Turing (1950) p. 441.
32. Turing et al. (1952), p. 495.
33. Turing (1950), p. 449.
34. Turing (1951), pp. 485, 486.
35. Turing (1950) p. 449.
36. Turing et al. (1952), p. 495.
37. M. Davis, 'Foreword to the Centenary Edition', in S. Turing, *Alan M. Turing: Centenary Edition*, Cambridge University Press (2012), p. xvi.
38. *The Washington Post*, 'A bot named "Eugene Goostman" passes the Turing test . . . kind of' (9 June 2014) (https://www.washingtonpost.com/blogs/compost/wp/2014/06/09/a-bot-named-eugene-goostman-passes-the-turing-test-kind-of/).
39. BBC News, 'Computer AI passes Turing test in "world first" ' (9 June 2014) (http://www.bbc.com/news/technology-27762088).
40. University of Reading, 'Turing test success marks milestone in computing history' (8 June 2014) (http://www.reading.ac.uk/news-and-events/releases/PR583836.aspx).
41. See, for example, A. Hodges, *Alan Turing: the Enigma*, Vintage (1992), p. 415, and R. French, 'The Turing test: the first 50 years', *Trends in Cognitive Sciences*, 4 (2000), 115–22, p. 115.
42. Turing et al. (1952), p. 494.
43. Turing et al. (1952), p. 494.
44. See, for example, N. Block, 'Psychologism and behaviorism', *Philosophical Review*, 90 (1981), 5–43.
45. N. Block, 'The mind as the software of the brain' in E. E. Smith and D. N. Osherson (eds), *An Introduction to Cognitive Science*, 2nd edn, Vol. 3, MIT Press (1995), p. 381.
46. Searle first presented his famous Chinese room thought experiment in J. R. Searle, 'Minds, brains, and programs', *Behavioural and Brain Sciences*, 3 (1980), 417–24, 450–6, and also in J. R. Searle, *Minds, Brains and Science: the 1984 Reith Lectures*, Penguin (1989); see also J. Preston and M. Bishop (eds) *Views Into the Chinese Room*, Oxford University Press (2002).
47. For further discussion of Searle's Chinese room argument, see B. J. Copeland, *Artificial Intelligence*, Blackwell (1993), Chapter 6, and B. J. Copeland, 'The Chinese Room from a logical point of view', in Preston and Bishop (Note 46).

CHAPTER 26 TURING'S MODEL OF THE MIND (SPREVAK)

1. Turing (1950), Turing et al. (1952), Turing (1948), Turing (1950), and Turing (c.1951).
2. Turing (1951) and Turing (1948).
3. Turing (1948), pp. 420, 412.
4. Letter from Turing to W. Ross Ashby (no date), Woodger Archive, M11/99; a digital facsimile is in *The Turing Archive for the History of Computing* (http://www.AlanTuring.net/turing_ashby/).
5. Turing (1950).
6. Turing (1936), p. 59.
7. Turing (1936), pp. 75–6.
8. H. Putnam, 'Minds and machines' and 'The mental life of some machines', in *Mind, Language and Reality: Philosophical Papers*, Vol. 2, Cambridge University Press (1975), 362–87 and 408–28, respectively.
9. H. Putnam, 'Philosophy and our mental life', in *Mind, Language and Reality: Philosophical Papers*, Vol. 2, Cambridge University Press (1975), 291–303.
10. See A. Clark, 'Whatever next? Predictive brains, situated agents, and the future of cognitive science', *Behavioral and Brain Sciences*, 36 (2013), 181–253.
11. D. C. Dennett, *Consciousness Explained*, Little, Brown & Company (1991).
12. See A. Zylberberg, S. Dehaene, P. R. Roelfsema, and M. Sigman, 'The human Turing machine: a neural framework for mental programs', *Trends in Cognitive Sciences*, 15 (2011), 293–300, and J. Feldman, 'Symbolic representation of probabilistic worlds', *Cognition*, 123 (2012), 61–83.

CHAPTER 27 THE TURING TEST—FROM EVERY ANGLE (PROUDFOOT)

1. University of Reading, 'Turing Test success marks milestone in computing history' (8 June 2014) (http://www.reading.ac.uk/news-and-events/releases/PR583836.aspx); Kevin Warwick, quoted in the University of Reading press release; A. Griffin, 'Turing Test breakthrough as super-computer becomes first to convince us it's human', *The Independent* (8 June 2014); 'Turing Test bested, robot overlords creep closer to assuming control', *International Business Times* (9 June 2014); M. Masnick, 'No, a "supercomputer" did NOT pass the Turing test for the first time and everyone should know better', *Techdirt* (9 June 2014); Stevan Harnad quoted in I. Semple and A. Hern, 'Scientists dispute whether computer "Eugene Goostman" passed Turing test', *The Guardian* (9 June 2014); A. Mann, 'That computer actually got an F on the Turing test', *Wired* (9 June 2014); G. Marcus, 'What comes after the Turing test?', *The New Yorker* (9 June 2014).
2. Turing (1948), p. 431.
3. Turing (1945), p. 389.
4. A. Newell, J. C. Shaw, and H. A. Simon, 'Chess-playing programs and the problem of complexity', in E. A. Feigenbaum and J. Feldman (eds), *Computers and Thought*, MIT Press (1995), 39–70 (originally published in 1963 by McGraw-Hill); this is a reprint of their article in the *IBM Journal of Research and Development*, 2(4) (1958), 320–35.
5. The words 'hope of humanity' are from S. Levy, 'Man vs. machine', *Newsweek* (5 May 1997), p. 5; Kasparov in an interview with Daniel Sieberg 'Kasparov: "Intuition versus the brute force of calculation" ', *CNNAccess* (10 February 2003) (http://edition.cnn.com/2003/TECH/fun.games/02/08/cnna.kasparov); R. Wright, 'Can machines think?', *Time Magazine*, 147(13) (25 March 1996).
6. Kasparov's words are taken from the following: Garry Kasparov, in Vikram Jayanti's 2003 documentary film *Game Over: Kasparov and the Machine*, available at: http://topdocumentaryfilms.com/game-over-kasparov-and-the-machine/; ICC Kasparov Interview (22 November 1998), available at: http://www6.chessclub.com/resources/articles/interview_kasparov.html; G. Kasparov, *How Life Imitates Chess: Making the Right Moves, from the Board to the Boardroom*, Bloomsbury Publishing (2008), p. 161; Kasparov in *Game Over*; G. Kasparov, 'IBM owes mankind a rematch', *Time Magazine*, 149(21) (26 May 1997).
7. M. Gardner, 'Kasparov's defeat by Deep Blue', in *A Gardner's Workout: Training the Mind and Entertaining the Spirit*, A. K. Peters (2001), p. 98.

8. Turing (1950), pp. 442, 441, 443.

9. C. Thompson, 'The other Turing test', *Wired*, 13(7) (2005); C. Thompson, 'The female Turing test', *Collision Detection: Science, Technology, Culture* (1 April 2005).

10. Turing (1951), p. 484. On whether the imitation game is gendered, see B. J. Copeland, 'The Turing test', in J. H. Moor (ed.) *The Turing Test: The Elusive Standard of Artificial Intelligence*, Kluwer (2003), 1–21.

11. Turing (1950), p. 449; Turing (1948), p. 431; Turing et al. (1952), pp. 495, 503; Turing (1950), pp. 454, 443, 441.

12. Turing et al. (1952), p. 495; see Copeland (Note 10).

13. On strategic judging in the 2000 contest see Copeland (Note 10), p. 7; J. H. Moor, 'The status and future of the Turing test', in *The Turing Test* (Note 10), p. 204 (the results of the 2000 contest are set out on p. 205). For the outcomes of the 2003 contest, see http://www.loebner.net/Prizef/loebner-prize.html.

14. Turing (1950), p. 442.

15. H. R. T. Roberts, 'Thinking and machines', *Philosophy*, 33(127) (1958), p. 356; R. M. French, 'Dusting off the Turing test', *Science*, 336 (2012), 164–5, p. 164; T. W. Polger, 'Functionalism as a philosophical theory of the cognitive sciences', *Wiley Interdisciplinary Reviews: Cognitive Science*, 3(3) (2012), 337–48, p. 338.

16. Turing (1936) p. 75; Turing (c.1951), p. 472.

17. A. D. Ritchie and W. Mays, 'Thinking and machines', *Philosophy*, 32(122) (1957), 258–61, p. 261.

18. For more on Turing's own words concerning his imitation game, see D. Proudfoot, 'What Turing himself said about the imitation game', *IEEE Spectrum*, 52(7) (2015), 42–7.

19. Turing (1948), p. 431.

20. S. Harnad, 'Other bodies, other minds: a machine incarnation of an old philosophical problem', *Minds and Machines*, 1 (1991), 43–44, p. 44.

21. XPRIZE: http://www.xprize.org/prize-development; http://www.xprize.org/TED.

22. British Computer Society Machine Intelligence Competition: http://www.bcs-sgai.org/micomp/index.html.

23. Excerpts from transcripts provided in K. Warwick and H. Shah, 'Good machine performance in Turing's imitation game', *IEEE Transactions on Computational Intelligence and AI in Games*, 99 (2013), 289–99; J. Vincent, 'Turing test: what is it—and why isn't it the definitive word in artificial intelligence?', *The Independent* (9 June 2014); http://www.reading.ac.uk/news-and-events/releases/PR583836.aspx.

24. http://n4g.com/news/1092379/unreal-bots-beat-turing-test-ai-players-are-officially-more-human-than-gamers; M. Polceanu, 'MirrorBot: using human-inspired mirroring behavior to pass a Turing test', *IEEE Conference on Computational Intelligence in Games (CIG), 11–13 August 2013* (doi: 10.1109/CIG.2013.6633618).

25. 'Marvin Minsky on Singularity 1 on 1: the Turing test is a joke!' Interview with Nikola Danaylov, 2013 (http://www.youtube.com/watch?v=3PdxQbOvAlI); A. Sloman, 'The mythical Turing test' (26 May 2010) (http://www.cs.bham.ac.uk/research/projects/cogaff/misc/turing-test.html); D. McDermott, 'Don't improve the Turing test, abandon it': address to the *Towards a Comprehensive Turing Test (TCIT)* workshop, 2010 AISB Convention (http://www.youtube.com/watch?v=zXEB9ctWJp8).

26. Turing et al. (1952), p. 495; Turing (1950), p. 448; see B.J. Copeland and D. Proudfoot, 'The legacy of Alan Turing', *Mind*, 108 (1998), 187–95.

27. R. M. French, 'Subcognition and the limits of the Turing test', *Mind*, 99 (1990), 53–65, p. 53; D. Michie, 'Turing's test and conscious thought', *Artificial Intelligence*, 60 (1993), 1–22, p. 9; P. R. Cohen, 'If not Turing's test, then what?', *AI Magazine*, 26(4) (2005), 61–67, p. 62; N. Block, 'The mind as the software of the brain', in E. E. Smith and D. N. Osherson (eds), *An Invitation to Cognitive Science, Vol. 3 Thinking*, 2nd edn, MIT Press (1995), p. 378; P. J. Hayes and K. M. Ford, 'Turing test considered harmful', *IJCAI'95 Proceedings of the 14th International Joint Conference on Artificial Intelligence* (Montreal, August 1995), Vol. 1, Morgan Kaufman (1995), 972–7, pp. 972, 976; R. M. French, 'Moving beyond the Turing test', *Communications of the ACM*, 55(12) (2012), 74–7.

28. Hayes and Ford (Note 27), p. 974; G. Fostel, 'The Turing test is for the birds', *ACM SIGART Bulletin*, 4(1) (1993), 7–8, p. 8.

29. K. M. Ford and P. J. Hayes, 'On computational wings: rethinking the goals of artificial intelligence', *Scientific American Presents*, 9(4) (1998), p. 79.
30. French 'Subcognition and the limits of the Turing test' (Note 27), p. 53; French, 'Dusting off the Turing test' (Note 15), p. 165.
31. P. Millican, 'The philosophical significance of the Turing machine and the Turing test', in B. Cooper and J. van Leeuwen (eds), *Alan Turing: His Work and Impact*, Elsevier (2013), p. 599; 'Artificial stupidity', *The Economist*, 324(7770) (1 August 1992), p. 14.
32. See Copeland (Note 10), pp. 10–11.
33. Turing (1950), p. 442.
34. The phrase 'intellectual statues' is from B. Whitby, 'The Turing test: AI's biggest blind alley?', in P. Millican and A. Clark (eds), *The Legacy of Alan Turing, Vol. I: Machines and Thought*, Oxford University Press (1996), p. 58; the notion of an 'intelligence amplifier' is from D. B. Lenat, 'Building a machine smart enough to pass the Turing test: could we, should we, will we?', in R. Epstein, G. Roberts, and G. Beber (eds), *Parsing the Turing Test: Philosophical and Methodological Issues in the Quest for the Thinking Computer*, Springer (2008), p. 281.
35. G. Jefferson, 'The mind of mechanical man', *British Medical Journal*, 1(4616) (1949), 1105–1110, p. 1110. On anthropomorphism in AI, see D. Proudfoot, 'Anthropomorphism and AI: Turing's much misunderstood imitation game', *Artificial Intelligence*, 175(5–6) (2011), 950–7.
36. G. Kasparov, 'The chess master and the computer', *New York Review of Books*, 57(2) (11 February 2010); 'Computers and chess: not so smart', *The Economist* (1–7 February 2003), p. 13; Douglas Hofstadter as quoted in B. Weber, 'Mean chess-playing computer tears at meaning of thought', *New York Times* (19 February 1996).
37. DeepBlue: http://www.research.ibm.com/deepblue/meet/html/d.3.3a.html#whychess; Watson: http://www.ibm.com/smarterplanet/us/en/ibmwatson/ (for detailed information on Watson's architecture see D. Ferrucci, E. Brown, J. Chu-Carroll, J. Fan, D. Gondek, A. A. Kalyanpur, A. Lally, J. W. Murdock, E. Nyberg, J. Prager, N. Schlaefer, and C. Welty, 'Building Watson: an overview of the DeepQA project', *AI Magazine*, 31(3) (2010), 59–79); 'Artificial Intelligence: the Difference Engine: the answering machine', *Economist* (18 February 2011); R. Kurzweil, 'Kurzweil: why IBM's Jeopardy victory matters', *PCMag* (20 January 2011).
38. IBM Watson AI XPRIZE: 'Announcing the IBM Watson AI XPRIZE; a cognitive computing competition' (http://www.xprize.org/ai).
39. J. Searle, 'Watson doesn't know it won on "Jeopardy"', *Wall Street Journal* (23 February 2011).
40. Howard Yu, quoted in S. Borowiec, 'Computer beats man in final Seoul matchup', *Los Angeles Times* (16 March 2016); S. Mundy, 'AlphaGo seals 4–1 win over Korea grandmaster', *Financial Times* (16 March 2016). AlphaGo's structure is set out in D. Silver et al., 'Mastering the game of Go with deep neural networks and tree search', *Nature*, 529 (2016), 484–9.
41. Google DeepMind: AlphaGo (https://deepmind.com/alpha-go).
42. David Silver quoted in 'AlphaGo vs Lee Sedol: history in the making', *Chess News* (13 March 2016) (http://en.chessbase.com/post/alphago-vs-lee-sedol-history-in-the-making); J. Naughton, 'Can Google's AlphaGo really feel it in its algorithms?', *The Observer* (31 January 2016).
43. T. Chatfield, 'How much should we fear the rise of artificial intelligence?', *The Guardian* (18 March 2016).
44. G. Johnson, 'Recognizing the artifice in artificial intelligence', *The New York Times* (5 April 2016).
45. 'Google's AlphaGo might have bested the world Go champ—but Chinese netizens say it's not smart enough to win at mahjong', *South China Morning Post* (15 March 2016).
46. Letter from Anthony Oettinger to Jack Copeland (19 June 2000); see Copeland (Note 10).
47. 'Computer says Go' [editorial], *The Times* (10 March 2016).
48. N. Sanders, 'Computer says Go' [letter to the Editor], *The Times* (11 March 2016).
49. A. Oettinger, 'Programming a digital computer to learn', *Philosophical Magazine*, 43, (1952), 1243–63, p. 1250. See B. J. Copeland and D. Proudfoot, 'Alan Turing—father of the modern computer', *The Rutherford Journal for the History and Philosophy of Science and Technology*, 4 (2011–2012) (http://www.rutherfordjournal.org/article040101.html).

50. Eric Brown, at a NEOSA Tech Week event in Cleveland (18 April 2012) (http://www.youtube.com/watch?v=bfLdgDYjC6A).

51. K. Jennings, 'Watson, Jeopardy and me, the obsolete know-it-all', TEDxSeattleU, filmed February 2013 (http://www.ted.com/talks/ken_jennings_watson_jeopardy_and_me_the_obsolete_know_it_all); John Searle is quoted in Weber (Note 36); A. Levinovitz, 'The mystery of Go, the ancient game that computers still can't win', *Wired* (12 May 2014).

52. D. Michie (Note 27), p. 20; G. Jefferson (Note 35), p. 110 (Turing quotes Jefferson's words in Turing (1950), p. 451).

53. Turing (1950), p. 452; Turing (1947), p. 394.

54. Turing (1951), p. 485; Turing (1948), p. 410; Turing (1950), p. 450.

55. For Turing's attitude to this reaction, see D. Proudfoot, 'Mocking AI panic', *IEEE Spectrum*, 52(7) (2015), 46–7.

56. H. L. S. (Viscount) Samuel, *Essay in Physics*, Blackwell (1951), pp. 133–4; H. Cohen, 'The status of brain in the concept of mind', *Philosophy*, 27(102) (1952), 195–210, p. 206; Turing (1950), pp. 455, 459; Turing et al. (1952), p. 500.

57. Turing (1952), p. 459.

CHAPTER 28 TURING'S CONCEPT OF INTELLIGENCE (PROUDFOOT)

1. The central argument in this paper is developed in D. Proudfoot, 'Rethinking Turing's test', *Journal of Philosophy*, 110 (2013), 391–411; see also D. Proudfoot, 'Anthropomorphism and AI: Turing's much misunderstood imitation game', *Artificial Intelligence*, 175 (2011), 950–7.

2. It is the *capacity for* or *tendency to* 'thinking' behaviour that is the thinking, according to the behaviourist. All references to behaviourism in the text should be understood in this way.

3. See W. Mays and D. G. Prinz, 'A relay machine for the demonstration of symbolic logic', *Nature*, 165(4188) (1950), 197–8.

4. See W. Mays, 'Can machines think?', *Philosophy*, 27(101) (1952), 148–62, pp. 151, 160, 151. Like Mays, many read Turing as claiming that if the machine's behaviour *isn't* indistinguishable from that of the human contestant, it *doesn't* think. But Turing made it clear that an intelligent machine might do badly in his game.

5. P. Ziff, 'About behaviourism', *Analysis*, 18(6) (1958), 132–6, p. 132.

6. Mays (Note 4), pp. 149, 162, and 150.

7. Mays (Note 4), p. 158.

8. Turing (1948), p. 431.

9. Turing et al. (1952), p. 495; Turing (1950), p. 441; Copeland (in *The Essential Turing*, p. 436) points out Turing's protocol for scoring the game.

10. Turing et al. (1952), pp. 495, 503.

11. The 1952 game is played by both machine and human contestants, but only one contestant is interviewed at a time.

12. Turing et al. (1952), p. 496.

13. Except insofar as the machine must be able to produce the appropriate behaviour in real time: see Proudfoot 2013 (Note 1), pp. 400–1.

14. Turing (1948), p. 411; Turing (1948), p. 431; and Turing et al. (1952), p. 500.

15. Turing (1948), p. 431.

16. A 'paper machine' is a human being 'provided with paper, pencil, and rubber, and subject to strict discipline', carrying out a set of rules (Turing (1948), p. 416).

17. Why did Turing design a 3-player game? See Proudfoot 2013 (Note 1), pp. 409–10.

18. R. A. Brooks, 'Intelligence without reason', in L. Steels and R. A. Brooks (eds), *The Artificial Life Route to Artificial Intelligence*, Lawrence Erlbaum (1995), 25–81, p. 57.

19. Turing (1948), p. 412.

20. Turing (1950), p. 449; Turing et al. (1952), p. 495.

21. Turing et al. (1952), p. 495; Turing (1950), p. 442.

22. In Proudfoot (2013) (Note 1), pp. 401–2, the schema I propose is world-relativized.
23. G. Jefferson, 'The mind of mechanical man', *British Medical Journal*, 1(4616) (1949), 1105–10, pp. 1109–10 and 1107.
24. Turing (1951), p. 484.
25. Turing (1951), p. 485.
26. The argument in the last section is developed in in D. Proudfoot, 'Turing and free will: a new take on an old debate', in J. Floyd and A. Bokulich (eds), *Philosophical Explorations of the Legacy of Alan Turing*, Boston Studies in the Philosophy and History of Science (in press).

CHAPTER 29 CONNECTIONISM: COMPUTING WITH NEURONS (COPELAND AND PROUDFOOT)

1. This chapter consists of part of our article 'Alan Turing's forgotten ideas in computer science', *Scientific American*, 280 (April 1999), 99–103, with minor additions and modifications. We are grateful to the editors of *Scientific American* for permission to reproduce this material here.
2. D. E. Rumelhart and J. L. McClelland, 'On learning the past tenses of English verbs', in J. L. McClelland, D. E. Rumelhart, and the PDP Research Group, *Parallel Distributed Processing: Explorations in the Microstructure of Cognition, Vol. 2: Psychological and Biological Models*, MIT Press (1986), 216–71.
3. F. Rosenblatt, 'The Perceptron, a perceiving and recognizing automaton', *Report no. 85–460–1*, Cornell Aeronautical Laboratory (1957); 'The Perceptron: a probabilistic model for information storage and organisation in the brain', *Psychological Review*, 65 (1958), 386–408; *Principles of Neurodynamics*, Spartan (1962).
4. Turing (1948).
5. Robin Gandy in an interview with Jack Copeland, October 1995; minutes of the Executive Committee of the National Physical Laboratory for 28 September 1948, National Physical Laboratory Library; a digital facsimile is available in *The Turing Archive for the History of Computing* (http://www.AlanTuring.net/npl_minutes_sept1948).
6. In C. R. Evans and A. D. J. Robertson, *Key Papers: Cybernetics*, Butterworth (1968).
7. Turing (1948), p. 422.
8. Turing (1948), p. 424.
9. Rosenblatt (1958) (Note 3), p. 387.
10. Copeland, 'Enigma', in *The Essential Turing*, pp. 254–5.
11. Turing (1948), p. 428.
12. B. G. Farley and W. A. Clark, 'Simulation of self-organizing systems by digital computer', *Institute of Radio Engineers Transactions on Information Theory*, 4 (1954), 76–84; W. A. Clark and B. G. Farley, 'Generalization of pattern recognition in a self-organizing system', in *AFIPS '55 (Western): Proceedings of the March 1–3, 1955, Western Joint Computer Conference*, ACM (1955), 86–91.
13. For further information about Turing's connectionist networks, see: B. J. Copeland and D. Proudfoot, 'On Alan Turing's anticipation of connectionism', *Synthese*, 108 (1996), 361–77; B. J. Copeland and D. Proudfoot, 'Turing and the computer', in Copeland et al. (2005); C. Teuscher, *Turing's Connectionism: an Investigation of Neural Network Architectures*, Springer (2002).

CHAPTER 30 CHILD MACHINES (PROUDFOOT)

1. Turing (1950), p. 460.
2. Turing (1948), pp. 431, 422. Turing considered two options for constructing a child machine: build as simple a child machine as possible, or build one with a great deal of programming, such as 'a complete system of logical inference' (Turing (1950), p. 461). In the latter case, the machine would have 'imperatives'—for example, if the teacher said 'Do your homework now', the machine would do the homework (Turing (1950), pp. 461–2).
3. Turing (1950), pp. 460–1; Turing (1948), pp. 429, 430; Turing (1947), p. 393; Turing (1948), p. 412.
4. Turing (1948), p. 420; Turing (1950), p. 463; Turing (1948), pp. 431–2.

5. Turing (1950), pp. 460–1; Turing (c.1951), p. 473.

6. Turing (1947), p. 393; Turing (1950), p. 458; Turing (c.1951), p. 473; Turing (1947), p. 393; Turing et al. (1952), p. 497.

7. Turing (1948), pp. 416, 424; Turing (1950), p. 460.

8. Turing (1948), pp. 418, 424, 422.

9. Turing (1948), p. 425; Turing (1950), p. 461; Turing (1948), p. 426.

10. Anthony Oettinger in an interview with Jack Copeland (January 2000).

11. Letter from Oettinger to Copeland (19 June 2000); A. Oettinger, 'Programming a digital computer to learn', *Philosophical Magazine*, 43 (1952), 1243–63, p. 1247.

12. See further B. J. Copeland and D. Proudfoot, 'Turing and the computer', in Copeland et al. (2005), 107–48, p. 126.

13. Letter from Christopher Strachey to Turing (15 May 1951) in the King's College Archive. This letter is quoted courtesy of the Camphill Village Trust.

14. D. Michie, 'Recollections of early AI in Britain: 1942–1965' (2002), transcript of talk given by Donald Michie at the Computer Conservation Society's Artificial Intelligence: Recollections of the Pioneers seminar on 11 October 2002, The Artificial Intelligence Applications Institute (http://www.aiai. ed.ac.uk/events/ccs2002/CCS-early-british-ai-dmichie.pdf); D. Michie, 'The Turing Institute: an experiment in cooperation', *Interdisciplinary Science Reviews*, 14(2) (1989), 117–19, p. 118; D. Michie, 'Edinburgh will set the pace', *The Scotsman* (17 February 1966), p. 11.

15. MENACE was reported in D. Michie, 'Trial and error', in S. A. Barnett and A. McLaren (eds), *Science Survey 1961: Part 2*, Penguin (1961), 129–45; reprinted in D. Michie, *On Machine Intelligence*, 2nd edn, Ellis Horwood (1986), 11–23. The first FREDDY robot was built in 1969 and reported in H. G. Barrow and S. H. Salter, 'Design of low-cost equipment for cognitive robot research', in B. Meltzer and D. Michie (eds), *Machine Intelligence 5*, Edinburgh University Press (1969), 555–66.

16. D. Michie, 'Strong AI: an adolescent disorder', *Informatica*, 19 (1995), 461–8.

17. D. Michie, 'Return of the imitation game' [revised version], *Linköping Electronic Articles in Computer and Information Science*, 6(28) (2001), 1–17, on pp. 1, 4, 6, 7, 17. Michie and Claude Sammut's conversational program 'Sophie' (at one time installed at the Sydney Powerhouse Museum) established rapport with a human user, switching from 'goal mode', where the program provided information about the exhibits, to 'chat mode' where the program's responses prompted the user to speak about themselves (ibid. p. 8).

18. Turing (1950), p. 460; Turing (1948), p. 421.

19. Turing (c.1951), pp. 474–5.

20. Turing (1948), pp. 428, 416, 428, 427, 428; Turing (1950), p. 461; Turing et al. (1952), pp. 428, 497.

21. Turing (1948), p. 428.

22. Turing (1948), p. 420; Turing (c.1951), p. 473; Turing (1950), pp. 460–1; Turing (1948), p. 421.

23. Turing (1948), p. 420. On 'infest the countryside' see B. Meltzer and D. Michie (eds) (Note 15), p. B; Turing (1948), pp. 420–1.

24. http://news.mit.edu/1998/cog-0318.

25. Turing (c.1951), p. 473; Turing (1950), p. 462.

26. Personal communication from Javier Movellan.

27. Personal communication from Javier Movellan; the research on reaching behaviour is due to Dan Messinger and Tingfan Wu.

28. Turing (1951), p. 486.

29. The discussion of social robots in this and the following section is drawn from the discussion that I gave in D. Proudfoot, 'Can a robot smile? Wittgenstein on facial expression', in T. P. Racine and K. L. Slaney (eds), *A Wittgensteinian Perspective on the Use of Conceptual Analysis in Psychology*, Palgrave Macmillan (2013), 172–94.

30. Turing (1950), p. 442.

31. See T. Wu, N. J. Burko, P. Ruvulo, M. S. Bartlett, and J. R. Movellan, 'Learning to make facial expressions', *2009 IEEE 8th International Conference on Development and Learning*, available at: https://www.researchgate.net/profile/Javier_Movellan/publication/228395247_Learning_to_make_facial_expressions/links/0deec5227d33038e6c000000.pdf.

32. Turing et al. (1952), p. 486.
33. Personal communication from Javier Movellan; the research on smiling is due to Dan Messinger and Paul Ruvolo.
34. Michie, it seems, believed that rapport is a psychological state, and that it suffices if the human observer *feels* rapport with the machine: see Michie (Note 17), p. 8.
35. On Turing's conception of intelligence, see D. Proudfoot, 'Anthropomorphism and AI: Turing's much misunderstood imitation game', *Artificial Intelligence*, 175(5–6) (2011), 950–7; D. Proudfoot, 'A new interpretation of the Turing test', *The Rutherford Journal for the History and Philosophy of Science and Technology*, 1 (2005) (http://rutherfordjournal.org/article010113.html); D. Proudfoot, 'Rethinking Turing's test', *Journal of Philosophy*, 110(7) (2013), 391–411.
36. Turing (1948), pp. 411, 431.
37. Turing (1950), p. 461; C. Breazeal, 'Regulating human–robot interaction using "emotions", "drives" and facial expressions', *Proceedings of the 1998 Autonomous Agents Workshop, Agents in Interaction—Acquiring Competence Through Imitation*, Minneapolis (1998), 14–21, available at: http://robotic.media.mit.edu/wp-content/uploads/sites/14/2015/01/Breazeal-Agents-98.pdf.
38. C. Breazeal, *Sociable Machines: Expressive Social Exchange Between Humans and Robots*, DSc dissertation, MIT (2000) (http://groups.csail.mit.edu/lbr/mars/pubs/phd.pdf), p. 190.
39. Rodney Brooks, interviewed in E. Guizzo and E. Ackermann, 'The rise of the robot worker', *IEEE Spectrum*, 49(10) (2012), 34–41, p. 41.
40. On anthropomorphism in AI, see Proudfoot (2011) (Note 35).
41. Turing (1948), p. 429: Turing said that we could also produce initiative in a machine by beginning with a 'fully disciplined' (that is, programmed) machine, rather than an unorganized machine, and allowing it to make more and more 'choices' (Turing (1948), pp. 429–30).
42. Turing (1950), p. 453; Turing (c.1951), p. 474; Turing (1948), p. 429.
43. Turing (1947), p. 394. For more on Turing's view that a machine can be the ultimate origin of its behaviour, see D. Proudfoot, 'Turing and free will: a new take on an old debate', in J. Floyd and A. Bokulich (eds), *Philosophical Explorations of the Legacy of Alan Turing*, Boston Studies in the Philosophy and History of Science (in press).
44. Turing (1951), p. 486; Turing et al. (1952), p. 495.

CHAPTER 31 COMPUTER CHESS—THE FIRST MOMENTS (COPELAND AND PRINZ)

1. Golombek quoted in A. G. Bell, *The Machine Plays Chess?*, Pergamon (1978), p. 15.
2. Letter from A. C. Pigou to Sara Turing (26 November 1956), King's College Archive, catalogue reference A 10.
3. J. W. von Goethe, *Gotz von Berlichingen mit der eisernen Hand*, 1773, Act 2, Scene 1, in *Goethes Werke*, Vol. 8, Böhlau, p. 50. With thanks to Klaus Weimar and Giovanni Sommaruga for advice concerning the English equivalent of Goethe's '*ein Probirstein des Gehirns*'.
4. Michie interviewed by Copeland, February 1998.
5. Good interviewed by Copeland, February 2004.
6. Michie interviewed by Copeland, October 1995.
7. J. von Neumann, 'Zur Theorie der Gesellschaftsspiele' [On the theory of games], *Mathematische Annalen*, 100 (1928), 295–320.
8. Good quoted in Bell (Note 1), p. 14.
9. Turing (1945), p. 389.
10. B. J. Copeland and G. Sommaruga, 'The stored-program universal computer: did Zuse anticipate Turing and von Neumann?' in G. Sommaruga and T. Strahm, *Turing's Revolution*, Birkhauser/Springer (2015).
11. Turing (1947), p. 393.
12. Turing (1948), p. 431.
13. Extract from a letter that Champernowne wrote to *Computer Chess*, 4 (January 1980), 80–1.

14. Turing (1953), pp. 574–5.
15. Turing et al. (1952), p. 503.
16. Turing (1953), p. 575.
17. A. L. Samuel, 'Some studies in machine learning using the game of checkers', *IBM Journal of Research and Development*, 3 (1959), 211–29; reprinted in E. A. Feigenbaum and J. Feldman (eds), *Computers and Thought*, McGraw-Hill (1963).
18. For the sake of smooth flow, the piece values have been relocated from their position a few paragraphs earlier in Turing's typescript, and the heading 'Piece values' has been added.
19. Glennie quoted in Bell (Note 1), pp. 17–18.
20. Turing (1953), p. 573.
21. With thanks to Jonathan Bowen and Eli Dresner for supplying the translations into modern notation.
22. From *The Essential Turing*, pp. 573–4.
23. B. V. Bowden (ed.), *Faster Than Thought*, Pitman (1953).
24. The chapter 'Digital computers applied to games' also contained extensive contributions by Audrey Bates, Vivian Bowden, and Christopher Strachey. Unfortunately the whole article was mistakenly attributed to Turing alone in D. C. Ince (ed.), *Collected Works of A. M. Turing: Mechanical Intelligence*, Elsevier (1992).
25. Game 2 is from Bowden (Note 23), p. 293.
26. Bell (Note 1), pp. 20–1. Bell also introduced typos.
27. Dresner in conversation with Copeland, January 2016.
28. D. Michie, 'Game-playing and game-learning automata', in L. Fox (ed.), *Advances in Programming and Non-numerical Computation*, Pergamon (1966), p. 189.
29. See further *The Essential Turing*, pp. 564–5.
30. C. Gradwell, 'Early days', reminiscences in a newsletter 'For those who worked on the Manchester Mk I computers', April 1994.
31. D. G. Prinz, *Introduction to Programming on the Manchester Electronic Digital Computer*, Ferranti Ltd, Moston, Manchester (28 March 1952). A digital facsimile is in *The Turing Archive for the History of Computing* (http://www.AlanTuring.net/prinz).
32. Prinz programmed the *Würfelspiel* in collaboration with David Caplin. See C. Ariza, 'Two pioneering projects from the early history of computer-aided algorithmic composition', *Computer Music Journal*, 35 (2011), 40–56.
33. Letter from Dietrich Prinz to Alex Bell, 16 August 1975.
34. Letter from Prinz to Bell (Note 33).
35. W. E. Kühle, D. G. Prinz, and F. Herriger, 'Magnetronröhre' [The magnetron tube], German patent 639,572, filed February 1934, issued December 1936; W. E. Kühle, F. Herriger, W. Ilberg, and D. G. Prinz, 'Röhrenanordnung unter Verwendung eines durch Nebenschluss regelbaren permanenten Magneten' [Setting up tubes by using a permanent magnet adjustable by a shunt], German patent 660,398, filed February 1934, issued May 1938; F. Herriger and D. G. Prinz, 'Magnetronröhre' [The magnetron tube], German patent 664,735, filed March 1934, issued September 1938; D. G. Prinz, 'Magnetron', US patent 2,099,533, filed July 1935, issued November 1937; W. E. Kühle, D. G. Prinz, and F. Herriger, 'Magnetron', US patent 2,031,778, filed August 1935, issued February 1936. (With thanks to Giovanni Sommaruga for advice on translating the titles of Prinz's German patents.)
36. See, for example, R. W. Burns, *Communications: an International History of the Formative Years*, Institution of Engineering and Technology, London (2004), p. 594.
37. 'Collar the lot' was attributed to Churchill (speaking about Italians in the UK) in a memo of June 1940 by Nigel Ronald of the Foreign Office (NA FO371/25197, 'Detention of enemy aliens'); see also P. and L. Gillman, *'Collar the Lot!' How Britain Interned and Expelled its Wartime Refugees*, Quartet (1980), pp. 153, 309.
38. G. Hamilton, 'How Jewish "enemy aliens" overcame a "traumatic" stint in Canadian prison camps during the Second World War', *National Post* (Canada) (7 February 2014).
39. Turing, (1940), pp. 320–1.

40. There is a modest amount of personal information about Prinz in his patent 'Improvements relating to control apparatus for equalizing the temperatures of two bodies remote from each other', UK patent 637,312, draft filed May 1947, issued May 1950.

41. W. Mays and D. G. Prinz, 'A relay machine for the demonstration of symbolic logic', *Nature*, 165(4188) (4 February 1950), 197–8; D. G. Prinz and J. B. Smith, 'Machines for the solution of logical problems', in Bowden (Note 23), Chapter 15.

42. D. W. Davies, 'A theory of chess and noughts and crosses', *Science News*, 16 (1950), 40–64.

43. Davies (Note 42), p. 62.

44. Bowden (Note 23), p. 295.

45. D. G. Prinz, 'Robot chess', *Research*, 5 (1952), 261–6.

46. Prinz quoted in Bell (Note 1), p. 27.

47. Bowden (Note 23), p. 296.

48. Bowden (Note 23), p. 296.

49. Prinz quoted in Bell (Note 1), p. 28.

50. Bowden (Note 23), p. 297.

51. Correspondence between Feist and Copeland, 2016.

52. Correspondence between Feist and Copeland (Note 51).

53. https://en.chessbase.com/post/alan-turing-plays-garry-kasparov-at-che-58-years-after-his-death. The Turing Engine is looking two ply ahead. (With thanks to Mathias Feist and Jonathan Bowen for transcribing the game.)

54. G. P. Collins, 'Claude E. Shannon: founder of information theory', *Scientific American*, 14 October 2002 (http://www.scientificamerican.com/article/claude-e-shannon-founder/).

55. Seldom noticed, the date October 8 1948 is included at the foot of the published version.

56. C. E. Shannon, 'Programming a computer for playing chess', *Philosophical Magazine*, 41 (1950), 256–75; see also C. E. Shannon, 'A chess-playing machine', *Scientific American*, 182 (1950), 48–51.

57. Shannon, 'Programming a computer for playing chess' (Note 56), p. 257.

58. Good interviewed by Copeland (Note 5).

59. I. Althöfer and M. Feist, 'Chess exhibition match between Shannon Engine and Turing Engine', *Preliminary Report*, Version 5 (17 April 2012) (http://www.althofer.de/shannon-turing-exhibition-match.pdf).

60. Correspondence between Feist and Copeland (Note 51).

61. *Manchester Evening News* (25 June 2012).

CHAPTER 32 TURING AND THE PARANORMAL (LEAVITT)

1. In D. R. Hofstadter, *Gödel, Escher, Bach: an Eternal Golden Braid*, Vintage (1990), p. 599; the author writes: 'Objection (9) I find remarkable. I have seen the Turing paper reprinted in a book—but with objection (9) omitted—which I find equally remarkable'.

2. Turing (1950), pp. 457–8.

3. A. Gauld, *The Founders of Psychical Research*, Schocken (1968), pp. 88, 138. For an incisive overview of psychic research in the twentieth century, see P. Lamont, *Extraordinary Beliefs: a Historical Approach to a Psychological Problem*, Cambridge University Press (2013).

4. 'Report of the committee on thought-transference', *Proceedings of the Society for Psychical Research*, 2 (1884), 2–3.

5. For more on crisis apparitions and the census of hallucinations, see D. Blum, *Ghost Hunters: William James and the Search for Psychic Proof of Life After Death*, Penguin (2006).

6. For an account of the Lady Wonder episode, see M. Gardner, *Fads and Fallacies in the Name of Science* (1957), pp. 351–2.

7. C. D. Broad, 'Introduction to Mr. Whately Carington's and Mr. Soal's papers', *Proceedings for the Society of Psychical Research*, 46 (1940–41), 27.

8. J. B. Rhine, *New Frontiers of the Mind*, Farrar & Rinehart (1937), p. 55.

9. J. B. Rhine, *Extra-Sensory Perception*, Boston Society for Psychic Research (1934). This is only the broadest outline of how Rhine conducted his tests: in fact, his methods varied from trial to trial. For a detailed account of his work, see S. Horn, *Unbelievable*, Ecco (2009). For Rhine's own account of his experiments in dice rolling, see his *The Reach of the Mind*, William Sloan (1947); Faber & Faber published the UK version of this book in 1948.

10. Rhine (Note 8), pp. 58, 165.

11. S. G. Soal and F. Bateman, *Modern Experiments in Telepathy*, Yale University Press (1954), pp. 104–19. On pp. 338–9, Soal and Bateman cite the opening sentences of Turing's argument 9 as an 'example of the outlook of the scientific materialist. According to this view, human beings are just material aggregates which behave according to the laws of quantum theory. The brain is, in this view, no more than an electrical switchboard of amazing complexity, but without an operator'. Clearly they had not read the rest of the article from which they were quoting.

12. Soal and Bateman (Note 11), pp. 123–5. See also S. G. Soal and W. Whately Carington, 'Experiments in non-sensory cognition', *Nature* (9 March 1940), 389–390. Soal is a fascinating and elusive figure. A putative adept at automatic writing, in 1927 he held the pencil in a series of séances whose result was a sequence of letters ostensibly written by the dead Oscar Wilde. In *Oscar Wilde from Purgatory*, her own account of the episode, Mrs. Travers Smith called Soal 'Mr. V.' and explained that he 'wrote with Mrs. T. S.'s hand resting on his. When she took her hand off, the pencil only tapped and did not continue . . . Mr. V. is a mathematical scholar and had no special interest in Oscar Wilde'. Subsequently, Soal himself published a rather Jamesian essay on the séances in which he revealed that the Wilde letters were cobbled together from books in his own possession. The account read in part: 'If, however, it should eventually turn out that in cases where the communicators are shown to be purely fictitious characters the supernormal selection of material is as varied and ingenious as in the apparently spiritistic cases, then we should have at least a presumption in favour of the view that in these latter also the supernormal selection may be the work of living minds'. H. Travers Smith, *Oscar Wilde from Purgatory*, T. Werner Laurie (1924), p. 7; S. G. Soal, 'Note on the "Oscar Wilde" script', *Journal of the Society for Psychical Research*, 23 (1926), 110–12.

13. Soal and Bateman (Note 11), p. 124.

14. Soal and Bateman (Note 11), p. 132.

15. Three of Soal's four agents were women. One, Rita Elliott, he also identified in his book as 'Mrs. S. G. Soal'. According to the General Register Office, however, Soal's wife, whom he married in 1942, was named Beatrice Potter.

16. Soal and Bateman (Note 11), pp. 134–7.

17. Soal and Bateman (Note 11), p. 139.

18. S. G. Soal and K. M. Goldney, 'Experiments in precognitive telepathy', *Proceedings of the Society for Psychical Research*, 47 (1942–45), 21–150.

19. C. D. Broad, 'The experimental establishment of precognition', *Philosophy*, 19(74) (November 1944), p. 261. In 1948 Soal's research received an even stronger endorsement from G. Evelyn Hutchinson, a professor of biology at Yale. Writing in *American Scientist*, Hutchinson characterized Soal's experiments as 'the most carefully conducted investigations of the kind ever to have been made' and quoted from T. S. Eliot's 'Burnt Norton':
    Time present and time past
    Are both perhaps present in time future,
    And time future contained in time past.
    'Summing up this general consideration', Hutchinson concluded, 'it would seem to the present writer that, either we must suppose that the experiments are to be taken at face value, and that reality is not exactly what it is generally supposed to be in scientific work . . . or we must reject the experiments and accept that, without being precognitive, we are justified by past experience in setting limits to what is possible'. G. Evelyn Hutchinson, 'Marginalia', *American Scientist*, 36(2) (April 1948), pp. 291–5.

20. 'Evidence of telepathy', *The Times* (2 September 1949), 2.

21. R. Braddon, *The Piddingtons: Their Exclusive Story*, Werner Laurie (1950).

22. 'Telepathy', *The Times* (15 September 1949), 5.
23. The Piddingtons' disinclination to be tested at the society's headquarters might well have been due to the treatment that the 'vaudeville telepath' Frederick Marion had received when he had submitted to similar testing a decade earlier. Soal had conducted the tests. His conclusion, published in *Nature* in 1938, was that Marion was 'hyperaesthetic'—that is, he possessed visual, auditory, and tactical acuity of an unusually high order, but no telepathic capacities. For instance, when he identified a face-down playing card as the queen of diamonds, what he was really recognizing 'was the actual piece of pasteboard which he had previously touched, and not the value on the face of the card'. Not surprisingly, Marion took umbrage at this diagnosis. S. G. Soal, 'Scientific testing of a vaudeville telepathist', *Nature* (26 March 1938), 565–6; Soal and Bateman (Note 11), p. 96.
24. Turing (1950), p. 450.
25. Turing (1950), p. 458.
26. Turing (1950), p. 458.
27. Here and following, I quote from 'Nature of spirit', Turing Digital Archive, AMT/C/29; where necessary, I have inserted commas to make the sentences easier to follow.
28. Hodges (1983), p. 45.
29. Letter from A. M. Turing to S. Turing (16 February 1930), Turing Digital Archive, AMT/K/1/20.
30. Turing (1950), p. 458.
31. M. Campbell-Kelly, 'Programming the Mark I: early programming activity at the University of Manchester', *Annals of the History of Computing*, 2(2) (April 1980), p. 136.
32. Turing (1950), p. 458.
33. W. James, 'The confidences of a "psychical researcher"', *The American Magazine* (October 1909), 580–9.
34. Turing (1950), p. 454.
35. Turing et al. (1952), p. 495.
36. Turing (1950), p. 458.
37. Turing (1947), p. 394.
38. Turing (*c*.1951), p. 475.
39. Turing (*c*.1951), p. 475.
40. Turing (1951), p. 484.
41. 'Nature of spirit' (Note 27).
42. C. Scott and P. Haskell, '"Normal" explanation of the Soal–Goldney experiments in extrasensory perception', *Nature*, 245 (7 September 1973), 52–3. For a detailed account of the manifold other ways in which Soal committed fraud, see C. E. M. Hansel, *The Search for Psychic Power: ESP & Parapsychology Revisited*, Prometheus (1989), pp. 100–15. For an investigation of the use and abuse of statistical methods in parapsychology, see J. E. Alcock, *Parapsychology: Science or Magic?*, Pergamon (1981).

CHAPTER 33 PIONEER OF ARTIFICIAL LIFE (BODEN)

1. Turing (1952).
2. This chapter draws on Chapter 15 of M. A. Boden's *Mind as Machine: a History of Cognitive Science*, Clarendon Press (2006). Parts of it were given as the Fifth Turing Memorial Lecture (2009) at Bletchley Park.
3. William Blake, *Songs of Experience* (1794).
4. Boden (Note 2), Sections 4.i–ii, 10.i.f, 12.i.b, 16.ii.
5. See J. A. Anderson and E. Rosenfeld (eds), *Talking Nets: an Oral History of Neural Networks*, MIT Press (1998), p. 118.
6. Turing (1952), p. 522.
7. Turing (1952), Abstract.
8. E. T. Brewster, *Natural Wonders Every Child Should Know*, Grosset & Dunlap (1912).
9. M. A. Boden, 'D'Arcy Thompson: a grandfather of A-Life', in P. N. Husbands, O. Holland, and M. W. Wheeler (eds), *The Mechanical Mind in History*, MIT Press (2008), 41–60.
10. D. W. Thompson, *On Growth and Form*, Cambridge University Press (1917) (expanded 2nd edn published 1942).

11. Turing (1948).

12. Turing (1952), p. 521.

13. Turing (1952), p. 552; Turing's hand-drawings appeared on pp. 546 and 551.

14. F. Jacob and J. Monod, 'Genetic regulatory mechanisms in the synthesis of proteins', *Journal of Molecular Biology*, 3 (1961), 318–56.

15. See Boden (Note 2), Chapters 4.viii and 15.

16. See C. G. Langton, 'Artificial life', in C. G. Langton (ed.), *Artificial Life (Proceedings of an Interdisciplinary Workshop on the Synthesis and Simulation of Living Systems, September 1987)*, Addison-Wesley (1989), 1–47; revised version in M. A. Boden (ed.), *The Philosophy of Artificial Life*, Oxford University Press (1996), 39–94.

17. Thompson (Note 10), p. 1026.

18. Thompson (Note 10), p. 1090.

19. Turing (1950).

20. S. Wolfram, 'Statistical mechanics of cellular automata', *Review of Modern Physics*, 55 (1983), 601–44; 'Cellular automata as models of complexity', *Nature*, 311 (1984), 419–24; 'Universality and complexity in cellular automata', *Physica D*, 10 (1984), 1–35; *Theory and Applications of Cellular Automata*, World Scientific (1986).

21. Turing (1950), p. 463.

22. See C. G. Langton, 'Life at the edge of chaos', in C. G. Langton, C. Taylor, J. D. Farmer, and S. Rasmussen (eds), *Artificial Life II*, Addison-Wesley (1992), 41–91; S. A. Kauffman, *The Origins of Order: Self-Organization and Selection in Evolution*, Oxford University Press (1993), pp. 29ff.

23. See Kauffman (Note 22); R. Sole and B. C. Goodwin, *Signs of Life: How Complexity Pervades Biology*, Basic Books (2000).

24. Turing (1948), p. 431.

25. Turing (1950), pp. 460, 463.

26. Turing (1950), p. 446.

27. G. Turk, 'Generating textures on arbitrary surfaces using reaction–diffusion', *Computer Graphics*, 25 (1991), 289–98.

28. B. C. Goodwin and P. T. Saunders (eds), *Theoretical Biology: Epigenetic and Evolutionary Order from Complex Systems*, Edinburgh University Press (1989); Kauffman (Note 22).

29. See N. H. Shubin and P. Alberch, 'A morphogenetic approach to the origin and basic organization of the tetrapod limb', *Evolutionary Biology*, 20 (1986), 319087; G. F. Oster, N. Shubin, J. D. Murray, and P. Alberch, 'Evolution and morphogenetic rules: the shape of the vertebrate limb in ontogeny and phylogeny', *Evolution*, 42 (1988), 862–84; Turing (1952), p. 39.

30. C. H. Waddington, *Organisers and Genes*, Cambridge University Press (1940).

31. See B. C. Goodwin and L. E. H. Trainor, 'Tip and whorl morphogenesis in *Acetabularia* by calcium-regulated strain fields', *Journal of Theoretical Biology*, 117 (1985), 79–106; C. Briere and B. C. Goodwin, 'Geometry and dynamics of tip morphogenesis in *Acetabularia*', *Journal of Theoretical Biology*, 131 (1988), 461–75; B. C. Goodwin and C. Briere, 'A mathematical model of cyto-skeletal dynamics and morphogenesis in *Acetabularia*', in D. Menzel (ed.), *The Cytoskeleton of the Algae*, CRC Press (1992), pp.219–38; B. C. Goodwin, *How the Leopard Changed its Spots: the Evolution of Complexity*, Weidenfeld & Nicolson (1994), pp. 88–103; a new Preface by the author is added in the 2nd edn (Princeton University Press, 2001).

32. Turing (1952), p. 556.

33. Goodwin (Note 31), p. 94.

34. Ironically, much the same thing happened to the founder of morphology, Johann von Goethe. In 1853 his work on this topic was lavishly praised by the leading scientist of the time, Hermann von Helmholtz, who wrote 'On Goethe's scientific researches' (transl. by H. W. Eve, in H. Helmholtz, *Popular Lectures on Scientific Subjects*, new edn, Longmans Green (1884), pp.29–52). Helmholtz said that Goethe's work offered 'ideas of infinite fruitfulness' and had earned 'immortal renown'. But its renown, immortal or not, very soon went into hibernation, being overshadowed by the publication of Darwin's *On the Origin of Species* only 6 years later. Biologists' interests turned from timeless questions about form to historical questions about evolution.

35. See Turing (1952), p. 68.
36. P. T. Saunders (ed.), *Morphogenesis: Collected Works of A. M. Turing*, Vol. 3, Elsevier Science (1992).
37. 'Countless times', because citation indices do not normally consult the bibliographies in books, often omit non-scientific journals, and never search the 'public' media.
38. See Turing (1952), p. 557.
39. Saunders (Note 36).
40. B. Richards, 'The morphogen theory of phyllotaxis: III—a solution of the morphogenetic equations for the case of spherical symmetry', in Saunders (Note 36).
41. *Turing Digital Archive*, AMT/C/27 (http://www.turingarchive.org).
42. H. O'Connell and M. Fitzgerald, 'Did Alan Turing have Asperger's syndrome?', *Irish Journal of Psychiatric Medicine*, 20(1) (2003), 28–31, p. 29.
43. Quoted in W. Allaerts, 'Fifty years after Alan M. Turing: an extraordinary theory of morphogenesis', *Belgian Journal of Zoology*, 133(1) (2003), 3–14, p. 8.

CHAPTER 34 TURING'S THEORY OF MORPHOGENESIS (WOOLLEY, BAKER, AND MAINI)

1. Turing (1952).
2. E. Kreyszig, *Advanced Engineering Mathematics* Wiley-India (2007).
3. J. D. Murray, *Mathematical Biology II: Spatial Models and Biomedical Application*, Springer-Verlag (2003).
4. A. Gierer and H. Meinhardt, 'A theory of biological pattern formation', *Biological Cybernetics*, 12 (1972), 30–9.
5. V. Castets, E. Dulos, J. Boissonade, and P. De Kepper, 'Experimental evidence of a sustained standing Turing-type nonequilibrium chemical pattern', *Physical Review Letters*, 64 (1990), 2953–6.
6. Q. Ouyang and H. L. Swinney, 'Transitions from a uniform state to hexagonal and striped Turing patterns', *Nature*, 352 (1991), 610–12.
7. I. Lengyel and I. R. Epstein, 'Modeling of Turing structures in the chlorite–iodide–malonic acid–starch reaction system', *Science*, 251 (1991), 650–2.
8. S. Sick, S. Reinker, J. Timmer, and T. Schlake, 'WNT and DKK determine hair follicle spacing through a reaction-diffusion mechanism', *Science*, 314 (2006), 1447–50.
9. A. D. Economou, A. Ohazama, T. Porntaveetus, P. T. Sharpe, S. Kondo, M. A. Basson, A. Gritli-Linde, M. T. Cobourne, and J. B. A. Green, 'Periodic stripe formation by a Turing mechanism operating at growth zones in the mammalian palate', *Nature Genetics*, 44 (2012), 348–51.
10. S. W. Cho, S. Kwak, T. E. Woolley, M. J. Lee, E. J. Kim, R. E. Baker, H. J. Kim, J. S. Shin, C. Tickle, P. K. Maini, and H. S. Jung, 'Interactions between Shh, Sostdc1 and Wnt signaling and a new feedback loop for spatial patterning of the teeth', *Development*, 138 (2011), 1807–16.
11. R. Sheth, L. Marcon, M. F. Bastida, M. Junco, L. Quintana, R. Dahn, M. Kmita, J. Sharpe, and M. A. Ros, 'Hox genes regulate digit patterning by controlling the wavelength of a Turing-type mechanism', *Science*, 338 (2012), 1476–80.
12. S. Kondo and R. Asai, 'A reaction-diffusion wave on the skin of the marine angelfish *Pomacanthus*', *Nature*, 376 (1995), 765–8.
13. J. D. Murray, 'Parameter space for Turing instability in reaction diffusion mechanisms: a comparison of models', *Journal of Theoretical Biology*, 98 (1982), 143.
14. J. Bard and I. Lauder, 'How well does Turing's theory of morphogenesis work?', *Journal of Theoretical Biology*, 45 (1974), 501–31.
15. E. J. Crampin, E. A. Gaffney, and P. K. Maini, 'Reaction and diffusion on growing domains: scenarios for robust pattern formation', *Bulletin of Mathematical Biology*, 61 (1999), 1093–120.
16. E. A. Gaffney and N. A. M. Monk, 'Gene expression time delays and Turing pattern formation systems', *Bulletin of Mathematical Biology*, 68 (2006), 99–130.
17. T. E. Woolley, R. E. Baker, E. A. Gaffney, P. K. Maini, and S. Seirin-Lee, 'Effects of intrinsic stochasticity on delayed reaction-diffusion patterning systems', *Physical Review E*, 85 (2012), 051914.

18. A. Nakamasu, G. Takahashi, A. Kanbe, and S. Kondo, 'Interactions between zebrafish pigment cells responsible for the generation of Turing patterns', *Proceedings of the National Academy of Sciences of the United States of America*, 106 (2009), 8429–34.

19. T. E. Woolley, R. E. Baker, E. A. Gaffney, and P. K. Maini, 'Stochastic reaction and diffusion on growing domains: Understanding the breakdown of robust pattern formation', *Physical Review E*, 84 (2011), 046216.

20. G. E. P. Box and N. R. Draper, *Empirical Model-building and Response Surfaces*, Wiley (1987).

21. Turing (1952).

CHAPTER 35 RADIOLARIA: VALIDATING THE TURING THEORY (RICHARDS)

1. D. W. Thompson, *On Growth and Form*, Cambridge University Press (1917) (expanded 2nd. edn published 1942).

2. Turing (1952).

3. B. Richards, 'The morphogenesis of Radiolaria', MSc Thesis, University of Manchester (1954).

4. P. T. Saunders, *Morphogenesis*, North-Holland (1992).

CHAPTER 36 INTRODUCING TURING'S MATHEMATICS (WHITTY AND WILSON)

1. J. L. Britton (ed.), *The Collected Works of A. M. Turing: Pure Mathematics*, North-Holland (1992).

2. I. J. Good, 'Studies in the history of probability and statistics, XXXVII: A. M. Turing's statistical work in World War II', *Biometrika*, 66(2) (1979), 393–6.

3. A. M. Turing, 'On the Gaussian error function', King's College Fellowship Dissertation (1935).

4. S. L. Zabell, 'Alan Turing and the central limit theorem', *American Mathematical Monthly*, (June–July 1995), 483–94.

5. W. Burnside, *The Theory of Groups of Finite Order*, 2nd edn, Cambridge University Press (1911).

6. Britton (Note 1).

7. P. S. Novikov, 'On the algorithmic unsolvability of the word problem in group theory', *Trudy Matematicheskogo Instituta imeni V. A. Steklova*, 44, Academy of Sciences of the USSR (1955), 3–143.

8. J. Stillwell, 'The word problem and the isomorphism problem for groups', *Bulletin of the American Mathematical Society*, 6(1) (1982), 33–56.

9. Britton (Note 1).

10. A. E. Ingham, *The Distribution of Prime Numbers*, Cambridge Mathematical Tracts 30, Cambridge University Press (1932).

11. D. Zagier, 'The first 50 million prime numbers', *Mathematical Intelligencer*, (1997), 221–4.

12. B. Riemann, 'Ueber die Anzahl der Primzahlen unter einer gegebenen Grösse' [On the number of primes less than a given magnitude], *Monatsberichte der Berliner Akademie*, (November 1859), 671–80.

13. E. C. Titchmarsh, 'The zeros of the Riemann zeta-function', *Proceedings of the Royal Society of London. Series A, Mathematical and Physical Sciences*, 157 (1936), 261–3.

14. Turing (1943).

15. Turing (1953a).

16. A. R. Booker, 'Turing and the Riemann hypothesis', *Notices of the American Mathematical Society*, 53(10) (2006), 1208–11.

17. D. A. Hejhal and A. M. Odlyzko, 'Alan Turing and the Riemann zeta function', in S. B. Cooper and J. van Leeuwen (eds), *Alan Turing—His Work and Impact*, Elsevier (2013), 265–79.

CHAPTER 37 DECIDABILITY AND THE ENTSCHEIDUNGSPROBLEM (WHITTY)

1. Turing (1936).

2. J. J. Gray, *The Hilbert Challenge*, Oxford University Press (2000).

3. P. Smith, *An Introduction to Gödel's Theorems*, Cambridge University Press (2007).

4. B. J. Copeland, 'Computable numbers: a guide', *The Essential Turing*, pp. 5–57.
5. J. P. Jones, H. Wada, D. Sato, and D. Wiens, 'Diophantine representation of the set of prime numbers', *American Mathematical Monthly*, 83 (1976), 449–64.

CHAPTER 38 BANBURISMUS REVISITED: DEPTHS AND BAYES (SIMPSON)

1. Hinsley & Stripp (1993).
2. R. Whelan, 'The use of Hollerith equipment in Bletchley Park', NA, HW25/22.
3. Sybil Cannon (nee Griffin), private communication.
4. Whelan described this visit with gusto (Note 2).
5. C. H. O'D. Alexander, 'Cryptographic history of work on the German Naval Enigma', (*c*.1945), NA, HW25/1 and *The Turing Archive for the History of Computing* (http://www.AlanTuring. net/alexander_naval_enigma).
6. What I call 'alignment', Alexander called 'distance'.
7. Alexander (Note 5).
8. Alexander (Note 5).
9. I. J. Good, *Probability and the Weighing of Evidence*, Griffin (1950).
10. Christine Brose (nee Ogilvie-Forbes), private communication.
11. A. M. Turing, 'Mathematical theory of ENIGMA machine', (*c*.1940) (also known as 'Turing's treatise on the Enigma'), NA, HW25/3 and *The Turing Archive for the History of Computing* (http://www. AlanTuring.net/profs_book).
12. A. M. Turing, 'Visit to National Cash Register Corporation of Dayton, Ohio' (1942), *The Turing Archive for the History of Computing* (http://www.AlanTuring.net/turing_ncr).
13. Joan Clarke in Hinsley & Stripp (1993).
14. Turing (Note 11).
15. Brose (Note 10).
16. Eileen Johnson (nee Plowman), private communication.
17. C. H. O'D. Alexander, 'The factor method' (*c*.1945), NA, HW43/26, Study 1.
18. E. H. Simpson, 'Bayes at Bletchley Park', *Significance* (June 2010).

CHAPTER 39 TURING AND RANDOMNESS (DOWNEY)

1. É. Borel, 'Les probabilités dénombrables et leurs applications arithmétiques', *Rendiconti del Circolo Matematico di Palermo*, 27 (1909), 247–71.
2. Mathematically, saying that a number is normal means that the collection of absolutely normal numbers has a 'Lebesgue measure' of 1; this corresponds to saying that if we throw a dart at the real line, then with probability 1 it would hit an absolutely normal number.
3. A. Copeland and P. Erdős, 'Note on normal numbers', *Bulletin of the American Mathematical Society*, 52 (10) (1946), 857–60. Their proof relies on the 'density' of primes in base 10.
4. Turing (*c*. 1936).
5. C. Schnorr, 'A unified approach to the definition of a random sequence', *Mathematical Systems Theory*, 5 (1971), 246–58.
6. It turns out that if this feature is not allowed and you can only bet in discrete amounts with a minimum bet, a completely different notion of randomness comes about called 'integer-valued randomness'. It is a question of physics as to whether this is the correct notion of randomness for the universe, since it depends upon whether space–time is a continuum or discrete.
7. Turing (*c*. 1936).
8. V. Becher, S. Figueira, and R. Picchi, 'Turing's unpublished algorithm for normal numbers', *Theoretical Computer Science*, 377 (2007), 126–38.
9. V. Becher, 'Turing's normal numbers: towards randomness', in S. B. Cooper, A. Dawar, and B. Löwe (eds), *CiE 2012*, Springer Lecture Notes in Computer Science 7318, Springer (2012), 35–45.

10. H. Lebesgue, 'Sur certaines démonstrations d'existence', *Bulletin de la Société Mathématique de France*, 45 (1917), 132–44; W. Sierpiński, 'Démonstration élémentaire du théorème de M. Borel sur les nombres absolument normaux et détermination effective d'un tel nombre', *Bulletin de la Société Mathématique de France*, 45 (1917), 127–32.

11. V. Becher and S. Figueira. 'An example of a computable absolutely normal number', *Theoretical Computer Science*, 270 (2002), 947–58.

12. Turing (1936).

13. See the chapters by R. Soare, A. Nerode, and W. Sieg in Downey, R. (ed.), *Turing's Legacy*, Cambridge University Press (2015).

14. Turing (1950).

15. Turing (1950).

16. This is easily seen for polynomials of one variable. For example, a non-zero cubic polynomial expression such as $ax^3 + bx^2 + cx + d$ can have at most three roots, so it can take the value zero at no more than three values of x. It is unlikely that a random choice would select one of these three values!

17. B. J. Copeland and O. Shagrir, 'Turing versus Gödel on computability and the mind', in B. J. Copeland, C. Posy, and O. Shagrir (eds), *Computability: Gödel, Turing, Church and Beyond*, MIT Press (2013), 1–33.

18. Turing (c.1951).

19. R. Impagliazzo and A. Wigderson, 'P = BPP if E requires exponential circuits: derandomizing the XOR lemma', in *Proceedings of the 29th Annual ACM Symposium on the Theory of Computing (STOC '97)*, ACM (1997), 220–9.

20. M. Agrawal, N. Kayal, and N. Saxena, 'PRIMES is in P', *Annals of Mathematics*, 160 (2004), 781–93.

21. C. Schnorr and H. Stimm, 'Endliche Automaten und Zufallsfolgen', *Acta Informatica*, 1 (1972), 345–59.

22. E. Mayordomo, 'Construction of an absolutely normal real number in polynomial time', Preprint (November 2012); V. Becher, P. Heiber, and T. A. Slaman, 'A polynomial-time algorithm for computing absolutely normal numbers', *Information and Computation*, 232 (2013), 1–9.

23. For general introductions to the area of randomness and its relation to computability, see R. Downey and D. Hirschfeldt, *Algorithmic Randomness and Complexity*, Springer-Verlag (2010); A. Nies, *Computability and Randomness*, Oxford University Press (2009). For a more informal discussion as to the general theory of randomness, see H. Zenil (ed.), *Randomness Through Computation: Some Answers, More Questions*, World Scientific (2011); this has essays by leading experts in the area. For more on the mathematical (particularly, logical) developments stemming from Turing's work, see R. Downey (ed.), *Turing's Legacy*, Cambridge University Press (2014).

CHAPTER 40 TURING'S MENTOR, MAX NEWMAN (GRATTAN-GUINNESS)

1. The history of modern mathematical analysis is well covered; see, for example, N. H. Jahnke (ed.), *A History of Analysis*, American Mathematical Society (2003). A similar story obtains for complex-variable analysis; see U. Bottazzini, *The Higher Calculus. A History of Real and Complex Analysis from Euler to Weierstrass*, Springer (1986).

2. See G. H. Moore, *Zermelo's Axiom of Choice*, Springer (1982).

3. See C. S. Roero and E. Luciano, 'La scuola di Giuseppe Peano', in C. S. Roero (ed.), *Peano e la sua Scuola. Fra Matematica, Logica e Interlingua. Atti del Congresso Internazionale di Studi (Torino, 2008)*, Deputazione Subalpina di Storia Patria (2010), 1–212.

4. See I. Grattan-Guinness, *The Search For Mathematical Roots, 1870–1940. Logics, Set Theories and the Foundations of Mathematics from Cantor through Russell to Gödel*, Princeton University Press (2000), Chapters 8 and 9.

5. See V. Peckhaus, 'Hilbert, Zermelo und die Institutionalisierung der mathematischen Logik in Deutschland', *Berichte zur Wissenschaftsgeschichte*, 15 (1992), 27–38; W. Sieg, 'Hilbert's programs: 1917–1922', *Bulletin of Symbolic Logic*, 5 (1999), 1–44.

6. See W. Aspray, 'Oswald Veblen and the origins of mathematical logic at Princeton', in T. Drucker (ed.), *Perspectives on the History of Mathematical Logic*, Birkhäuser (1991), 54–70.

7. See I. Grattan-Guinness, 'Re-interpreting "λ": Kempe on multisets and Peirce on graphs, 1886–1905', *Transactions of the C. S. Peirce Society*, 38 (2002), 327–50.

8. See M. Hallett, *Cantorian Set Theory and Limitation of Size*, Clarendon Press (1984), Chapter 8.

9. A. Tarski, *Introduction to Logic and to the Methodology of the Deductive Sciences* (transl. O. Helmer), 1st edn, Oxford University Press (1941), 125–30.

10. See I. Grattan-Guinness, 'The reception of Gödel's 1931 incompletability theorems by mathematicians, and some logicians, up to the early 1960s', in M. Baaz, C. H. Papadimitriou, H. W. Putnam, D. S. Scott, and C. L. Harper (eds), *Kurt Gödel and the Foundations of Mathematics. Horizons of Truth*, Cambridge University Press (2011), 55–74.

11. Hodges (1983), pp. 109–14.

12. Turing (1936).

13. See I. Grattan-Guinness, 'The mentor of Alan Turing: Max Newman (1897–1984) as a logician', *Mathematical Intelligencer*, 35(3) (September 2013), 54–63.

14. See also B. J. Copeland, 'From the *Entscheidungsproblem* to the Personal Computer', in M. Baaz, C. H. Papadimitriou, H. W. Putnam, D. S. Scott, and C. L. Harper (eds), *Kurt Gödel and the Foundations of Mathematics. Horizons of Truth*, Cambridge University Press (2011), 151–84.

15. The Newman Archive is in St John's College, Cambridge; thanks to David Anderson much of it is available in digital form at http://www.cdpa.co.uk/Newman/. Individual items are cited in the style 'NA, [box] a- [folder] b- [document] c'; here 2–12–3.

16. A Mathematical Tripos course in 'logic' was launched in 1944 by S. W. P. Steen; the Moral Sciences Tripos continued to offer its long-running course on the more traditional parts of 'logic'.

17. For Turing's teaching, see Hodges (1983), pp. 153, 157, and the Faculty Board minutes for 29 May 1939.

18. M. H. A. Newman, 'Stratified systems of logic', *Proceedings of the Cambridge Philosophical Society*, 39 (1943), 69–83; M. H. A. Newman and A. Turing, 'A formal theorem in Church's theory of types', *Journal of Symbolic Logic*, 7 (1943), 28–33.

19. M. H. A. Newman, 'On theories with a combinatorial definition of "equivalence"', *Annals of Mathematics*, 43 (1942), 223–43. For a discussion see J. R. Hindley, 'M. H. Newman's typability algorithm for lambda-calculus', *Journal of Logic and Computation*, 18 (2008), 229–38.

20. W. Newman, 'Max Newman—mathematician, codebreaker, and computer pioneer', in Copeland et al. (2006), 176–88.

21. See J. F. Adams, 'Maxwell Herman Alexander Newman, 7 February 1897–22 February 1984', *Biographical Memoirs of Fellows of the Royal Society*, 31 (1985), 436–52.

22. M. Gardiner, *A Scatter of Memories*, Free Association Books (1988), pp. 61–8.

23. The Wirtinger letter is in the Newman Archive (Note 15), item 2–1–2.

24. K. Gödel, 'Über formal unentscheidbare Sätze der Principia Mathematica und verwandter Systeme I', *Monatshefte für Mathematik und Physik*, 38 (1931), 173–98.

25. For the Vienna Circle see F. Stadler, *The Vienna Circle*, Springer (2001). For Hahn see K. Sigmund, 'A philosopher's mathematician: Hans Hahn and the Vienna Circle', *Mathematical Intelligencer*, 17(4) (1995), 16–19.

26. M. H. A. Newman, 'On approximate continuity', *Transactions of the Cambridge Philosophical Society*, 23 (1923), 1–18. For the context see F. A. Medvedev, *Scenes from the History of Real Functions* (transl. R. Cooke), Birkhäuser (1991).

27. M. H. A. Newman, 'The foundations of mathematics from the standpoint of physics', manuscript dated 1923, Newman Archive (Note 15), item F 33.1.

28. D. Hilbert, 'Die logischen Grundlagen der Mathematik', *Mathematische Annalen*, 88 (1922), 151–65 (also in *Gesammelte Abhandlungen*, Vol. 3, Springer (1935), 178–91).

29. L. E. J. Brouwer, 'Begründung der Mengenlehre unabhängig vom logischen Satz von ausgeschlossenen Dritten' Erster Teil, Allgemeine Mengenlehre, *Verhandlingen der Koninklijke Akademie van Wetenschappen te Amsterdam*, 12(5) (1918), 1–43; Zweite Teil, Theorie der Punktmengen, 7 (1919), 1–33; also in *Collected Works*, Vol. 1, North-Holland (1975), 150–221; C. H. H. Weyl. 'Über

die neue Grundlagenkrise der Mathematik', *Mathematische Zeitschrift*, 10 (1921), 39–79 (also in *Gesammelte Abhandlungen*, Vol. 2, Springer (1968), 143–80).

30. See the Penrose Papers, University College London Archives, especially boxes 20–1 and 26–8.
31. See H. Harris, 'Lionel Sharples Penrose. 1898–1972', *Biographical Memoirs of Fellows of the Royal Society*, 19 (1973), 521–61 (also in *Journal of Medical Genetics*, 11 (1974), 1–24).
32. B. A. W. Russell, *The Analysis of Matter*, Kegan Paul (1927).
33. M. H. A. Newman, 'Mr. Russell's "Causal theory of perception" ', *Mind* (new ser.), 37 (1928), 137–48.
34. See also W. Demopoulos and M. Friedman, 'The concept of structure in *The Analysis of Matter*', in A. D. Irvine (ed.), *Bertrand Russell: Language, Knowledge and the World*, Routledge (1999), 277–94.
35. I. Grattan-Guinness, 'Logic, topology and physics: points of contact between Bertrand Russell and Max Newman', *Russell* (new ser.), 32 (2012), 5–29.
36. M. H. A. Newman, 'Alan Mathison Turing, 1912–1954', *Biographical Memoirs of Fellows of the Royal Society*, 1 (November 1955), 253–263.
37. Information comes from the Royal Society Archives, and the Newman Archive (Note 15), items 2–15–10 to –13.
38. There seems to be no Newman material in the Frank Plumpton Ramsey papers at the University of Pittsburgh.
39. On this ironic situation see Grattan-Guinness (Note 4), pp. 327–8, 388–91, and 592–3. 'I remember talking to you about Gödel's proof soon after it appeared', Newman recalled to Russell on 25 September 1966, Newman Archive (Note 15), item 2–15–11.
40. On Turing's circle of Cambridge connections during and after graduation, see Hodges (1983), Chapter 4.

CHAPTER 41 IS THE WHOLE UNIVERSE A COMPUTER? (COPELAND, SPREVAK, AND SHAGRIR)

1. Turing (1950), p. 446.
2. B. J. Copeland and R. Sorensen, 'Multiple realizability: the Copeland–Sorensen optical universal computing machine', in B. J. Copeland, G. Piccinini, D. Proudfoot, and O. Shagrir (eds), *The Philosophy of Computing* (forthcoming).
3. Turing (1936), p. 59.
4. L. Wittgenstein, *Remarks on the Philosophy of Psychology*, Vol. 1, Blackwell (1980), section 1096.
5. Turing (1948), p. 416.
6. Turing (1947), pp. 387, 391.
7. Turing (1950), p. 444.
8. A. M. Turing, *Programmers' Handbook for Manchester Electronic Computer Mark II*, p. 1; a digital facsimile is in *The Turing Archive for the History of Computing* (http://www.AlanTuring.net/ programmers_handbook).
9. We canvas alternative answers in our forthcoming chapter 'Zuse's thesis, Gandy's thesis, and Penrose's thesis', in M. Cuffaro and S. Fletcher (eds), *Physical Perspectives on Computation, Computational Perspectives on Physics*, Cambridge University Press.
10. A. M. Turing, 'Proposed electronic calculator', in Copeland et al. (2005), p. 386
11. J. Searle, 'Is the brain a digital computer?', *Proceedings and Addresses of the American Philosophical Association*, 64 (1990), 21–37 (pp. 25–27); J. Searle, *The Rediscovery of the Mind*, MIT Press (1992), pp. 205–9; H. Putnam, *Representation and Reality*, MIT Press (1988), pp. 121–5.
12. B. J. Copeland, 'What is computation?', *Synthese*, 108 (1996), 335–59; D. J. Chalmers, 'Does a rock implement every finite-state automaton?', *Synthese*, 108 (1996), 309–33.
13. J. von Neumann, 'The general and logical theory of automata', in A. H. Taub (ed.), *Collected Works*, Vol. 5, Pergamon Press (1963).
14. B. J. Copeland and G. Sommaruga, 'The stored-program universal computer: did Zuse anticipate Turing and von Neumann?' in G. Sommaruga and T. Strahm (eds), *Turing's Revolution*, Birkhauser/ Springer (2015), pp. 99–100.

15. K. Zuse, 'Some remarks on the history of computing in Germany', in Metropolis et al. (1980), p. 611.

16. Konrad Zuse interviewed by Uta Merzbach in 1968 (Computer Oral History Collection, Archives Centre, National Museum of American History, Washington, DC).

17. Copeland and Sommaruga (Note 14).

18. Zuse in conversation with Brian Carpenter at CERN on 17 June 1992 (Copeland is grateful to Carpenter for sending some brief notes on the conversation which Carpenter made at the time); K. Zuse, *Der Computer—Mein Lebenswerk* [The computer—my life's work], 4th edn, Springer (2007), p. 101.

19. Donald Davies interviewed by Christopher Evans in 1975 ('The pioneers of computing: an oral history of computing', Science Museum, London; copyright Board of Trustees of the Science Museum).

20. Robin Gandy and Wilfried Sieg view CA as parallel computers that differ from Turing machines in allowing changes in *arbitrarily many* cells, whereas in a Turing machine only the content of one of the tape's cells can be altered at each step; R. Gandy, 'Church's thesis and principles of mechanisms', in J. Barwise, H. J. Keisler and K. Kunen (eds), *The Kleene Symposium*, North-Holland (1980), 123–48; W. Sieg, 'On computability', in A. Irvine (ed.), *Handbook of the Philosophy of Mathematics*, Elsevier (2009), 535–630.

21. S. Wolfram, *Theory and Applications of Cellular Automata*, World Scientific (1986), p. 8.

22. For more on this, see W. Poundstone, *The Recursive Universe*, William Morrow & Company (1985).

23. K. Zuse, *Rechnender Raum*, Friedrich Vieweg & Sohn (1969).

24. 'Looking at life with Gerardus 't Hooft'. *Plus Magazine* (January 2002) (https://plus.maths.org/content/looking-life-gerardus-t-hooft).

25. For a good summary, see J. D. Bekenstein, 'Information in the holographic universe', *Scientific American*, 17 (April 2007), 66–73.

26. See M. Moyer, 'Is space digital?' *Scientific American*, 23 (August 2014), 104–11.

27. A. Cho, 'Controversial test finds no sign of a holographic universe', *Science*, 350 (2015), 1303.

28. M. Tegmark, *Our Mathematical Universe*, Knopf (2014).

29. For a critical discussion of Tegmark's proposal, see the commentary by Scott Aaronson at http://www.scottaaronson.com/blog/?p=1753.

30. L. Wittgenstein, *Tractatus Logico-Philosophicus*, Kegan Paul, Trench and Trubner (1922), proposition 7.

31. A. Church, On the concept of a random sequence, *Bulletin of the American Mathematical Society*, 46 (1940), 130–5.

32. For a detailed examination of Turing's various formulations of his thesis, see B. J. Copeland, 'The Church–Turing thesis', in E. Zalta (ed.), *The Stanford Encyclopedia of Philosophy* (http://www.plato.stanford.edu/entries/church-turing).

33. A. Church, 'An unsolvable problem of elementary number theory', *American Journal of Mathematics*, 58 (1936), 345–63; Turing (1936), pp. 88–90.

34. See S. C. Kleene, 'Origins of recursive function theory', *Annals of the History of Computing*, 3 (1981), 52–67 (on pp. 59, 61); K. Gödel, 'Some basic theorems on the foundations of mathematics and their implications' (1951), in S. Feferman et al. (eds), *Collected Works*, Vol. 3, Oxford University Press (1995), pp. 304–5.

35. Searle mistakenly calls this 'Church's thesis': J. Searle, *The Rediscovery of the Mind*, MIT Press (1992), pp. 200–1; see also J. Searle, *The Mystery of Consciousness*, New York Review (1997), p. 87.

36. B. J. Copeland, 'The broad conception of computation', *American Behavioral Scientist*, 40 (1997), 690–716.

37. S. Guttenplan, *A Companion to the Philosophy of Mind*, Blackwell (1994), p. 595.

38. P. M. Churchland and P. S. Churchland, 'Could a machine think?'. *Scientific American*, 262 (January 1990), 26–31 (on p. 26).

39. M. B. Pour-El, and I. Richards 'The wave equation with computable initial data such that its unique solution is not computable', *Advances in Mathematics*, 39 (1981), 215–39.

40. I. Stewart, 'Deciding the undecidable', *Nature*, 352 (1991), 664–5; B. J. Copeland, 'Even Turing machines can compute uncomputable functions', in C. Calude, J. Casti, and M. Dinneen (eds), *Unconventional Models of Computation*, Springer-Verlag (1998), pp. 150–64; B. J. Copeland, 'Super Turing-machines', *Complexity*, 4 (1998), 30–2; B. J. Copeland, 'Accelerating Turing machines', *Minds and Machines*, 12

(2002), 281–300. The term 'accelerating Turing machine' was introduced by Copeland in lectures in 1997. The variant term 'accelerated Turing machine' (see e.g. C. S. Calude and L. Staiger, 'A note on accelerated Turing machines', *Centre for Discrete Mathematics and Theoretical Computer Science Research Reports* (2009) (http://hdl.handle.net/2292/3857); L. G. Fearnley, 'On accelerated Turing machines', Honours Thesis in Computer Science (2009), University of Auckland; P. H. Potgieter and E. Rosinger, 'Output concepts for accelerated Turing machines', *Centre for Discrete Mathematics and Theoretical Computer Science Research Reports* (2009) (http://hdl.handle.net/2292/3858)) originated in Copeland's 'Even Turing machines can compute uncomputable functions'.

41. B. A. W. Russell, *Our Knowledge of the External World as a Field for Scientific Method in Philosophy*, Open Court (1915), pp. 172–3.
42. B. J. Copeland and O. Shagrir, 'Do accelerating Turing machines compute the uncomputable?' *Minds and Machines*, 21 (2011), 221–39.
43. The conference papers were published in 1990: I. Pitowsky, 'The physical Church thesis and physical computational complexity', *Iyyun*, 39 (1990), 81–99.
44. H. Andréka, I. Németi, and P. Németi, 'General relativistic hypercomputing and foundation of mathematics', *Natural Computing*, 8 (2009), 499–516; I. Németi and G. Dávid, 'Relativistic computers and the Turing barrier', *Applied Mathematics and Computation*, 178 (2006), 118–42; G. Etesi and I. Németi, 'Non-Turing computations via Malament–Hogarth space-times', *International Journal of Theoretical Physics*, 41 (2002), 341–70. David Malament (in private communications) and Mark Hogarth have described essentially similar setups: M. L. Hogarth, 'Does general relativity allow an observer to view an eternity in a finite time?', *Foundations of Physics Letters*, 5 (1992), 173–81, M. L. Hogarth, 'Non-Turing computers and non-Turing computability', *PSA: Proceedings of the Biennial Meeting of the Philosophy of Science Association*, 1 (1994), 126–38.
45. Andréka, Németi, and Németi (Note 44), p. 501.
46. Andréka, Németi, and Németi (Note 44), p. 511.
47. M. Bunge and R. Ardila, *Philosophy of Psychology*, Springer-Verlag (1987), p. 109.
48. D. Deutsch, 'Quantum theory, the Church–Turing principle and the universal quantum computer', *Proceedings of the Royal Society of London. Series A, Mathematical and Physical Sciences*, 400 (1985), 97–117 (p. 99).
49. R. Penrose, *Shadows of the Mind: a Search for the Missing Science of Consciousness*, Oxford University Press (1994), p. 21.
50. A. Hodges, *Alan Turing: the Enigma*, Vintage (1992), p. 109.
51. A. Hodges, 'What would Alan Turing have done after 1954?', in C. Teuscher (ed.), *Alan Turing: Life and Legacy of a Great Thinker*, Springer-Verlag (2003), p. 51.
52. B. J. Copeland and D. Proudfoot, 'Alan Turing's forgotten ideas in computer science', *Scientific American*, 280 (1999), 99–103.
53. A. M. Turing (1951), 'Can digital computers think?', first published in B. J. Copeland, 'A lecture and two radio broadcasts on machine intelligence by Alan Turing', in K. Furukawa, D. Michie, and S. Muggleton (eds), *Machine Intelligence 15*, Oxford University Press (1999), pp. 445–76; also in *The Essential Turing*.
54. Copeland (Note 53), pp. 451–2. See also B. J. Copeland, 'Turing and the physics of the mind', in S. B. Cooper and J. van Leeuwen (eds), *Alan Turing: His Work and Impact*, Elsevier (2013), 651–66.
55. A. Hodges, 'What would Alan Turing have done after 1954?', Lecture at the Turing Day, Lausanne, Switzerland (June 2002).
56. Turing (1951), p. 483.
57. A. Hodges, 'Beyond Turing's machines', *Science*, 336 (13 April 2012), 163–4.

CHAPTER 42 TURING'S LEGACY (BOWEN AND COPELAND)

1. 'The great minds of the century', *Time*, 153(12) (29 March 1999) (http://content.time.com/time/magazine/article/0,9171,990608,00.html); P. Gray, 'Computer scientist: Alan Turing', *Time*, 153(12) (29 March 1999).

2. Google Scholar, 'Alan Turing' (https://scholar.google.co.uk/citations?user=VWCHlwkAAAAJ).
3. Turing (1936), Turing (1950), Turing (1952).
4. 'A. M. Turing Award', Association for Computing Machinery (http://amturing.acm.org/).
5. C. Page and M. Richards, 'A letter from Christopher Strachey', *Resurrection: the Journal of the Computer Conservation Society*, 73 (Spring 2016), 22–4 (http://www.computerconservationsociety.org/resurrection/res73.htm#d).
6. P. Berma, G. D. Doolen, R. Mainieri, and V. I. Tsifrinovich, 'Turing machines', *Introduction to Quantum Computers*, World Scientific (1998), 8–12; E. Bernstein and U. V. Vazirani, 'Quantum complexity theory', *STOC '93 Proceedings of the Twenty-Fifth Annual ACM Symposium on Theory of Computing*, ACM (1993), 11–20; D. Aharonov and U. V. Vazirani, 'Is quantum mechanics falsifiable? A computational perspective on the foundations of quantum mechanics', in B. J. Copeland, C. Posy, and O. Shagrir (eds), *Computability: Gödel, Turing, Church and Beyond*, MIT Press (2013).
7. P. Rendell, 'Game of Life—universal Turing machine', YouTube (2010, uploaded 2012) (http://www.youtube.com/watch?v=My8AsV7bA94).
8. F. L. Morris and C. B. Jones, 'An early program proof by Alan Turing', *IEEE Annals of the History of Computing*, 6(2) (1984), 139–43.
9. R. Highfield, 'What to think about machines that think', Science Museum, London (11 December 2015) (http://blog.sciencemuseum.org.uk/what-to-think-about-machines-that-think/).
10. M. Boden, 'Grey Walter's anticipatory tortoises', *The Rutherford Journal*, 2 (2006–7) (http://rutherfordjournal.org/article020101.html).
11. D. Rooney, 'Codebreaker—Alan Turing's life and legacy', Science Museum, London, YouTube (19 June 2012) (http://www.youtube.com/watch?v=I3NkVMHh0_Q); R. Highfield, 'Codebreaker wins Great Exhibition award', Science Museum, London (17 December 2012) (http://blog.sciencemuseum.org.uk/codebreaker-wins-great-exhibition-award).
12. 'Alan Turing at 100', *Nature*, 482 (25 February 2012), 450–65 (http://www.nature.com/news/specials/turing).
13. *Breaking the Code*, IMDb (1996) (http://www.imdb.com/title/tt0115749).
14. *Turing: a Staged Case History*, written and directed by Maria Elisabetta Marelli (https://vimeo.com/channels/712706).
15. http://www.thehopetheatre.com/productions/lovesong-electric-bear/; https://www.thestage.co.uk/reviews/2016/to-kill-a-machine-review-at-kings-head-theatre-london/.
16. https://georgezarkadakis.com/2011/10/28/notes-on-the-imitation-game/; www.scientificamerican.com/article/alan-turing-comes-alive.
17. http://www.jamesmccarthy.co.uk/blog/codebreaker-an-introduction.
18. Letter from Turing to Sara Turing (16 February 1930), in the King's Collage Archive, catalogue reference K1/20. Turing's 1930 letters to Morcom's mother are in Hodges (1983).
19. *The Imitation Game*, IMDb (2014) (http://www.imdb.com/title/tt2084970).
20. S. McKay, 'Turing, Snow White and the poisoned apple', *The Spectator* (9 May 2015) (http://www.spectator.co.uk/2015/05/turing-snow-white-and-the-poisoned-apple/).
21. S. Kettle, 'Alan Turing' (http://www.stephenkettle.co.uk/turing.html).
22. 'Bletchley Park marks Turing bombe stamp', BBC News (8 November 2014) (http://www.bbc.co.uk/news/uk-england-beds-bucks-herts-29950356); 'Bletchley Park marks Alan Turing centenary with stamp issue', BBC News (24 January 2012) (http://www.bbc.co.uk/news/uk-england-beds-bucks-herts-16688942).
23. 'A song for Alan Turing', YouTube (16 October 2011) (http://www.youtube.com/watch?v=ksUyhJRkvNk).
24. http://www.scribd.com/document/65056089/For-Alan-Turing-solo-piano. *Alan Turing's 100th Birthday Party* at King's College, Cambridge was organized by Jack Copeland and Mark Sprevak (https://sites.google.com/site/turingace2012/).
25. J. McCarthy, 'Codebreaker: an introduction' (http://www.jamesmccarthy.co.uk/blog/codebreaker-an-introduction).
26. McCarthy (Note 25).

27. J. Rogers, ' "We wrote it for Alan": Pet Shop Boys take their Turing opera to the Proms', *The Guardian* (20 July 2014) (https://www.theguardian.com/music/2014/jul/20/pet-shop-boys-alan-turing-enigma-proms-tribute-interview). (Thanks to James Gardner.)

28. 'Nico Muhly takes on Alan Turing: "No one wants a gay martyr oratorio" ', *The Guardian* (6 June 2015) (http://www.theguardian.com/music/2015/jun/06/nico-muhly-takes-on-alan-turing-no-one-wants-a-gay-martyr-oratorio).

29. R. Collins, 'Steve Jobs review: "manically entertaining" ', *The Telegraph* (12 November 2015) (http://www.telegraph.co.uk/film/steve-jobs/review).

30. H. Frith, 'Unraveling the tale behind the Apple logo', CNN (7 October 2011) (http://www.edition.cnn.com/2011/10/06/opinion/apple-logo); A. Robinson, 'Film: reality and check', *The Lancet*, 386 (21 November 2015), 2048.

31. S. McKay, 'How Alan Turing's secret papers were saved for the nation', *The Telegraph* (30 July 2011) (http://www.telegraph.co.uk/lifestyle/8668156/How-Alan-Turings-secret-papers-were-saved-for-the-nation.html).

32. S. Hickey, 'Alan Turing notebook sells for more than $1m at New York auction', *The Guardian* (13 April 2015) (https://www.theguardian.com/science/2015/apr/13/alan-turings-notebook-sells-for-more-than-1m-at-new-york-auction).

33. R. Lewin, *Ultra Goes to War*, Grafton (1978), p. 64.

34. A. Hodges, 'Oration at Alan Turing's birthplace' (http://www.turing.org.uk/publications/oration.html).

35. 'PM apology after Turing petition', BBC News (11 September 2013) (http://news.bbc.co.uk/1/hi/technology/8249792.stm); G. Brown, 'Gordon Brown: I'm proud to say sorry to a real war hero', *The Telegraph* (10 September 2009) (http://www.telegraph.co.uk/news/politics/gordon-brown/6170112/Gordon-Brown-Im-proud-to-say-sorry-to-a-real-war-hero.html).

36. 'Government rejects pardon request for Alan Turing', BBC News (8 March 2012), (http://www.bbc.co.uk/news/technology-16919012).

37. S. Swinford, 'Alan Turing granted Royal pardon by the Queen', *The Telegraph* (24 September 2013) (http://www.telegraph.co.uk/history/world-war-two/10536246/Alan-Turing-granted-Royal-pardon-by-the-Queen.html).

38. 'Royal pardon for codebreaker Alan Turing', BBC News (24 December 2013) (http://www.bbc.co.uk/news/technology-25495315).

39. A. Cowburn, *The Independent* (20 October 2016) (http://www.independent.co.uk/news/uk/politics/alan-turing-law-government-pardon-rachel-barnes-historic-crimes-a7370621.html).

40. R. Dawkins, *The God Delusion*, Bantam Press (2006), p. 289.

INDEX